T0372387

Collected Papers of S.R.S. Varadhan

Collected Papers of S.R.S. Varadhan

Volume I: Limit Theorems, Review Articles
Volume II: PDE, SDE, Diffusions, Random Media
Volume III: Large Deviations
Volume IV: Particle Systems and Their Large Deviations

S.R.S. Varadhan

Collected Papers I

Limit Theorems, Review Articles

Editors
Rajendra Bhatia
Abhay Bhatt
K. R. Parthasarathy

Author:
Professor S.R.S. Varadhan
New York University
New York, USA

Volume Editors:
Professor Rajendra Bhatia
Indian Statistical Institute
New Delhi, India

Professor Abhay Bhatt
Indian Statistical Institute
New Delhi, India

Professor K. R. Parthasarathy
Indian Statistical Institute
New Delhi, India

A co-publication with the Hindustan Book Agency, New Delhi, licensed for sale in all countries
outside of India. Sold and distributed within India by the Hindustan Book Agency, P 19 Green Park Extn.,
New Delhi 110 016, India
© Hindustan Book Agency 2012
HBA ISBN 978-93-80250-39-7 (4 volume set)

ISBN 978-3-642-33231-9 (4 volume set)
ISBN Volume I 978-3-642-33544-0
Springer Heidelberg Dordrecht London New York

Library of Congress Control Number: 2012947217

Math. Subj. Class (2010): 60-06, 60F99, 60J60, 60G44, 60F10, 60K35

Cover design: WMXDesign GmbH, Heidelberg, Germany

Printed on acid-free paper

Springer is part of Springer Science+Business Media (www.springer.com)

Printed and bound by Gopsons Papers Ltd., Noida

Preface

Srinivasa Varadhan began his research career at the Indian Statistical Institute (ISI), Calcutta, where he came in 1959 as a graduate student. His first paper appeared in *Sankhyā*, the Indian Journal of Statistics in 1962. The erudite founder of the Institute, and of the Journal, traced the origins of probabilistic thinking to the Indian-Jaina philosophy of *syādvāda*, or the "assertion of possibilities". The school of logic that propagated this was active around the fourth century BCE, and a full exposition of this theory is given in a sixth century text called *Saptabhaṅgīnaya* (the dialectic of seven-fold prediction).

In the middle of the twentieth century, research work on all aspects of statistics, practical and theoretical, was started in all earnestness at ISI. Varadhan was there in the golden period of the Institute. Among his teachers were C. R. Rao and R. R. Bahadur, and just a little older than him were fellow students V. S. Varadarajan, R. Ranga Rao and K. R. Parthasarathy. Together with this latter group Varadhan began the study of probability on topological groups and on Hilbert spaces, and quickly made an international reputation.

At this time Varadhan realised that there are strong connections between Markov processes and differential equations, and in 1963 he came to the Courant Institute in New York where he has stayed ever since. (According to him "here you do not have to make an effort to learn PDE, you just have to breathe it".) Here he came under the influence of probabilists Monroe Donsker and Marc Kac, and of a young graduate student Daniel Stroock. With Stroock he wrote a series of papers on the Martingale Problem and Diffusions, and with Donsker another series of papers on Large Deviations. With this work Varadhan's reputation as one of the leading mathematicians of the time was firmly established. Since then he has contributed to several other topics in probability, analysis and physics, and collaborated with several distinguished mathematicians.

These Collected Works contain all his research papers over the half-century from 1962 to the time of going to print in early 2012. It is befitting that they are being published in India. We are happy and proud to be associated with this project.

Editors

Contents

Volume 1

Limit Theorems

Contents

Review Articles

Volume 2

PDE, SDE, Diffusions

Contents

Random Media

Contents

Volume 3

Large Deviations

Contents

Volume 4

Particle Systems and their Large Deviations

Contents

Autobiography: S. R. S. Varadhan

According to my school records I was born on Jan 2, 1940, in the city of Madras, in the state of Madras in India, which was then a British colony. The city is now called Chennai and the state has become Tamil Nadu in the Republic of India.

My father was born in the last year of the nineteenth century, in 1899 and he married my mother in 1917, when he was eighteen and she was ten. I am an only child and my parents had been married for nearly twenty five years when I was born. Both my parents were the eldest siblings in rather large families and I have always received special attention from all my uncles, aunts, cousins, grandmothers and other assorted relatives. I was born so late that I did not really get to know either of my grandfathers.

As a child I grew up in several small towns not far from Madras. My father was a high school teacher and later principal in the county school system and was periodically transferred from one town to another within the county or district as it is called in India. That was both good and bad for me. I was treated with consideration by all my teachers. But I could not do any mischief at school without my father finding out right away.

Growing up in these small towns was fun. There was plenty of time after school to play with friends on the riverbed that was mostly dry, or play indoors on rainy days. There was very little homework, and in fact there was only minimal learning at school. I did well relative to my class, but was not challenged in any sense. It was only in the last year of the high school, that my mathematics teacher took a special interest in a small group of us and would ask us to come to his house on weekends to do some mathematics problems for fun. More than anything else he taught me that solving mathematical problems or puzzles can be fun.

I had some vague ideas of becoming a doctor as a child. But once, with a group of fellow students from the high school, I went to a medical exhibition at the local

S.R.S. Varadhan (✉)
Courant Institute of Mathematical Sciences, New York University, New York, NY 10012, USA
e-mail: varadhan@cims.nyu.edu

H. Holden, R. Piene (eds.), *The Abel Prize*,
DOI 10.1007/978-3-642-01373-7_12, © Springer-Verlag Berlin Heidelberg 2010

(a) As a seven year old in traditional Brahmin attire

(b) As a nine year old in a group with teachers in school. I am there because my father was the principal. I am sitting at his feet. He is the one with the head dress (turban)

medical college where the medical students demonstrated their surgical skills on cadavers. That cured me of the desire to enter the medical profession.

In 1954, I moved out to live with my uncle in a suburb of Madras to attend a local college where we were required to study for two years before entering the University. This was a difficult transition. From a rather parochial school system where everything was taught in Tamil, the vernacular, I had to grasp new ideas that were being taught in English. Some of the professors were from UK and I could not quite understand their accent. I studied English and Tamil as my two languages. These were the harder subjects at college since we were expected to write critical essays, where as in school we were only required to remember facts and memorize poetry.

In addition to languages I studied Mathematics, Physics and Chemistry. I did well in all. I liked Physics the most, and I began slowly to see the connections between them.

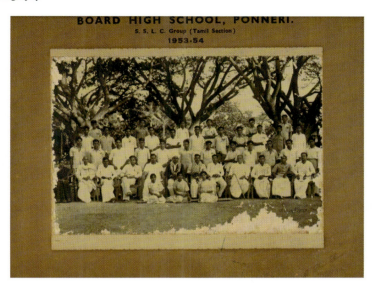

(c) High school graduation picture. I am standing on the top row. My father as principal is there too, sitting in the front row wearing a turban

It was time to enter the University and the seats were limited and the competition was tough. I came from a Brahmin community, viewed by the government as privileged, and there was reverse discrimination. The goal was to get into a Honors program that led directly to a Master's degree in three years and saved you a year.

It was at this time that I heard of Statistics for the first time. I was told that there was only one college in the state that offered an Honours program in Statistics and they admitted only fourteen students each year. My father knew somebody who knew somebody who could help and luckily I was admitted. I had a choice between Physics and Statistics and I opted for Statistics.

Basically it was a three year program devoted to pure mathematics, probability and statistics. The Presidency College where the program was offered, was situated on the beach. You looked out the window and it was a sandy beach that was as nice as Ipanema. Meanwhile, my father had retired and moved out to the same town as my uncle and I stayed with my parents and commuted for a year. I wanted my parents to let me stay in the dorm for the last two years and they did, after much persuasion on my part.

I really enjoyed those two years. Long walks on the beach with a dozen or so close friends from my class. Lots of movies, arguments, discussions and a generally carefree student life. We are still a close group and see each other from time to time. Although I am the only academic, the others have gone on to do well in government, business and other areas. The studies themselves were relatively easy. I was learning a lot of new concepts and they all felt natural and I could do exceptionally well without much effort. The trouble with the system was that the expectations were not high and it was too rigid. We had only set courses and no electives.

(d) Picture with Kolmogorov during his visit to India (1962). I am third from the left behind Kolmogorov. With me are K.R. Parthasarathy (graduate student), B.P. Adhikari (Professor), me, J. Sethuraman (graduate student), C.R. Rao (Head of the Institute and my advisor) and P.K. Pathak (graduate student)

I graduated after three years having established a new record for grades obtained in the final examination where you are tested in one week on everything that you have learned in three years.

The usual career path at this point is to write a competitive civil service examination that is used to recruit high level government officials. My parents expected me to sit for this. If you succeed, you are set for life. I wanted to go for research. When my father saw that I was firm, he supported me and I went off to Calcutta to study at the Indian Statistical Institute. I had no idea what I was supposed to do. When I arrived in August, 1959, they gave me a desk and expected me to write a thesis in three years. No graduate courses were offered. There were seminars that were optional. I took one on point set topology from Varadarajan and one on measure theory from Bahadur. I did not know what to do and so learned to play bridge and I played a lot. But slowly over the year I met up with a few other graduate students who were there already for a year or two, and we organized our own seminars and programs of study. We lectured to each other, formulated our own problems and tried to solve them. It was a wonderful learning experience and I started to do "Research".

The atmosphere at ISI was very stimulating. We had a steady stream of distinguished visitors, during the pleasant winter months. Sir Ronald Fisher came every year. I had just missed Norbert Wiener, who had come the year before I went there. We had tea, twice a day, when we all met at the tea room and talked informally. J.B.S. Haldane was a regular at tea, constantly puffing on his cigar. There was always some excitement in the air and some thing or other was always happening. During my second and third year I learned a lot of functional analysis and more or less finished my dissertation. I had begun to learn Markov Processes and wanted to start working on it. During my third year Kolmogorov visited us for two months and I gave a seminar on my thesis and he was in the audience asking questions. He spoke no English, none of us knew any Russian, and so we talked through an interpreter who knew French. A group of graduate students accompanied him on a two week tour of parts of India. He was a member of my thesis committee and brought a copy of my thesis with him, to Moscow, promising to send a report from there. The report came after six months, but only after Parthasarathy, a colleague who went to Moscow on an exchange program, provided a steady daily reminder.

At this point a year had gone by after my dissertation was finished, I formally got my Ph.D., and it was time to go abroad, for a postdoctoral study. In those post Sputnik days that meant USA. Varadarajan, who had just returned to India, after three years in the US, suggested that I go to NYU, to Courant Institute. He wrote a letter on my behalf that went unacknowledged for nearly three months. In the end I was offered a postdoctoral position and I arrived in New York in the Fall of 1963.

There was an international organization that greeted foreign students upon arrival, and they offered to find a place for me to stay for my first night. They met me at the old Idlewild airport and arranged for me to stay in a hotel on Times Square. When I showed up at Courant the next day and told them where I was staying, they were horrified and found me a place to stay on Tenth and University Place. Later I found myself a studio in the West Village.

Before leaving India I became engaged to be married to Vasundara and the wedding was to take place the following June, in Madras. The year went by very fast. I met a small group of Indian academics and over the years we have remained close.

Work went well at Courant. I saw a lot of Mony Donsker, who was the principal probabilist at Courant. There were lots of visitors, seminars and other activities that kept me busy during the week and I socialized with my Indian group during weekends.

Donsker and I, discussed a lot. He had many students working on an assortment of problems. This provided me with a broad perspective and I was beginning to view things from a slightly different angle that made a lot of sense to me. Donsker was interested in specific problems that he wanted to be solved and I was always interested in developing techniques that would enable one to solve a class of problems. We complemented each other and worked closely for over twenty-five years. I went to India to get married in the summer and we found a nice apartment in Washington Square Village. My wife, who was only sixteen, started undergraduate studies at Washington Square College. When my post doctoral fellowship was renewed for two more years it proved ideal for both of us. Then, I joined the faculty in the Fall of 1966, as an assistant professor.

Donsker had a student named Schilder who did a very fine thesis on Laplace asymptotics on Wiener space. I thought a lot about it, and felt that one could develop it

(e) This picture is at the airport on my way   (f) This picture is at the airport, returning to
to Courant as a postdoc (Fall 64)   USA with Vasu immediately after our wed-
                                                      ding in the summer of 1964

considerably. This has been a major thrust of my research over the years and goes by the name of Large Deviation Theory. It is the technique of estimating precisely, in a logarithmic scale, how small the probabilities of certain rare events are. I have found this to be a problem that crops up in many different contexts, and have returned to it again and again during my career.

We used to have a joint seminar with Rockefeller Institute, where Mark Kac was a professor. We would go up to 68th and York often. Once on the way back, in a taxi, someone commented on a result by Cieselski, a Polish mathematician and I immediately saw connections with large deviations and diffusions. The diffusion processes are Markov processes with continuous paths, closely related to second order elliptic partial differential operators. I was able to work out this connection and it proved to be a nice result.

There was a graduate student at Rockefeller University, Dan Stroock, who came by to talk to me about this and we became close friends. He came to Courant in 1966 and stayed for six years. We worked very closely during this period and I believe we changed the way Markov processes were viewed. We introduced systematically methods based on martingales that have become more or less standard today. We jointly wrote a book that appeared in 1979 and has been received well.

In the fall of 1973, we (the family at this time included two sons Gopal born in 69 and we lost him on 9/11/01, and Ashok born 1972) had just returned after a sabbatical from Sweden and India. Donsker as usual was interested in a specific problem about a probabilistic explanation of the variational formula for the ground state energy of the Schrödinger operator. I remember going to Durham to give a talk at Duke, and sitting in their library before my talk, I saw the connection. This led to a whole series of results on large deviation theory that Donsker and I worked on, till his untimely death in 1991.

I had become an Associate professor in 1968, won an Alfred P. Sloan Fellowship in 1969, and became a Professor in 1972. This was also a tough time at NYU, involving a financial crisis and the sale of the Bronx campus. We moved to Stanford for a year long visit in 1976–77, enjoyed the outdoor life and the open skies. But professionally I could not wait to return to New York. I also started working occasionally with George Papanicolau, who had lots of applied problems that often required new techniques. I was invited to give a talk at the International Congress of Mathematicians, in Helsinki in 1978 which was a kind of recognition.

I served for four years from 1980–84 as Director of Courant and survived the experience. I was surprised to find that the central administration consisted for the most part of talented individuals doing their best at a difficult task. I began to appreciate for the first time that we at Courant were part of a larger University. My research slowed down some due to the demands of administrative work, but I was able to continue with it. I had started to attract a steady stream of graduate students by this time and enjoyed working with them. I became recognized as an expert on Large Deviations, lectured often on it and wrote a set of lecture notes on the topic.

This was also the time when Wall Street discovered probability. The Black and Scholes model was suddenly very popular. I started some consulting for a small company that managed pension funds and worked closely with Harry Markowitz, who was also a consultant. Harry went on to win the Nobel Prize in Economics. Richard Brignoli, who was the CEO of the company was a maverick and that made the experience all the more enjoyable. It lasted several years, until the company went under. Not because it performed poorly, but the partners had a fallout, fought in court and the lawyers ended up getting everything. Our son Gopal got his first experience working for the company in the summer and later part time during his senior year at college. He chose to stay with the financial industry and his younger brother Ashok has followed him.

I took a sabbatical in 84–85, after serving for four years as Director of the Courant Institute supported partly on a Guggenheim fellowship. Around this time I went to a conference in Marseilles and we usually walked down to the sea and back up the hill, to the campus at Luminy, after lunch. During one of the walks, George explained to me a problem of establishing what he called bulk diffusion under rescaling for interacting diffusions. The problem intrigued me and I thought it would be a simple problem. I spent my sabbatical year thinking about it and found it rather difficult. I found out later that there were a whole class of such problems generally referred to as problems of hydrodynamic scaling. I worked on the problem a lot, but made no progress. Although I had some ideas, they were not sufficient to solve the problem. We had a seminar at which the speaker Josef Fritz from Budapest, proved a result of similar type for a different model. I thought about the new model and found to my surprise, that my ideas worked well for this model and could provide a better result than Fritz. George, Guo who was a student of George, and I worked feverishly on this model and completed our work. This led me to a whole set of problems requiring new methods and I worked closely with a very talented younger colleague, H.T. Yau. He has since developed the subject and has taken it in a variety of different directions.

The period from 1984–94 was personally very hard for me. My parents grew old and infirm. Since I had no siblings, I had the responsibility for their welfare. This meant frequent and prolonged trips back and forth between New York and Madras. This was hard on my wife, who had stayed in school part time, while raising the two boys, and earned a Ph.D. degree (her fourth degree at NYU) in 1985. My father died in 1990 and my mother in 1994. My wife lost her mother in 1991. The generation above us started thinning out slowly.

There was a bit of a problem at the office in 1992. George, who was to have been the director, left to go to Stanford and there was no time to find a replacement. I was drafted to serve for two years during which period a search committee was appointed and found an excellent choice of director in Dave McLaughlin.

This was also a time when I was receiving honors and recognition and that was gratifying. I was elected a Fellow of American Academy of Arts and Sciences and an Associate Fellow of the Third World Academy of Sciences in 1988, a Fellow of the Institute of Mathematical Statistics in 1991, a member of National Academy of Sciences in 1995, and a Fellow of The Royal Society in 1998. I received the George D. Birkhoff Prize from AMS and SIAM in 1994 and the Leroy P. Steele prize of the AMS that I shared with Dan Stroock in 1996. NYU gave me the Margaret and Herman Sokol award in 1995. I was invited to give a plenary address at the International Congress of Mathematicians in Zurich, in 1994. I was appointed as Frank J. Gould Professor of Science at NYU. In 2001, I was elected to serve as the President of Institute of Mathematical Sciences for 2002–03. I received honorary degrees from University of Paris in 2003 and Indian Statistical Institute in 2004.

Tragedy hit on 9/11/01 when we lost our older son Gopal who was working at the World Trade Center that day. I was in Paris, visiting IHP for a month and was told of the crash by my wife who called me from New York. It took four days to get back to New York. We miss him. Our grandson Gavin, born to our younger son Ashok and his wife Maggie helps take some of the edge off our grief.

Professionally I consider myself to have been fortunate, in that my career spanned a period, when science and mathematics were generously funded by the public. I had a stimulating ambience at Courant that has traditionally provided a very supportive environment for their younger faculty. I have enjoyed working closely with my colleagues as well as everyone of my nearly thirty doctoral students. Finally, I hope to have several more years of productive academic life.

Varadhan with M. D. Donsker

Varadhan with G. C. Papanicolaou

Varadhan with D. W. Stroock

Varadhan receiving the Abel Prize from HM King Harald V, Oslo, 2007

Varadhan receiving the National Medal for Science from Mr. Barack Obama,
President of the United States, 2011

Introduction

S. R. S. Varadhan

These volumes of collected works begin with the work done at the Indian Statistical Institute, with R. Ranga Rao and K. R. Parthasarathy. We were learning functional analysis, particularly topics around weak convergence of probability distributions, an area that had been investigated by V.S. Varadarajan earlier at ISI. We wanted to apply it to study convolution properties of probability distributions on topological groups. While some initial work was carried out directly, other parts dealing with locally compact abelian groups needed the use of characteristic functions. These methods were later modified to cover the infinite dimensional case of Hilbert Space, which was a significant part of my Ph.D thesis. Later developments include the work with D. Stroock [12] on the analog of the Central Limit Theorem on Lie Groups and extension to limit theorems of products of independent random elements on a Lie Group carried out by P. Feinsilver [4] in his thesis. These works use martingale techniques developed with Stroock in the sixties at the Courant Institute.

The work done at the Courant Institute falls mainly into four interrelated areas; large deviations, diffusion processes, interacting particle systems and processes in random environments. The work on large deviations began in 1964 when, Michael Schilder, a student of Monroe Donsker proved in his thesis [11] a result that can be interpreted as a large deviation result for Brownian motion. He showed that if P is the Wiener measure on $C[0, 1]$, then for any bounded continuous function $V(\cdot)$

$$\lim_{\epsilon \to 0} \epsilon \log E^P \left[\exp[\frac{1}{\epsilon} \int_0^1 V(\sqrt{\epsilon}x(s))ds] \right] = \sup_{\substack{f(\cdot) \\ f(0)=0}} \left[\int_0^1 V(f(t))dt - \frac{1}{2} \int_0^1 [f'(t)]^2 dt \right].$$

My interest in large deviation theory began with trying to interpret this result as the combination of a Laplace formula and a large deviation result directly in $C[0, 1]$. This point of view was developed in [14]. Some time in 1965 while returning after a seminar from Rockefeller University in a taxi, some one remarked about a theorem of Czieselski [1] that compared the ratio of the two fundamental solutions of the heat equation

$$r(t, x, y) = \frac{p_D(t, x, y)}{p(t, x, y)}$$

where p_D corresponds to the process killed at the boundary of a domain D and p to the whole space. The ratio $r \to 1$ as $t \to 0$ for all $x, y \in D$, if and only if

D is essentially convex. The explanation is that if a Brownian path goes from x to y, in a short time, it would most likely go along the straight line from x to y. The analogue for diffusions would be that the straight line has to be replaced by the geodesic in the Riemannian metric associated to the diffusion. This, along with my interest in large deviations, led to the study of small time behavior of diffusions. Michael Glass in his Ph.D. thesis [7], extended this to small second order perturbations of vector fields. This topic was developed profoundly and independently by Freidlin and Wentzell in their work [5] on small random perturbations of dynamical systems.

One of the things that Dan Stroock, who was then finishing up his thesis at Rockefeller University, and I noticed was that in proving the large deviation results for diffusions the main tool was some exponential martingales that were associated with the diffusion process. We then started wondering if the martingales characterized the process. We quickly worked out that if they did, then one can use the machinery of weak convergence to take a fresh look at many of the convergence theorems in Markov Processes. For the next few years Dan and I explored many aspects of Markov Process theory through their associated martingales. This continued till Dan left for Colorado in 1972, and we later wrote our book on multidimensional diffusion processes [13]. Martingale methods have since become an important tool in proving limit theorems for stochastic processes.

I was on a sabbatical leave during 1972-73, and when I returned, Donsker asked me a question. Can we explain the variational formula

$$\lambda(V) = \inf_{\|f\|_2 = 1} \left[\int V(x)[f(x)]^2 dx + \frac{1}{2} \int |\nabla f|^2 dx \right],$$

for the principal eigenvalue $\lambda(V)$ of $-\frac{1}{2}\Delta + V$ as the Laplace formula that corresponds to some large deviation result? After all there is the Feynman-Kac formula and its consequence

$$\lambda(V) = -\lim_{t \to \infty} \frac{1}{t} \log E[\exp[-\int_0^t V(x(s))ds]]$$

where E is the expectation with respect to Brownian motion. I was thinking of this sitting in the library of the Mathematics Department at Duke University, where I was visiting to give a talk, and I saw how there was a simple answer. We can rewrite the variational formula in terms of $g = f^2$ as

$$\lambda(V) = \inf_{\substack{g \geq 0 \\ \int g\, dx = 1}} \left[\int V(x)g(x)dx + \frac{1}{8} \int \frac{|\nabla g|^2}{g} dx \right],$$

16

and the Feynman-Kac formula as

$$\lambda(V) = -\lim_{t\to\infty} \frac{1}{t} \log E[\exp[-t \int_R V(x)\, L_t(dx)],$$

where $L_t(A) = \frac{1}{t} \int_0^t \mathbf{1}_A(x(s))ds$ is the empirical distribution, or occupation time, i.e., the proportion of time spent by the Brownian path in the set A. This suggests there is a large deviation result for the behavior of the distribution of L_t on the space $\mathcal{M}$ of probability distributions in R, with the rate function for the distribution with density g being

$$I(g) = \frac{1}{8} \int \frac{|\nabla g|^2}{g} dx.$$

We, in fact, established such a result for Brownian motion [2] and for other Markov processes. At this time Mark Kac raised a problem about what he named the Wiener Sausage: if $|D_\epsilon(t)|$ is the volume of the range of a Brownian path in R^d enlarged by a neighborhood of size ϵ, show that

$$\lim_{t\to\infty} \frac{1}{t^{\frac{d}{d+2}}} \log E[\exp[-\nu D_\epsilon(t)]] = -k(d,\nu)$$

exists and calculate it explicitly. This prompted us to investigate the role of topology in large deviations. While the usual theory of large deviations involved weak topology for measures, to answer the question raised by Kac, we need the strong or L_1 topology. The ϵ neighborhood provides a certain level of smoothing, and the issue was to show that was enough. Mark had suggested that a similar result for random walk on the number of distinct sites visited could be more tractable. But it turns out that the random walk case is proved by a combination of the central limit theorem and the result for Brownian motion [3].

Similar large deviation results were obtained independently by Jurgen Gärtner [6]. It turns out that one can prove the existence of the limit

$$\lambda(V) = \lim_{t\to\infty} \frac{1}{t} \log E\Big[\exp[\int_0^t V(x(s))ds]\Big]$$

for a wide class of Markov processes. While according to the ergodic theorem L_t converges to the invariant distribution μ_0, the dual

$$I(\mu) = \sup_V \Big[\int V(x)\mu(dx) - \lambda(V)\Big]$$

of $\lambda(V)$ is the rate function for large deviations of L_t from μ_0.

We adopted a different method where the upper bound was established by the use of exponential martingales, and the lower bound was obtained through "tilting" which provides a direct connection to entropy that controls, in some sense, all large deviations probabilities. The idea is that if we have a model and a rare phenomenon, whose probability we want to estimate, we can change the model so that the the rare phenomenon is normal in the new model. The amount of change is measured by relative entropy. Since there may be many possible "tilts" that can produce the phenomenon in question, it is natural to choose the optimal tilt, i.e., one with the least relative entropy. This then ends up as the large deviation rate.

Over the years I have worked with George Papanicolaou on problems involving random media of different kinds. These problems have their origins in problems of homogenization of periodic and random media that arise in the study of composite materials. The theory of interacting particle systems is closely related, in the sense that the random medium felt by any one particle keeps changing randomly due to the movement of other particles. Some of these problems are related to the study of non-equilibrium statistical mechanics, where a system with multiple equilibria evolves from non-equilibrium. Local equilibria are quickly reached, parametrized by variables that correspond to conserved quantities, that evolve in space and time when suitably rescaled. The limiting equations that describe their evolution are often a non-linear system of conservation laws. While attending a conference in Luminy, near Marseille, during a walk to the calanque after lunch, George posed the question of studying a large collection of $N\rho$ interacting Brownian motions, rescaled as

$$dx_i(t) = -N \sum V'(N(x_i(t) - x_j(t)) + d\beta_i(t)$$

in a periodic one dimensional space, i.e., the unit circle. In the limit the local density of particles $\rho(t, x)$ will evolve according to a nonlinear heat equation

$$\rho_t = (D(\rho)\rho_x)_x.$$

$D(\rho)$ is determined through the statistical mechanical equilibrium distribution, or the Gibbs state, μ_ρ at density ρ. I thought about this problem for a while. Initially I was confident it could be solved easily, but that did not turn out to be so. I did make some progress but was stuck at a crucial point. Some time later Josef Fritz from Hungary was giving a seminar at Courant in which he was proving a similar result for a different model. To my surprise I realized that the difficulty I had for the model involving interacting Brownian motions, could be resolved easily in this model dealing with interacting random fields at lattice locations. This resulted in a joint paper with Guo and Papanicolaou [8]. This involved field variables at lattice locations that represented charges that were randomly exchanged between neighboring sites. The random exchange was driven by a nonlinear force, modified by a random noise, which was Brownian motion

$$d\xi_i(t) = [\phi'(x_{i-1}(t)) - 2\phi'(x_i(t)) + \phi(x_{i+1}(t))]\, dt + [d\beta_{i-1,i}(t) - d\beta_{i,i+1}(t)].$$

The rescaled limiting equation for the averages of $\xi_i(t)$ is again a nonlinear diffusion equation of the form

$$m_t = (D(m)m_x)_x$$

with an explicit formula for D in terms of ϕ. I was then curious to see how some inhomogeneity will change the answer. Replace $d\beta_{i-1,i}(t)$ by $\sigma(x_{i-1}(t), x_i(t))d\beta_{i-1,i}(t)$. To my surprise I found the answer to be much more complicated, and it involved some very heavy analysis in infinite dimensional spaces [15]. In the end I could get a variational formula for $a(\rho)$ and the limiting equation

$$m_t = (a(m)D(m)m_x)_x.$$

I remember being in L'Aquila, Italy, and explaining to Herbert Spohn what I could do. He was quick to recognize the significance of what I had done, and explained to me the difference between gradient and non-gradient models. The diffusion coefficient $a(m)$ is, in general, given by a Green-Kubo formula, in terms of the entire evolution in time of the equilibrium, and it simplifies only in gradient models. What I had done was perhaps the first example of a non-gradient model whose diffusive scaling limit had been rigorously established. I quickly understood the significance and with a group of students, started working on problems of this elk, referred to as hydrodynamic scaling limits. Around this time I was lucky to have H. T. Yau, join our group. He had an excellent background in mathematical physics (and everything else!) and together with many students and postdocs we worked on several problems in this area. It turned out that there is a subtle connection between these problems and large deviations that proved to be useful in sorting out many things.

During the last decade I have been studying large deviations for random walks in a random environment. I was always of the opinion that one should consider the environment as seen by the moving particle, and with many students we explored this. It turned out that certain nonlinear homogenization problems are related to large deviations of random walks or diffusions in a random environment. With Rezakhanlou and Kosygina we worked on these issues [9],[10].

I have been lucky to have found excellent collaborators, throughout my career. The Mathscinet lists nearly thirty coauthors. I have enjoyed working with all of them: from fellow graduate students Ranga Rao and Parthasarathy to colleagues Donsker, Stroock, Papanicolaou, Nirenberg, Yau and Chatterjee at Courant; visitors and post-docs Sznitman, Olla, Kipnis and Kifer. There is even an unpublished but widely circulated manuscript with P. L. Lions. I have continued to work with many of my past students, Rezakhanlou, Quastel, Kosygina and Sethuraman. All of them have enriched my professional life and I am greatly indebted to them.

References

[1] Ciesielski, Z. Heat conduction and the principle of not feeling the boundary. Bull. Acad. Polon. Sci. Sr. Sci. Math. Astronom. Phys. 14 1966 435–440.

[2] Donsker, M. D.; Varadhan, S. R. S. Asymptotic evaluation of certain Wiener integrals for large time. Functional Integration and its Applications (Proc. Internat. Conf., London, 1974), pp. 15–33. Clarendon Press, Oxford, 1975.

[3] Donsker, M. D.; Varadhan, S. R. S. Asymptotics for the Wiener sausage. Comm. Pure Appl. Math. 28 (1975), no. 4, 525–565.

[4] Feinsilver, Philip. Processes with independent increments on a Lie group. Trans. Amer. Math. Soc. 242 (1978), 73–121.

[5] Freidlin, Mark I.; Wentzell A.D. On Small random perturbations of dynamical systems. Russian Mathematical Surveys Volume 25(1970) , number 1 pages 1-56

[6] Gärtner, Jurgen. Large deviation theorems for a class of random processes. (Russian) Teor. Verojatnost. i Primenen. 21 (1976), no. 1, 95–106.

[7] Glass, Michael Steven. Perturbation of a first order equation by a small diffusion. Thesis (Ph.D.)New York University. ProQuest LLC, Ann Arbor, MI, 1970. 53 pp..

[8] Guo, M. Z.; Papanicolaou, G. C.; Varadhan, S. R. S. Nonlinear diffusion limit for a system with nearest neighbor interactions. Comm. Math. Phys. 118 (1988), no. 1, 31–59.

[9] Kosygina, Elena; Rezakhanlou, Fraydoun; Varadhan, S. R. S. Stochastic homogenization of Hamilton-Jacobi-Bellman equations. Comm. Pure Appl. Math. 59 (2006), no. 10, 1489–1521.

[10] Kosygina, Elena; Varadhan, S. R. S. Homogenization of Hamilton-Jacobi-Bellman equations with respect to time-space shifts in a stationary ergodic medium. Comm. Pure Appl. Math. 61 (2008), no. 6, 816–847.

[11] Schilder, M. Some asymptotic formulas for Wiener integrals. Trans. Amer. Math. Soc. 125 (1966) 63–85.

[12] Stroock, Daniel W.; Varadhan, S. R. S. Limit theorems for random walks on Lie groups. Sankhya Ser. A 35 (1973), no. 3, 277–294.

[13] Stroock, Daniel W.; Varadhan, S. R. S. Multidimensional diffusion processes. Grundlehren der Mathematischen Wissenschaften [Fundamental Principles of Mathematical Sciences], 233. Springer-Verlag, Berlin-New York, 1979.

[14] Varadhan, S. R. S. Asymptotic probabilities and differential equations. Comm. Pure Appl. Math. 19 (1966) 261–286.

[15] Varadhan, S. R. S. Nonlinear diffusion limit for a system with nearest neighbor interactions. II. Asymptotic problems in probability theory: stochastic models and diffusions on fractals (Sanda/Kyoto, 1990), 75–128, Pitman Res. Notes Math. Ser., 283, Longman Sci. Tech., Harlow, 1993.

Prize Citations

National Medal of Science, 2011.

To Srinivasa S.R. Varadhan, for his work in probability theory, especially his work on large deviations from expected random behavior, which has revolutionized this field of study during the second half of the twentieth century and become a cornerstone of both pure and applied probability. The mathematical insights he developed have been applied in diverse fields including quantum field theory, population dynamics, finance, econometrics, and traffic engineering.

Abel Prize, 2007.

To S. R. S. Varadhan, for his fundamental contributions to probability theory and in particular for creating a unified theory of large deviations.

Probability theory is the mathematical tool for analyzing situations governed by chance. The law of large numbers, discovered by Jacob Bernoulli in the eighteenth century, shows that the average outcome of a long sequence of coin tosses is usually close to the expected value. Yet the unexpected happens, and the question is: how? The theory of large deviations studies the occurrence of rare events. This subject has concrete applications to fields as diverse as physics, biology, economics, statistics, computer science, and engineering.

The law of large numbers states that the probability of a deviation beyond a given level goes to zero. However, for practical applications, it is crucial to know how fast it vanishes. For example, what capital reserves are needed to keep the probability of default of an insurance company below acceptable levels? In analyzing such actuarial "ruin problems", Harald Cramér discovered in 1937 that standard approximations based on the Central Limit Theorem (as visualized by the bell curve) are actually misleading. He then found the first precise estimates of large deviations for a sequence of independent random variables. It took 30 years before Varadhan discovered the underlying general principles and began to demonstrate their tremendous scope, far beyond the classical setting of independent trials.

In his landmark paper "Asymptotic probabilities and differential equations" in 1966 and his surprising solution of the polaron problem of Euclidean quantum field theory in 1969, Varadhan began to shape a general theory of large deviations that was much more than a quantitative improvement of convergence rates. It addresses a fundamental question: what is the qualitative behaviour of a stochastic system if it deviates from the ergodic behaviour predicted by some law of large numbers or if it arises as a small perturbation of a deterministic system? The key to the answer is a powerful variational principle that describes the unexpected behaviour in terms of a new probabilistic mode minimizing a suitable entropy distance to the initial probability measure. In a series of joint papers with Monroe D. Donsker exploring the hierarchy of large deviations in the context of Markov processes, Varadhan demonstrated the relevance and the power of this new approach. A striking application is their solution of a conjecture of Mark Kac concerning large time asymptotics of a tubular neighbourhood of the Brownian motion path, the so-called "Wiener sausage".

Varadhan's theory of large deviations provides a unifying and efficient method for clarifying a rich variety of phenomena arising in complex stochastic systems, in fields as diverse as quantum field theory, statistical physics, population dynamics, econometrics and finance, and traffic engineering. It has also greatly expanded our ability to use computers to simulate and analyze the occurrence of rare events. Over the last four decades, the theory of large deviations has become a cornerstone of modern probability, both pure and applied.

Varadhan has made key contributions in several other areas of probability. In joint work with Daniel W. Stroock, he developed a martingale method for characterizing diffusion processes, such as solutions of stochastic differential equations. This new approach turned out to be an extremely powerful way of constructing new Markov processes, for example infinite-dimensional diffusions arising in population genetics.

Another major theme is the analysis of hydrodynamical limits describing the macroscopic behavior of very large systems of interacting particles. A first breakthrough came in joint work with Maozheng Guo and George C. Papanicolaou on gradient models. Varadhan went even further by showing how to handle non-gradient models, greatly extending the scope of the theory. His ideas also had a strong influence on the analysis of random walks in a random environment. His name is now attached to the method of "viewing the environment from the travelling particle", one of the few general tools in the field.

Varadhan's work has great conceptual strength and ageless beauty. His ideas have been hugely influential and will continue to stimulate further research for a long time.

Leroy P. Steele Prize for Seminal Contribution to Research, 1996.

To Daniel Stroock and S.R.S. Varadhan for their four papers: *Diffusion processes with continuous coefficients I and II*, Comm. Pure Appl. Math. 22 (1969), 345-400, 479-530; *On the support of diffusion processes with applications to the strong maximum principle*, Sixth Berkeley Sympos. Math. Statist. Probab., vol. III, 1970, pp. 333-360; *Diffusion processes with boundary conditions*, Comm. Pure Appl. Math. 34 (1971), 147-225; *Multidimensional diffusion processes*, Springer-Verlag, 1979, in which they introduced the new concept of a martingale solution to a stochastic differential equation, enabling them to prove existence, uniqueness, and other important properties of solutions to equations which could not be treated before by purely analytic methods; their formulation has been widely used to prove convergence of various processes to diffusions.

Birkhoff Prize in Applied Mathematics (AMS-SIAM George David Birkhoff Prize), 1994.

To S. R. S. Varadhan for important contributions to the martingale characterization of diffusion processes, to the theory of large deviations for functionals of occupation times of Markov processes, and to the study of random media.

Varadhan's Work on Diffusion Theory

Daniel W. Stroock*

Given a second order, (possibly degenerate) elliptic differential operator

$$L = \frac{1}{2} \sum_{i,j=1}^{N} a_{ij}(x)\partial x_i \partial x_j + \sum_{i=1}^{N} b_i(x)\partial x_i \qquad (1)$$

acting on $C^2(\mathbb{R};\mathbb{C})$ and a Borel probability measure μ on $\mathbb{R}^N$, what does it mean for a Borel probability measure $\mathbb{P}_\mu$ on $C([0,\infty);\mathbb{R}^N)$ to be a diffusion process determined by L with initial μ? When Varadhan and I asked ourselves this question in the mid 1960's, we found no answer which satisfied us.

It was not that there were no answers. Indeed, already in the 1930's Kolmogorov (cf. [6]) had mapped out a path which started from L and ended at a measure $\mathbb{P}_\mu$. Namely, he began by solving what is now called Kolmogorov's forward equation

$$\partial_t p(t, x, \cdot) = L^* p(t, x, \cdot) \quad \text{with } p(0, x, \cdot) = \delta_x, \qquad (2)$$

where L^* is the formal adjoint of L given by

$$L^*\psi(y) = \frac{1}{2} \sum_{i,j=1}^{N} \partial_{y_i}\partial_{y_j}\big(a_{ij}(y)\psi(y)\big) - \sum_{i=1}^{N} \partial_{y_i}\big(b_i(y)\psi(y)\big).$$

He then showed that, under suitable regularity and non-degeneracy conditions on the coefficients a and b, there is one and only one solution to (2) which is the density of a probability measure on $\mathbb{R}^N$. Further he showed that $p(t, x, y)$ is a continuous function of $(t, x, y) \in (0, \infty) \times \mathbb{R}^N \times \mathbb{R}^N$ and that it satisfies the Chapman–Kolmogorov equation

$$p(s + t, x, y) = \int p(t, \xi, y)p(s, x, \xi) \, d\xi. \qquad (3)$$

*Department of Mathematics, Massachusetts Institute of Technology, Cambridge, MA 02139

Knowing (3), he could check that, for any $\mu \in M_1(\mathbb{R}^N)$, the family of probability measures given by

$$P_\mu\big((t_1, \ldots, t_n); \Gamma\big) = \int_\Gamma \cdots \int \prod_{m=1}^n p\big(t_m - t_{m-1}, y_{m-1}, y_m\big)\, \mu(dy_0) dy_1 \cdots dy_n$$

$$\text{for } n \geq 0,\ 0 = t_0 < t_1 < \cdots < t_n, \text{ and } \Gamma \in \mathcal{B}_{(\mathbb{R}^N)^{n+1}}$$

is consistent in the sense that if $\{t_1', \ldots, t_{n'}'\} \subset \{t_1, \ldots, t_n\}$ then $P_\mu\big((t_1', \ldots, t_{n'}'); \cdot\big)$ is the marginal distribution of $P_\mu\big((t_1, \ldots, t_n); \cdot\big)$ on the coordinates corresponding to $\{t_1', \ldots, t_{n'}'\}$. Having established consistency, Kolmogorov applied his Consistency Theorem to produce a measure $\tilde{\mathbb{P}}_\mu$ on $(\mathbb{R}^N)^{[0,\infty)}$ whose finite dimensional marginals are given by

$$\tilde{\mathbb{P}}_\mu\big(\{\omega : \big(\omega(t_0), \ldots, \omega(t_n)\big) \in \Gamma\}\big) = P_\mu\big((t_1, \ldots, t_n); \Gamma\big).$$

Finally, using his continuity criterion, he showed that $C\big([0, \infty); \mathbb{R}^N\big)$ has outer measure 1 under $\tilde{\mathbb{P}}_\mu$ and therefore that the restriction of $\tilde{\mathbb{P}}_\mu$ to $C\big([0, \infty); \mathbb{R}^N\big)$ determines a unique Borel probability measure $\mathbb{P}_\mu$ there.

The trail which Kolmogorov blazed was widened and smoothed by many people, including Wm. Feller [4], J. Doob[2], G. Hunt [5], R. Blumenthal and R. Getoor [11], E.B. Dynkin [3], and a host of others. What emerged, especially after the work of Hunt, from this line of research was an isomorphism between Markov processes and potential theory. However, none of these really addressed the question which was bothering Varadhan and me. Namely, from our point of view, the route mapped out by Kolmogorov was too circuitous: one started with L, constructed from it either a transition probability function or resolvent operator, and only then characterized the resulting Markov processes in terms of these ancillary objects. What we were looking for is a characterization of $\mathbb{P}_\mu$ directly in terms of L, one from which it would be clear that if, for each $n \geq 1$, $\mathbb{P}_n$ is related to L_n with initial distribution μ_n, and if $\mu_n \longrightarrow \mu$ weakly and $L_n\varphi \longrightarrow L\varphi$ for smooth φ's with compact support, then $\mathbb{P}_n$ tends weakly to $\mathbb{P}_\mu$. Obviously, the best way to achieve this goal is to phrase the characterization in terms $\mathbb{P}_\mu$-integrals involving $L\varphi$.

Formulating such a characterization was relatively easy. Namely, no matter how one goes about its construction, $\mathbb{P}_\mu$ should have the property that[1]

$$\mathbb{E}^{\mathbb{P}_\mu}\big[\varphi\big(\omega(t_2)\big) \,\big|\, \mathcal{B}_{t_1}\big] - \varphi\big(\omega(t_1)\big) = \mathbb{E}^{\mathbb{P}_\mu}\left[\int_{t_1}^{t_2} L\varphi\big(\omega(\tau)\big)\, d\tau \,\bigg|\, \mathcal{B}_{t_1}\right]$$

$$\text{for all } \varphi \in C_c^\infty(\mathbb{R}^N; \mathbb{C}), \tag{4}$$

---

[1]Throughout, I will be somewhat casual about integrability questions. In the following, no integrability problems arise unless the coefficients are unbounded, in which case special considerations have to be taken into account.

where $\mathcal{B}_t$ is the σ-algebra generated by $\{\omega(\tau) : \tau \in [0,t]\}$. More succinctly,

$$\left(\varphi\big(\omega(t)\big) - \int_0^t L\varphi\big(\omega(\tau)\big)\, d\tau, \mathcal{B}_t, \mathbb{P}_\mu \right)$$

is a martingale for all $\varphi \in C_c^\infty(\mathbb{R}^N; \mathbb{C})$. $\hspace{2cm}$ (5)

Thus, we said that $\mathbb{P}_\mu$ *solves the martingale problem for L with initial distribution μ* if μ is the $\mathbb{P}_\mu$-distribution of $\omega \rightsquigarrow \omega(0)$ and (5) holds.

Having formulated the martingale problem, our first task was to prove that, in great generality, solutions exist. Of course, seeing as all roads lead to solutions to the martingale problem, in situations to which they applied, we could have borrowed results from earlier work (e.g., of either the Kolmogorov or Itô schools). However, not only would that have made no contribution to known existence results, it would have been very inefficient. At least when a and b are bounded and continuous, existence of solutions to the martingale problem is an easy application of compactness, completely analogous to the proof of existence of solutions to ordinary differential equations with bounded continuous coefficients. Our second task was to show that, under appropriate conditions, solutions are unique, and we knew that uniqueness posed a challenge of an entirely different order from the one posed by existence. Indeed, if all roads lead to solutions to the martingale problem, then uniqueness would prove that all roads must have the same destination.

Our approach to uniqueness depended on the assumptions being made about the coefficients a and b. One approach turned on the observation that any solution to the martingale is the distribution of a solution to an Itô stochastic integral equation. When the coefficients are reasonably smooth, this observation allowed us to prove uniqueness for the martingale problem as a consequence of uniqueness for the associated stochastic integral equation. However, when the coefficients are not smooth, Itô's theory cannot cope. Specifically, because his theory ignores ellipticity properties, it is difficult to see how one goes about bolstering it with powerful results from the vast and beautiful P.D.E. literature.

The key to overcoming the problem just raised is to realize that uniqueness for the martingale problem is a purely distributional question, whereas the uniqueness found in Itô's theory is a path-by-path statement about a certain transformation (the "Itô map") of Brownian paths. Thus, to take advantage of the P.D.E. existence theory, we abandoned Itô and adopted an easy duality argument which shows that uniqueness for the martingale problem requires only a sufficiently good existence statement about the initial value problem for the backward equation $\partial_t u = Lu$. Namely, if u is a classical solution[2] to the

---

[2] By "classical solution," I mean that the solution is twice continuously differentiable in space and once in time.

backward equation with initial data f, then one can show that, for any solution $\mathbb{P}$ to the martingale problem for L,

$$\left(u\big((t_2 - t)^+, \omega(t)\big), \mathcal{B}_t, \mathbb{P}\right)$$

is a martingale and therefore that, for all $0 \leq t_1 < t_2$,

$$\mathbb{E}^{\mathbb{P}}\big[f\big(\omega(t_2)\big) \,\big|\, \mathcal{B}_{t_1}\big] = u\big(t_2 - t_1, \omega(t_1)\big).$$

Hence, if such classical solutions exist for enough f's, then the conditional distribution of $\omega \rightsquigarrow \omega(t_2)$ under $\mathbb{P}$ given $\mathcal{B}_{t_1}$ is uniquely determined for all $0 \leq t_1 < t_2$, and from this, together with the condition that the $\mathbb{P}$-distribution of $\omega \rightsquigarrow \omega(0)$ is μ, it follows that there is only one solution to the martingale problem. This argument works without a hitch when, for example, a and b are bounded and uniformly Hölder continuous and a is uniformly elliptic (i.e., $a \geq \epsilon I$ for some $\epsilon > 0$), since in that case P.D.E. theorists had long ago provided the necessary existence statement about the backward equation.

When uniform Hölder continuity is replaced by simple uniform continuity, the existence theory for the backward equation is less satisfactory. To be precise, although solutions continue to exist, they are no longer classical. Instead, their second derivatives exist only in $L^p(\mathbb{R}^N; \mathbb{C})$ for $p \in (1, \infty)$. As a consequence, the preceding duality argument breaks down unless one can show *a priori* that every solution to the martingale problem is sufficiently regular to tolerate limit procedures involving $L^p(\mathbb{R}^N)$-convergence of the integrands. Without question, the proof of such an *a priori* regularity result was the single most intricate part of our program. In broad outline, our proof went as follows. First, we developed machinery which allowed us to reduce everything to the case when $b = 0$ and a is an arbitrarily small perturbation of the identity. Second, we again used the connection with Itô stochastic integral equations to show that any solution $\mathbb{P}$ can be approximated by $\mathbb{P}_n$'s which, conditioned on $\mathcal{B}_{\frac{m}{n}}$, are Gaussian during the time interval $\left[\frac{m}{n}, \frac{m+1}{n}\right]$ and therefore have the required regularity properties. Finally, as an application of the Calderon–Zygmund theory of singular integral operators, for each $p \in (1, \infty)$ we were able to show that, when the perturbation is sufficiently small, the $L^p(\mathbb{R}^N)$-regularity properties of the approximating $\mathbb{P}_n$'s can be controlled independent of n and are therefore enjoyed by $\mathbb{P}$ itself.

The steps outlined above led us to a proof of the following theorem, which should be considered the centerpiece of [12].

Theorem. *Let L be given by* (1), *where a and b are bounded, a is continuous, b is Borel measurable, and $a(x)$ is symmetric and strictly positive definite for each $x \in \mathbb{R}^N$. Then, for each Borel probability measure μ on $\mathbb{R}^N$, there is precisely* *one solution to the martingale for L with initial distribution μ. Moreover, if δ_x is the unit mass at x and $\mathbb{P}_x = \mathbb{P}_{\delta_x}$, then $x \rightsquigarrow \mathbb{P}_x$ is weakly continuous and $\{\mathbb{P}_x : x \in \mathbb{R}^N\}$ is a strong Markov family.*

Concluding Remarks: There are several additional comments which may be helpful.

(1) Although, with 20-20 hindsight, the characterization given by (4) is the most obvious one, it is not the one which we chose initially. Instead, because we were influenced at the time both by Henry McKean's treatment in [9] of Itô's theory of stochastic integral equations and by a possible analogy with characteristic functions, we chose to characterize $\mathbb{P}_\mu$ by saying that

$$\left(\exp\left[\left(\xi, \omega(t) - \int_0^t b\big(\omega(\tau)\big)\, d\tau \right)_{\mathbb{R}^N} - \tfrac{1}{2} \int_0^t \big(\xi, a(\tau)\xi \big)_{\mathbb{R}^N}\, d\tau \right], \mathcal{B}_t, \mathbb{P}_\mu \right)$$

is a martingale for all $\xi \in \mathbb{C}^N$. $\qquad\qquad (6)$

The equivalence of (4) and (5) is a quite easy application of a elementary Fourier analysis and the observation that, aside from integrability issues,

$$\left(M(t)V(t) - \int_0^t M(\tau)\, dV(\tau), \mathcal{B}_t, \mathbb{P} \right)$$

is a martingale when $\big(M(t), \mathcal{B}_t, \mathbb{P} \big)$ is a continuous martingale and $V(t)$ is a continuous, $\{\mathcal{B}_t : t \geq 0\}$-adapted process of locally bounded variation.

(2) The preceding theorem is not the one for which we had originally hoped that we would have to settle. When we started, we had hoped for a statement in which a need only be Borel measurable and locally uniformly elliptic. Using special, heavily dimension dependent arguments, we did prove such a result when $N \in \{1,2\}$, and we somewhat naïvely assumed that it was only the limits on our own technical powers which prevented us from doing so in higher dimensions. However, a few years ago Nikolai Nadirashvili gave in [?] a beautiful example which showed that the limits on our powers were not responsible for our inability to go to higher dimensions. Namely, he produced a uniformly elliptic, bounded, Borel measurable a on $\mathbb{R}^3$ for which the associated martingale problem admits more than one solution.

(3) In retrospect, there are many aspects of this work which look rather crude and unnecessarily cumbersome. For instance, around the same time, first P.A. Meyer [?] and then N. Kunita and S. Watenabe [8] gave a much better and cleaner treatment of the stochastic integral calculus which we developed. More significant, had we been aware of the spectacular work being done by N. Krylov, in particular the estimate in [7], we could have vastly simplified the regularity proof just described.

(**4**) In a later work, we extended the martingale problem approach to cover diffusions with boundary conditions. The resulting paper [13] has, for good reasons, been much less influential: it is nearly unreadable. On the other hand, it contains information which, so far as I know, is available nowhere else. In particular, I do not think the theorems there dealing with approximations by discrete Markov chains have ever been improved.

(**5**) Varadhan and my joint work on diffusion theory culminated in our papers [14] and [15] on degenerate diffusions. The first of these contains our initial version of "the support theorem," which has been given several other proofs over the years. Among other things, the second paper contains an extension of "the support theorem" to cover situations in which the diffusion matrix a is smooth but does not admit a smooth square root. It seems that most probabilists are not particularly impressed by this extension, but we were very proud of it at the time.

References

[1] R.M. Blumenthal and R.K. Getoor, *Markov Processes and Potential Theory*, Academic Press, 1961.

[2] J.L. Doob, *Stochastic Processes*, J. Wiley, N.Y., 1950.

[3] E.B. Dynkin, *Markov Processes, volumes I & II*, Springer–Verlag translation of 1963 Russian edition, 1965.

[4] Wm. Feller, *Zur Theorie der Stochastichen Prozesse*, Math. Ann. **133** (1936) 113–160.

[5] G. Hunt, *Markov processes and potentials I, II, III*, Illinois J. Math. **1** (1957) 44-93; **1** (1957) 316-369; **2** (1958) 151-213.

[6] A.N. Kolmogorov, *Uber die Analytischen Methoden in der Wahrscheinlichkeitsrechnung*, Math. Ann. **104** (1931) 415–458.

[7] Krylov, N., *On quasi diffusion processes*, Theory Prob. and Appl. **11** (1966) 373-389.

[8] Kunita, H. and Watanabe, S. , *On square integrable martingales*, Nagoya Math. J. **30** (1967) 209–245.

[9] H.P. McKean, *Stochastic Integrals*, Academic Press and now available in the AMS Chelsea Series, 1973.

[10] 0P.A. Meyer, *Probability and Potentials*, Blaisdell, Boston, 1966.

[11] 1N.S. Nadirashvili, *Nonuniqueness in the martingale problem and the Dirichlet problem*, Ann. Scuola Norm. Sup. Pisa Cl. Sci. (4) **24** (1997) 537–550.

[12] Stroock, D. and Varadhan, S. , *Diffusion processes with continuous coefficients, I and II*, CPAM **XXII** (1969) 345–400 and 479–30.

[13] _____, *Diffusion processes with boundary conditions*, CPAM **XXIV** (1971) 147–225.

[14] _____, *On the support of diffusion processes, with applications to the strong maximum principle*, Proceedings of the Sixth Berkeley Symposium on Probability and Statistics, 1970333–359 .

[15] _____, *On degenerate elliptic-parabolic operators of the second order and their associated diffusions*, CPAM **XXV** (1972) 651–713.

Varadhan's Work on Large Deviations

Daniel W. Stroock[*]

Suppose that $\{\mu_n : n \in \mathbb{N}\}$ is a sequence of Borel probability measures on a Polish space Ω, and assume that, as $n \to \infty$, μ_n degenerates to the point mass δ_{ω_0} at $\omega_0 \in \Omega$. Then it is reasonable to say that, at least for large n's, neighborhoods of ω_0 represent "typical" behavior and that their complements represent "deviant" behavior, and it is often important to know how fast their complements are becoming deviant. Finding a detailed solution to such a problem usually entails rather intricate analysis. However, if one's interest is in behavior which is "highly deviant" in the sense that it is dying out at an exponential rate, and if one is satisfied with finding the exponential rate at which it is disappearing, then one is studying *large deviations* and life is much easier. Indeed, instead of trying to calculate the asymptotic limit of quantities like[1] $\mu_n\big(B(\omega_0, r)^c\big)$, one is trying to calculate $\lim_{n\to\infty} \frac{1}{n} \log \mu_n\big(B(\omega_0, r)^c\big)$, which is an inherently simpler task.

To develop intuition for this type of analysis, remember that we are dealing here with probability measures. Thus, the only way that the μ_n's can be degenerating to δ_{ω_0} is that more and more of their mass is getting concentrated in a neighborhood of ω_0. In the nicest situation, this concentration is taking place because $\mu_n(d\omega) = \frac{1}{Z_n} e^{-nI(\omega)} \lambda(d\omega)$, where $I(\omega) > I(\omega_0) \geq 0$ for $\omega \neq \omega_0$ and λ is some reference measure. Indeed, assuming that $\lim_{n\to\infty} \frac{1}{n} \log Z_n = 0$, then

$$\lim_{n\to\infty} \frac{1}{n} \log \mu_n(\Gamma) = \lim_{n\to\infty} \log \|1_\Gamma e^{-I}\|_{L^n(\lambda)}$$
$$= \log \|1_\Gamma e^{-I}\|_{L^\infty(\lambda)} = -\operatorname*{essinf}_{\omega \in \Gamma} I(\omega).$$

That is,

$$\lim_{n\to\infty} \frac{1}{n} \log \mu_n(\Gamma) = -\operatorname*{essinf}_{\omega \in \Gamma} I(\omega). \tag{1}$$

Of course, for many applications (e.g., to number theory, geometry, or statistical mechanics), the non-appearance of Z_n in the answer would mean that one has thrown out the baby with the wash. On the other hand, because it is so crude,

---

[*]Department of Mathematics, Massachusetts Institute of Technology, Cambridge, MA 02139

[1]Here, $B(\omega, r)$ denotes the ball of radius r centered at ω.

the type of thinking used in the preceding predicts correct results even when it has no right to. To wit, suppose that $\Omega = \{\omega \in C([0,1]; \mathbb{R}) : \omega(0) = 0\}$ and μ_n is the distribution of $\omega \in \Omega \longmapsto n^{-\frac{1}{2}}\omega \in \Omega$ under standard Wiener measure. Clearly the μ_n's are degenerating to the point mass at the path $\mathbf{0}$ which is identically 0. Moreover, Feynman's representation of μ_n is

$$\mu_n(d\omega) = \frac{1}{Z_n} e^{-nI(\omega)} \lambda(d\omega),$$

where

$$I(\omega) = \frac{1}{2} \int_0^1 |\dot{\omega}(t)|^2 \, dt$$

and λ is Lebesgue measure on Ω. Ignoring the fact that none of this is mathematically kosher and proceeding formally, one is led to the guess that (1) may nonetheless be true, at least after one has taken into account some of its obvious flaws. In particular, there are two sources of concern. The first of these is the almost sure non-differentiability of Wiener paths. However this objection is easily overcome by simply defining $I(\omega) = \infty$ unless ω has a square integrable derivative. The second objection is that λ does not exist and therefore the "ess" before the "inf" has no meaning. This objection is more serious, and its solution requires greater subtlety. In fact, it was his solution to this problem which was one of Varadhan's seminal contributions to the whole field of large deviations. Namely, our derivation of (1) was purely measure theoretic: we took no account of topology. On the other hand, not even the sense in which the μ_n's are degenerating can be rigorously described in purely measure theoretic terms. The best that one can say is that they are tending weakly to $\delta_\mathbf{0}$. Thus, one should suspect that (1) must be amended to reflect the topology of Ω and that topology should appear in exactly the same way that it does in weak convergence. With this in mind, one can understand Varadhan's answer that (1) should be replaced by

$$- \inf_{\omega \in \Gamma^\circ} I(\omega) \leq \varliminf_{n \to \infty} \frac{1}{n} \log \mu_n(\Gamma)$$
$$\leq \varlimsup_{n \to \infty} \frac{1}{n} \log \mu_n(\Gamma) \leq - \inf_{\omega \in \overline{\Gamma}} I(\omega). \tag{2}$$

Monroe Donsker provided the original inspiration for this type of analysis of rescaled Wiener measure, and his student Schilder was the first to obtain rigorous results. However, it was Varadhan in [1] who first realized that Schilder's work could be viewed in the context of large deviations, and it was he who gave and proved the validity of the formulation in (2). Indeed, a strong case can be made for saying that the modern theory of large deviations was born in [1]. In particular, (2) quickly became the archetype for future results, and families

$\{\mu_n : n \in \mathbb{N}\}$ for which (2) are now said to satisfy *the large deviation principle with rate function I.* In addition, it was in the same article that Varadhan proved how to pass from (2) to the sort of results which Schilder had proved. Namely, he proved that if (2) holds with a rate function I which has compact level sets[2] and $F : \Omega \longrightarrow \mathbb{R}$ is a bounded, continuous function, then

$$\lim_{n\to\infty} \mathbb{E}^{\mu_n}\left[e^{nF}\right] = \sup_{\omega\in\Omega}\left(F(\omega) - I(\omega)\right). \tag{3}$$

This result, which is commonly called *Varadhan's Lemma*, is exactly what one would expect from the model case when $\mu_n(d\omega) = \frac{1}{Z_n}e^{-nI}\,\lambda(d\omega)$, and its proof in general is quite easy, but one would be hard put to overstate its importance. Not only is it a practical computational tool, it provides a link between the theory of large deviations and convex analysis. Specifically, when Ω is a closed, convex subset of a topological vector space E, then, under suitable integrability assumptions, Varadhan's Lemma combined with the inversion formula for Legendre transforms often can be used to identify the rate function I as the Legendre transform

$$\Lambda^*(\omega) = \sup_{\lambda\in E^*}\left\{\langle\omega,\lambda\rangle - \Lambda(\lambda) : \lambda \in E^*\right\},$$

where

$$\lambda \in E^* \longrightarrow \Lambda(\lambda) \equiv \lim_{n\to\infty}\frac{1}{n}\log\mathbb{E}^{\mu_n}\left[e^{n\lambda(\omega)}\right].$$

Had he only laid the foundation for the field, Varadhan's impact on the study of large deviations would have been already profound. However, he did much more. Perhaps his deepest contributions come from his recognition that large deviations underlie and explain phenomena in which nobody else even suspected their presence. The depth of his understanding is exemplified by his explanation of Marc Kac's famous formula for the principal eigenvalue of a Schrödinger operator. Donsker had been seeking such an explanation for years, but it was not until he joined forces with Varadhan that real progress was made on the problem. Prior to their article [2], all applications (Schilder's theorem, including its extensions and improvements by Varadhan, as well as the many beautiful articles by Freidlin and Wentcel) of large deviations to diffusion theory had been based on the observation that, during a short time interval, "typical" behavior of a diffusion is given by the solution to an ordinary differential equation. Thus, the large deviations in these applications come from the perturbation of an ordinary differential equation by a Gaussian-like noise term. The large deviations in [2] have an entirely different origin. Instead of short time behavior of the diffusion paths themselves, the quantity under consideration is

---

[2]That is, $\{\omega : I(\omega) \leq R\}$ is compact for each $R \in [0, \infty)$.

the long time behavior of their empirical distribution. In this case, "typical" behavior is predicted by ergodic theory, and the large deviations are those of the empirical distribution from ergodic behavior. The situation in [2] is made more challenging by the fact that there really is no proper ergodic behavior of Brownian motion on the whole of space since, in so far as possible, the empirical distribution of a Brownian path is trying to become normalized Lebesgue measure. What saves the day is the potential term in the Schrödinger operator, whose presence penalizes paths which attempt to spread out too much.

The upshot of Donsker and Varadhan's analysis in [2] is a new variational formula for the principal eigenvalue. Although their formula reduces to classical one in the case of self-adjoint operators, it has the advantage that it relies entirely on probabilistic reasoning (i.e., the minimum principle) and, as they showed in [3], is therefore equally valid for operators which are not self-adjoint. More important, it launched a program which produced a spectacular sequence of articles. The general theory was developed in [3] and [5], each one raising the level[3] of abstraction and, at the same time, revealing more fundamental principles. However, they did not content themselves with general theory. On the contrary, they applied their theory to the solve a remarkably varied set of problems, ranging from questions about the range of a random walk in [6] to questions coming from mathematical physics about function space integrals in [5] and [7], and each abstraction was designed to tackle a specific problem.

As is nearly always the case when breaking new ground, the applications required ingenious modifications of the general theory. To give an indication of just how ingenious these modifications had to be, consider the "Wiener sausage" calculation in [4]. The problem there, which grew out of a question posed by the physicist J. Luttinger, was to find the asymptotic volume of the tubular neighborhood of a Brownian path as the time goes to infinity and the diameter of the neighborhood goes to 0. If one thinks about it, one realizes that this volume can be computed by looking at a neighborhood in the space of measures of the empirical distribution. However, the neighborhood that one needs is the one determined by the variation norm, whereas their general theory deals with the weak topology. Thus, except in one dimension where local time comes to the rescue, they had to combine their general theory with an intricate approximation procedure in order to arrive at their goal. Their calculation in [5] is a true *tour de force*, only exceeded by their solution to the polaron problem in [7].

---

[3]Among their disciples, these levels are designated as levels I, II, and III. Because it reminds me, depending on my mood, of the voice one hears on the elevator as one ascends from "boys clothing" to "women's lingerie" at Macy's or of the levels of hell through which Dante descended, I have never subscribed to this terminology.

In conclusion, it should be emphasized that Varadhan's contributions to the theory of large deviations were to both its foundations and its applications. Because of his work, the subject is now seen as basic tool of analysis, not simply an arcane branch of probability and statistics. With twenty-twenty hindsight, it has become clear that large deviations did not always provide the most efficient or best approach to some of the problems which he solved, but there can be no doubt that his insights have transformed the field forever.

References

[1] Varadhan, S.R.S., *Diffusion processes in a small time interval* , CPAM **20** (1967) 659–685.

[2] Donsker, M. and Varadhan, S.R.S., *On a variational formula for the principal eigenvalue for operators with maximum principle*, PNAS **72** (1975) 780–783.

[3] _____, *Asymptotic evaluation of certain Markov process expectations for large time, I & II*, CPAM **28** (1975) 1–47 & 279–301.

[4] _____, *Asymptotics for the Wiener sausage* CPAM **28#4** (1975) 525–565.

[5] _____, *Asymptotic evaluation of certain Markov process expectations for large time, III* , CPAM **29** #4 (1976) 389–461.

[6] _____, *On the number of distinct sites visited by a random walk*, CPAM **32** #6 (1979) 721–747.

[7] _____, *The polaron problem and large deviations* , Physics Reports. A Review Section of Physics Letters **77** (No. 3) (1981) 235–237.

Varadhan's Work on Large Deviations and Homogenization

Fraydoun Rezakhanlou*

The main theme of this expository article is to review some of Raghu Varadhan and his collaborators' contributions to the question of homogenization for the following stochastic models:

- Stationary Hamilton-Jacobi (HJ) and Hamilon-Jacobi-Bellman (HJB) equation.

- Random walk in random environment (RWRE).

- Simple exclusion process (SEP).

All the above models share similar scaling behaviors and in some sense represent evolving height functions which are governed by local and random growth rules. In fact the law of a RWRE satisfies an equation which resembles a discrete HJB equation, and growth rates of the particle currents in SEP are described by a nonlinear function of the height differences. Reviewing Raghu Varadhan's fundamental contributions sheds light on some universal behavior of stochastic growth models.

1. Hamilton-Jacobi and Hamilon-Jacobi-Bellman Equation

To introduce the basic idea behind homogenization, we first consider (inhomogeneous) Hamilton–Jacobi (HJ) equation

$$u_t = H(x, u_x), \tag{1.1}$$

where H is stationary and ergodic in the first variable x. More precisely, we have a probability space $(\Omega, \mathcal{F}, \mathbb{P})$ with $\mathcal{F}$ a Borel σ-field on Ω and $\mathbb{P}$ a probability measure on $(\Omega, \mathcal{F})$, which is invariant with respect to a family of translation operators, i.e., for every $x \in \mathbb{R}^d$, there exists a measurable function $\tau_x : \Omega \to \Omega$

*This work is supported in part by NSF grant DMS07-07890. Department of Mathematics, UC Berkeley

so that $\tau_x \circ \tau_y = \tau_{x+y}$ and $\mathbb{P}(\tau_x A) = \mathbb{P}(A)$ for every $A \in \mathcal{F}$ and $x, y \in \mathbb{R}^d$. We also assume that τ_x is ergodic; i.e., $\tau_x A = A$ for all $x \in \mathbb{R}^d$ implies that either $\mathbb{P}(A) = 1$ or 0.

Now $H(x, p, \omega) = H_0(\tau_x \omega, p)$ where $H_0 : \Omega \times \mathbb{R}^d \to \mathbb{R}$ is a measurable function. We think of (x, t, u) as the *microscopic coordinates* with the graph of $u(\cdot, t)$ representing a random interface. To switch to *macroscopic coordinates*, we set

$$u^\varepsilon(x, t; \omega) = \varepsilon u\left(\frac{x}{\varepsilon}, \frac{t}{\varepsilon}; \omega\right). \tag{1.2}$$

We now have

$$u_t^\varepsilon = H\left(\frac{x}{\varepsilon}, u_x^\varepsilon\right). \tag{1.3}$$

We note that the right-hand side of (1.3) fluctuates greatly over macroscopic shifts in position x. The huge fluctuation in H, though, does not necessarily imply correspondingly huge fluctuations in u^ε. This is the homogenization phenomenon, i.e., we expect $u^\varepsilon \to \bar{u}$ as $\varepsilon \to 0$ with $\bar{u}$ solving a *homogenized HJ equation*

$$\bar{u}_t = \bar{H}(\bar{u}_x), \tag{1.4}$$

where $\bar{H} : \mathbb{R}^d \to \mathbb{R}$ is the *homogenized Hamiltonian* and does not depend on ω.

As our second example, we consider Hamilton–Jacobi–Bellmann equation

$$u_t = H(x, u_x) + \frac{1}{2}\Delta u \tag{1.5}$$

with $H(x, p) = H(x, p, \omega)$ as before. We define u^ε as in (1.2) and (1.3) becomes

$$u_t^\varepsilon = H\left(\frac{x}{\varepsilon}, u_x^\varepsilon\right) + \frac{\varepsilon}{2}\Delta u^\varepsilon. \tag{1.6}$$

Again we expect to have $u^\varepsilon \to \bar{u}$ with $\bar{u}$ satisfying an equation of the form (1.4) for a different homogenized Hamiltonian $\bar{H}$. Indeed the homogenization for both (1.3) and (1.6) have been achieved by Souganidis [S], Rezakhanlou–Tarver [RT], Lions-Souganidis and Kosygina–Rezakhanlou–Varadhan [KRV] provided that $H(x, p)$ is convex in p, and satisfies suitable technical assumptions that we do not elaborate in this article. (See also Kosygina-Varadhan [KR] when H is allowed to depend on the time variable.) Most notably [KRV] obtains a variational formula for $\bar{H}$. In the case of (1.6), $\bar{H}$ is given by

$$\bar{H}(p) = \inf_g \operatorname*{esssup}_\omega \left[H_0(p + g(\omega), \omega) + \frac{1}{2}\nabla \cdot g(\omega) \right] \tag{1.7}$$

where essential supremum is taken with respect to the probability measure $\mathbb{P}$ and the infimum is taken over functions $g : \Omega \to \mathbb{R}^d$ such that $\mathbb{E}g = 0$ and

$\nabla \cdot g = 0$ weakly. Here ∇ is the generator of the group $\{\tau_x\}$. That is,

$$\nabla f(\omega) \cdot v = \lim_{t \to 0} \frac{1}{t}(f(\tau_{tv}\omega) - f(\omega)), \tag{1.8}$$

whenever the limit exists. We expect a similar formula to hold in the case of (1.3), namely

$$\bar{H}(p) = \inf_g \operatorname*{esssup}_w H_0(p + g(\omega), \omega). \tag{1.9}$$

Before we turn to our next model, let us make an observation regarding the homogenization of (1.6). Note that if

$$H(x, p, \omega) = \frac{1}{2}|p|^2 + b(x, \omega) \cdot p + V(x, \omega), \tag{1.10}$$

and u is a solution of (1.5), then by Hopf–Cole transform, the function $w = e^u$ solves

$$w_t = \frac{1}{2}\Delta w + b(x, \omega) \cdot \nabla w + V(x, \omega)w(x, \omega). \tag{1.11}$$

By Feynmann–Kac Formula, there is a probabilistic representation for w using a diffusion with a drift b. More precisely, if $X(t, x; \omega)$ denotes the solution to

$$dX(t) = b(X(t), \omega)dt + d\beta(t), \tag{1.12}$$
$$X(0) = x,$$

then

$$w(x, t; \omega) = E^\omega w(X(t, x; \omega), 0)e^{\int_0^t V(X(s, x; \omega), \omega)ds}. \tag{1.13}$$

Here β is a standard Brownian motion and E^ω denotes the expected value for the process $X(t)$. The function V is the potential and if $V \leq 0$, then $-V$ may be interpreted as a killing rate for the diffusion X. With this interpretation, $w(x, t; \omega)$ is the expected value of $w(\hat{X}(t), 0)$ with $\hat{X}$ denoting the diffusion with the killing. We now would like to use our probabilistic representation to rewrite u^ε. If

$$u^\varepsilon(x, 0; \omega) = f(x) \tag{1.14}$$

for a deterministic initial condition f, then

$$u^\varepsilon(x, t; \omega) = \varepsilon \log E^\omega \exp\left[\varepsilon^{-1}f(\varepsilon X(t/\varepsilon, x/\varepsilon; \omega)) + \int_0^{t/\varepsilon} V(X(s, x/\varepsilon; \omega), \omega)ds\right].$$
$$\tag{1.15}$$

In particular

$$u^\varepsilon(0, 1; \omega) = \varepsilon \log E^\omega e^{\varepsilon^{-1}f(X(\varepsilon^{-1}; \omega)) + \int_0^{\varepsilon^{-1}} V(X(s; \omega), \omega)ds} \tag{1.16}$$

where $X(s;\omega) := X(s,0;\omega)$ is the diffusion starting from the origin. On the other hand, since $\bar{H}$ is convex (which is evident from (1.7)) we may use Hope–Lax–Oleinik formula to write

$$\bar{u}(x,t) = \sup_y \left(f(y) - t\bar{L}\left(\frac{y-x}{t}\right) \right) \tag{1.17}$$

where $\bar{L}$ is the convex conjugate of $\bar{H}$. In particular,

$$\lim_{\varepsilon \to 0} u^\varepsilon(0,1;\omega) = \bar{u}(0,1) = \sup_y (f(y) - \bar{L}(y)). \tag{1.18}$$

By a celebrated lemma of Varadhan, (1.18) is equivalent to saying that for almost all ω, the diffusion $\hat{X}$ satisfies a *large deviation principle* with the rate function $\bar{L}$. When $b \equiv 0$ and

$$-V(x,\omega) = \sum_{j \in I} V_0(x - x_j) \tag{1.19}$$

with $\omega = \{x_j : j \in I\}$ a Poisson point process and V_0 a continuous function of compact support, the large deviation principle for $\hat{X}$ was earlier established by Sznitman [Sz].

In words, large deviation principle for the diffusion $\hat{X}(\cdot;\omega)$ is equivalent to homogenization for the equation (1.6). Let us write P^ω for the law of the process $\hat{X}(\cdot;\omega)$. What we have in (1.18) is an example of a *quenched* large deviation principle. We may also consider the annealed law

$$\bar{P} = \int P^\omega \, \mathbb{P}(d\omega) \tag{1.20}$$

and wonder whether an annealed large deviation principle is true for the process $\hat{X}$. More precisely, whether or not

$$\lim_{\varepsilon \to 0} \varepsilon \log \int E^\omega \exp\left[\varepsilon^{-1} f(X(\varepsilon^{-1},\omega)) + \int_0^{\varepsilon^{-1}} V(X(s,\omega),\omega)ds \right] \mathbb{P}(d\omega)$$
$$= \sup_y (f(y) - J(y)) \tag{1.21}$$

for a suitable rate function J. In terms of u^ε, this is equivalent to saying

$$\lim_{\varepsilon \to 0} \varepsilon \log \int e^{\varepsilon^{-1} u^\varepsilon(0,1;\omega)} \mathbb{P}(d\omega) = \sup_y (f(y) - J(y)). \tag{1.22}$$

This would follow if we can establish a large deviation principle for the convergence of u^ε to $\bar{u}$. That is, if we can find a function $K_f(y;x,t)$ such that

$$\lim_{\varepsilon \to 0} \varepsilon \log \int e^{\varepsilon^{-1} \lambda u^\varepsilon(x,t;\omega)} \mathbb{P}(d\omega) = \sup_y (\lambda y - K_f(y;x,t)). \tag{1.23}$$

The annealed large deviation (1.22) in the case of $b \equiv 0$ and V of the (1.19) can be found in the manuscript [S] but (1.23) remains open even when $b = 0$.

It is worth mentioning that there is also a variational description for the large deviation rate function $\bar{L}$, namely

$$\bar{L}(v) = \inf_{a} \inf_{\mu \in \Gamma_{a,v}} \int L_0(\omega, a(\omega)) \mu(d\omega) \qquad (1.24)$$

where $L_0(\omega, v)$ is the convex conjugate of $H_0(\omega, p)$ and $\Gamma_{a,v}$ is the set of invariant measures for the diffusions $\mathcal{A}_a = a(\omega) \cdot \nabla + \frac{1}{2}\Delta$ with $\int a(\omega)\mu(d\omega) = v$. In the case of (1.3) the generator $\mathcal{A}_a$ takes the form $a \cdot \nabla$ and when H is periodic in x (i.e., Ω is the d-dimensional torus with $\mathbb{P}$ the uniform measure), the formula (1.24) is equivalent to a formula of Mather for the averaged Lagrangian and our homogenization is closely related to the weak KAM theory. See Fathi [F] and Evans–Gomez [EG] for more details.

2. Random Walk in Random Environment

As our second class of examples, we consider a discrete version of the diffusion (1.12). This is simply a random walk in a random environment (RWRE). To this end, let us write $\mathcal{P}$ for the space of probability densities on the d-dimensional lattice $\mathbb{Z}^d$, i.e., $p \in \mathcal{P}$ if $p : \mathbb{Z}^d \to [0, 1]$ with $\sum_z p(z) = 1$. We set $\Omega = \mathcal{P}^{\mathbb{Z}^d}$ and $\omega \in \Omega$ is written as

$$\omega = (p_a : a \in \mathbb{Z}^d).$$

Given $\omega \in \Omega$, we write $X(n, a; \omega)$ to denote a random walk at time n with starting point $a \in \mathbb{Z}^d$ and transition probabilities $p_a, a \in \mathbb{Z}^d$. More precisely

$$P^\omega(X(n+1) = y \mid X(n) = x) = p_x(y - x).$$

Given a function $g : \mathbb{Z}^d \to \mathbb{R}$, we write

$$T_n g(x) = E^\omega g(X(n, x; \omega)),$$

so that

$$T_1 g(x) = \sum_{y \in \mathbb{Z}^d} g(y) p_x(y - x).$$

To compare with (1.11) in the case of $V \equiv 0$, we also write

$$w(x, n) = T_n g(x)$$

for a given initial g. This trivially solves

$$w(x, n+1) = (T_1 w(\cdot, n))(x).$$

To compare with (1.5), we set $u = \log w$ so that

$$u(x, n+1) - u(x, n) = (Au(\cdot, n))(x)$$

where

$$Ag(x) = \log T_1 e^g(x) - g(x) = \log \sum_z e^{g(x+z)-g(x)} \, p_x(z).$$

Now homogenization means that we are interested in

$$\bar{u}(x, t) = \lim_{\varepsilon \to 0} \varepsilon u\left(\left[\frac{x}{\varepsilon}\right], \left[\frac{t}{\varepsilon}\right]; \omega\right)$$

provided that ω is distributed according to an ergodic stationary probability measure $\mathbb{P}$ where

$$\tau_x \omega = (p_{y+x} : y \in \mathbb{Z}^d).$$

(Here by $[a]$ we mean the integer part of a.) Again $\bar{u}$ solves (1.4) provided that $\lim_{\varepsilon \to 0} u^\varepsilon(x, 0) = \bar{u}(x, 0) = f(x)$ exists initially. The function $\bar{L}$, the convex conjugate of $\bar{H}$, is the quenched large deviation rate function for $X(n; \omega)$. More precisely, for any bounded continuous $f : \mathbb{R}^d \to \mathbb{R}$,

$$\lim_{n \to \infty} n^{-1} \log E e^{f(n^{-1} X(n, 0; \omega))} = \sup_y (f(y) - \bar{L}(y)).$$

This has been established under an ellipticity condition on p_x by Varadhan [V1]. See Bolthausen and Sznitman [BS] for a survey on earlier results. The analog of (1.7) is the following formula of Rosenbluth [Ro] :

$$\bar{H}(p) = \inf_g \operatorname*{esssup}_\omega \sum_y p_0(y) e^{p \cdot z + g(\omega, z)} \tag{2.1}$$

with infimum over functions $(g(\cdot, z) : \Omega \to \mathbb{R} : z \in \mathbb{Z}^d)$ such that $\mathbb{E}g(\cdot, z) = 0$ and g is a "closed 1-form". By the latter we mean that for every loop $x_0, x_1, \ldots, x_{k-1}, x_k = x_0$, we have that $\sum_{r=0}^{k-1} g(\tau_{x_r} \omega, x_{k+1} - x_k) = 0$.

We now turn to the annealed large deviations for a RWRE. For this, we need to select a tractable law for the environment. Pick a probability measure β on $\mathcal{P}$ and set $\mathbb{P}$ to be the product of β to obtain a law on $\mathcal{P}^{\mathbb{Z}^d}$. The annealed measure $\bar{P} = \int P^\omega \, \mathbb{P}(d\omega)$ has a simple description. For this, we write $Z(n) = X(n+1) - X(n)$ for the jump the walk performs at time n. We also define

$$N_{x,z}(n) = \#\{i \in \{0, 1, 2, \ldots, n\} : X(i) = x, Z(i) = z\}.$$

We certainly have

$$P(X(1; \omega) = x_1, \ldots, X(n, \omega) = x_n) = \prod_{z, x \in \mathbb{Z}^d} (p_x(z))^{N_{x,z}(n)},$$

$$\bar{P}(X(1) = x_1, \ldots, X(n) = x_n) = \prod_{x \in \mathbb{Z}^d} \int \prod_z (p(z))^{N_{x,z}(n)} \beta(dp),$$

where now

$$N_{x,z}(n) = \#\{i \in \{0, 1, \ldots, n-1\} : x_i = x, x_{i+1} - x_i = z\}.$$

Evidently $\bar{P}$ is the law of a non-Markovian walk in $\mathbb{Z}^d$. Varadhan in [V1] establishes the annealed large deviations principle under a suitable ellipticity condition on β. The method relies on the fact that the environment seen from the walker is a Markov process for which Donsker–Varadhan Theory may apply if we have enough control on the transition probabilities.

If we set

$$W_n = (0 - X(n), X(1) - X(n), \ldots, X(n-1) - X(n), X(n) - X(n))$$
$$= (s_{-n}, \ldots, s_{-1}, s_0 = 0)$$

for the chain seen from the location $X(n)$, then we obtain a walk of length n that ends at 0. The space of such walks is denoted by $\mathbf{W}_n$. Under the law $\bar{P}$, the sequence $W_1, W_2, \ldots$ is a Markov chain with the following rule:

$$\bar{P}(W_{n+1} = T_z W_n \mid W_n) = \frac{\bar{P}(T_z W_n)}{\bar{P}(W_n)} = \frac{\int_{\mathcal{P}} p(z) \prod_a p(a)^{N_{0,a}} \beta(dp)}{\int_{\mathcal{P}} \prod_a p(a)^{N_{0,a}} \beta(dp)}, \qquad (2.2)$$

where $N_{0,a} = N_{0,a}(W_n)$ is the number of jumps of size a from 0 for the walk W_n. Here $T_z W_n$ denotes a walk of size $n + 1$, which is formed by translating the walk W_n by $-z$ so that it ends at $-z$ instead of 0, and then making the new jump of size z so that it ends at 0.

We wish to establish a large deviation principle for the Markov chain with transition probability $q(W, z)$ given by (2.2) where $W = W_n \in \bigcup_{m=0}^{\infty} \mathbf{W}_m$ and z is the jump size. We assume that with probability one, the support of $p_0(\cdot)$ is contained in the set $D = \{z : |z| \leq C_0\}$. Naturally q extends to those infinite walks $W \in \mathbf{W}_\infty$ with $N_{0,a} < \infty$ for every $a \in D$. If we let $\mathbf{W}_\infty^{tr}$ denote the set of transient walks, then the expression $q(W, z) = \mathbf{q}(W, T_z W)$ given by (2.2) defines the transition probability for a Markov chain in $\mathbf{W}_\infty^{tr}$. Donsker–Varadhan Theory suggests that the empirical measure

$$\frac{1}{n} \sum_{m=0}^{n-1} \delta_{W_m}$$

satisfies a large deviation principle with a rate function

$$\mathcal{I}(\mu) = \int_{\mathbf{W}_\infty^{tr}} \mathbf{q}_\mu(W, z) \log \frac{\mathbf{q}_\mu(W, z)}{\mathbf{q}(W, z)} \mu(dW)$$

43

where μ is any T-invariant measure on $\mathbf{W}_\infty^{tr}$ and $\mathbf{q}_\mu(W, z)$ is the conditional probability of a jump of size z, given the past history. We then use the contraction principle to come up with a candidate for the large deviation rate function

$$H(v) = \inf \left\{ I(\mu) : \int z_0 \mu(dW) = v \right\}$$

where $(z_j : j \in \mathbb{Z})$ denotes the jumps of a walk W. Several technical difficulties arise as one tries to apply Donsker–Varadhan Theory because of non-compactness of the state space and the fact that the transition probabilities are not continuous. These issues are handled masterfully in [V1].

3. Simple Exclusion Process

We now turn to our final model. This time our environment $\omega = (p_i(t) : i \in \mathbb{Z})$ is a collection of independent Poisson clocks. More precisely p_i, $i \in \mathbb{Z}$ are independent and each p_i is a Poisson process of rate 1; $p_i(t) = k$ for $t \in [\tau_1^i + \cdots + \tau_k^i, \tau_1^i + \cdots + \tau_{k+1}^i)$ with τ_j^i independent mean 1 exponential random variables. Given a realization of ω and an initial height function $h^0 \in \Gamma = \{h : \mathbb{Z} \to \mathbb{Z}$ such that $0 \le h(i+1) - h(i) \le 1\}$, we construct $h(i, t) = h(i, t; \omega)$ such that $h(\cdot, t; \omega) \in \Gamma$ for all t. More precisely, at each Poisson time $t = \tau_1^i + \cdots + \tau_k^i$, the height $h(i, t)$ increases by one unit provided that the resulting height function h^i belongs to Γ; otherwise the increase is suppressed.

The process $h(\cdot, t)$ is a Markov process with the following rules: $h \to h^i$ with rate $\eta(i+1)(1 - \eta(i))$ where $\eta(i) = h(i) - h(i-1)$. The process $(\eta(i, t; \omega) : i \in \mathbb{Z})$ is also Markovian with the interpretation that $\eta(i, t) = 1$ if the site i is occupied by a particle and $\eta(i, t) = 0$ if the site i is vacant. Now the growth $h \to h^i$ is equivalent to jumping a particle from site $i + 1$ to i provided that the site i is vacant. Since $h \in \Gamma$ is nondecreasing, we may define its inverse $x \in \Gamma'$, where

$$\Gamma' = \{x : \mathbb{Z} \to \mathbb{Z} \text{ such that } x(h + 1) > x(h)\}.$$

Since h increases at a site $i + 1$ if the site i is occupied by a particle, we may regard $x(h)$ as the position of a particle of label h. Equivalently we may interpret $h(i)$ as the label of a particle at an occupied site i.

The process $x(h, t; \omega)$ is also a Markov process with the following rules: $x(h, t) \to x(h, t) - 1$ with rate $\mathbb{1}(x(h, t) - x(h-1, t) > 1)$. In words $x(h)$ decreases by one unit with rate 1 provided that the resulting configuration x_h is still in Γ'. For the construction of $x(h, t; \omega)$ we may use the clocks ω or equivalently we may use clocks that are assigned to sites $h \in \mathbb{Z}$. More precisely, if $\omega' = (p_h'(t) : h \in \mathbb{Z})$ is a collection of independent Poisson processes of rate 1, then we decrease $x(h)$ by one unit when the clock p_h' rings. The processes $x(h, t; \omega)$ and $x(h, t; \omega')$ have the same distribution. If we define $\zeta(h,t) = x(h,t) - x(h-1,t) - 1$, then $\zeta(h,t)$ represents the gap between the h-th and $(h-1)$-th particles in the exclusion process. The process $(\zeta(h,t) : h \in \mathbb{Z})$ is the celebrated *zero range process* and can be regarded as the occupation number at site h. The ζ-process is also Markovian where a ζ-particle at site h jumps to site $h+1$ with rate $\mathbb{1}(\zeta(h) > 0)$.

As in the previous sections, we set

$$u^\varepsilon(x,t;\omega) = \varepsilon h\left(\left[\frac{x}{\varepsilon}\right], \frac{t}{\varepsilon};\omega\right), \quad x^\varepsilon(u,t;\omega) = \varepsilon x\left(\left[\frac{u}{\varepsilon}\right], \frac{t}{\varepsilon};\omega\right)$$

and as a homogenization we expect to have $u^\varepsilon \to \bar{u}$, $x^\varepsilon \to \bar{x}$ with $\bar{u}$ and $\bar{x}$ deterministic solutions to Hamilton-Jacobi equations

$$\bar{u}_t = \bar{H}_1(\bar{u}_x) = \bar{u}_x(1 - \bar{u}_x) \tag{3.1}$$

$$\bar{x}_t = \bar{H}_2(\bar{x}_u) = \frac{1}{\bar{x}_u} - 1. \tag{3.2}$$

(See [R].) As for the large deviations, we will be interested in

$$\lim_{\varepsilon \to 0} \varepsilon \log \mathbb{P}(u^\varepsilon(x,t) \geq u) = \lim_{\varepsilon \to 0} \varepsilon \log \mathbb{P}(x^\varepsilon(u,t) \geq x) =: -W(x,u,t). \tag{3.3}$$

Evidently $W(x,u,t) = 0$ if $u \leq \bar{u}(x,t)$ or $x \leq \bar{x}(u,t)$. However we have that $W(x,u,t) > 0$, whenever $u > \bar{u}(x,t)$ or $x > \bar{x}(u,t)$. As it turns out,

$$\lim_{\varepsilon \to 0} \varepsilon \log \mathbb{P}(u^\varepsilon(x,t) \leq u) = -\infty \tag{3.4}$$

for $u < \bar{u}(x,t)$ because for such a number u,

$$\liminf_{\varepsilon \to 0} \varepsilon^2 \log \mathbb{P}(u^\varepsilon(x,t) \leq u) > 0$$

as was demonstrated by Jensen and Varadhan [JV]. Quoting from [JV], the statement (3.3) has to do with the fact that one may slow down $x(h,t)$ for $h \leq h_0$ in a time interval of order $O(\varepsilon^{-1})$ by simply slowing down $x(h_0,t)$. This can be achieved for an entropy price of order $O(\varepsilon^{-1})$. However for $x^\varepsilon(u,t) \leq \bar{x}(u,t) - \delta$, $\delta > 0$, we need to speed up $O(\varepsilon^{-1})$ many particles for a time interval of order $O(\varepsilon^{-1})$. This requires an entropy price of order $O(\varepsilon^{-2})$.

As was observed by Seppäläinen [S1], both h and x processes enjoy a strong monotonicity property. More precisely, if we write $x(h,t;\omega) = T_t^\omega x^0(h)$ for the x-process starting from the initial configuration $x^0 \in \Gamma'$, then $T_t^\omega(\sup_\alpha x_\alpha^0) = \sup_\alpha T_t^\omega(x_\alpha^0)$. In words, if the initial height $x^0 = \sup_\alpha x_\alpha^0$ is the supremum of a family of height functions x_α^0, then it suffices to evolve each x_α^0 separately for a given realization of ω and take the supremum afterwards. From this, it is not hard to show that such a strong monotonicity must be valid for W and this in turn implies that W solves a H-J equation of the form

$$W_t = K(W_x, W_u). \tag{3.5}$$

Here the initial data $W(x, u, 0)$ is the large deviation rate function at initial time. Of course we assume that there is a large deviation rate function initially and would like to derive a large deviation principle at later times. In the case of exclusion or zero-range process, it is not hard to guess what K is because when the process is at equilibrium, then the height function at a given site has a simple description. To construct the equilibrium measures for the x-process, let us pick a number $b \in (0, 1)$ and define a random initial height function $x(\cdot, 0)$ by the requirement that $x(\cdot, 0) = 0$ and that $(x(h + 1, 0) - x(h, 0) - 1 : h \in \mathbb{Z})$ are independent geometric random variables of parameter b. That is $x(h + 1, 0) - x(h, 0) = k + 1$ with probability $(1 - b)b^k$. Let us write $\mathbb{P}^b$ for the law of the corresponding process $x(h, t; \omega)$. Using Cramer's large-deviation theorem, we can readily calculate that for u positive,

$$W(x, u, 0) = -\lim_{\varepsilon \to 0} \varepsilon \log \mathbb{P}^b(x^\varepsilon(u, 0) \geq x) = u \left(I_1 \left(\frac{x}{u} + 1, b \right) \right)^+ \tag{3.6}$$

where $I_1(r, b) = r \log \frac{r}{b(1+r)} - \log[(1 - b)(1 + r)]$. As is well-known (see for example Chapter VIII, Corollary 4.9 of Liggett [L]), $-x(0, t)$ is a Poisson process which decreases one unit with rate b. Again Cramer's theorem yields

$$W(x, 0, t) = bt \left(I_2 \left(\frac{-x}{bt} \right) \right)^+ \tag{3.7}$$

where $I_2(r) = r \log r - r + 1$. The expressions (3.5)–(3.7) provide us with enough information to figure out what K is. We refer to [S2] for a large deviation principle of the form (3.3) for a related particle system known as Hammersley's model.

Alternatively, we may study the large derivation of the particle densities. For this purpose, we define the empirical measure by

$$\pi^\varepsilon(t, dx) = \pi^\varepsilon(t, dx; \omega) = \varepsilon \sum_i \delta_{\varepsilon i}(dx)\eta(i, t/\varepsilon; \omega).$$

We regard π^ε as an element of the Skorohod space $X = \mathcal{D}([0, T]; M)$ where M is the space of locally bounded measures. The law of $\omega \mapsto \pi^\varepsilon(t, dx; \omega)$ induces a probability measure $\mathcal{P}^\varepsilon$ on X.

Hydrodynamic limit for the exclusion process means that $\mathcal{P}^\varepsilon \Rightarrow \mathcal{P}$ where $\mathcal{P}$ is concentrated on the single *entropy* solution of

$$\bar{\rho}_t = (\bar{\rho}(1 - \bar{\rho}))_x, \tag{3.8}$$

46

for a given initial data $\bar{\rho}(x,0) = \bar{\rho}^0(x)$. The function $\hat{\rho}$ is related to the macroscopic height function $\bar{u}$ by $\bar{\rho} = \bar{u}_x$. In Jensen–Varadhan [JV], a large deviation principle has been established for the convergence of $\mathcal{P}^\varepsilon$. Roughly

$$\mathcal{P}^\varepsilon \left(\pi^\varepsilon(t, dx) \text{ is near } \mu(t, dx) \right) \approx e^{-\varepsilon^{-1}\mathcal{I}(\mu)} \qquad (3.9)$$

with the following rate function $\mathcal{I}$. First $\mathcal{I}(\mu) = +\infty$ unless $\mu(t, dx) = m(x, t)dx$ and m is a weak solution of (3.8). However when $0 < \mathcal{I}(m) < \infty$, then m is a non-entropic solution of (3.8). In fact $\mathcal{I}(\mu) = \mathcal{I}_0(\mu) + \mathcal{I}_{dym}(\mu)$ where $\mathcal{I}_0(\mu)$ is the large-deviation rate function coming from the initial deviation and depends only on our choice of initial configurations, and $\mathcal{I}_{dym}(\mu)$ is the contribution coming from dynamics and quantitatively measures how the entropy condition is violated. By the entropy condition we mean that for a pair (ϕ, q) with ϕ convex and $\phi' \bar{H}_1' = q'$ for $\bar{H}_1(p) = p(1 - p)$, we have

$$\phi(\bar{\rho})_t + q(\bar{\rho})_x \leq 0 \qquad (3.10)$$

in the weak sense. The left-hand side is a negative distribution which can only be a negative measure. As our discussions around (3.5) and (3.6) indicate, the invariant measures play an essential role in determining the large-deviations rate function. As it turns out the relevant ϕ to choose is simply the large deviation rate function for the invariant measure which is given by

$$\phi(m) = m \log m + (1 - m) \log(1 - m) + \log 2.$$

Here for the invariant measure we choose a Bernoulli measure ν under which $(\eta(i) : i \in \mathbb{Z})$ are independent and $\nu(\eta(i) = 1) = 1/2$. To measure the failure of the entropy solution, we take a weak solution m for which the corresponding

$$\phi(m)_t + q(p)_x = \gamma = \gamma^+ - \gamma^-$$

is a measure with γ^+ and γ^- representing the positive and negative parts of γ. We now have

$$\mathcal{I}_{dyn}(\mu) = \gamma^+(\mathbb{R} \times [0, T]).$$

It is customary in equilibrium statistical mechanics to represent a state as a probability measure with density $\frac{1}{Z}e^{-\beta H}$ with H some type of energy and Z the normalizing constant. In non-equilibrium statistical mechanics a large deviation principle of the form (3.9) offers an analogous expression with $\mathcal{I}(\mu)$ playing the role of "effective" energy (or rather potential). What we learn from [JV] is that after the entropy solution, the most frequently visited configurations are those associated with non-entropic solutions and the entropic price for such visits are measured by the amount the inequality (3.10) fails. Even though the entropy solutions for scalar conservation laws are rather well-understood, our understanding of non-entropic solutions is rather poor perhaps because we had no reason to pay attention to them before. The remarkable work [JV] urges us to look more deeply into non-entropic solutions for gaining insight into the way the microscopic densities deviate from the solution to the macroscopic equations.

References

[BS]   Bolthausen, E., Sznitman, A-S., Ten lectures on random media. DMV Seminar, 32. Birkhuser Verlag, Basel, 2002.

[EG]   Evans, L. C., Gomes, D., Effective Hamiltonians and averaging for Hamiltonian dynamics. I. Arch. Ration. Mech. Anal. 157 (2001), 1–33.

[F]   Fathi,A., The Weak Kam Theorem in Lagrangian Dynamics (Cambridge Studies in Advanced Mathematics), to appear.

[JV]   Jensen, L. H., and Varadhan, S. R. S., Large deviations of the asymmetric exclusion process in one dimension, preprint.

[KR]   Kosygina, E., Varadhan S.R.S. Homogenization of Hamilton-Jacobi-Bellman equations with respect to time-space shifts in a stationary ergodic medium, Comm. Pure Appl. Math. **61** (2008), 816847.

[KRV] Kosygina, E., Rezakhanlou, F., and Varadhan, S. R. S., Homogenization of stochastic Hamilton–Jacobi–Bellmann equations, Comm. Pure Appl. Math. **59**(2006), 1489–1521.

[LS]   Lions, P-L,, Souganidis, P. E., Homogenization of "viscous" Hamilton-Jacobi equations in stationary ergodic media, Comm. Partial Differential Equations **30** (2005), no. 1-3, 335–375.

[L]   Liggett,T. "Interacting particle systems." Springer-Verlag, Berlin, 2005.

[R]   Rezakhanlou, F., Hydrodynamic limit for attractive particle systems on $\mathbb{Z}^d$, Comm. Math. Phys. **140** (1991), 417–448.

[RT]   Rezakhanlou, F., and Tarver, J. E., Homogenizations for stochastic Hamilton–Jacobi equations, Arch. Ration. Mech. Anal. **151**(4) (2000), 227–309.

[Ro]   Rosenbluth, J., Quenched large deviations for multidimensional random walk in random environment: A variational formula. Ph.D thesis, New York University, 2006.

[S1]   Seppäläinen, T., Existence of hydrodynamics for the totally asymmetric simple K-exclusion process, Ann. Probab. **27** (1999), 361–415.

[S2]   Seppäläinen, T., Large deviations for increasing sequences on the plane, Probab. Theory Related Fields, **112** (1998), 221–244.

[S]   Souganidis, P. E., Stochastic homogenization of Hamilton–Jacobi equations and some applications, Asymptot. Anal. **20**(1) (1999), 1–11.

[Sz]   Sznitman, A. S., Brownian Motion, Obstacles and Random Media, Springer, 1998.

[V1]   Varadhan, S. R. S., Large deviations for the asymmetric exclusion process, 1–27, Adv. Stud. Pure Math., **39**, Math. Soc. Japan, Tokyo, 2004.

[V2]   Varadhan, S. R. S., Large deviations for random walks in a random environment, Comm. Pure Appl. Math. **56** (2003), 1222-1245.

Varadhan's Work on Hydrodynamical Limits

Jeremy Quastel* and Horng-Tzer Yau†

In the centuries since the discovery of Newton equation, the quest to solve the many-body problem has been one of the most persistent endeavours of mathematics and physics. Although progress was made in approximating it when the number of particles is small, a solution for large numbers of particles in any useful form seems simply impossible. The fundamental observation of Boltzmann was that the typical behavior for classical Hamiltonian systems in equilibrium is governed by ensemble averages – Gibbs states in today's language. This avoided the difficulty of directly solving the Newton equations by postulating statistical ensembles, and lead to modern statistical physics and ergodic theory. Boltzmann's formulation concerned systems in equilibrium; in other words, behavior of systems as the time approaches infinity.

At the other end is the kinetic theory for short time behavior. Classical dynamics are exactly solvable when there is no interactions among particles, i.e., free dynamics. In short time classical dynamics can be understood by supplementing the free dynamics with collisions. The fundamental observation of kinetic theory, the idealization of the collision processes, is again due to Boltzmann in his celebrated work on the Boltzmann equation.

For systems neither in equilibrium nor near free dynamics, i.e., for time scales too short for equilibrium theory, but longer than the scale for kinetic theory, the most useful descriptions are still the classical macroscopic equations, e.g., the Euler and Navier-Stokes equations. These are continuum formulations of conservation of mass and momentum, and also contain some phenomenological concepts such as viscosity. They are equations for macroscopic quantities such as density, velocity (momentum) and energy, while the Boltzmann equation is an equation for the probability density of finding a particle at a fixed position and velocity. The classical Hamiltonian plays no active role in either formulation, and all the microscopic effects are summarized by the viscosity in the Navier-Stokes equations, or the collision operator in the Boltzmann equation. But the central theoretical question, i.e., understanding the connection

---

*Departments of Mathematics and Statistics, University of Toronto, Toronto, Canada M5S 2E4

†Department of Mathematics, Harvard University, Cambridge, MA 02138

between large particle systems and their continuum approximations, remained unsolved and is still one of the fundamental questions in non-equilibrium statistical physics.

Since classical dynamics of large systems are all but impossible to solve, a more feasible goal is to replace the classical dynamics with stochastic dynamics. From the sixties to early eighties, tremendous effort was made by Dobrushin, Lebowitz, Spohn, Presutti, Spitzer, Liggett, and others to understand large stochastic particle systems. A key focus was to derive rigorously the classical phenomenological equations from the interacting particle systems in suitable scaling limits. The methods at the time were based on coupling and perturbative arguments; the systems which can be treated rigorously were restricted to special one dimensional systems, perturbations of the symmetric simple exclusion or mean-field type interactions. Together with Guo and Papanicoloau, Varadhan [3] introduced the first general approach, the entropy method. The key ideas are the dissipative nature of the entropy, and large deviations. As long as the equilibrium measures of the dynamics are known, and the scaling is diffusive, this approach is very effective and now has been applied to many systems. We now sketch the approach in [3]. The method is most transparent in a model (the hydrodynamical limit of this model was first proved under some more restrictive assumptions in [1]) where the particle number is replaced by a real valued scalar field $\phi_x \in \mathbb{R}$, $x \in \{1, \cdots, N\}$ with periodic boundary condition so that $N + 1 = 1$. These evolve as interacting diffusions. Denote by μ_0 the product measure such that the law of ϕ_x is given by $e^{-V(\phi_x)}$. Let $f_t \mu_0$ be distribution of the field ϕ at the time t. The dynamics of ϕ will be given by the evolution equation for f_t:

$$\partial_t f_t = L f_t, \quad L = \sum_{j=1}^{N} \left(\frac{\partial}{\partial \phi_j} - \frac{\partial}{\partial \phi_{j+1}} \right)^2$$

$$- \sum_{j=1}^{N} (V'(\phi_j) - V'(\phi_{j+1})) \left(\frac{\partial}{\partial \phi_j} - \frac{\partial}{\partial \phi_{j+1}} \right) \quad (1)$$

This dynamics is reversible with respect to the invariant measure μ_0 and the Dirichlet form is given by

$$D(f) := - \int f L f \, d\mu_0 = \sum_{j=1}^{N} \int \left(\frac{\partial f}{\partial \phi_j} - \frac{\partial f}{\partial \phi_{j+1}} \right)^2 d\mu_0 \quad (2)$$

We rescale the time diffusively so that the evolution equation becomes

$$\partial_t f_t = \varepsilon^{-2} L f_t, \quad (3)$$

where $\varepsilon = N^{-1}$ is the scaling parameter.

The dynamics (3) can be written as a conservation law,

$$d\phi_i = (w_{i+1} - w_i)dt + dM_i, \qquad w_i = N^2 V'(\phi_i) - N^2 V'(\phi_{i-1}) \qquad (4)$$

where M_i are martingales and w_i is the microscopic current. The current is itself a gradient and our dynamics is formally diffusive, i.e., for any smooth test function J

$$dN^{-1} \sum_i J(\varepsilon i)\phi_i \sim N^{-1} \sum_i J''(\varepsilon i)V'(\phi_i)dt + dM. \qquad (5)$$

Denote by $\rho(x, t)$ the local average of ϕ around $x = \varepsilon i$,

$$\rho(x, t) = \lim_{\delta \to 0} \lim_{\varepsilon \to 0} \frac{1}{2\varepsilon\delta + 1} \sum_{|j-i| \leq \delta\varepsilon^{-1}} \phi_j \qquad (6)$$

Let $p(\lambda)$ denote the pressure

$$p(\lambda) = \log \int_{\mathbb{R}} d\phi e^{\lambda\phi - V(\phi)} \qquad (7)$$

and $h(m)$ denote the free energy, the Legendre transform of p. It is not hard to check that the martingale term in (5) vanishes in the limit and hence the main task to establish the hydrodynamic limit of the form

$$\partial_t \rho = \partial_{xx} h'(\rho) \qquad (8)$$

is to prove that we can replace $V'(\phi_i)$ in (5) by $h'(\rho(x, t))$ in the sense of a law of large numbers with respect to the distribution f_t satisfying (3).

Consider the local Gibbs measure with a chemical potential λ,

$$d\mu_\lambda(\phi_1, \cdots, \phi_N) = \exp\left[\sum_j \lambda(\varepsilon j)\phi_j\right] d\mu_0 \qquad (9)$$

where the chemical potential λ is allowed to depend on the site slowly. If f_t is a local Gibbs state, then certainly we can replace $V'(\phi_i)$ in (5) by $h'(\rho(x, t))$ in the sense of law of large numbers. The key observation of GPV is to consider the evolution of the entropy:

$$\partial_t S(f_t) = -D(\sqrt{f_t}), \qquad S(f) = \int f \log f d\mu_0 \qquad (10)$$

For a typical system of N variables given by a density f with respect to μ_0, the entropy $S(f)$ is of order N. This implies that

$$\int_0^t D(\sqrt{f_s})ds \leq CN \qquad (11)$$

This information alone is sufficient to establish that the solution of (3) is close enough to a local Gibbs state that the law of large numbers continues to hold.

Systems where the current is itself a gradient of some other function are known as *gradient systems*. For such systems, [3] provides a general framework to establish the hydrodynamical limit. However many systems are of nongradient type. A simple illustrative example is to modify the dynamics (1) so that the generator L becomes the symmetric operator with Dirichlet form,

$$D(f) := -\int fLf d\mu_0 = \sum_{j=1}^{N} \int a(\phi_j, \phi_{j+1}) \left(\frac{\partial f}{\partial \phi_j} - \frac{\partial f}{\partial \phi_{j+1}} \right)^2 d\mu_0 \qquad (12)$$

The current can be computed easily and is given (up to scale factors) by

$$w_j = -\frac{\partial a(\phi_j, \phi_{j+1})}{\partial \phi_j} + \frac{\partial a(\phi_j, \phi_{j+1})}{\partial \phi_{j+1}} + a(\phi_j, \phi_{j+1})[\phi_1 - \phi_{j+1}]. \qquad (13)$$

To establish the hydrodynamical limit, it is now required to prove that in some sense w_j can be replaced by $(x = \varepsilon j)$

$$w_j = D(\rho(x,t))[\phi_{j+1} - \phi_j] \qquad (14)$$

for some function D, which will be the diffusion coefficient of the hydrodynamical equation. The key observation of Varadhan's work on nongradient systems is that

$$w_j = D(\rho(x,t))[\phi_{j+1} - \phi_j] + LF \qquad (15)$$

where F is some local function of ϕ_j and L is the generator given by (12). The idea is that functions of the type LF represent incoherent rapid fluctuations which vanish over the long time scale of the hydrodynamic limit. This fluctuation is indeed in the system and the hydrodynamical limit can be established only if we properly account for its effect.

The sense in which (15) holds is in the H_{-1} sense, corresponding to the vanishing of the variance in the central limit theorem for the corresponding additive functional. This goes back to Varadhan's earlier work on *tagged particles* [4]. The problem of proving the convergence of tagged particles to appropriate diffusions is somewhat complementary to the hydrodynamic limit. Varadhan introduced the martingale method in this context so that the idea of viewing the system from the point of view of the particle can be implemented. These ideas have had broad influence not only in hydrodynamic limits, but also in homogenization theory and in random walk in random environment. In fact, there is even a more explicit connection between the tagged particle problems and the non-gradient systems. Suppose that one gives each particle one of m different labels, and watches the evolution of the m different densities in the hydrodynamic limit. The corresponding particle systems are usually of non-gradient form [8] as long as $m \geq 2$. This is a weak form of tagging, and the large m limit of this system is a (weak) way to keep track of individual particles. It can be proved, via nongradient system method, that each species of particles evolves according to a diffusion equation and thus the hydrodynamical limit of tagged particles in nonequilibrium is established [9]. The advantage of this approach is that it can be done in nonequilibrium, identifying the collective drift imposed by the flow of the bulk towards equilibrium. However, it is strictly speaking not the behavior of a single tagged particle, but the average behavior of tagged particles with vanishing density.

The equation (15) is quite difficult to solve as it involves the full generator L. In order to solve it, Varadhan developed a method which can be viewed as an infinite dimensional version of Hodge theory. This is a deep theory and we shall only attempt to give some flavor. First note that because of the entropy bound (11), one only has to solve (15) in equilibrium. So the diffusion coefficient can be treated as a constant. Now for simplicity take $j = 0$. The current w_0 has the property that

$$\int w_0 d\mu_\lambda = 0 \qquad (16)$$

for any Gibbs state with constant chemical potential λ. The space of functions with this property corresponds to a space of *closed forms*. A subspace of *exact forms* corresponds to the fluctuation terms LF. The deep result is that the exact forms are of codimension one in the space of closed forms with orthogonal complement corresponding to $\phi_{j+1} - \phi_j$. This solves (15). This approach, as it stands, is based on the integration by parts nature of L and applies only to reversible dynamics. It is possible to formulate it also for nonreversible dynamics and the formal analogy between this equation and the Hodge theory can be strengthened [15].

The two fundamental papers [3, 16] of Varadhan ushered in an era of hydrodynamical limits based on the idea of entropy. The developments following these two papers are astonishing and we shall only mention a few. The approach of [3] was successfully applied to many systems, including interacting Brownian motions [14], interacting Ornstein-Uhlenbeck processes [6] and Ginzburg-Landau models [11]. The interacting Brownian motions and interacting Ornstein-Uhlenbeck processes are continuum systems with no lattice structure. The hydrodynamical limit for the Ginzburg-Landau models was proved for all temperatures, including the phase transition region—a remarkable result. Furthermore, the approach of GPV was successfully applied to kinetic scaling and led to the derivation of the Boltzmann equation from stochastic particle systems [13]. The idea that the solution of (3) is heuristically a local Gibbs states goes back many decades. But the estimates obtained in GPV are in fact strong enough to prove it. It is observed in [17] that one can bypass many technical difficulties in GPV and prove directly that the local Gibbs states are in fact an approximate solution to (3) in the sense of relative entropy. The assumptions needed in this approach are: 1. Some ergodic properties of the dynamics. 2. Smoothness of solutions to the hydrodynamical equations. This method is more restrictive than GPV for diffusive systems, but it essentially relies only on the identification of the invariant measures of the dynamics and it applies also to hyperbolic systems before the formation of shocks. It was adapted in [7] to derive the classical Euler equation from Hamiltonian systems with vanishing noise. This is the most significant advance since Morrey stated this problem in the 60's. Once the hyperbolic equations develop shocks, a very different method is needed, see [12, 2] for references and related results.

Varadhan's work on nongradient systems requires a spectral gap of order ℓ^{-2} for the system in a box of side length ℓ. This inspired work on the estimates of spectral gaps of conservative dynamics and it led to the development of martingale methods for estimating spectral gap for conservative dynamics. Using this spectral estimate, Varadhan and his coworker [15] established the hydrodynamical limit of lattice gas in the high temperature phase.

The idea that the current can be decomposed into a dissipative term $\phi_j - \phi_{j+1}$ and a fluctuation term LF is a deep idea and is really a rigorous statement of the so-called fluctuation-dissipation theorem in physics. In a sense, the insight that this equation is fundamental to the hydrodynamical limit is at least as significant as the solution of this equation for the specific model considered in [16]. Although the fluctuation-dissipation equation was solved in [16] only for reversible dynamics, it was realized that one can develop a method to solve this equation for nonreversible dynamics provided that the spatial dimension is larger than two [5]. This led to the derivation of the incompressible Navier-Stokes (INS) equations from stochastic lattice gases for dimension $d = 3$ [10]. The result obtained in [10] is very strong; it identifies the large deviation rate that the hydrodynamical equation is not a Leray solution and does not assume that the INS equations have classical solutions. The physical significance is the following: The first principles equation governing a classical fluid is the Newton equation, which is time reversible and has no dissipation. The INS equations possess viscosity and are time irreversible. Therefore, a derivation of the INS equations from classical mechanics would have to answer the fundamental question relating to the origin of dissipation and the breaking of the time reversibility in classical dynamics. Although the underlying dynamics in [10] is stochastic, it was proved that the viscosity in the INS equations was *strictly* larger than the original viscosity of the underlying stochastic dynamics. In other words, the deterministic part of the dynamics makes a nontrivial contribution to the viscosity. We remark that the condition $d = 3$ is critical. For dimension $d = 2$, it was proved that the hydrodynamic limit equations for such lattice gas models are not the INS equations. Indeed, even the diffusive scaling is incorrect and there are logarithmic corrections. Although these works do not answer directly the fundamental question regarding the derivation of the incompressible Navier-Stokes equations from the classical dynamics, it is the first time we understand the generation of the viscosity from many particle dynamics. These developments are largely attributed to Varadhan's insight of the importance of the fluctuation-dissipation equations (15).

References

[1] Fritz, J.: On the hydrodynamic limit of a scalar Ginzburg-Landau lattice model: the resolvent approach. Hydrodynamic behavior and interacting particle systems (Minneapolis, Minn., 1986), 75–97, IMA Vol. Math. Appl., 9, Springer, New York, 1987.

[2] Fritz, J.; Tooth, B.: Derivation of the Leroux system as the hydrodynamic limit of a two-component lattice gas. Comm. Math. Phys. 249 (2004), no. 1, 1–27.

[3] Guo, M. Z.; Papanicolaou, G. C.; Varadhan, S. R. S. Nonlinear diffusion limit for a system with nearest neighbor interactions. Comm. Math. Phys. 118 (1988), no. 1, 31–59.

[4] Kipnis, C.; Varadhan, S. R. S.: Central limit theorem for additive functionals of reversible Markov processes and applications to simple exclusions. Comm. Math. Phys. 104 (1986), no. 1, 1–19.

[5] Landim, C.; Yau, H. T. Fluctuation-dissipation equation of asymmetric simple exclusion processes. Probab. Theory Related Fields 108 (1997), no. 3, 321–356.

[6] Olla, Stefano; Varadhan, S. R. S.: Scaling limit for interacting Ornstein-Uhlenbeck processes. Comm. Math. Phys. 135 (1991), no. 2, 355–378.

[7] Olla, S.; Varadhan, S. R. S.; Yau, H.-T. Hydrodynamical limit for a Hamiltonian system with weak noise. Comm. Math. Phys. 155 (1993), no. 3, 523–560.

[8] Quastel, J.: Diffusion of color in the simple exclusion process. Comm. Pure Appl. Math. 45 (1992), no. 6, 623–679

[9] Quastel, J.; Rezakhanlou, F.; Varadhan, S. R. S. Large deviations for the symmetric simple exclusion process in dimensions $d \geq 3$. Probab. Theory Related Fields 113 (1999), no. 1, 1–84.

[10] Quastel, J.; Yau, H.-T. : Lattice gases, large deviations, and the incompressible Navier-Stokes equations. Ann. of Math. (2) 148 (1998), no. 1, 51–108.

[11] F. Rezakhanlou: Hydrodynamic limit for a system with finite range interactions, Comm. Math. Phys. 129 (1990) 445–480.

[12] Rezakhanlou, F. : Hydrodynamic limit for attractive particle systems on Z^d. Comm. Math. Phys. 140 (1991), no. 3, 417–448.

[13] Rezakhanlou, F.: Boltzmann-Grad limits for stochastic hard sphere models. Comm. Math. Phys. 248 (2004), no. 3, 553–637.

[14] Varadhan, S. R. S. Scaling limits for interacting diffusions. Comm. Math. Phys. 135 (1991), no. 2, 313–353.

[15] Varadhan, S. R. S.; Yau, H.-T.: Diffusive limit of lattice gas with mixing conditions. Asian J. Math. 1 (1997), no. 4, 623–678.

[16] Varadhan, S. R. S. Nonlinear diffusion limit for a system with nearest neighbor interactions. II. Asymptotic problems in probability theory: stochastic models and diffusions on fractals (Sanda/Kyoto, 1990), 75–128, Pitman Res. Notes Math. Ser., 283, Longman Sci. Tech., Harlow, 1993.

[17] Yau, H.T.: Relative entropy and hydrodynamics of Ginsburg-Landau models, Letters Math. Phys., **22**, 63–80,(1991)

Book Review

BULLETIN (New Series) OF THE
AMERICAN MATHEMATICAL SOCIETY
Volume 2, Number 3, May 1980
© 1980 American Mathematical Society
0002-9904/80/0000-0219/$02.75

Multidimensional diffusion processes, by D. W. Stroock and S. R. S. Varadhan, Die Grundlehren der mathematischen Wissenschaften, vol. 233, Springer-Verlag, Berlin and New York, 1979, xii + 338 pp., $34.80.

1. *Is it best to think of a 'diffusion' as meaning* (i) *a continuous strong Markov process,* (ii) *a strong solution of an Itô stochastic differential equation, or* (iii) *a solution of a martingale problem?* Both the Markov-process approach and the Itô approach (which holds so special a place in the hearts of probabilists after the appearance of McKean's wonderful book [7]) have been immensely successful in diffusion theory. The Stroock-Varadhan book, developed from the historic 1969 papers by its authors, presents the martingale-problem approach as a more powerful–and, in certain regards, more intrinsic–means of studying the foundations of the subject.

The martingale-problem method has been applied with great success to other problems in Markov-process theory, both 'pure' (Stroock [10], . . .) and applied (Holley and Stroock [3], [4], . . .). It has conditioned our whole way of thinking about still-more-general processes (Jacod [5], . . .). Moreover, the method's ideas and results now feature largely in work on filtering and control (Davis [1], . . .).

I 'batter' you with the preceding paragraph because the authors make the uncompromising decision not 'to proselytize by intimidating the reader with myriad examples demonstrating the full scope of the techniques', but rather to persuade the reader 'with a careful treatment of just one problem to which they apply'. Halmos's doctrine 'More is less, and less is more' is thereby thoroughly tested; but if one had to choose a single totally-integrated piece of work which in depth and importance shows that probability theory has 'come of age', it would surely be the theorem towards which so much of this book is directed–or perhaps Stroock's extension of it [10]. Most of the main tools of stochastic-process theory are used, after first having been honed to a sharper edge than usual. But it is the formidable combination of probability theory with analysis (in the form of deep estimates from the theory of singular integrals) which is the core of the work.

The book's importance has persuaded me to write a review accessible to the general reader–I even define *continuous martingale*! I therefore concentrate on background material, with just a few clues (some for the sharp-eyed!) on how the book's results are proved.

2. The following purely-analytic (and very special) corollary of the main Stroock-Varadhan theorem can help set the scene; and the fact that, in spite of so much research, no analytic proof of its uniqueness assertion has been discovered, can serve as a first indication of the depth of the theorem.

Let L be a second-order elliptic operator on $C_K^\infty(\mathbf{R}^d)$ of the form

$$L = \frac{1}{2} \sum_{i,j \leq d} a_{ij}(x)\partial_i\partial_j + \sum_{i \leq d} b_i(x)\partial_i, \qquad (2.1)$$

where ∂_i denotes $\partial/\partial x_i$, $\{a_{ij}(x)\}$ is a strictly positive-definite real symmetric matrix for each x, and where each $a_{ij}(\cdot)$ and each $b_i(\cdot)$ is a bounded continuous real-valued function on $\mathbf{R}^d$. No Lipschitz conditions are assumed. Then there is one and only one Feller-Dynkin semigroup $\{P_t: t \geq 0\}$[1] with infinitesimal generator extending L. (Chapter 10 of the book examines closely the extent to which the 'boundedness' condition on a and b may be relaxed. It cannot be relaxed completely because one must preclude 'explosion'.)

3. With a general Feller-Dynkin semigroup $\{P_t\}$ is associated an $\mathbf{R}^d$ valued Markov process $X = \{X_t: t \geq 0\}$ with *right-continuous* paths, such that the law P^x of X started from x is given by the usual recipe: for $0 \leq t_1 \leq t_2 \leq \cdots \leq t_n$, we have[2]

$$P^x\big[X_{t_1} \in dx_1; X_{t_2} \in dx_2; \ldots ; X_{t_n} \in dx_n\big] \qquad (3.1)$$
$$= P_{t_1}(x, dx_1)P_{t_2-t_1}(x_1, dx_2) \ldots P_{t_n-t_{n-1}}(x_{n-1}, dx_n).$$

But if $\{P_t\}$ has generator extending our operator L, then, because L is *local*, we can take X to be (path-)*continuous*.[3]

---

[1] A 'Feller-Dynkin' semigroup $\{P_t: t \geq 0\}$ is a family of bounded linear maps $P_t: C_0(\mathbf{R}^d) \to C_0(\mathbf{R}^d)$, where $C_0(\mathbf{R}^d)$ is the Banach space of continuous functions on $\mathbf{R}^d$ which 'vanish at ∞', such that

$$f \geq 0 \Rightarrow P_t f \geq 0; \quad P_t 1 = 1; \quad P_s P_t = P_{s+t}; \quad P_t f \to f \quad (\varepsilon \downarrow 0).$$

('Pointwise' and 'strong-topology' convergence in the last statement are equivalent under the remaining hypotheses.) The statement that $\{P_t\}$ has *infinitesimal generator extending* L amounts to the statement that

$$P_t f - f = \int_0^t P_s L f \, ds, \quad \forall f \in C_K^\infty(\mathbf{R}^d). \qquad (2.2)$$

Note that the differentiated form $(P_t f)' = P_t L f$ of (2.2) corresponds to the Fokker-Planck equation.

[2] Here, the transition probability measure $P_t(x, \cdot)$ is of course that associated with the map $f \mapsto P_t f(x)$ on $C_0(\mathbf{R}^d)$ by the Riesz representation theorem.

[3] Conversely, if each P^x measure associated with a Feller-Dynkin semigroup $\{P_t\}$ assigns mass 1 to the set of continuous paths, then the generator G of $\{P_t\}$ must be *local* and must obey the *maximum principle*: if $f \in D(G)$, then $Gf \leq 0$ at a maximum of f. Hence if $C_K^\infty(\mathbf{R}^d) \subset D(G)$, then G acts on $C_K^\infty(\mathbf{R}^d)$ as a (possibly singular) *second-order* elliptic operator with continuous coefficients. The fact that it can happen that $D(G) \cap C_K^\infty(\mathbf{R}^d) = \{0\}$ is one reason why the definition of a diffusion as a continuous strong Markov process, though intrinsic, is too wide.

A Markov process is a much richer structure than its transition semigroup, and it may be subjected to all sorts of transformations which have no analytic counterpart. This explains why probability theory can obtain analytic results which escape analytic proof (and the literature contains many very striking instances of this). But in the richer setting, the purely-analytic questions are of secondary interest.

4. Continuous P-martingales. We discuss martingale theory only in the special context appropriate to diffusion theory.

Let $\Omega = C([0, \infty), \mathbf{R}^d)$ be the space of all continuous functions ω from $[0, \infty)$ to $\mathbf{R}^d$. For $t \in [0, \infty)$ and $\omega \in \Omega$, write X for the coordinate process with $X_t(\omega) = \omega(t)$. Put $\mathcal{F}_\infty^\circ = \sigma\{X_s: s \geqslant 0\}$, the smallest σ-algebra of subsets of Ω such that every map X_s from Ω to $\mathbf{R}^d$ is $\mathcal{F}_\infty^\circ$ measurable; and, for $t \geqslant 0$, put $\mathcal{F}_t^\circ = \sigma\{X_s: s \leqslant t\}$.

Let M be a continuous adapted real-valued process. Thus, the map M: $[0, \infty) \times \Omega \to \mathbf{R}$ is such that $t \mapsto M_t(\omega)$ is continuous for every ω, and M is $\{\mathcal{F}_t^\circ\}$ *adapted* in that for each t, $M_t(\cdot)$ is $\mathcal{F}_t^\circ$ measurable.[4]

Let P be a probability measure on $(\Omega, \mathcal{F}_\infty^\circ)$. Then M is called a (continuous) P-*martingale* if $M_t \in \mathcal{L}^1(\Omega, \mathcal{F}_t^\circ, P)$, $\forall t$, and, whenever $s \leqslant t$ and $\Lambda \in \mathcal{F}_s^\circ$,

$$\int_\Lambda M_t(\omega) P(d\omega) = \int_\Lambda M_s(\omega) P(d\omega).[5]$$

5. Martingale problems for L. Let L be as at (2.1). Let P be a probability measure on $(\Omega, \mathcal{F}_\infty^\circ)$. Let $x \in \mathbf{R}^d$.

DEFINITION. We say that P *solves the martingale problem for L starting from* x if $P[X_0 = x] = 1$ and, $\forall f \in C_K^\infty(\mathbf{R}^d)$, the process C^f, where

$$C_t^f = f(X_t) - \int_0^t Lf(X_s)\, ds$$

is a P-martingale.

Assume for the moment that a Feller-Dynkin semigroup $\{P_t\}$ exists with infinitesimal generator extending L. As explained in §3, it is possible to define a probability measure P^x on $(\Omega, \mathcal{F}_\infty^\circ)$ via (3.1). It is almost trivial to use (2.2) to show that P^x solves the martingale problem for L starting from x.

In the special context corresponding to §2, the main Stroock-Varadhan theorem takes the following form.

THEOREM. *Let L be a second-order elliptic operator on $C_K^\infty(\mathbf{R}^d)$ satisfying the hypotheses described at (2.1). Then, for each x in $\mathbf{R}^d$, there is one and only one solution of the martingale problem for L starting from x. Moreover, this solution is of the Markovian form P^x derived as at (3.1) from the unique Feller-Dynkin semigroup $\{P_t\}$ with generator extending L.*

In fact, for the existence-and-uniqueness part of the result, Stroock and Varadhan allow the matrices $a(\cdot)$ and $b(\cdot)$ to be *time-dependent*, and impose

---

[4]The intuitive thrust of this requirement is that $M_t(\omega)$ is determined when the values $X_s(\omega)$ for $s \in [0, t]$ are known.

[5]This is the full statement of the requirement that the conditional P-expectation of M_t, given the information $\mathcal{F}_s^\circ$ about what has happened up to time s, is M_s: under P, the 'game' M is fair.

only a *measurability* requirement on *b*. This generalisation is not merely an academic matter: it is just what engineers need (see Davis [1], . . .).

Before thinking briefly about how the theorem is proved, let us spend some time in getting an intuitive feeling for continuous martingales and semi-martingales.[6]

6. Continuous *P*-semimartingales.

Again let *P* be a probability measure on $(\Omega, \mathcal{F}_\infty^\circ)$. The integrability condition $(M_t \in \mathcal{L}^1)$ in the definition of 'martingale' is a nuisance. Call a continuous adapted process *M* a (continuous) *local P-martingale* if, for every *n*, M^n is a *P*-martingale, where

$$M_t^n = M_{t \wedge \tau(n)} - M_0, \quad \text{and} \quad \tau(n) = \inf\{s: |M_s - M_0| > n\}.$$

(This is one of many instances where I use simple definitions and/or results which work only for continuous processes. Meyer [8] has the 'correct'–that is, *generalisable*–versions).

We now arrive at one of the central concepts of Strasbourg theory. Call a continuous adapted process *Z* a (continuous) *P-semimartingale* if *Z* may be written as follows:

$$Z_t = Z_0 + M_t + V_t, \tag{6.1}$$

where *M* is a continuous local *P*-martingale with $M_0 = 0$, and *V* is a (continuous adapted) process with paths of finite variation and with $V_0 = 0$. It is surprising that *the decomposition* (6.1) *is P-unique*: if $Z = Z_0 + M_t^* + V_t^*$, then $P[M_t = M_t^*, \forall t] = 1$.[7]

7. Generalised Itô formula: continuous case.

Let *Z* be a continuous *P*-semimartingale with decomposition as at (6.1). You can see that *V* represents the 'drift' of *Z* away from the fair (local martingale) situation.

By the celebrated *Meyer decomposition theorem*, there is a *P*-unique continuous adapted process $\langle M, M \rangle$ with nondecreasing paths such that $\langle M, M \rangle_0 = 0$ and $M^2 - \langle M, M \rangle$ is a local *P*-martingale. Moreover, by a theorem of Kunita, Watanabe, and Doléans, there exists a sequence (n_k) along which, with *P*-probability 1,

$$\langle M, M \rangle_t = \lim \sum_i \left[M_{t \wedge i2^{-n}} - M_{t \wedge (i-1)2^{-n}} \right]^2 \quad \forall t,$$

---

[6]The theory of martingales, and of the still-more-important semimartingales, has been developed to a very advanced level by the French school of probabilists led by Meyer. (But we must not forget massive Japanese contributions.) I myself have on occasion been mischievous in comments on 'Strasbourg theory', and I think that Stroock and Varadhan are perhaps just a little mischievous in this regard. Strasbourg publications are generally inclined to be somewhat abstract in spirit. However, they do contain a profound analysis of intuitive thinking; and, with sufficient poetic licence, they may be regarded as in the direct tradition of Lévy's work. Meyer [8] is the definitive work on the Strasbourg theory of stochastic integrals. Dellacherie, Doléans, Letta, and Meyer [2] is a most helpful 'Strasbourg' look at part of the Stroock-Varadhan theory. These two papers are my guides for the next few sections. Incidentally, I am using the long-thought-about Strasbourg notation throughout the review.

[7]An immediate consequence of the *P*-uniqueness just asserted is that a *continuous* local *P*-martingale *M* cannot have paths of finite variation unless it is constant. (*You* supply the 'almost surely' qualification.) This explains why we need a *stochastic integral*. It is important that the mildest regularity requirements on a 'stochastic integral' force us to consider only stochastic integrals relative to (possibly discontinuous) semimartingales.

the limit existing uniformly on compact t-intervals. Infinitesimally,

$$(dZ_t)^2 = (dM_t)^2 = d\langle M, M \rangle_t$$

(in some sense!), so Taylor's theorem leads us to anticipate the generalised Itô formula for the continuous case:

$$df(Z) = f'(Z)\, dZ + \tfrac{1}{2} f''(Z) d\langle M, M \rangle. \tag{7.1}$$

The moral is that a C^2 function f of a continuous P-semimartingale Z is again a continuous P-semimartingale with decomposition

$$f(Z_t) = f(Z_0) + \int_0^t f'(Z)\, dM$$

$$+ \left\{ \int_0^t f'(Z_s)\, dV_s + \frac{1}{2} \int_0^t f''(Z_s)\, d\langle M, M \rangle_s \right\}, \tag{7.2}$$

the first of the integrals, a stochastic integral, yielding a local P-martingale.[8] The final two integrals in (7.2) are Stieltjes integrals over s for each ω.

8. Diffusions as semimartingales. If Z is our continuous P-semimartingale as at (6.1), and f is a C_K^∞ function on **R**, then, by (7.2),

$$f(Z_t) - \left\{ \int_0^t f'(Z_s)\, dV_s + \frac{1}{2} \int_0^t f''(Z_s)\, d\langle M, M \rangle_s \right\}$$

is a local P-martingale. (More general martingale problems loom into view; but back to diffusions . . . .)

Take $d = 1$, and $L = \tfrac{1}{2} a(x) d^2/dx^2 + b(x) d/dx$, where $a(\cdot)$ and $b(\cdot)$ are continuous (but 'measurable' will do!) and $a(\cdot) > 0$. We see that *a measure P on $(\Omega, \mathscr{F}_\infty^\circ)$ solves the martingale problem for L starting from x if and only if $P[X_0 = x] = 1$ and the coordinate process X is a P-semimartingale with decomposition $X = X_0 + M + V$ satisfying*

$$V_t = \int_0^t b(X_s)\, ds, \qquad \langle M, M \rangle_t = \int_0^t a(X_s)\, ds.$$

'Vector' generalisations to $d > 1$ are obvious. See footnote 8!

9. The martingale problem for Wiener measure. Continue with $d = 1$. Take $b \equiv 0$, $a \equiv 1$, $x = 0$. If P satisfies the corresponding martingale problem, then $P[X_0 = 0] = 1$, and X is a local P-martingale with $\langle X, X \rangle_t = t$. A trivial modification of (7.1) shows that $dY_t^\theta = i\theta Y_t^\theta\, dX$, where $Y_t^\theta = \exp(i\theta X_t + \tfrac{1}{2}\theta^2 t)$. Hence Y is a local P-martingale, and indeed a P-*martingale* because Y_t^θ is bounded by $\exp(\tfrac{1}{2}\theta^2 t)$. It is immediate to anyone who recognizes the characteristic function of the normal distribution that *X is an $(\{\mathscr{F}_t^\circ\}, P)$ Brownian motion*: X_t is $\mathscr{F}_t^\circ$ measurable $(\forall t)$, and, under P, $X_{t+h} - X_t$ is independent of $\mathscr{F}_t^\circ$ and has the normal distribution of mean 0 and variance h.

---

[8]With U standing for $f'(Z)$, the stochastic integral $\int U\, dM$ is a limit of Riemann sums $\Sigma U_s(M_{s+h} - M_s)$, with each M-increment pointing into the future of the U-value. This is why $\int U\, dM$ is a local martingale. The super-elegant modern (Kunita, Motoo, Watanabe, . . . , Meyer) theory *defines* $\int U\, dM$ as that P-unique continuous local P-martingale starting at 0 such that for every continuous local P-martingale N with $N_0 = 0$, $\langle \int U\, dM, N \rangle_t = \int_0^t U_s d\langle M, N \rangle$ up to P-uniqueness, where $\langle M, N \rangle = \tfrac{1}{2}(\langle M + N, M + N \rangle - \langle M, N \rangle - \langle N, N \rangle)$. See Meyer [8].

Thus P is forced to be Wiener measure. This result, discovered by Lévy (who else?!), is what gave hope for the martingale problem method. Splendid as the Kunita-Watanabe proof just given is, it is tied to this 'freak' case, and so is superseded by the proof in §12 below.

10. Martingale problems and stochastic differential equations. Again take $d = 1$, and assume that P solves the martingale problem starting from x for $L = \frac{1}{2}a(x)d^2/dx^2 + b(x)d/dx$, where a and b are continuous (or just measurable) and $a(\cdot) > 0$.

Put $\sigma(x) = a(x)^{1/2}$, and write

$$dB = \sigma(X)^{-1} dM = \sigma(X)^{-1}[dX - b(X) dt], \qquad B_0 = 0. \qquad (10.1)$$

Then B is a local P-martingale and $(dB)^2 = a(X)^{-1}d\langle M, M \rangle = dt$. By Lévy's theorem, B is an $(\{\mathcal{F}_t^\circ\}, P)$ Brownian motion. Moreover, the formally-obvious consequence:

$$dX = \sigma(X) dB + b(X) dt \qquad (10.2)$$

of (10.1) is easily proved. (It is obvious from the modern theory in footnote 8.) Note that we constructed B from X.

This is the reverse of the situation in Itô theory, where the aim is to study (10.2) as an equation for an 'unknown' X for a given Brownian motion B. Under Lipschitz conditions on a and b, Itô solved (10.2) via successive approximation, Picard-style, obtaining X as a B-adapted process (X_t is $\sigma\{B_s: s \leqslant t\}$ measurable). It is not known whether a B-adapted solution X of (10.2) exists when a mere continuity hypothesis is imposed on a and b: X may require more randomness than B can provide. To understand what might be going on here, study the still-rather-mystifying *Tsirel'son counterexample* in Liptser and Shiryayev [6].

Many aspects of the relationship between Itô theory and S-V theory are subtle and difficult. Chapter 8 of the S-V book is a careful study, following on from important work of Watanabe and Yamada. One of the main results is that 'Itô uniqueness' (watch the formulation!) implies uniqueness in the martingale-problem sense.

11. Why has the martingale-problem method succeeded where others have failed? In regard to *existence theorems*, the reason lies in the 'solidarity' of probability measures on Polish spaces under 'weak' convergence, and in the fact that *the martingale-problem method is ideally suited for establishing 'weak' compactness of families of measures*. This is because martingale inequalities automatically provide modulus-of-continuity properties for application of Arzéla-Ascoli-Prohorov criteria. (That 'practical' weak-convergence results are handled effectively is evidenced by the book's Chapter 11.) Itô's results are 'dense', and we can take limits!

We now turn to the deeper *uniqueness results*. These are deeper than analysis can *formulate*! But the cunning martingale-problem method *stays sufficiently close to analysis to be able to utilise its estimates*. By contrast, the more probabilistic *Itô theory is too subtle for analysis to be of any help*!

12. Thumbnail sketch of proof of the uniqueness result. Let L be as at (2.1) but with $b \equiv 0$. The *Cameron-Martin-Girsanov change-of-measure theorem*

63

allows us to 'add in (measurable) b' later. Since $a(\cdot)$ is continuous, it is locally almost constant. *Localisation techniques* (*probabilistic*–with no analytic counterpart) make it enough to deal with the case: $a(\cdot)$ *is always close to the identity matrix, so L is 'appoximately'* $\frac{1}{2}\Delta$. You can very easily derive directly from the definition that if P^x (*not* known to be Markovian) solves the martingale problem for L starting from x, then, for $\lambda > 0$, and $f \in C_K^\infty(\mathbf{R}^d)$,

$$R_\lambda^x(\lambda f - Lf) = f(x) \quad \text{where } R_\lambda^x(h) = \int_0^\infty dt \, e^{-\lambda t} \int_\Omega h \circ X_t(\omega) P^x(d\omega).$$

(12.1)

The idea of perturbation theory is to try to show that R_λ^x is therefore determined because $R_\lambda^x g = (\lambda - L)^{-1} g(x)$, where we try to define

$$(\lambda - L)^{-1} = V_\lambda (I - K_\lambda)^{-1},$$

(12.2)

where $K_\lambda = \frac{1}{2}\Sigma\Sigma[a_{ij}(\cdot) - \delta_{ij}]\partial_i\partial_j V_\lambda$, and $V_\lambda = (\lambda - \frac{1}{2}\Delta)^{-1}$ is the usual Brownian resolvent. The key fact with which we can work is provided by Littlewood-Paley theory (see Meyer [9] for a fine probabilistic proof), namely: *for $p > 1$, the map $\partial_i\partial_j V_\lambda$ from $C_K^\infty(\mathbf{R}^d)$ to $L^p(\mathbf{R}^d)$ extends* (*uniquely*) *to a bounded linear operator on $L^p(\mathbf{R}^d)$.* By holding $a(\cdot)$ everywhere close to the identity matrix, we can arrange that, for all λ in an interval, $\|K_\lambda\|_p < 1$. While (12.2) is now meaningful in $L^p(\mathbf{R}^d)$ terms, we need to make a much more careful analysis (not done in this review!) to show that *for $p > \frac{1}{2}d$, R_λ^x is a bounded linear functional on $L^p(\mathbf{R}^d)$.* Only then can we say that R_λ^x is determined.[9] By 'inversion of Laplace transforms', $P^x \circ X_t^{-1}$ *is uniquely determined for all t* (*and this for every x*). *On conditioning the martingale-problem formulation relative to $\mathfrak{F}_t^\circ$ in the sense of regular conditional probabilities,* we find that

$$\left[P^x \circ X_{t+h}^{-1} | \mathfrak{F}_t^\circ \right] = P^{X(t)} \circ X_h^{-1} \quad (P^x \text{ almost surely}).$$

The 'full' uniqueness of P^x and its Markov property (*the two are inextricably linked*) *follow immediately.* (For *Markov selection principles* in the presence of nonuniqueness, see Chapter 12 of the book.)

For their more general and much fuller results, Stroock and Varadhan use much deeper inequalities due to Jones, Fabes, and Rivière. The book's Appendix proves these estimates, and links such results to the theory of *BMO*, etc.

13. I have emphasised the great importance of the Stroock-Varadhan book. It contains a lot more than I have indicated; in particular, its many exercises contain much interesting material.

For immediate confirmation of the subject's sparkle, virtuosity, and depth, see Mozart–sorry, I mean *McKean*–[7]. The Stroock-Varadhan book proceeds on its inexorable way like a massive Bach fugue. 'Too much counterpoint; and, what is worse, *Protestant* counterpoint', said Beecham of Bach. But old

---

[9]Argue (via localisation) that we can assume that $a(\cdot)$ is constant far out. Hence we can approximate $a(\cdot)$ uniformly by smooth $a_k(\cdot)$. Now it is fairly easy to show that if $g \in C_K^\infty$, then we can find f_n in C_K^∞ with $\lambda f_n - Lf_n \to g$ in L^p. Thus, since R_λ^x is continuous on L^p, $R_\lambda^x g$ may be determined from (12.1).

J. S. can be something of a knockout if his themes get hold of you. And his influence on what followed was (you may say) substantial.

REFERENCES

1. M. H. A. Davis, *A pathwise solution of the equations of non-linear filtering*, Teor. Veroyatnost i Primenen (to appear).

2. C. Dellacherie, C. Doléans-Dade, G. Letta and P. A. Meyer, *Diffusions à coefficients continus d'après D. W. Stroock et S. R. S. Varadhan*, Séminaire de Probabilités IV, Lecture Notes in Math., vol. 124, Springer-Verlag, Berlin and New York, 1970, pp. 241–282.

3. R. A. Holley and D. W. Stroock, *A martingale approach to infinite systems of interacting processes*, Ann. Probability **4** (1976), 195–228.

4. _____, *Nearest neighbor birth and death processes on the real live*, Acta Math. **140** (1978), 103–154.

5. J. Jacod, *Calcul stochastique et problèmes de martingales*, Lecture Notes in Math., vol. 714, Springer-Verlag, Berlin and New York, 1979.

6. R. S. Liptser and A. N. Shiryayev, *Statistics of random processes*. I: *general theory* (English translation from the Russian), Applications of Mathematics, vol. 5, Springer-Verlag, Berlin and New York, 1977.

7. H. P. McKean, *Stochastic integrals*, Academic Press, New York, 1969.

8. P. A. Meyer, *Un cours sur les intégrales stochastiques*, Séminaire de Probabilités X, Lecture Notes in Math., vol. 511, Springer-Verlag, Berlin and New York, 1976, pp. 245–400.

9. _____, *Démonstration probabiliste de certaines inégalités de Littlewood-Paley*, Séminaire de Probabilités X, Lecture Notes in Math., vol. 511, Springer-Verlag, Berlin and New York, 1976, pp. 125–183.

10. D. W. Stroock, *Diffusion processes associated with Lévy generators*, Z. Wahrscheinlichkeitstheorie ver. Gebiete **32** (1975), 209–244.

DAVID WILLIAMS

Limit Theorems

Sankhya Ser. A 24 (1962) 213-238

LIMIT THEOREMS FOR SUMS OF INDEPENDENT RANDOM VARIABLES WITH VALUES IN A HILBERT SPACE

By S. R. S. VARADHAN

Indian Statistical Institute

SUMMARY. In this article distributions on a Real Separable Hilbert Space are considered. Limit distributions are derived for sums of infinitesimal random variables.

A representation similar to the Levy-Khintchine representation is derived for infinitely divisible distributions. Necessary and sufficient conditions for compactness are obtained in terms of the quantities occurring in the representation.

1. INTRODUCTION

The celebrated Levy-Khintchine representation for the characteristic function $\phi(t)$ of an infinitely divisible distribution on the real line is given by

$$\phi(t) = \exp\left[i\nu t - \frac{\sigma^2 t^2}{2} + \int_{-\infty}^{\infty} \left(e^{itx} - 1 - \frac{itx}{1+x^2} \right) \frac{1+x^2}{x^2} \, dG(x) \right] \quad \ldots \quad (1.1)$$

where ν and σ are real constants, $\sigma \geqslant 0$ and $G(x)$ is a bounded non-decreasing function of x which is continuous at the origin. ν, σ and G are uniquely determined by $\phi(t)$. Conversely, any function of the type (1.1) is the characteristic function of an infinitely divisible distribution. Khintchine and Bawly went further and proved that the limit distributions of sums of uniformly infinitesimal random variables are infinitely divisible and that they can be obtained as limits of certain accompanying infinitely divisible distributions. For all historical and other details concerning these we refer to Gnedenko and Kolmogorov (1954). Attempts have been made by several authors to extend these results to more general situations. We mention, in particular, the works of Bochner (1955) and (1958), Hunt (1956), Kloss (1961), Levy (1939), Takano (1955) and Vorobev (1954). Takano has generalised both the representation and the limit theorems to the case of a finite dimensional vector space. Levy has considered the circle group (the multiplicative group of complex numbers of modulus unity). The axiomatic development of the concept of characteristic functions by Bochner (1958) throws some light on the nature of the problem. Hunt has obtained the representation for one parametric semigroups of distributions on a Lie group. Vorobev has considered finite groups and Kloss a wide class of compact groups.

The case of a general locally compact abelian group was considered very recently by K. R. Parthasarathy, R. Rangarao and the present author (1962a). The results obtained are mentioned briefly in the next section and we will have occasion to use them in the later sections.

The case of a non-locally compact group offers considerable difficulty. However, the particular case of a separable Hilbert space can be tackled using some compactness criteria of Prohorov (1956). In this paper we will consider the case of a Hilbert space, and the analysis will throw some light on the precise nature of the difficulties in extending these results to non-locally compact groups.

2. The locally compact case

In the present section we will briefly mention the results obtained for the locally compact case. We will be using some of these results in the latter sections.

Let X be a locally compact abelian group which is separable metric and let Y be its character group. For $x \epsilon X$ and $y \epsilon Y$ we write (x, y) to denote the value of the character y at the point x.

Let $\mathcal{M}$ denote the convolution semi-group of probability measures on X, endowed with the weak topology. A sequence μ_n in $\mathcal{M}$ is shift compact whenever the sequence formed by some translates of these distributions is compact. Throughout the article $*$ denotes the convolution operation. When x is an element of the group and λ a distribution $\lambda * x$ denotes the convolution of λ with the distribution degenerate at x. λ^n denotes the n-th convolution of λ. $|\lambda|^2$ denotes $\lambda * \bar{\lambda}$ where $\bar{\lambda}$ is the distribution defined by $\bar{\lambda}(A) = \lambda(-A)$. $-A$ consists of the set of inverses of A. The following theorem was proved by Parthasarathy et al (1962).

Theorem 2.1 : Let $\mu_n = \alpha_n * \beta_n$ for each n. If μ_n is compact then α_n and β_n are shift compact.

Definition 2.1 : A distribution μ is said to be infinitely divisible if for each n there are elements x_n in X and $\lambda_n \epsilon \mathcal{M}$ such that

$$\mu = \lambda_n^n * x_n.$$

Remark 2.1 : This modification of the classical definition is intended to avoid the consequence of the presence of elements that are not divisible. In a divisible group this definition is equivalent to the classical definition as can be easily seen.

Theorem 2.2 : The totality of infinitely divisible distributions is a closed sub-semigroup of $\mathcal{M}$.

Definition 2.2 : If F is any totally finite measure on X the distribution $e(F)$ associated with it is defined as follows.

$$e(F) = e^{-F(X)} \left[1 + F + \frac{F^2}{2!} + \dots \right].$$

For $\mu \epsilon \mathcal{M}$ we denote by $\mu(y)$ its characteristic function. We then have

$$e(F)(y) = [\int [(x, y) - 1] dF].$$

Theorem 2.3 : If μ is infinitely divisible and if its characteristic function vanishes at some point then μ has an idempotent factor.

Theorem 2.4 : Let $\mu_n = e(F_n)$. Then the necessary and sufficient conditions that

(a) μ_n is shift compact,

(b) if μ is any limit of shifts of μ_n then μ has no idempotent factors, are

(i) for each neighbourhood N of identity e the family F_n restricted to $X - N$ is weakly conditionally compact;

(ii) for each $y \epsilon Y$

$$\sup_n \int [1 - \text{Real } (x, y)] dF_n < \infty.$$

LIMIT DISTRIBUTION FOR RANDOM VARIABLES

Definition 2.3 : A double sequence $\alpha_{nj} : j = 1, 2, \ldots, k_n$; $n = 1, 2, \ldots$ of distributions is said to be uniformly infinitesimal if

$$\underset{n \to \infty}{\text{lt}} \sup_{1 \leqslant j \leqslant k_n} \sup_{y \epsilon K} |\alpha_{nj}(y) - 1| = 0$$

for every compact subset $K \subseteq Y$.

Lemma 2.1 : *There exists a real valued function $g(x, y)$ on $X \times Y$ which is continuous in both the variables x and y and such that*

(i) $g(x, y_1 + y_2) = g(x, y_1) + g(x, y_2)$ *for each $x \epsilon X$ and $y_1, y_2 \epsilon Y$,*

(ii) *for any compact subset $K \subseteq Y$*

$$\sup_{x \epsilon X} \sup_{y \epsilon K} |g(x, y)| < +\infty,$$

(iii) *for each compact subset $K \subseteq Y$ there exists a neighbourhood N_K of the identity in X such that*

$$(x, y) = e^{ig(x,y)}$$

holds for $x \epsilon N_K$ and $y \epsilon K$,

(iv) $g(x, y) = 0$ *whenever $x = e$ for any y.*

Theorem 2.5 : *Let α_{nj} be a uniformly infinitesimal sequence and let*

$$\mu_n = \prod_{j=1}^{k_n} \alpha_{nj}.$$

Suppose μ_n is shift compact and that no limit of shifts of μ_n has an idempotent factor. Let $\beta_{nj} = e[\alpha_{nj} * g_{nj}]$ where g_{nj} is that element of the group X defined by the equality

$$(g_{nj}, y) = \exp\{-i \int g(x, y) d\alpha_{nj}\}.$$

If

$$\lambda_n = \prod_{j=1}^{k_n} \beta_{nj} * g_n \text{ where } g_n = - \sum_j g_{nj},$$

then

$$\underset{n \to \infty}{\overline{\lim}} \sup_{y \epsilon K} |\lambda_n(y) - \mu_n(y)| = 0$$

for every compact set K of Y.

Corollary 2.1 : *Limit distribution of sums infinitesimal independent random variables is infinitely divisible.*

Definition 2.4 : A distribution μ is said to be Gaussian if it has the following properties,

(i) μ is infinitely divisible,

(ii) $\mu = e(F) * \alpha$, where α is infinitely divisible implies that F vanishes outside the identity.

Therem 2.6 : *A distribution μ is Gaussian if and only if $\mu(y)$ is of the form*

$$\mu(y) = (x, y) \exp[-\phi(y)]$$

where x is a fixed element of X and $\phi(y)$ a continuous non-negative function of y satisfying the equation

$$\phi(y_1 + y_2) + \phi(y_1 - y_2) = 2[\phi(y_1) + \phi(y_2)]$$

for every pair y_1, y_2 in Y.

SANKHYĀ : THE INDIAN JOURNAL OF STATISTICS : SERIES A

Theorem 2.7 : *A distribution μ on X is infinitely divisible without an idempotent factor if and only if $\mu(y)$ is of the form*

$$\mu(y) = (x_0, y) \exp\left[\int [(x, y) - 1 - ig(x, y)]dM - \phi(y)\right]$$

where

(i) x_0 *is a fixed element of X,*

(ii) $\phi(y)$ *is a function as in Theorem 2.6,*

(iii) M *is a σ-finite measure giving finite mass outside every neighbourhood of the identity,*

(iv) $\int [1 - \text{Real}\,(x, y)]dM < +\infty$ *for each $y \in Y$.*

Remark 2.2 : This representation is not in general unique. Examples of non-uniqueness and conditions for uniqueness are discussed by Parthasarathy *et al* (1962a) . It turns out that if the character group is connected then the representation is unique.

Corollary 2.2 : *Every one parametric weakly continuous semigroup μ_t has a unique representation*

$$\mu_t(y) = (x_t, y)\ \exp[t \int [(x, y) - 1 - ig(x, y)]dM - t\phi(y)]$$

where x_t is a continuous semigroup in X and M and ϕ are as in Theorem 2.7.

Remark 2.3 : Of the results mentioned in this section Theorems 2.1, 2.2 and the necessity part of Theorem 2.3 are valid in any complete separable metric group. We will also need the following corollary and theorem deducible from Theorem 2.2.

Corollary 2.3 : λ_n *is shift compact if and only if $|\lambda_n|^2$ is compact.*

Theorem 2.8 : *Let λ_n be a sequence such that λ_n is a factor of λ_{n+1} for all n. Then if λ_n is shift compact there are translates λ'_n of λ_n which converge.*

Remark 2.4 : We can assume instead that λ_{n+1} is a factor of λ_n for each n and then also the theorem will be valid.

3. PRELIMINARIES

X is a real separable Hilbert space, (x, y) denotes the inner product and $\|x\|$ the norm. With vector addition as group operation X becomes a complete separable metric group. We will denote by $\mathcal{M}$ the semigroup of all distributions.

For every $\mu \in \mathcal{M}$ its characteristic function is defined on X by the formula

$$\mu(y) = \int e^{i(x,\, y)}\ d\mu(x).$$

We will, in this section, mention some results obtained by Prohorov (1956) concerning compactness criteria for distributions on X.

Definition 3.1 : A positive semi-definite Hermitian operator A is called an *S*-operator if it has finite trace. The class of sets of the type $[x :(Sx, x) < t]$ where S runs over *S*-operators and t over positive numbers forms a neighbourhood system at the origin for a certain topology which is called the *S*-topology. A net x_a converges to zero in *S*-topology if and only if (Sx_a, x_a) converges to zero for every *S*-operator *S*.

LIMIT DISTRIBUTION FOR RANDOM VARIABLES

We have the following theorem concerning characteristic functions and S-topology which was obtained by Sazanov (1958).

Theorem 3.1 : *In order that a function $\mu(y)$ may be the characteristic function of a distribution on X it is necessary and sufficient that $\mu(0) = 1$, $\mu(y)$ be positive definite and continuous at $y = 0$ in the S-topology. (Here and elsewhere in the article 0 will denote the null element of X or the identity of the group and will be called the origin).*

We also have the following theorem of Prohorov (1956).

Theorem 3.2 : *In order that a positive definite function $\mu(y)$ with $\mu(0) = 1$ be the characteristic function of a distribution on X it is necessary and sufficient that for every $\epsilon > 0$ there exists an S-operator S_ϵ such that*

$$1 - R\mu(y) \leqslant (S_\epsilon y, y) + \epsilon$$

where R denotes the real parts.

Definition 3.2 : Let μ be a distribution on X for which $\int (x, y)^2 d\mu < \infty$ for each y. Then the covariance operator S of μ is that Hermitian operator for which

$$(Sy, y) = \int (x, y)^2 d\mu(x).$$

This operator S will be positive semi-definite and will be an S-operator if and only if

$$\int \|x\|^2 \, d\mu(x) < \infty.$$

Definition 3.3 : A sequence S_n of S-operators will be called compact if and only if the following two conditions are satisfied

$$\text{(i)} \quad \sup_n \text{Trace } (S_n) < \infty$$

$$\text{(ii)} \quad \underset{N \to \infty}{\text{lt}} \sup_n \sum_{j=N}^{\infty} (S_n e_j, e_j) \to 0$$

for some orthonormal sequence $e_1, e_2, \ldots e_j, \ldots$

When S is the covariance operator of a distribution μ on X for which

$$\int \|x\|^2 d\mu(x) < \infty$$

we have the relation

$$\sum_{j=N}^{\infty} (Se_j, e_j) = \int r_N^2(x) d\mu(x)$$

where

$$r_N^2(x) = \sum_{j=N}^{\infty} (x, e_j)^2$$

and $e_1, e_2, \ldots$ any orthonormal basis.

The following theorems concerning conditions for compactness of a sequence μ_n of distributions on X were obtained by Prohorov (1956).

Theorem 3.3 : *In order that a sequence μ_n of distributions on X be weakly conditionally compact it is necessary and sufficient that for every $\epsilon > 0$, there exists a compact sequence S_n^ϵ of S-operators such that*

$$1 - R \mu_n(y) \leqslant (S_n^\epsilon y, y) + \epsilon$$

for all n and y. Here $\mu_n (y)$ is the characteristic function of μ_n.

SANKHYĀ : THE INDIAN JOURNAL OF STATISTICS : Series A

Theorem 3.4 : *Let μ_n be a sequence of distributions for which the covariance operators S_n exist and are S-operators. Let further S_n be compact. Then μ_n is weakly conditionally compact.*

Theorem 3.5 : *In order that a sequence μ_n of distributions on X may be weakly conditionally compact it is necessary that for any $\epsilon > 0$*

$$\text{lt}_{N \to \infty} \sup_n \mu_n[r_N^2(X) > \epsilon] = 0.$$

In defining the covariance operator in Definition 3.2. we assumed that μ is a distribution. Actually, if M is any σ-finite measure for which

$$\int \|x\|^2 \, dM < \infty$$

then S is well defined as an S-operator by the relation

$$(Sy, y) = \int (x, y)^2 \, dM(x).$$

We will need also the following.

Theorem 3.6 : *(Rangarao, 1960). Let $\mu_n \to \mu$. Let $\{f_\alpha\}$ be a family of continuous functions equicontinuous and uniformly bounded. Then*

$$\text{lt}_{n \to \infty} \sup_\alpha \left| \int f_\alpha d\mu_n - \int f_\alpha d\mu \right| = 0.$$

4. An estimate of the variance

Let $X_1, X_2, \ldots, X_n$, be n symmetric independent random variables in the Hilbert space X. We will give in this section an estimate for the variance $E\|X_1 + \ldots X_n\|^2$ when each X_j is bounded uniformly in norm by a constant C independent of j.

To this end we introduce the concentration function $Q_\mu(t)$ following Levy [(1954), page 138].

Definition 4.1 : The concentration function $Q_\mu(t)$ of a distribution μ in the Hilbert space X is defined for $0 < t < \infty$ as

$$Q_\mu(t) = \sup_{x \epsilon X} \mu(S_t + x)$$

where S_t denotes the sphere $[x : \|x\| \leqslant t]$ and $S_t + x$ its translate by the element x of X.

We now list a few elementary properties of these functions.

Theorem 4.1 : (1) $Q_\mu(t)$ *is a nondecreasing function of t and* $\lim_{t \to \infty} Q_\mu(t) = 1$.

(2) *If* $\mu_1 * \mu_2 = \mu$ *then, for every* t, $Q_\mu(t) \leqslant \min [Q_{\mu_1}(t), Q_{\mu_2}(t)]$. (3) *If* μ_n *is shift compact then*

$$\lim_{t \to \infty} \inf_n Q_{\mu_n}(t) = 1.$$

Proof : (1) and (2) are obvious and (3) is a consequence of tightness.

Theorem 4.2 : *Let $X_1, X_2, \ldots, X_n$ be n mutually independent symmetric random variables. Let $S_j = X_1 + \ldots + X_j$. Further, let $Q(t)$ denote the concentration function of the sum $S_n = X_1 + \ldots X_n$. If T is defined as*

$$T = \sup_{1 \leqslant j \leqslant n} \|S_j\|$$

then one has for any $t > 0$ $\qquad P\{T > 4t\} \leqslant 2[1 - Q(t)].$

LIMIT DISTRIBUTION FOR RANDOM VARIABLES

Proof : This remarkable result of Levy, which is a refinement of Kolmogorov's inequality in terms of the variance, although proved by him for the real line, offers no difficulty for generalisation. Let the events E_k be defined as follows :

$$E_k = [\|S_1\| \leqslant 4t, \ldots, \|S_{k-1}\| \leqslant 4t, \|S_k\| > 4t]$$
$$\{T > 4t\} = E_1 \cup E_2 \ldots \cup E_n, \quad E_i \cap E_j = \phi.$$

By $Pr\{\ \}$ we will denote the probability of the event within the brackets given that the event E_r has occurred. We then have

$$Pr\{\|S_n\| \leqslant 2t\} \leqslant Pr\{\|S_n - S_r\| > 2t\} = P\{\|S_n - S_r\| > 2t\}. \qquad \ldots (4.1)$$

This is because E_r and $\|S_n\| \leqslant 2t$ imply that $\|S_n - S_r\| > 2t$ and $S_n - S_r$ is distributed independently of E_r. Let us further suppose that $Q(t) > \frac{1}{2}$. We now consider the distribution μ_{rn} of $S_n - S_r$ whose concentration function is denoted by $Q_{rn}(t)$. Since μ_{rn} is a factor of μ we have $Q_{rn}(t) > \frac{1}{2}$. This implies the existence of a point x in the space X such that

$$\mu_{rn}(S_t + x) > Q_{rn}(t) - \epsilon > \frac{1}{2}. \qquad \ldots (4.2)$$

Since μ_{rn} is a symmetric distribution

$$\mu_{rn}(S_t - x) = \mu_{rn}(-S_t + x) = \mu_{rn}(S_t + x) > \frac{1}{2}.$$

Therefore, $\qquad\qquad S_t - x \cap S_t + x \neq \phi.$

In other words there exists a point y such that

$$\|x + y\| \leqslant t, \quad \|x - y\| \leqslant t.$$

These two imply that $\|x\| \leqslant t$ and hence

$$x + S_t \subseteq S_{2t}. \qquad \ldots (4.3)$$

From (4.2) and (4.3) follows $\quad \mu_{rn}(S_{2t}) > Q_{rn}(t) - \epsilon.$

Since ϵ is arbitrary we have $\quad \mu_{rn}(S_{2t}) \geqslant Q_{rn}(t) \qquad\qquad \ldots (4.4)$

which is the same as $\quad P\{\|S_n - S_r\| > 2t\} \leqslant 1 - Q_{rn}(t) \leqslant 1 - Q(t). \qquad \ldots (4.5)$

We have further, using (4.1) and (4.5)

$$P\{T > 4t, \|S_n\| \leqslant 2t\} = \sum_{r=1}^{n} Pr\{\|S_n\| \leqslant 2t\} P[E_r] \leqslant (\sum_{r=1}^{n} P[E_r])(1 - Q(t)) = P\{T > 4t\}(1 - Q(t)),$$
$$\ldots (4.6)$$

we also have $\quad P\{T \leqslant 4t, \|S_n\| \leqslant 2t\} \leqslant P\{T \leqslant 4t\} = 1 - P\{T > 4t\}. \qquad \ldots (4.7)$

Adding (4.6) and (4.7) we get $\quad P\{\|S_n\| \leqslant 2t\} \leqslant 1 - P\{T > 4t\} Q(t). \qquad \ldots (4.8)$

From (4.4) putting $r = 0$, we obtain

$$P\{\|S_n\| \leqslant 2t\} \geqslant Q(t), \qquad \ldots (4.9)$$

(4.8) and (4.9) imply, since $Q(t) > \frac{1}{2}$,

$$P\{T > 4t\} \leqslant \frac{1 - Q(t)}{Q(t)} \leqslant 2[1 - Q(t)]. \qquad \ldots (4.10)$$

However, if $Q(t) \leqslant \frac{1}{2}$ the theorem is trivially true. Thus the proof of the theorem is complete.

We have the following theorem which gives an estimate of the variance. This is well known for the real line. We reproduce the proof for the real line, which can be found in Halmos [(1950) page 198], replacing however $|x|$ by $\|x\|$.

Theorem 4.3 : *Let X_1, X_2, ..., X_n be n independent random variables in a Hilbert space, uniformly bounded by a constant c in norm. Let each X_i have zero expectation. In addition let*

$$P\{\sup_{1 \leqslant j \leqslant n} \|S_j\| \leqslant d\} \geqslant \epsilon > 0$$

where

$$S_j \doteq X_1 + \dots X_j$$

Then

$$E\ \|S_n\|^2 \leqslant \frac{d^2 + (c+d)^2}{\epsilon}\ .$$

Proof : Let the sets E_k be defined as follows

$$E_k = \sup_{1 \leqslant j \leqslant k} \|S_j\| \leqslant d.$$

Then $E_1 \supset E_2 \dots \supset E_n$ and $P(E_n) \geqslant \epsilon > 0$. By P we mean the product measure of $X_1, X_2, ..., X_n$. Let us define

$$F_k = E_{k-1} - E_k$$

and

$$\alpha_k = \int_{E_k} \|S_k\|^2\, dP.$$

We will take E_0 as the whole space and α_0 as zero. We then have

$$\alpha_k - \alpha_{k-1} = \int_{E_k} \|S_k\|^2\, dP - \int_{E_{k-1}} \|S_{k-1}\|^2\, dP$$

$$= \int_{E_{k-1}} \|S_k\|^2\, dP - \int_{F_k} \|S_k\|^2\, dP - \int_{E_{k-1}} \|S_{k-1}\|^2\, dP$$

$$= \int_{E_{k-1}} \|S_{k-1}\|^2\, dP - 2\int_{E_{k-1}} (S_{k-1}, X_k) dP + \int_{E_{k-1}} \|X_k\|^2\, dP$$

$$- \int_{F_k} \|S_k\|^2\, dP - \int_{E_{k-1}} \|S_{k-1}\|^2\, dP$$

$$= \int_{E_{k-1}} \|X_k\|^2\, dP - \int_{F_k} \|S_k\|^2\, dP$$

$$\geqslant P(E_{k-1}) E\|X_k\|^2 - (c+d)^2 P(F_k). \qquad \dots (4.11)$$

We notice here that $\int_{E_{k-1}} (S_{k-1}, X_k)\, dP = 0$ since X_k is independent of S_{k-1}, has zero expectation and is independent of E_{k-1} as well. Moreover, $F_k \subset E_{k-1}$ hence $\|S_k\| \leqslant \|S_{k-1}\| + \|X_k\| \leqslant (c+d)$ over F_k. Since $P(E_n) \leqslant P(E_k)$ for any k we have

$$\alpha_k - \alpha_{k-1} \geqslant P(E) E\|X_k\|^2 - (c+d)^2 P(F_k) \quad \text{for } k = 1, 2, \dots \qquad \dots (4.12)$$

Since $F_1, F_2, ..., F_n$ are all disjoint adding (4.12) for $k = 1, 2, ..., n$, we have

$$\alpha_n \geqslant P(E) E\ \|S_n\|^2 - (c+d)^2. \qquad \dots (4.13)$$

However, since $\|S_n\| \leqslant d$ on E_n $\alpha_n = \int_{E_n} \|S_n\|^2\, dP \leqslant d^2,$ $\qquad \dots (4.14)$

(4.13) and (4.14) give

$$E\|S_n\|^2 \leqslant \frac{(d+d)^2 + d^2}{P(E)} \leqslant \frac{(c+d)^2 + d^2}{\epsilon}\ .$$

This completes the proof.

LIMIT DISTRIBUTION FOR RANDOM VARIABLES

Theorem 4.4 : *Let X_1, X_2, ..., X_n be symmetric independent random variables in the Hilbert space such that $\| X_i \| \leqslant c$ for $i = 1, 2, ..., n$. Let $Q(t)$ denote the concentration function of $S_n = X_1 + ... X_n$. Then*

$$E\|S_n\|^2 \leqslant \frac{16t^2 + (c + 4t)^2}{2Q(t) - 1}$$

for any t such that $Q(t) > \frac{1}{2}$.

Proof : This follows at once from Theorems 4.2 and 4.3 and the fact that a bounded symmetric random variable has zero expectation.

5. INFINITELY DIVISIBLE DISTRIBUTIONS

In this section we will obtain a representation for infinitely divisible distributions. The definition of an infinitely divisible distribution is the same as Definition 2.1, but since the Hilbert space as a group is divisible it is equivalent to the classical definition which requires that μ be written as λ_n^n for each n.

As we have already remarked at the end of Section 2 that some of the results mentioned in that Section are valid for the Hilbert space we will now state them. We will keep in mind also that the Hilbert space has no nontrivial compact subgroups and hence there are no idempotent distributions.

Theorem 5.1 : *The infinitely divisible distributions form a closed subsemigroup among all distributions.*

Theorem 5.2 : *If μ is an infinitely divisible distribution and $\mu(y)$ its characteristic function then $\mu(y)$ is nonvanishing.*

For every finite measure F the infinitely divisible distribution $e(F)$ is associated in the same way as in Definition 2.2. We then have the following theorem.

Theorem 5.3 : *Let $\mu_n = e(F_n)$. In order that μ_n may be shift compact it is necessary that*

(i) *for each neighbourhood N of the identity F_n restricted to N' is weakly conditionally compact,*

(ii) $\sup_n \int [1 - \cos (x, y)] dF_n < \infty$ *for each y.*

Theorem 5.4 : *Let $\mu_n \Longrightarrow \mu$. Then $\mu_n(y) \to \mu(y)$ uniformly over every bounded sphere.*

Proof : Since the set of functions (x, y) as y varies over a bounded sphere forms a equicontinuous family of functions in x the theorem follows.

Theorem 5.5 : *Let μ_n be shift compact and $\mu_n (y) \to \mu(y)$ uniformly over bounded spheres. Then $\mu_n \Longrightarrow \mu$.*

Proof : Since μ_n is shift compact let x_n in X be chosen such that $\mu_n * x_n$ is compact. We will now show that x_n in X is compact. If x_n is not compact then we can produce a subsequence from x_n which has no further convergent subsequence. We will denote the subsequence by x_n itself. Since $\mu_n * x_n$ is weakly compact it has a subsequence $\mu_{n_j} * x_{n_j}$ converging weakly. Thus $\mu_{n_j}(y)e^{i(x_{n_j}, y)}$ as well as $\mu_{n_j}(y)$ converge uniformly over bounded spheres. Since from the compactness of $|\mu_n|^2$ one can conclude the existence of a sphere S such that $\inf_n |\mu_n(y)| \geqslant \epsilon > 0$ for all y in S, it follows that $e^{i(x_{n_j}, y)}$ converges uniformly over S and hence x_{n_j} converges in norm. This proves the theorem.

Before obtaining the representation we will show that if $e(F_n)$ is shift compact then

$$\sup_n \int_{\|x\| \leqslant 1} \|x^2\| dF_n < \infty.$$

To this end we consider the following lemma.

Lemma 5.1 : *Let $f(y)$ be a non-negative function on X such that $f(2y) \leqslant 4f(y)$ for all values of y. If $f(y) \leqslant \epsilon$ whenever $(Sy, y) \leqslant \delta$ where S is some S-operator then*

$$f(y) \leqslant (S_1 y, y) + \epsilon \quad \text{for all } y$$

where $\quad S_1 = 4\epsilon\delta^{-1}S.$

Proof : Defining $S_0 = \epsilon\delta^{-1}S$,

we see that $\qquad f(y) \leqslant \epsilon$ whenever $(S_0 y, y) \leqslant \epsilon.$ ... (5.1)

Further, if $(S_0 y, y) \leqslant 4^n\epsilon$ where n is a positive integer, then denoting by y_n the element $2^{-n} y$, we have

$$(S_0 y_n, y_n) = 4^{-n}(S_0 y, y) \leqslant \epsilon$$

consequently $f(y_n) \leqslant \epsilon$. But since $f(2y) \leqslant 4f(y)$,

we have $\qquad f(y) \leqslant 4^n f(y_n) < 4^n\epsilon.$

So from (5.1) we have

$$f(y) \leqslant 4^n\epsilon \text{ whenever } (S_0 y, y) \leqslant 4^n\epsilon. \quad \text{... (5.2)}$$

Let y be any element of X and let $(S_0 y, y) = t.$

Case (i) : Let $t > \epsilon$. If n is a non-negative integer such that

$$4^n \epsilon < t \leqslant 4^{n+1}\epsilon. \quad \text{... (5.3)}$$

We have, since $(S_0 y, y) = t \leqslant 4^{n+1}\epsilon$, using (5.2) and (5.3)

$$f(y) \leqslant 4^{n+1} \epsilon \leqslant 4t = 4(S_0 y, y). \quad \text{... (5.4)}$$

Case (ii) : Let $t \leqslant \epsilon$. Then from (4.1) we have

$$f(y) \leqslant \epsilon. \quad \text{... (5.5)}$$

(5.4) and (5.5) give at once $\quad f(y) \leqslant \max[\epsilon, (4S_0 y, y)] \leqslant \epsilon + (S_1 y, y).$

We shall need while proving the next theorem the following inequality. If $a_1, \ldots a_m$ are any m real numbers such that $|a_j| \leqslant 1$ for $1 \leqslant j \leqslant m$, then

$$1 - a_1 a_2 \ldots a_m \leqslant \sum_{j=1}^m (1 - a_j). \quad \text{... (5.6)}$$

This inequality is proved by induction if all the a's are positive. If at least one of them say a_r is negative

$$1 - a_1 \ldots a_m \leqslant 1 + |a_1 \ldots a_m| \leqslant 1 + |a_r| = 1 - a_r \leqslant \sum_{j=1}^m (1 - a_j).$$

We will now prove the following theorem.

LIMIT DISTRIBUTION FOR RANDOM VARIABLES

Theorem 5.6 : *Let F_n be a sequence of finite measures such that $e(F_n)$ is shift compact. Then*

$$\sup_n \int_{\|x\| \leqslant 1} \|x\|^2 dF_n < \infty.$$

Proof : We assume without any loss of generality that each F_n vanishes outside the unit sphere. Otherwise we can consider the restriction of F_n to the unit sphere instead of F_n. Let $M_n = F_n + \bar{F}_n$. Then $e(M_n) = |e(F_n)|^2 = \lambda_n$ is compact. We will show that $\int \|x\|^2 dM_n$ is uniformly bounded. To this end we assume that the total mass of M_n is an integer for every n. If this were not so we can write $M_n = M_n^{(1)} + M_n^{(2)}$ where $M_n^{(1)}$ is symmetric with an integral total mass and $M_n^{(2)}$ has total mass less than unity. Consequently

$$\int_{\|x\| \leqslant 1} \|x\|^2 dM_n^{(2)} \leqslant 1.$$

Since our aim is to prove that $\qquad \sup_n \int_{\|x\| \leqslant 1} \|x\|^2 dM_n < \infty$

it suffices to show that $\qquad \sup_n \int_{\|x\| \leqslant 1} \|x\|^2 dM_n^{(1)} < \infty.$

Now since the total mass of M_n is an integer say k_n we will write

$$M_n = F_{n1} + \ldots F_{nk_n}$$

where F_{nj} for $j = 1, 2, \ldots, k_n$, $n = 1, 2, \ldots$ is a symmetric probability measure in the unit sphere. Let us now denote by μ_n the convolution

$$\mu_n = F_{n1} * F_{n2} \ldots * F_{nk_n}.$$

Since each F_{nj} is symmetric and has zero expectation

$$\int \|x\|^2 d\mu_n = \sum_{j=1}^{k_n} \int \|x\|^2 dF_{nj} = \int \|x\|^2 dM_n.$$

Hence it suffices to show that $\qquad \sup_n \int \|x\|^2 d\mu_n < \infty.$

If $Q_n(t)$ denotes the concentration function of μ_n from Theorem 4.4 we have

$$\int \|x\|^2 d\mu_n \leqslant \frac{16t^2 + (4t+1)^2}{2Q_n(t) - 1}$$

whenever $Q_n(t) > \frac{1}{2}$. Therefore, it is enough to prove that

$$\inf_n Q_n(t) \geqslant \tfrac{3}{4} \quad \text{for some } t$$

which will follow from Theorem 4.1 if we prove that μ_n is weakly conditionally compact.

Since each $F_{nj}(y)$ is a real characteristic function and

$$\mu_n(y) = \prod_{j=1}^{k_n} F_{nj}(y) \qquad \qquad \ldots \quad (5.7)$$

it follows from (5.6) that $\quad 1 - \mu_n(y) \leqslant \sum_{j=1}^{k_n} [1 - F_{nj}(y)] = \sum_{j=1}^{k_n} \int [1 - \cos(x, y)] dF_{nj}(x)$

$$= \int [1 - \cos(x, y)] dM_n(x) = f_n(y) \text{ say.} \qquad \ldots \quad (5.8)$$

SANKHYĀ : THE INDIAN JOURNAL OF STATISTICS : Series A

We also have $\lambda_n(y) = \exp[-f_n(y)]$ and hence for any given ϵ there exists a δ depending only on ϵ such that

$$f_n(y) \leqslant \epsilon \quad \text{if} \quad 1 - \lambda_n(y) \leqslant \delta. \qquad \qquad \dots (5.9)$$

Since λ_n is compact we have from Theorem 3.5 for any given $\delta > 0$ a compact sequence S_n of S-operators depending on δ only such that

$$1 - \lambda_n(y) \leqslant (S_n y, y) + \frac{\delta}{2}. \qquad \qquad \dots (5.10)$$

From (5.9) and (5.10) it follows that whenever $(S_n y, y) \leqslant \delta/2, f_n(y) \leqslant \epsilon$ and hence from Lemma 5.1 we have

$$f_n(y) \leqslant (S_n' y, y) + \epsilon \qquad \qquad \dots (5.11)$$

where $\qquad\qquad S_n' = 8\epsilon\delta^{-1}S_n.$

Since S_n is compact so is S_n' and Theorem 3.3 and (5.11) imply that μ_n is weakly conditionally compact. The proof of the theorem is now complete.

We will denote by $K(x, y)$ the following function

$$K(x, y) = e^{i(x,y)} - 1 - \frac{i(x, y)}{1 + \|x\|^2}.$$

Theorem 5.7 : Let μ_n for each n be of the form $e(F_n)$ where F_n is a finite measure. Let $\mu_n * x_n \Longrightarrow \mu$ for some suitably chosen elements x_n in X. We further assume that F_n is increasing. Then F_n increases to a measure F which may be σ-finite but gives finite mass outside every neighbourhood of the origin and for which

$$\int_{\|x\| \leqslant 1} \|x\|^2 dF < \infty.$$

In addition $\qquad\qquad \mu(y) = \exp[i(x_0, y) + \int K(x, y) dM(x)]$

where x_0 is a fixed element of X.

Proof : Let $\lambda_n(y)$ be defined as

$$\lambda_n(y) = \exp[\int K(x, y) dF_n].$$

Then $\lambda_n(y)$ is the characteristic function of λ_n which is the shift of μ_n by the element

$$Z_n = -\int \frac{x}{1 + \|x\|^2} dF_n.$$

We will now show that $\lambda_n(y)$ converges uniformly in y over bounded spheres. For this purpose we write

$$\int K(x, y) dF_n = \int_{\|x\| \leqslant 1} K(x, y) dF_n + \int_{\|x\| > 1} K(x, y) dF_n. \qquad \dots (5.12)$$

Let F be the limit of F_n. From Theorems 5.3 and 5.6 it follows that F is finite outside every neighbourhood of the origin and

$$\int_{\|x\| \leqslant 1} \|x\|^2 dF < \infty.$$

Since $\qquad\qquad |K(x, y)| \leqslant 2 + \frac{\|x\| \, \|y\|}{1 + \|x\|^2} \leqslant 2 + \|y\|$

LIMIT DISTRIBUTION FOR RANDOM VARIABLES

it follows that

$$\sup_{y \in S} \Big| \int_{||x||>1} K(x, y)dF_n - \int_{||x||>1} K(x, y)dF \Big| \to 0 \text{ as } n \to \infty \qquad \dots \quad (5.13)$$

for every bounded sphere S. On the other hand if $||x|| \leqslant 1$

$$|K(x, y)| \leqslant |e^{i(x, y)} - 1 - i(x, y)| + \frac{|(x, y)| \, ||x||^2}{1 + ||x||^2}$$

$$\leqslant \tfrac{1}{2}(x, y)^2 + |(x, y)| \, ||x||^2 \leqslant \tfrac{1}{2} ||x||^2 ||y||^2 + ||y|| \, ||x||^2.$$

Since $\int_{||x|| \leqslant 1} ||x||^2 \, dF$ is finite it follows that

$$\sup_{y \in S} \Big| \int_{||x|| \leqslant 1} K(x, y)dF_n - \int_{||x|| \leqslant 1} K(x, y)dF \Big| \to 0 \text{ as } n \to \infty \qquad \dots \quad (5.14)$$

for every bounded sphere S. (5.12), (5.13) and (5.14) imply that

$$\sup_{y \in S} |\lambda_n(y) - \lambda(y) | \to 0 \quad \text{as } n \to \infty$$

for every bounded sphere S where $\lambda(y) = \exp [\int K(x, y)dF]$.

Since λ_n is shift compact from Theorem 5.5 it follows that $\lambda_n \Longrightarrow \lambda$ and λ has to be a shift of μ. Hence the theorem follows.

Theorem 5.8 : *Let $\mu(y)$ be a function of the form*

$$\mu(y) = \exp [\int K(x, y)dF(x)]$$

where F is a σ-finite measure giving finite mass outside every neighbourhood of the identity and for which

$$\int_{||x|| \leqslant 1} ||x||^2 \, dF < \infty.$$

Then $\mu(y)$ is the characteristic function of an infinitely divisible distribution.

Proof : Let N_n denote the sphere of radius $\dfrac{1}{n}$ around the origin and F_n the restriction of F to N_n'. Then F_n increases to F. Let

$$\mu_n(y) = \exp [\int K(x, y)dF_n].$$

From the proof of Theorem 5.1 it follows that

$$\sup_{y \in V} |\mu_n(y) - \mu(y) | \to 0 \text{ as } n \to \infty$$

for every bounded sphere V. In view of Theorems 5.1 an l 5.5 it is enough to show that μ_n is shift compact. We will now show that $\lambda_n = |\mu_n|^2$ is compact.

$$\lambda_n = |\mu_n|^2 = |e(F_n)|^2 = e(M_n) \text{ where } M_n = F_n + \bar{F}_n.$$

Since F_n increases to F it follows that M_n increases to M where $M = F + \bar{F}$. Without any loss of generality we can assume that F and hence M vanishes outside the sphere $||x|| \leqslant 1$. We further have

$$1 - \lambda_n(y) = 1 - \exp [\int [\cos(x, y) - 1]dM_n] = 1 - \exp [-\int [1 - \cos(x, y)]dM_n]$$

$$\leqslant \int [1 - \cos(x, y)] \, dM_n \leqslant \int [1 - \cos(x, y)] \, dM$$

$$\leqslant \tfrac{1}{2} \int_{||x|| \leqslant 1} (x, y)^2 \, dM = \tfrac{1}{2}(Sy, y).$$

Since
$$\int_{\|x\| \leqslant 1} \|x\|^2 \, dM = 2 \int_{\|x\| \leqslant 1} \|x\|^2 \, dF < \infty$$

it follows that S is an S-operator. Since S is a fixed S-operator independent of n it follows from Theorem 3.5 that λ_n is compact. Consequently μ_n is shift compact and the theorem is proved.

Gaussian distributions are defined in the Hilbert Space in exactly the same manner as in Definition 2.4. We shall now prove

Theorem 5.9 : *A distribution μ on X is Gaussian if and only if $\mu(y)$ is of the form*

$$\mu(y) = \exp \{i(x_0, y) - (Sy, y)\}$$

where X_0 is a fixed element and S an S-operator.

Proof: Let us take a countable dense subset $y_1, y_2, ..., y_n ...$ in X and consider the map τ from X to Z^∞, the countable product of the circle groups, defined as

$$\tau(x) = [e^{i(x, y_1)}, ... \, e^{i(x, y_n)}, ...].$$

Let H be the image of X under τ in Z^∞. Then τ is a both ways measurable isomorphism of the two groups X and H. If μ is Gaussian on X then $\mu\tau^{-1}$ is Gaussian in H and hence in Z^∞. From Theorem 2.5 we have

$$\mu\tau^{-1}(\theta) = \theta(Z) \exp[-\phi(\theta)] \qquad \ldots \quad (5.15)$$

where θ is a character on Z^∞, Z is a fixed element of Z^∞ and ϕ a function with properties specified in Theorem 2.5. $\theta(Z)$ denotes the value of the character θ at the point Z. Further any character θ on Z^∞ is of the form

$$\theta(Z) = Z_1^{n_1} ... Z_k^{n_k} \qquad \ldots \quad (5.16)$$

where n_1, n_k are integer and $Z_1, ... Z_k$ the first k coordinates of Z in Z^∞.

Therefore,
$$\mu\tau^{-1}(\theta) = \mu(n_1 y_1 + ... \, n_k y_k)$$

where $\mu(y)$ is the characteristic function on X and $n_1, ... n_k$ are related to θ by the relation (5.16). Hence
$$|\mu\tau^{-1}(\theta)| = |\mu| \, (n_1 y_1 + ... n_k y_k) = e^{-\phi(\theta)}.$$

Since for any θ and θ'
$$\phi(\theta + \theta') + \phi(\theta - \theta') = 2[\phi(\theta) + \phi(\theta')]$$

it follows that
$$h(y + y') + h(y - y') = 2[h(y') + h(y)] \qquad \ldots \quad (5.17)$$

whenever y, y' are of the form $n_1 y_1 + ... n_k y_k$, where
$$h(y) = -\log |\mu(y)|.$$

Since $y_1, y_2, ...$ are dense and h is continuous in the norm topology of X it follows that the relation (5.17) is valid for any pair y, y'. This implies that $h(y)$ can be put as
$$h(y) = (Sy, y)$$

where S is a positive semi-definite Hermitian operator. From the continuity of $\mu(y)$ in the S-topology it follows that S is an S-operator. If we now consider the distribution λ on X defined by the equation
$$\lambda(y) = \exp[-(Sy, y)].$$

We have
$$\lambda\tau^{-1}(\theta) = e^{-\phi(\theta)}.$$

LIMIT DISTRIBUTION FOR RANDOM VARIABLES

Hence $\lambda\tau^{-1}$ is a shift of $\mu\tau^{-1}$. Since both $\lambda\tau^{-1}$ and $\mu\tau^{-1}$ give unit mass for the subgroup H of Z^∞ the element Z by which $\lambda\tau^{-1}$ is shifted to obtain $\mu\tau^{-1}$ belongs to H and hence $Z = \tau(x_0)$ for some $x_0 \epsilon X$. Consequently,

$$\mu = \lambda * x_0$$

or
$$\mu(y) = \exp\left[i(x_0, y) - (Sy, y)\right].$$

Conversely, if $\mu(y)$ is of the form $\quad \mu(y) = \exp\left[i(x_0, y) - (Sy, y)\right]$

the distribution $\mu\tau^{-1}$ in Z^∞ is Gaussian and hence so is μ.

We now prove the representation theorem for infinitely divisible distributions.

Theorem 5.10 : *A function $\mu(y)$ is the characteristic function of an infinitely divisible distribution μ on X if and only if it is of the form*

$$\mu(y) = \exp\left[i(x_0, y) - (Sy, y) + \int K(x, y) \, dM(x)\right] \qquad \text{... (5.18)}$$

where x_0 is a fixed element of X, S an S-operator and M a σ-finite measure giving finite-mass outside every neighbourhood of the origin and for which

$$\int_{\|x\| \leqslant 1} \|x\|^2 \, dM < \infty.$$

Here $K(x, y)$ is the function $\quad K(x, y) = \left[e^{i(x, y)} - 1 - \dfrac{i(x, y)}{1 + \|x\|^2} \right].$

The representation (5.18) of $\mu(y)$ is unique.

Proof : Let $\mu(y)$ be the characteristic function of an infinitely divisible distribution μ. Then in the same manner as in the proof of Theorem 7.1 of Parthasarathy *et al* (1962a) we can construct a sequence of distributions λ_n such that (i) $\lambda_n = e(M_n)$, (ii) M_n increases to M, (iii) $\lambda_n * x_n \Longrightarrow \lambda$, (iv) $\mu = \lambda * \mu_0$ where μ_0 is Gaussian.

Now using Theorems 5.7 and 5.9 we can complete the proof. Sufficiency is immediate from Theorems 5.8 and 5.9. Uniqueness is proved in an exactly same manner as Theorem 8.1 of Parthasarathy *et al* (1962a) but keeping in mind that the space X playing the role of the character group is connected.

6. COMPACTNESS CRITERIA

In the present section we will find out the necessary and sufficient conditions in order that a sequence μ_n of infinitely divisible distributions may be weakly conditionally compact.

If μ is any infinitely divisible distribution by $\mu = [x, S, M]$ we will mean that the three quantities occurring in the representation of μ according to Theorem 5.10 are respectively x, S and M. We will associate with any such $\mu = [x, S, M]$ another S-operator which we will denote by T.

$$(Ty, y) = 2(Sy, y) + \int_{\|x\| \leqslant 1} (x, y)^2 \, dM(x) \qquad \text{... (6.1)}$$

T is an S-operator since $\quad \displaystyle\int_{\|x\| \leqslant 1} \| x \|^2 \, dM(x) < \infty.$

Lemma 6.1 : *In order that a sequence μ_n of Gaussian distributions with covariance operators S_n be shift compact it is necessary and sufficient that S_n should be compact.* [*If μ is Gaussian with covariance operator S its characteristic function is* $\exp[i(x_0, y) - \frac{1}{2}(Sy, y)]$.

Proof : Sufficiency is immediate from Theorem 3.4. We will prove necessity. If μ_n is shift compact then $|\mu_n|^2$ is compact. But $|\mu_n|^2$ is Gaussian with mean zero and covariance operator $2S_n$.

$$|\mu_n|^2(y) = \exp[-(S_n y, y)].$$

From Theorem 3.3 it follows that there exists a compact sequence U_n of S-operators such that

$$1 - |\mu_n|^2(y) \leqslant (U_n y, y) + \epsilon.$$

Hence there is a δ such that for any n, $(U_n y, y) \leqslant \delta$ implies $(S_n y, y) \leqslant 1$. From this we can deduce that

$$(S_n y, y) \leqslant \delta^{-1}(U_n y, y).$$

Since δ is independent of n and U_n is compact, S_n is also compact.

Lemma 6.2 : *Let μ be a symmetric infinitely divisible distribution with $\mu = [0, 0, M]$. Further, let M be concentrated in the unit sphere. Then*

$$\int \|x\|^4 \, d\mu \leqslant \int \|x\|^4 \, dM + 3[\int \|x\|^2 \, dM]^2 < \infty.$$

Proof : It is enough to prove the theorem when M is finite since the other case can be obtained by limit transition. Let $M(x) = t$ and F be the distribution such that $F(X) = 1$ and $M = tF$. Then

$$\mu = e^{-t} \sum \frac{t^r F^r}{r!}$$

$$\int \|x\|^4 \, dF^r = E \|X_1 + \ldots X_r\|^4$$
$$= r E \|X_1\|^4 + r(r-1)[E \|X_1\|^2]^2 + 2r(r-1) E(X_1, X_2)^2$$
$$= rE \|X_1\|^4 + 3r(r-1)(E \|X_1\|^2)^2.$$

Expectation is with respect to F and $X_1, \ldots X_r$ are independent random variables in X with the same distribution F. Terms with zero expectation have been omitted.

$$\int \|x\|^4 \, d\mu \leqslant e^{-t} \sum \frac{rt^r \, E \|X_1\|^4}{r!} + 3 \, e^{-t} \sum \frac{r(r-1)t^r(E \|X_1\|^2)^2}{r!}$$
$$= tE \|X_1\|^4 + 3t^2(E \|X_1\|^2)^2$$
$$= t \int \|X\|^4 \, dF + 3t^2[\int \|X\|^2 \, dF]^2$$
$$= \int \|X\|^4 \, dM + 3[\int \|X\|^2 \, dM]^2.$$

Lemma 6.3 : *Let μ_n be symmetric infinitely divisible distributions such that*

$$\mu_n = [0, 0, M_n]$$

with M_n vanishing outside the sphere $\|X\| \leqslant 1$ for all n. If μ_n is compact then

$$\sup_n \int \|X\|^4 \, d\mu_n < \infty.$$

LIMIT DISTRIBUTION FOR RANDOM VARIABLES

Proof: Since M_n is concentrated in the unit sphere

$$\int \|X\|^4 \, dM_n \leqslant \int \|X\|^2 \, dM_n.$$

Theorem 5.6 implies that $\qquad \sup_n \int \|X\|^2 \, dM_n < \infty$

and hence an application of Lemma 6.2 will complete the proof.

Remark 6.1 : In the same manner as Lemma 6.2 it can be shown that if M is symmetric and vanishes outside the sphere $\|X\| \leqslant 1$ for

$$\mu = [0, 0, M].$$

We have $\qquad \int (x, y)^2 d\mu = \int (x, y)^2 dM$ for all y.

Lemma 6.4 : *Let μ_n be a weakly conditionally compact sequence of symmetric distribution such that*

$$\sup_n \int \|X\|^4 \, d\mu_n < \infty.$$

Then if S_n is the covariance operator of μ_n, $\{S_n\}$ is compact.

Proof: If θ_n is a sequence of distributions on the real line such that $\theta_n \Longrightarrow \theta$ and $\int x^2 \, d\theta_n$ is uniformly bounded then

$$\int x d\theta_n \to \int x d\theta \text{ as } n \to \infty.$$

Theorem 3.5 can be applied now and the lemma follows at once.

Theorem 6.1 : *Let μ_n be symmetric distributions that are infinitely divisible with representations*

$$\mu_n = [0, S_n, M_n].$$

Then in order that μ_n be compact it is necessary and sufficient that

(i) *M_n restricted outside any neighbourhood of the identity is weakly conditionally compact,*

(ii) *T_n as defined in (6.1) is compact.*

Proof: We will first prove sufficiency. Let unit sphere be chosen as the neighbourhood and let us write $M_n = M_n^{(1)} + M_n^{(2)}$ where $M_n^{(1)}$ and $M_n^{(2)}$ are respectively the restrictions of M_n inside the unit sphere and outside the unit sphere. Since $M_n^{(2)}$ is weakly conditionally compact and $F_n \Longrightarrow F$ implies $e(F_n) \Longrightarrow e(F)$ it is enough to show that the distributions

$$\lambda_n = [0, S_n, M_n^{(1)}]$$

form a compact sequence $\int (x, y)^2 d\lambda_n(x) = 2(S_n y, y) + \int (x, y)^2 \, dM_n^{(1)}(x)$

$$= 2(S_n y, y) + \int_{\|x\| \leqslant 1} (x, y)^2 \, dM_n(x)$$

$$= (T_n y, y).$$

Since T_n is compact sufficiency follows from Theorem 3.4. Necessity of (i) is a consequence of Theorem 5.3 and (ii) follows from Lemmas 6.1, 6.2 and 6.4.

Theorem 6.2 : *In order that a sequence μ_n of infinitely divisible distributions with representations*

$$\mu_n = [x_n, S_n, M_n]$$

be shift compact it is necessary and sufficient that

(i) *M_n restricted to the complement of any neighbourhood N of the origin is weakly conditionally compact,*

(ii) *T_n as defined in (6.1) is compact.*

Proof : Since μ_n is shift compact if and only $|\mu_n|^2$ is compact what we need are conditions for the compactness of

$$|\mu_n|^2 = [0, 2S_n, M_n + \bar{M}_n].$$

In addition we have $\int\limits_{\|x\| \leqslant 1} (x, y)^2 \, d(M_n + \bar{M}_n) = 2 \int\limits_{\|x\| < 1} (x, y)^2 dM_n.$

Hence the theorem follows from 6.1.

Theorem 6.3 : *In order that μ_n with the representations*

$$\mu_n = [x_n, S_n, M_n]$$

be weakly conditionally compact it is necessary and sufficient that in addition to the conditions of Theorem 6.2 x_n should be compact in X.

Proof : In order to prove the theorem it suffices to show that whenever

$$\mu_n = [x_n, S_n, M_n]$$

is shift compact, $\qquad \lambda_n = [0, S_n, M_n]$

is compact. Let $F_n \Longrightarrow F$. Then the following convergence takes place in norm.

$$\int \frac{x}{1 + \|x\|^2} \, dF_n \to \int \frac{x}{1 + \|x\|^2} \, dF.$$

Hence we can assume that M_n vanishes for all n outside the sphere $\|x\| \leqslant 1$. Let us now consider

$$f_n(y) = \int K(x, y) \, dM_n(x)$$

$$|f_n(y)| \leqslant \tfrac{1}{2} \int\limits_{\|x\| \leqslant 1} (x, y)^2 \, dM_n(x) + \int\limits_{\|x\| > 1} \frac{|(x, y)| \, \|x\|^2}{1 + \|x\|^2} \, dM_n(x)$$

$$\leqslant \tfrac{1}{2}(T_n y, y) + \int |(x, y)| \, \|x\|^2 \, dM_n(x)$$

$$\leqslant \tfrac{1}{2}(T_n y, y) + [\int (x, y)^2 \|x\|^2 \, dM_n \int \|x\|^2 \, dM_n]^{\frac{1}{2}}$$

$$\leqslant \tfrac{1}{2}(T_n y, y) + C(T_n y, y)^{\frac{1}{2}}.$$

Hence given any $\rho > 0$ there is a $\delta > 0$ such that $|f_n(y)| \leqslant \rho$ whenever $(T_n y, y) < \delta$. But since $\lambda_n(y) = \exp[f_n(y)]$, for any $\epsilon > 0$ there exists a $\rho > 0$ such that $1 - R\lambda_n(y) \leqslant \epsilon$ whenever $|f_n(y)| \leqslant \rho$. Consequently, for any $\epsilon > 0$ there exists a $\delta > 0$ such that whenever $(T_n y, y) \leqslant \delta$ it follows that $1 - R\lambda_n(y) \leqslant \epsilon$. Here R denotes the real part. Now Lemma 5.1 and Theorem 3.5 imply the validity of the theorem since $\{T_n\}$ is compact.

LIMIT DISTRIBUTION FOR RANDOM VARIABLES

Remark 6.2 :   In defining the operator T we could have taken any bounded sphere around the origin instead of the unit sphere.   When M_n restricted outside any neighbourhood of the origin is known to be weakly conditionally compact, the compactness of T_n when it is based on some sphere implies the compactness of T_n when it is based on any finite sphere.

Since the representation is unique one can give conditions for the convergence of

$$\mu_n = [x_n, S_n, M_n]$$

to the distribution $$\mu = [x, S, M]$$

in terms of $[x_n, S_n, M_n]$ and $[x, S, M]$.   However, we will mention only the following.

Theorem 6.4 :   *Let μ_n be a sequence such that μ_n has the representation*

$$\mu = [x_n, S_n, M_n].$$

If $\mu_n \Longrightarrow \mu$, μ is Gaussian if and only if

$$M_n(N') \to 0 \ as \ n \to \infty$$

for every neighbourhood N of the origin.

Proof: Let μ be Gaussian.   Since μ cannot be written as $e(F) * \lambda$ with λ infinitely divisible it follows that $M_n(N') \to 0$ as $n \to \infty$ for every neighbourhood N of the origin, conversely, if $M_n(N') \to 0$ for every neighbourhood N in exactly the same manner as in the proof of Theorem 6.1 of Parthasarathy *et al* (1962a) it can be shown that $\mu(y)$ is of the form

$$\mu(y) = \exp\left[i\,(x_0, y) - (Sy, y) \right]$$

which shows that μ is Gaussian.

7.   Accompanying laws

Definition 7.1 :   A sequence α_{nj} of distributions $j = 1, 2, ..., k_n$, $n = 1, 2, ...$ is said to be uniformly infinitesimal if for any neighbourhood N of the origin

$$\mathop{\mathrm{lt}}_{n \to \infty} \ \mathop{\inf}_{1 \leqslant j \leqslant k_n} \ \alpha_{nj}(N) = 1.$$

Theorem 7.1 :   *In order that α_{nj} be uniformly infinitesimal it is necessary that*

$$\mathop{\mathrm{lt}}_{\to \infty} \ \mathop{\sup}_{1 \leqslant j \leqslant k_n} \ \mathop{\sup}_{\|y\| \leqslant K} \ |\alpha_{nj}(y) - 1| = 0$$

for every constant K.

Proof :   This is immediate from Theorem 5.4.

Theorem 7.2 :   *Let α_{nj} be uniformly infinitesimal symmetric distributions with non-negative characteristic function.   Let*

$$\mu_n = \mathop{\Pi}_{j=1}^{k_n} \alpha_{nj}$$

and λ_n be defined as $$\lambda_n = \mathop{\Pi}_{j=1}^{k_n} e(\alpha_{nj}).$$

In order that $\mu_n \Longrightarrow \mu$ it is necessary and sufficient that $\lambda_n \Longrightarrow \mu$.

Proof : Let μ_n be compact. Since the inequality $e^{x-1} \geqslant x$ is valid for all x

we have $\qquad e(\alpha_{nj})(y) \geqslant \alpha_{nj}(y) \quad$ for $j = 1, 2, ..., k_n, \quad n = 1, 2 ...$

since $\alpha_{nj}(y)$ is non-negative $\qquad \lambda_n(y) \geqslant \mu_n(y)$.

Or $\qquad\qquad\qquad\qquad 1-\lambda_n(y) \leqslant 1-\mu_n(y)$.

From the compactness of μ_n and Theorem 3.3 it follows that λ_n is compact. Now let λ_n be compact. It follows from Theorem 6.2 and the remark made after Theorem 6.3 that

(i) F_n restricted to N' is weakly conditionally compact for every neighbourhood N;

(ii) the sequence S_n of operators defined by

$$(S_n\, y,\, y) = \int_{\|x\| \,\leqslant\, t} (x,\, y)^2\, dF_n(x)$$

is compact for every t.

Here F_n denotes the sum $\alpha_{n1}+\alpha_{n2}+...\alpha_{nk_n}$. We will now show that μ_n is compact. We have

$$1-\mu_n(y) \leqslant \sum_{j=1}^{k_n} [1-\alpha_{nj}(y)] = \int [1-\cos\,(x,\,y)]dF_n(x)$$

$$= \int_{\|x\|<t} [1-\cos\,(x,\,y)]dF_n(x) + \int_{\|x\|>t} [1-\cos(x,y)\,dF_n(x)$$

$$= \tfrac{1}{2} \int_{\|x\|\leqslant t} (x,\,y)^2\, dF_n(x) + 2F_n[\|x\| > t] = \tfrac{1}{2}(S_n y,\, y) + 2F_n[\|x\| > t].$$

Since F_n is weakly compact outside any neighbourhood we can choose t such that for all n

$$F_n[\|x\| > t] < \epsilon/2.$$

Since for that fixed t, S_n is compact Theorem 3.3 shows that μ_n is compact.

We will now complete the proof by showing that whenever λ_n is compact, for every costant k

$$\sup_{\|y\|\leqslant K} |\lambda_n(y) - \mu_n(y)| \to 0.$$

To this end it is enough to show that

$$\sup_{\|y\|\leqslant K} \sup_n \sum_{j=1}^{k_n} [1-\alpha_{nj}(y)] < \infty.$$

But the expression is equal to

$$\sup_{\|y\|\leqslant K} \sup_n \int [1-\cos\,(x,\,y)]dF_n \leqslant \sup_{\|y\|\leqslant K} \sup_n \int_{\|x\|\leqslant 1} [1-\cos\,(x,\,y)]dF_n + \sup_n 2F_n(\|x\|>1)$$

$$\leqslant \tfrac{1}{2}K^2 \sup_n \int_{\|x\|\leqslant 1} \|x\|^2 dF_n + 2\sup_n F_n(\|x\| > 1)$$

$$< \infty.$$

The last step follows since $\lambda_n = e(F_n)$ is compact.

LIMIT DISTRIBUTION FOR RANDOM VARIABLES

Lemma 7.1 : *Let α_{nj} be uniformly infinitesimal. Then if x_{nj} is defined by the relation*

$$x_{nj} = \int_{\|x\| \leqslant 1} x \, d \, \alpha_{nj}$$

then

$$\sup_{1 \leqslant j \leqslant k_n} \|x_{nj}\| \to 0 \text{ as } n \to \infty.$$

Proof : Let ϵ be arbitrary and V the sphere $\|x\| \leqslant \epsilon$. Then

$$\|x_{nj}\| \leqslant \| \int_V d\alpha_{nj}\| + \| \int_{\epsilon < \|x\| \leqslant 1} x d\alpha_{nj}\| \leqslant \epsilon + \alpha_{nj}(V').$$

Hence

$$\limsup_{n \to \infty} \sup_{1 \leqslant j \leqslant k_n} \|x_{nj}\| \leqslant \epsilon.$$

Since ϵ is arbitrary the lemma follows.

Lemma 7.2 : *Let α_{nj} be uniformly infinitesimal. Let x_{nj} be defined as in Lemma 4.6. Then if $\theta_{nj} = \alpha_{nj} * (-x_{nj})$ there exists a n_0 such that for all $1 \leqslant j \leqslant k_n$ and $n \geqslant n_0$ we have*

$$\| \int_{\|x\| \leqslant 1} x d\theta_{nj}\| \leqslant 2\theta_{nj}[\|x_{nj}\| > 1].$$

Proof : Let n_0 be so chosen that

$$\sup_{1 \leqslant j \leqslant k_n} \|x_{nj}\| \leqslant 1/4 \text{ for all } n > n_0.$$

Then

$$\int_{\|x\| \leqslant 1} x d\theta_{nj} = \int_{\|x - x_{nj}\| \leqslant 1} (x - x_{nj}) d\alpha_{nj}$$

$$= \int_{\|x - x_{nj}\| \leqslant 1} x d\alpha_{nj} - \int_{\|x\| \leqslant 1} x d\alpha_{nj} + x_{nj}\alpha_{nj}[\|x - x_{nj}\| > 1].$$

Therefore, for $n > n_0$ and $1 \leqslant j \leqslant k_n$,

$$\| \int_{\|x\| \leqslant 1} x d\theta_{nj}\| \leqslant \int_{[\|x - x_{nj}\| \leqslant 1 \Delta \|x\| \leqslant 1]} \|x\| d\alpha_{nj} + \|x_{nj}\| \theta_{nj} [\|x\| > 1]$$

$$\leqslant \int_{\frac{3}{4} \leqslant \|x\| \leqslant \frac{5}{4}} \|x\| d\alpha_{nj} + \tfrac{1}{4} \theta_{nj} [\|x\| > 1]$$

$$\leqslant \tfrac{5}{4} \alpha_{nj} [\|x\| > 3/4] + \tfrac{1}{4} \theta_{nj} [\|x\| > 1]$$

$$\leqslant \tfrac{5}{4} \theta_{nj} [\|x\| > 1] + \tfrac{1}{4} \theta_{nj} [\|x\| > 1]$$

$$\leqslant 2\theta_{nj} [\|x\| > 1].$$

Lemma 7.3 : *Let F_n be a sequence of σ-finite measures such that F_n restricted to N' is finite and weakly conditionally compact for every neighbourhood N of the origin. Then for any $\epsilon > 0$ there exists a compact set K such that*

$$F_n(K') \leqslant \epsilon \text{ for } n = 1, 2, \ldots$$

Proof : Let ϵ be any positive number. Choose a sequence N_r of neighbourhoods of the origin decreasing to the origin. Let A_r be defined as $N_{r-1} - N_r$, N_0 being taken as the whole space. From the conditions of the lemma it is possible to find a compact set K_r in A_r such that for all n,

$$F_n(A_r - K_r) \leqslant \frac{\epsilon}{2^r}.$$

Let K be defined as
$$K = \bigcup_{r=1}^{\infty} K_r \cup \{0\}.$$

Since
$$K \cap N_r' = K_1 \cup K_2 \ldots \cup K_r$$
and N_r decreases to the origin it follows that K is compact and

$$F_n(K') \leqslant \sum_{r=1}^{\infty} F_n(A_r - K_r) \leqslant \epsilon.$$

We now proceed to prove the main theorem of the section.

Let α_{nj} be uniformly infinitesimal sequence of distributions on X. Let μ_n, x_n, θ_{nj}, λ_n be defined as follows.

$$\mu_n = \prod_{j=1}^{k_n} \alpha_{nj}$$

$$x_{nj} = \int_{\|x\| \leqslant 1} x d\alpha_{nj}$$

$$\theta_{nj} = \alpha_{nj}*(-x_{nj})$$

$$\lambda_n = \prod_{j=1}^{k_n} e(\theta_{nj})* \left(\sum_{j=1}^{k_n} x_{nj} \right)$$

In what follows we will adopt the above notation.

Theorem 7.3 : *If μ_n is shift compact so is λ_n.*

Proof : Since μ_n is shift compact $|\mu_n|^2$ is compact. But

$$|\mu_n|^2 = \prod_{j=1}^{k_n} |\alpha_{nj}|^2.$$

It follows now from Theorem 6.2 that if one defines

$$F_n = \sum_{j=1}^{k_n} |\alpha_{nj}|^2 = \sum_{j=1}^{k_n} |\theta_{nj}|^2. \qquad \ldots \quad (7.1)$$

Then $e(F_n)$ is compact. We can now apply Theorem 6.2 and Lemma 7.3 and deduce that for any $\epsilon > 0$ there exists a compact set K such that

$$F_n(K') \leqslant \epsilon \quad \text{for } n = 1, 2, \ldots \qquad \ldots \quad (7.2)$$

Let us now define the S-operators T_n by the formula

$$(T_n y, y) = \int_{\|x\| \leqslant t} (x, y)^2 dF_n(x).$$

Then for any finite t the sequence T_n is compact.
Let us now define G_n as

$$G_n = \sum_{j=1}^{k_n} \theta_{nj}. \qquad \ldots \quad (7.3)$$

LIMIT DISTRIBUTION FOR RANDOM VARIABLES

In order to complete the theorem we have to show that $e(G_n)$ is shift compact or

(a) G_n is weakly conditionally compact when restricted outside any neighbourhood of the identity;

(b) if the S-operators S_n are defined as

$$(S_n y, y) = \int_{\|x\| \leqslant 1} (x, y)^2 dG_n$$

then S_n is compact.

Since α_{nj} is uniformly infinitesimal by Lemma 7.1 θ_{nj} are also uniformly infinitesimal. Hence for any $\epsilon > 0$ there exists a compact set C such that

$$\theta_{nj}(C) \geqslant 1-\epsilon \text{ for all } n \text{ and } 1 \leqslant j \leqslant k_n.$$

In the same manner as in the proof of Theorem 5.1 of Parthasarathy *et al* (1962a) we have for all n and $1 \leqslant j \leqslant k_n$

$$|\theta_{nj}|^2(K') = \int \theta_{nj}(K'+x)d\theta_{nj}(x) \geqslant \int_C \theta_{nj}(K'+x)d_{nj}(x)$$

$$\geqslant (1-\epsilon) \inf_{x \epsilon C} \theta_{nj}(K'+x) = (1-\epsilon)[1-\sup_{x \epsilon O} \theta_{nj}(K+x)]$$

$$= (1-\epsilon)\theta_{nj}(\overline{K+C'}) = (1-\epsilon)\,\theta_{nj}(K_1'). \qquad \dots (7.4)$$

where K_1 is another compact set.

In a similar manner it can be shown that if V and N are two neighbourhoods of the origin such that $V+V \subsetneq N$ and

$$\theta_{nj}(V) \geqslant 1-\epsilon \text{ for all } n \geqslant n_0, \ 1 \leqslant j \leqslant k_n \qquad \dots (7.5)$$

then for $\qquad n \geqslant n_0 \text{ and } 1 \leqslant j \leqslant k_n \ |\theta_{nj}|^2(V') \geqslant (1-\epsilon)\theta_{nj}(N') \qquad \dots (7.6)$

(7.2), (7.3) and (7.4) imply that for any $\epsilon > 0$ there exists a compact set K_1 such that

$$G_n(K_1') \leqslant \epsilon \text{ for all } n. \qquad \dots (7.7)$$

On the other hand (7.1), (7.3), (7.6) and the weak compactness of F_n when restricted to the complement of any neighbourhood of the origin imply that for any neighbourhood N of the origin,

$$\sup_n G_n(N') < \infty. \qquad \dots (7.8)$$

(7.7) and (7.8) prove that G_n is weakly conditionally compact when restricted outside any neighbourhood of the origin. To prove (b) let us consider

$$\int_{\|x\| \leqslant 2} (x, y)^2 \, d\,|\theta_{nj}|^2(x) = \iint_{\|x_1-x_2\| \leqslant 2} (x_1-x_2, y)^2 \, d\theta_{nj}(x_1)d\theta_{nj}(x_2)$$

$$\geqslant \iint_{\|x_1\| \leqslant 1, \ \|x_2\| \leqslant 1} (x_1-x_2, y)^2 \, d\theta_{nj}(x_1)d\theta_{nj}(x_2)$$

$$= 2\theta_{nj}(\|x\| \leqslant 1) \int_{\|x\| \leqslant 1} (x, y)^2 d\theta_{nj}(x) - 2[\int_{\|x\| \leqslant 1} (x, y)d\theta_{nj}]^2.$$

Since θ_{nj} is uniformly infinitesimal we can assume that $\theta_{nj}(\|x\| \leqslant 1) \geqslant \frac{1}{2}$ for all suitably large n and for all $1 \leqslant j \leqslant k_n$. Hence

$$(S_n y, y) \leqslant (T_n y, y) + 2(U_n y, y)$$

where
$$(U_n y, y) = \sum_{j=1}^{k_n} [\int (x, y) d\theta_{nj}(x)]^2.$$

Since we know that T_n is compact in order to show that S_n is compact it is enough to prove that U_n is compact. We will show now that trace of U_n tends to zero as $n \to \infty$, and this would complete the proof.

Let us put
$$y_{nj} = \int_{\|x\| \leqslant 1} x d\theta_{nj}$$

Then for the trace of U_n we have

trace of
$$U_n = \sum_{j=1}^{k_n} \|y_{nj}\|^2 \leqslant (\sup_{1 \leqslant j \leqslant k_n} \|y_{nj}\|) \left(\sum_{j=1}^{k_n} |y_{nj}| \right).$$

From Lemmas 7.1, 7.2 and 7.8 it follows that trace $U_n \to 0$ as $n \to \infty$.

Theorem 7.4 : *If λ_n is shift compact so is μ_n.*

Proof :
$$1 - |\mu_n|^2(y) = 1 - \prod_{j=1}^{k_n} |\theta_{nj}|^2(y) \leqslant \sum_{j=1}^{k_n} [1 - |\theta_{nj}|^2(y)]$$

$$\leqslant 2 \sum_{j=1}^{k_n} [1 - \operatorname{Real} \theta_{nj}(y)] = 2 \sum_{j=1}^{k_n} \int [1 - \cos (x, y)] d\theta_{nj}$$

$$= 2 \int [1 - \cos (x, y)] dG_n(x)$$

$$= 2 \int_{\|x\| \leqslant t} [1 - \cos (x, y)] dG_n(x) + 2G_n[\|x\| > t]$$

$$\leqslant \int_{\|x\| \leqslant t} (x, y)^2 dG_n(x) + 2G_n[\|x\| > t]$$

$$= (S_n y, y) + 2G_n[\|x\| > t]. \qquad \ldots (7.9)$$

From the shift compactness of λ_n for any $\epsilon > 0$ we can choose t such that

$$G_n[\|x\| > t] \leqslant \frac{\epsilon}{2} \text{ for all } n \qquad \ldots (7.10)$$

LIMIT DISTRIBUTION FOR RANDOM VARIABLES

and the sequence of operators S_n defined by

$$(S_n y, y) = \int\limits_{\|x\| \leqslant t} (x, y)^2 dG_n(x)$$

is compact. From (7.9), and (7.10) the compactness of S_n and Theorem 3.3 the present theorem follows.

Theorem 7.5 : *Let λ_n be shift compact. Then for any finite number t we have*

$$\text{lt} \sup_{n \to \infty} \sup_{\|y\| \leqslant t} |\lambda_n(y) - \mu_n(y)| = 0.$$

Proof : Since θ_{nj} is uniformly infinitesimal it follows from Theorem 7.1 that

$$\text{lt}_{n \to \infty} \sup_{\|y\| \leqslant t} \sup_{1 \leqslant j \leqslant k_n} |\theta_{nj}(y) - 1| = 0.$$

Hence it is enough to show that $\sup\limits_{n} \sup\limits_{\|y\| \leqslant t} \sum\limits_{j=1}^{k_n} |\theta_{nj}(y) - 1| < \infty.$ $\qquad$ ... (7.11)

To this end we have

$$\theta_{nj}(y) - 1 = \int [e^{i(x,y)} - 1] d\theta_{nj}$$

$$= \int\limits_{\|x\| \leqslant 1} [e^{i(x,y)} - 1 - i(x, y)] d\theta_{nj} + i \int\limits_{\|x\| \leqslant 1} (x, y) d\theta_{nj} + \int\limits_{\|x\| > 1} [e^{i(x,y)} - 1] d\theta_{nj}$$

$$|\theta_{nj}(y) - 1| \leqslant \tfrac{1}{2} \int\limits_{\|x\| \leqslant 1} (x, y)^2 d\theta_{nj} + \|y\| \, \| \int\limits_{\|x\| \leqslant 1} x d_{nj} \| + 2\theta_{nj}[\|x\| > 1].$$

(7.11) follows at once from the compactness of S_n, Lemma 7.2 and (7.8).

Theorem 7.6 : *In order that $\mu_n * x_n \Longrightarrow \mu$, where x_n are arbitrary points in X, it is necessary and sufficient that $\lambda_n * x_n \Longrightarrow \mu$.*

Proof : This is an immediate consequence of Theorems 7.3, 7.4, 7.5 and 5.5.

Theorem 7.7 : *Limit distribution of sums of independent uniformly infinitesimal random variables in X is infinitely divisible.*

Proof : Theorems 7.6 and 5.1 imply the present theorem since λ_n is infinitely divisible for each n.

Theorem 7.8 : *Let $\mu_n * x_n \Longrightarrow \mu$. In order that μ may be Gaussian it is necessary and sufficient that for each neighbourhood N*

$$\text{lt}_{n \to \infty} G_n(N') = 0,$$

where

$$G_n = \sum_{j=1}^{k_n} \theta_{nj}.$$

Proof : This follows immediately from Theorems 7.6 and 6.4.

SANKHYĀ : THE INDIAN JOURNAL OF STATISTICS : SERIES A

The author wishes to express his thanks to K. R. Parthasarathy and R. Rangarao fer several valuable comments and discussions.

REFERENCES

BOCHNER, S. (1955) : *Harmonic Analysis in Probability*, University of California Press.

———— (1958) : General analytical setting for the central limit theory of probability. Golden Jubilee Commemoration Volume of the Calcutta Mathematical Society, Part I, pp. 111-128.

GNEDENKO, B. V. and KOLMOGOROV, A. N. (1954) : *Limit Distributions for Sums of Independent Random Variables*, Addison-Wesley, Cambridge.

HALMOS, P. R. (1950) : *Measure Theory*, Van Nostrand, Princeton.

HUNT, G. A. (1956) : Semigroups of measures on Lie groups. *Trans. Amer. Math. Soc.*, 81, 264-293.

KLOSS, B. M. (1961) : Limiting distributions on bicompact abelian groups. *Teorya Veroyot i Premen.*, 6, 392-421.

LEVY, P. (1939) : L'addition des variables aleatoires des variables definies sur une cerconference. *Bull. Soc. Math.*, 67, 1-41.

———— (1954) : *Theorie de L'addition des Variables Aleatoires*, Gauthier—Villars.

PARTHASARATHY, K. R., RANGARAO, R. and VARADHAN, S. R. S. (1962) : On the category of indecomposable distributions on topological groups. *Trans. Amer. Math. Soc.*, 102, No. 2, 200-217.

———— (1962a) : On the semigroup of distributions on topological groups. To appear in *Illinois J. Math.*

PROHOROV, YU. V. (1956) : Convergence of random processes and limit theorems in the theory of probability. *Teorya Veroyot. i. Premen.*, 1, 177-238.

RANGARAO, R. (1960) : Some problems in probability theory. Thesis submitted to the Calcutta University (unpublished).

SAZANOV, V. (1958) : On characteristic functionals. *Teorya Veroyot i. Premen.*, 3, 201-205.

TAKANO, K. (1955) : Central convergence criterion in the multidimensional case. *Ann. Inst. Stat. Math.*, 7 95-102.

VARADARAJAN, V. S. (1958) : Weak convergence of measures on separable metric spaces. *Sankhyā*, 19, 15-22.

VOROBEV, N. N. (1954) : Addition of independent random variables on finite abelian groups. *Mat. Sbornik*, 34, 89-216.

Paper received : June, 1962.

Trans. Amer. Math. Soc. 102 (1962) 200-217

ON THE CATEGORY OF INDECOMPOSABLE DISTRIBUTIONS ON TOPOLOGICAL GROUPS

BY

K. R. PARTHASARATHY, R. RANGA RAO AND S. R. S. VARADHAN

1. **Introduction.** According to a theorem of A. I. Khinchine [1] any distribution on the real line can be written as the convolution of two distributions one of which is the convolution of a finite or a countable number of indecomposable distributions and the other is an infinitely divisible distribution without indecomposable factors. Further, any distribution which is not infinitely divisible has an indecomposable component. This result gives an indication of the existence of a large collection of indecomposable distributions. It is, however, not clear from this result alone that there exists a nonatomic or absolutely continuous indecomposable distribution. This question was raised by H. Cramér [2] and an answer in the affirmative was given by P. Lévy [3]. However, what is available is only a meagre supply of examples [3; 4; 5] even in the case of the real line. In this connection there arises naturally the question of the "size" of the class $\mathfrak{M}_1$, of indecomposable distributions. Or more precisely, what is the category of $\mathfrak{M}_1$? The object of this paper is to answer questions of this type.

These questions are studied here under the general framework of a complete separable metric group G. We consider three classes of distributions: (1) all indecomposable distributions, (2) all nonatomic indecomposable distributions and (3) all indecomposable distributions which are absolutely continuous with respect to the Haar measure in the case of a locally compact abelian group. It is shown that, under appropriate conditions on G, the indecomposable distributions form a dense G_δ in the class of all distributions in cases (1) and (2) and in the class of all absolutely continuous distributions in case (3). This lends substance to the statement that, in general, a distribution is indecomposable. It may be added that most of these results seem to be new even in the case when G is the real line.

In this connection it is interesting to compare these results with that of W. Rudin [6]. Rudin considers the convolution algebra $L_1(G)$ of all functions integrable with respect to the Haar measure in a locally Euclidean group G and shows that every element in $L_1(G)$ is decomposable. The decomposability of all elements is made possible by the fact that the factors are not confined to non-negative elements of $L_1(G)$ as we have done in our work.

Our analysis incidentally throws some light on the problem of existence of nonatomic measures in a separable metric space. It follows as a consequence of our work that the set of all nonatomic measures in a complete

Received by the editors May 15, 1961.

separable metric space without isolated point is a dense G_δ in the set of all measures under the weak topology.

2. **Preliminaries.** Throughout the paper we suppose that G denotes a complete separable metric group. Additional assumptions on G will be specifically mentioned as and when necessary. We employ the customary notation of denoting the group operation as xy, x, $y \in G$ in the case of general groups and as $x+y$ if G is abelian. e always denotes the unit in G. For any two subsets A, B of G we write $AB = [z: z = xy,\ x \in A,\ y \in B]$ and $A^{-1} = [z: z^{-1} \in A]$ (in case the group is abelian we use instead the symbols $A+B$ and $-A$ respectively).

The convolution operation. By a measure (or distribution) we mean a probability measure defined on the σ-field $\mathfrak{B}$ of Borel subsets of G. Let $\mathfrak{M}$ denote the collection of all probability measures on $\mathfrak{B}$. For $\mu, \nu \in \mathfrak{M}$ the convolution $\mu * \nu$ is defined as follows

$$(2.1) \qquad (\mu*\nu)(A) = \int \mu(Ax^{-1})d\nu(x).$$

With this operation $\mathfrak{M}$ becomes a semigroup which is abelian if and only if G is so. It should be noted that $\mu * \nu$ in (2.1) can be written in the equivalent form

$$(\mu*\nu)(A) = \int \nu(x^{-1}A)d\mu(x).$$

For each $g \in G$ and $\mu \in \mathfrak{M}$, $\mu * g$ denotes the right translate of μ by g i.e., the measure $\mu(Eg^{-1})$. $g * \mu$ is defined similarly. By a translate of μ we mean a measure of the form $\mu * g$ or $g * \mu$. Regarding the measure-theoretic terminology used but unaccompanied by explanation, we follow Halmos [7].

DEFINITION 2.1. A measure λ is *decomposable* if and only if there exist two nondegenerate measures μ and ν such that $\lambda = \mu * \nu$. In the contrary case λ is said to be *indecomposable*.

DEFINITION 2.2. A nondegenerate measure α is said to be a factor of a measure β if and only if there exists a measure γ such that either $\beta = \alpha * \gamma$ or $\beta = \gamma * \alpha$.

We shall denote by $\mathfrak{M}_0$ the set of all decomposable measures and $\mathfrak{M}_1$ the set of all indecomposable measures.

DEFINITION 2.3. The spectrum of a measure μ is the smallest closed set $A \subseteq G$ such that $\mu(A) = 1$.

The existence of the spectrum is well known and it is also easy to see that if A, B, and C are the spectra of the measures μ, ν and $\mu * \nu$ respectively then $C = $ closure of (AB).

Topologies in $\mathfrak{M}$. In the sequel we shall be mainly concerned with the weak topology in $\mathfrak{M}$. It is defined through convergence as follows:

DEFINITION 2.4. A sequence of measures $\{\mu_n\}$ converges weakly to a measure μ if and only if, for every real valued bounded continuous function f defined on G, $\int f d\mu_n \to \int f d\mu$.

It is clear that the class of subsets of $\mathfrak{M}$, of the form

$$\left[\mu : \left| \int f_i d\mu - \int f_i d\mu_0 \right| < \epsilon_i, i = 1, 2, \cdots, k \right]$$

where $(f_1, \cdots, f_k)$ is any finite set of bounded continuous functions and $(\epsilon_1, \cdots, \epsilon_k)$ is any finite set of positive numbers forms a neighbourhood system for the weak topology in $\mathfrak{M}$. It is useful to note that the sets of the type

$$[\mu : \mu(V_i) > \mu_0(V_i) - \epsilon_i, i = 1, 2, \cdots, k]$$

where $\epsilon_i > 0$ for all i and V_i are open subsets of G, are open in the weak topology.

Now we shall gather a few results about the weak topology in $\mathfrak{M}$ which we need in the sequel.

THEOREM 2.1 (PROHOROV [8], VARADARAJAN [9]). *If G is a complete separable metric space, the space $\mathfrak{M}$ of measures on G becomes a complete separable metric space under the weak topology.*

THEOREM 2.2 (PROHOROV [8]). *If G is a complete separable metric space, a subset $M \subset \mathfrak{M}$ is conditionally compact in the weak topology if and only if, for every $\epsilon > 0$, there exists a compact set $K_\epsilon \subset G$ such that $\mu(K_\epsilon) > 1 - \epsilon$ for every $\mu \in M$.*

THEOREM 2.3 (RANGA RAO [10]). *In a complete separable metric space G a sequence $\mu_n \in \mathfrak{M}$ converges weakly to $\mu \in \mathfrak{M}$ if and only if the following holds: For every class $\mathfrak{A}$ of continuous functions on G such that*
   (i) *$\mathfrak{A}$ is uniformly bounded,*
   (ii) *$\mathfrak{A}$ is compact in the topology of uniform convergence on compacta,*

$$\lim_{n \to \infty} \sup_{f \in \mathfrak{A}} \left| \int f d\mu_n - \int f d\mu \right| = 0.$$

All topological notions in $\mathfrak{M}$ used in §§1–6 refer to the weak topology. Only in the last section, we find it necessary to consider the strong topology induced by the norm $\|\mu\| = \sup_{A \in \mathfrak{B}} |\mu(A)|$, $\mu \in \mathfrak{M}$.

Indecomposable distributions on the real line. We now state two results due to P. Lévy [3] concerning absolutely continuous indecomposable distributions on the real line. We shall have occasion to use the latter one in the last section.

Let Z be any real-valued random variable which takes values in a bounded interval. We denote by $[Z]$ the integral part of Z and the conditional distribution of Z given that $[Z] = n$, by μ_n.

THEOREM 2.4. *Let Z be any real-valued random variable taking values in a bounded interval and satisfying the following properties:*
(a) *[Z] is even with probability one,*
(b) *the distribution of [Z] is indecomposable,*
(c) *the family of distributions μ_n, n running over possible values of [Z], has no common factor.*
Then the distribution of Z is indecomposable.

THEOREM 2.5. *Let μ_1, μ_2 denote the uniform distributions on the intervals $[a, b]$ and $[c, d]$ respectively. If $(b-a)/(d-c)$ is irrational then μ_1 and μ_2 have no common factor.*

3. **Shift compactness in $\mathfrak{M}$.** In the theory of sums of independent random variables we often come across the situation where a sequence of distributions fails to converge to any limit but actually does converge when suitably centered (see Gnedenko and Kolmogorov [11]). In this section we shall make a systematic analysis of this phenomenon in relation to the convolution operation between distributions on groups. To this end, it is convenient to introduce the following

DEFINTION 3.1. A family $\mathfrak{N}$ is said to be shift compact if, for every sequence $\mu_n \in \mathfrak{N}$ $(n = 1, 2, \cdots)$, there is a sequence of measures ν_n such that (1) ν_n is a translate of μ_n and (2) ν_n has a convergent subsequence.

The main result of this section is the following theorem which reveals an important structural property of the topological semigroup in relation to the notion of shift compactness. Its applications in the theory of factorisation of distributions in groups will be discussed in a separate paper.

THEOREM 3.1. *Let $\{\lambda_n\}$, $\{\mu_n\}$, $\{\nu_n\}$ be three sequences of measures on G such that*

$$(3.1) \qquad\qquad \lambda_n = \mu_n * \nu_n \qquad\qquad (n = 1, 2, \cdots).$$

If the sequence $\{\lambda_n\}$ is conditionally compact, then each one of the sequences $\{\mu_n\}$ and $\{\nu_n\}$ is shift compact.

As an immediate consequence we have the following

COROLLARY 3.1. *For any $\lambda \in \mathfrak{M}$, the family $\mathfrak{F}(\lambda)$ of all factors of λ is shift compact.*

Before proceeding to the proof of Theorem 3.1 we shall establish the following

LEMMA 3.1. *Let $\{\lambda_n\}$, $\{\mu_n\}$, $\{\nu_n\}$ be three sequences of measures on G such that $\lambda_n = \mu_n * \nu_n$ for each n. If the sequences $\{\lambda_n\}$ and $\{\mu_n\}$ are conditionally compact then so is the sequence $\{\nu_n\}$.*

Proof. Since the sequences λ_n and μ_n are conditionally compact it follows from Theorem 2.2 that, given $\epsilon > 0$, there exists a compact set K_ϵ such that

$$\lambda_n(K_\epsilon) > 1 - \epsilon, \qquad \mu_n(K_\epsilon) > 1 - \epsilon$$

for all n. Then we have

(3.2)
$$1 - \epsilon < \lambda_n(K_\epsilon) = \int \nu_n(x^{-1}K_\epsilon)d\mu_n(x)$$

$$\leq \int_{K_\epsilon} \nu_n(x^{-1}K_\epsilon)d\mu_n(x) + \epsilon$$

or

(3.3)
$$\int_{K_\epsilon} \nu_n(x^{-1}K_\epsilon)d\mu_n(x) > 1 - 2\epsilon.$$

(3.3) implies the existence of a point $x_n \in K_\epsilon$ with the property

$$\nu_n(x_n^{-1}K_\epsilon) > 1 - 3\epsilon$$

and consequently we have

$$\nu_n(K_\epsilon^{-1}K_\epsilon) > 1 - 3\epsilon$$

for all n. Since $K_\epsilon^{-1}K_\epsilon$ is compact and independent of n, another application of Theorem 2.2 leads to the fact that the sequence $\{\nu_n\}$ is conditionally compact. This completes the proof of the lemma.

Proof of Theorem 3.1. We choose a sequence ϵ_r of positive numbers such that $\sum \epsilon_r < \infty$. Then the conditional compactness of the sequence $\{\lambda_n\}$ implies, by Theorem 2.2, that there exists a sequence of compact sets K_r such that

$$\lambda_n(K_r) > 1 - \epsilon_r, \qquad\qquad r = 1, 2, \cdots$$

for all n. Now choose a positive sequence η_r descending to zero and satisfying

$$\sum_{r=1}^{\infty} \epsilon_r \eta_r^{-1} \leq \frac{1}{2}.$$

Let

(3.4)
$$E_{nr} = [x : \mu_n(K_r x^{-1}) > 1 - \eta_r], \qquad F_n = \bigcap_{r=1}^{\infty} E_{nr}.$$

Then, from (3.1) and (3.4), we have

$$1 - \epsilon_r \leq \lambda_n(K_r) = \int_{E_{nr}} \mu_n(K_r x^{-1})d\nu_n(x) + \int_{E'_{nr}} \mu_n(K_r x^{-1})d\nu_n(x)$$

$$\leq \nu_n(E_{nr}) + (1 - \eta_r)\nu_n(E'_{nr})$$

where E'_{nr} denotes the complement of the set E_{nr}. Thus we obtain

$$\nu_n(E'_{nr}) \leq \frac{\epsilon_r}{\eta_r}$$

and consequently

$$\nu_n(F'_n) \leq \sum_n \epsilon_r \eta_r^{-1} \leq \frac{1}{2}.$$

Hence $F_n \neq \phi$. Let x_n be any element in F_n. Then, from the definition of F_n, we have

(3.5) $$\mu_n(K_r x_n^{-1}) > 1 - \eta_r$$

for all n and all r.

We now write $\alpha_n = \mu_n * x_n$ (the right translate of μ_n by x_n) and $\beta_n = x_n^{-1} * \nu_n$. Then, obviously, $\lambda_n = \alpha_n * \beta_n$ and from (3.5) and Theorem 2.2 it follows that both the sequences λ_n and α_n are conditionally compact. Lemma 3.1 now implies that β_n is conditionally compact. The fact that α_n and β_n are translates of μ_n and ν_n respectively completes the proof of the theorem.

4. The class $\mathfrak{M}_1$ is a G_δ. The purpose of this section is simply to prove the following.

THEOREM 4.1. *Let G be a complete separable metric group. Then the class $\mathfrak{M}_1$ of all indecomposable distributions forms a G_δ.*

Before proceeding to the proof of this theorem we make a digression in order to pick up a few auxiliary facts.

Let $f_1, f_2, \cdots$, be a sequence of bounded functions on G with the following properties

(a) for each j, $f_j(x)$ is uniformly continuous in both the right and left uniformities of G.

(b) the sequence $\{f_j\}$ separates points of G.

The existence of such a sequence of functions may be seen as follows. Since G is a separable metric group there is a sequence of neighbourhoods $\{N_i\}$ of e such that $\bigcap_{i=1}^{\infty} N_i = \{e\}$. Then by a well known result (cf. A. Weil [16, pp. 13–14]) there exists a sequence of functions $f_i(x)$ $(i = 1, 2, \cdots)$ which are uniformly continuous in the two-sided uniformity (i.e., in both the right and left uniformities) and such that $f_i(e) = 0$ and $f_i(x) = 1$ for $x \notin N_i$. Let $\{x_n\}$ be a sequence dense in G. Then the countable family $S = \{\phi_{ij}\}$ where $\phi_{ij}(x) = f_i(xx_j)$ possesses both the properties. It is only necessary to prove that the family S separates points of G. If not let a and b be two distinct elements of G such that $\phi_{ij}(a) = \phi_{ij}(b)$ for all i and j or equivalently $f_i(ax_j) = f_i(bx_j)$ for all i and j. Since $\{x_j\}$ is dense it follows that $f_i(ab^{-1}) = f_i(e) = 0$ for all i. But $ab^{-1} \notin N_i$ for some i and hence $f_i(ab^{-1}) = 1$ for some i. This contradiction shows that $S = \{\phi_{ij}\}$ separates points of G.

In all that follows, $S = \{f_j\}$ is a fixed sequence with the above properties. It is then clear that a measure μ on G is degenerate if and only if the induced measure μf_j^{-1} on the real line is degenerate for each j. For any real valued bounded continuous function f and any measure μ, we write

$$(4.1) \quad V(f, \mu) = \sup_{a \in G} \max \left\{ \left[\int f^2(ax) d\mu - \left(\int f(ax) d\mu \right)^2 \right], \right.$$
$$\left. \left[\int f^2(xa) d\mu - \left(\int f(xa) d\mu \right)^2 \right] \right\}.$$

It is obvious that a measure μ is degenerate if and only if $V(f_j, \mu) = 0$ for all j.

LEMMA 4.1. *If f is bounded and uniformly continuous in both the right and left uniformities of the group G and μ_n is a sequence of measures converging weakly to μ, then*

$$\lim_{n \to \infty} V(f, \mu_n) = V(f, \mu).$$

Proof. For each f which is bounded and uniformly continuous in the right as well as the left uniformity, it is clear that each one of the families of functions $\{f(xa), a \in G\}$, $\{f^2(xa), a \in G\}$, $\{f(ax), a \in G\}$, $\{f^2(ax), a \in G\}$ is uniformly bounded and equi-continuous at each point of G. Consequently, they are conditionally compact in the topology of uniform convergence on compacta. The lemma is then an immediate consequence of Theorem 2.3.

Let now $E_{ij}(\epsilon)$ be defined as follows

$$(4.2) \quad E_{ij}(\epsilon) = [\mu : \mu = \alpha * \beta, \ V(f_i, \alpha) \geqq \epsilon, \ V(f_j, \beta) \geqq \epsilon]$$

where f_i and f_j are any two functions from S. Then we have

LEMMA 4.2. *For any $\epsilon > 0$ and each i, j the set $E_{ij}(\epsilon)$ is closed.*

Proof. Let μ_n be a sequence of measures in $E_{ij}(\epsilon)$ converging to some measure μ. Then by (4.2) there exist measures α_n and β_n such that

$$(4.3) \quad \mu_n = \alpha_n * \beta_n$$
$$V(f_i, \alpha_n) \geqq \epsilon, \qquad V(f_j, \beta_n) \geqq \epsilon.$$

From Theorem 3.1, it follows that there exists a sequence $a_n \in G$ such that the sequences of measures $\{\alpha_n * a_n\}$ and $\{a_n^{-1} * \beta_n\}$ are conditionally compact. Thus we can choose subsequences $\alpha_{n_k} * a_{n_k}$ and $a_{n_k}^{-1} * \beta_{n_k}$ converging to some measures α_0 and β_0 respectively. Since μ_n converges to μ and $\mu_{n_k} = \alpha_{n_k} * a_{n_k} * a_{n_k}^{-1} * \beta_{n_k}$ we have $\mu = \alpha_0 * \beta_0$. It is clear from the definition of $V(f, \mu)$ (see (4.1)) that

$$(4.4) \quad V(f, \alpha_{n_k} * a_{n_k}) = V(f, \alpha_{n_k}).$$

From Lemma 4.1 it follows immediately that

$$\lim_{k \to \infty} V(f_i, \alpha_{n_k}) = V(f_i, \alpha_0).$$

Similarly

$$\lim_{k \to \infty} V(f_j, \beta_{n_k}) = V(f_j, \beta_0).$$

Thus from (4.3) we have

$$V(f_i, \alpha_0) \geq \epsilon, \qquad V(f_j, \beta_0) \geq \epsilon$$

or $\mu \in E_{ij}(\epsilon)$. This completes the proof.

We shall now prove Theorem 4.1 by showing that the set of all decomposable measures is an F_σ. In fact

(4.5)
$$\mathfrak{M}_0 = \bigcup_{j=1}^{\infty} \bigcup_{i=1}^{\infty} \bigcup_{r=1}^{\infty} E_{ij}(r^{-1}).$$

It is clear that any measure belonging to the right side of (4.5) is decomposable and hence belongs to $\mathfrak{M}_0$. Let now μ be any measure in $\mathfrak{M}_0$. Then there exist two nondegenerate measures α and β such that $\mu = \alpha * \beta$. Since α and β are nondegenerate it follows from the remarks made earlier that there exist two functions f_i and f_j belonging to S, with the property

$$V(f_i, \alpha) > 0, \qquad V(f_j, \beta) > 0.$$

Let $\epsilon = \min [V(f_i, \alpha), V(f_j, \beta)]$. Then for $r > 1/\epsilon$, $\mu \in E_{ij}(r^{-1})$. Thus $\mathfrak{M}_0$ is contained in the right side of (4.5). An application of Lemma 4.2 completes the proof of the theorem.

REMARKS. 1. Let $\mathfrak{M}_c$ denote the class of nonatomic measures. It may be noted that the class of all indecomposable nonatomic measures is a G_δ in $\mathfrak{M}_c$ under the relative topology. At the moment, it is not clear whether even a single nonatomic measure exists in G. These points will be clarified in §6.

2. Let G be a locally compact abelian separable metric group and $\mathfrak{a}(G)$ the class of all absolutely continuous distributions in G. Since the norm topology in $\mathfrak{a}(G)$ is stronger than the weak topology it is clear that the set of all indecomposable absolutely continuous distributions is a G_δ in $\mathfrak{a}(G)$ (the relevant topology being the norm topology).

To determine the category of the various classes of indecomposable distributions, it is thus sufficient to find their closures. We note one case where the class $\mathfrak{M}_1$ is of first category. This is the situation when the group G is finite, as is implied in the work of Vorobev [12].

In the rest of the paper we will study the closures of three classes of indecomposable distributions

(1) the general case—the class $\mathfrak{M}_1$ itself;
(2) the nonatomic case—the class of all nonatomic measures in $\mathfrak{M}_1$;

(3) the absolutely continuous case—the class of indecomposable distributions absolutely continuous with respect to the Haar measure in a locally compact abelian group G.

5. The general case. Before we state the main theorem of this section we begin with some lemmas, the purpose of which is to construct indecomposable distributions in G. To this end we introduce the following definition.

DEFINITION 5.1. A subset $A \subset G$ is said to be *decomposable* if there exist two sets A_1, $A_2 \subset G$ such that (a) each of A_1, A_2 contains at least two elements and (b) $A_1 A_2 = A$; a set $A \subset G$ is said to be *indecomposable* if it is not decomposable.

LEMMA 5.1. *Let A be any countable indecomposable set and μ a measure such that $\mu(A) = 1$ and $\mu(\{g\}) > 0$ for every $g \in A$. Then μ is indecomposable.*

Proof. Let us suppose that μ is decomposable. Then $\mu = \mu_1 * \mu_2$ where μ_1 and μ_2 are nondegenerate measures with mass concentrated at a countable or finite number of points. Let

$$A_i = [g: g \in G, \mu_i(g) > 0], \qquad i = 1, 2.$$

From the conditions of the lemma it follows that $A = A_1 A_2$, which contradicts the fact that A is indecomposable.

LEMMA 5.2. *Let B be an infinite countable set $\{g_1, g_2, \cdots\}$ with the following property: $g_r g_s^{-1} \neq g_t g_u^{-1}$ for every set of distinct integers r, s, t, u no two of which are equal. If F is any finite subset of G then the set $B \cup F$ is indecomposable.*

Proof. Suppose the lemma is not true. Then there exist two sets A_1, A_2, at least one of which contains an infinite number of elements and such that $B \cup F = A_1 A_2$. Let $A_1 = (x_1, x_2 \cdots x_n \cdots)$ and $y_1, y_2 \in A_2$. Since the elements $x_r y_1, r = 1, 2, \cdots$ are all distinct all but a finite number of them belong to B. Thus there exists a finite set N of integers such that

$$x_r y_1 \in B \qquad \text{for } r \notin N.$$

Take any integer $m \notin N$. For at most one integer s, say $s = k_1$, $x_s y_1$ can be equal to $x_m y_2$. Similarly, for at most one integer, say $s = k_2$, $x_s y_2$ can be equal to $x_m y_1$. Choose any integer $n \notin N$ and different from k_1 and k_2. Then $x_m y_1$, $x_m y_2$, $x_n y_1$ and $x_n y_2$ are all distinct and belong to B. But

$$x_m y_1 (x_n y_1)^{-1} = x_m y_2 (x_n y_2)^{-1}$$

which contradicts the defining property of B.

LEMMA 5.3. *If G is an infinite group, then there exists a set B with the property described in Lemma 5.1.*

Proof. Let g_1, g_2, g_3 be any three distinct elements of the group G. Suppose g_1, g_2, $\cdots$, g_n have been chosen. Consider the set A_n of all elements of the form $g_{i_1}^{\pm 1} g_{i_2}^{\pm 1} g_{i_3}^{\pm 1}$ where i_1, i_2, i_3 are any three positive integers less than or equal to n. Since A_n is finite and the group G is infinite A_n' is nonempty. Choose any element g_{n+1} from A_n'. The sequence g_1, g_2, $\cdots$ chosen in this way has the required property.

THEOREM 5.1. *If the group G is infinite, then $\mathfrak{M}_1$ is a dense G_δ.*

Proof. Any measure in G is a weak limit of measures concentrated at a finite number of points. From Lemmas 5.1–5.3 it is clear that any measure with a finite spectrum is a weak limit of indecomposable distributions. Thus indecomposable distributions are dense in $\mathfrak{M}$. In view of Theorem 4.1, this completes the proof.

6. **The nonatomic case.** To start with we shall investigate the existence of a nonatomic measure in an arbitrary complete separable metric space. This, in itself, seems to be an interesting problem (cf. [13]).

THEOREM 6.1. *Let X be any complete separable metric space without any isolated points. Then there exists a nonatomic measure on X.*

Proof. Let $\mathfrak{M}$ be the class of all probability measures on X. Then by Theorem 2.1 $\mathfrak{M}$ is a complete separable metric space under the weak topology. For any given $\epsilon > 0$, we denote by $C(\epsilon)$ the class of all measures which have at least one atom of mass greater than or equal to ϵ. Then the class of all measures with atomic components can be represented as $\bigcup_{r=1}^{\infty} C(1/r)$. If there does not exist any nonatomic measure, we have

$$\mathfrak{M} = \bigcup_{r=1}^{\infty} C\left(\frac{1}{r}\right).$$

It is not difficult to verify by making use of Theorem 2.2 that $C(\epsilon)$ is closed in the weak topology. Thus by Baire's category theorem, at least one $C(1/r)$ has interior. Hence there exists a measure μ_0 with an atom of positive mass $> \delta > 0$ such that, whenever a sequence of measures μ_n converges weakly to μ_0, μ_n has an atom of mass at least δ for sufficiently large n. Since measures with finite spectrum (i.e., for which spectrum is a finite set) are everywhere dense we can, without loss of generality, assume that μ_0 is a measure with masses p_1, p_2, $\cdots$, p_k at the points x_1, x_2, $\cdots$, x_k respectively.

Let $N_n(x_1)$, $N_n(x_2)$, $\cdots$, $N_n(x_k)$ be sequences of neighbourhoods shrinking to x_1, x_2, $\cdots$, x_k respectively. We can and do assume that these neighbourhoods are disjoint. Since by assumption X has no isolated points each of these neighbourhoods contains an infinite number of points. We distribute the mass p_i among the points of $N_n(x_i)$ such that the mass at each point is less than $\delta/2$. By doing this for every i and every n we obtain a sequence of measures μ_n converging weakly to μ_0 and such that the mass of μ_n at any point is $\leq \delta/2$. This contradicts the defining property of μ_0 and shows that $C(1/r)$ has no interior for any r. The proof is complete.

COROLLARY 6.1. *Let X be any complete separable metric space with an uncountable number of points. Then there exists a nonatomic measure on X.*

Proof. Let Y denote the set of all accumulation points of X. Then it is well known (cf. Hausdorff [14, p. 146]) that X can be written in the form $X = Y \cup N$ where

(a) Y is closed and dense in itself,

(b) N is countable.

By Theorem 6.1 there exists a nonatomic measure on Y and hence on X.

The proof of Theorem 6.1 actually yields something more. In fact we have

COROLLARY 6.2. *Let X be a complete separable metric space without isolated points. Then the set of all nonatomic measures which give positive mass to each open subset of X is a dense G_δ in $\mathfrak{M}$.*

If G is any nondiscrete complete metric group, then it is clear that it is necessarily uncountable and cannot have any isolated point. Consequently we have the following:

THEOREM 6.2. *Let G be any nondiscrete complete separable metric group (not necessarily abelian). Then the set of all nonatomic indecomposable distributions which give positive mass to each open set is a dense G_δ in $\mathfrak{M}$.*

7. **The absolutely continuous case.** In this section we suppose that G is a locally compact abelian group and consider measures absolutely continuous with respect to the Haar measure on G. Let $\mathfrak{a} = \mathfrak{a}(G)$ denote the collection of these measures. The convergence notion that is appropriate for $\mathfrak{a}$ is the norm convergence of measures or the L_1 convergence of their densities. The main object here is to show that in the sense of this convergence the indecomposable measures in $\mathfrak{a}$ are dense.

In the first instance we develop in the following lemmas a general method of constructing absolutely continuous indecomposable distributions in G.

LEMMA 7.1. *Let A_1, A_2, A_3 be three closed subsets of G satisfying the following conditions:*

(1) $\quad (A_i - A_i) \cap (A_j - A_k) = \varnothing \qquad$ *for* $i = 1, 2, 3$ *and* $j \neq k$;

(2) $\quad (A_1 - A_2) \cap (A_2 - A_3) = (A_2 - A_3) \cap (A_3 - A_1)$

$$= (A_3 - A_1) \cap (A_1 - A_2) = \varnothing.$$

Let μ_1, μ_2, μ_3 be measures with $\mu_i(A_i) = 1$ and $\lambda = p_1\mu_1 + p_2\mu_2 + p_3\mu_3$ where $p_i > 0$ $(i = 1, 2, 3)$ and $p_1 + p_2 + p_3 = 1$. Then λ is decomposable if and only if μ_1, μ_2, μ_3 have a nondegenerate common factor.

Proof. If μ_1, μ_2, μ_3 have a nondegenerate common factor, it is obvious that λ is decomposable. Conversely let us suppose that λ is decomposable. Then there exist two nondegenerate measures α and β such that $\lambda = \alpha * \beta$. Let C

and D denote the spectra of α and β respectively. It is obvious that

(7.1) $$C + D \subset A_1 \cup A_2 \cup A_3 = A.$$

For each $c \in G$, we write

(7.2) $$D_i(c) = [d : d \in D \text{ and } c + d \in A_i] = D \cap (A_i - c)$$

for $i = 1, 2, 3$. The rest of the proof depends on an analysis of the nature of decomposition $\{D_i(c)\}$ of D. It is convenient to divide it into three steps.

I. The sets $D_i(c)$ possess the following properties.

(i) $\bigcup_{i=1}^{3} D_i(c) = D$ for each c.

(ii) $D_i(c) \cap D_j(c) = \varnothing$ for $i \neq j$.

(iii) For any two distinct c_1 and c_2, $D_i(c_1) = D_i(c_2)$ for some i, implies that $D_j(c_1) = D_j(c_2)$ for $j = 1, 2, 3$.

(iv) if $c_1 \neq c_2$, $D_i(c_1) \cap D_j(c_2) \neq \varnothing$ implies that $D_i(c_1) = D_j(c_2)$.

The first three properties are very simple. We shall prove (iv). Let us suppose that $D_i(c)$ and $D_j(c)$ have a common point d and $D_i(c_1) \neq D_j(c_2)$. Then there exists a point $d' \in D_i(c_1)$ which is not in $D_j(c_2)$. From (i) above it follows that there is a $k (\neq j)$ such that $d' \in D_k(c_2)$. From these facts we have

(7.3)
$$c_1 + d \in A_i, \qquad c_1 + d' \in A_i,$$
$$c_2 + d \in A_j, \qquad c_2 + d' \in A_k.$$

Consequently

$$d - d' \in (A_i - A_i) \cap (A_j - A_k).$$

This contradicts the assumption (1) of the lemma and proves (iv). It should be noted that the property (iv) implies that the decompositions $\{D_i(c)\}$ for $c \in C$ are only permutations of each other.

II. One of the following relations is always satisfied. Either

(a) for each $c \in C$, all but one of $D_i(c)$ are empty, i.e., $D_i(c) = D$ for some i, or

(b) for any two $c_1, c_2 \in C$, $D_i(c_1) = D_i(c_2)$ for $i = 1, 2, 3$.

The proof of this is quite straightforward and is similar to that of (iv) above. We shall not go into details.

III. Now we suppose that case (a) obtains. Let $C_i = [c : D_i(c) \neq \varnothing]$. It is then easily verified that (1) C_i's are mutually disjoint and their union is C, (2) $C_i + D \subset A_i$ for each i. Let the measures α_i ($i = 1, 2, 3$) be defined as follows:

$$\alpha_i(E) = \alpha(E \cap C_i)/\alpha(C_i).$$

(Note that $\alpha(C_i) > 0$.) It is then not difficult to verify that

$$\alpha_i * \beta = \mu_i \qquad \text{for } i = 1, 2, 3.$$

Thus for each i, β is a factor of μ_i.

In case (b), let $D_i = D_i(c)$. Obviously the D_i's are mutually disjoint and $C + D_i \subset A_i$ for each i. Writing $\beta_i(E) = \beta(E \cap D_i)/\beta(D_i)$ we get as before

$$\alpha * \beta_i = \mu_i \qquad \text{for } i = 1, 2, 3.$$

In this case α is the required common factor. This completes the proof of the lemma.

LEMMA 7.2. *Let G be a noncompact group. Then for any given compact set K, there exist elements g, h∈G such that the sets K, K+g, K+h satisfy the conditions* (1) *and* (2) *of Lemma* 7.1.

Proof. It may be verified that conditions (1) and (2) of Lemma 7.1 in this case reduce to choosing g and h such that none of the elements g, h, $g-h, g+h, 2g+h, 2h-g$ belong to the compact set $C = (K-K) - (K-K)$. Let

$$F = [x: x = 2y, y \in G].$$

Then there are two possibilities.

Case 1. F has compact closure. In this case we can choose an element g such that $g \notin C$ and $\overline{F} \cap (C+g) = \emptyset$. Since G is noncompact such elements exist. Let h be any element such that $h \notin C \cup (C+g) \cup (C-g) \cup (C+2g)$. The pair g, h satisfies our requirements.

Case 2. The closure of F is not compact. Let $g \notin C$ be arbitrary. Since $\overline{F}$ is not compact we can find an $h \in G$ such that $2h \notin C+g$ and $h \notin C \cup (C+g) \cup (C-g) \cup (C-2g) \cup (C+2g)$. As is easily verified the pair g, h serves our purpose. This completes the proof.

LEMMA 7.3. *Let G be an infinite compact metric abelian group. Let A be a subset such that*

(1) $$0 < \lambda(A) < 1,$$

(2) $$\int_A \chi_j(x) d\lambda(x) \neq 0 \qquad \text{for } j = 0, 1, 2, \cdots,$$

where λ is the normalized Haar measure on G and $\chi_0, \chi_1, \cdots$ are the characters of G. If λ_1, λ_2 are defined by

$$\lambda_1(E) = \lambda(E \cap A)/\lambda(A),$$
$$\lambda_2(E) = \lambda(E \cap A')/\lambda(A'),$$

then λ_1 and λ_2 do not have a common factor.

Proof. Let χ_0 be the identity character. Since for the Haar measure $\int \chi_j d\lambda = 0$ for $j \neq 0$ and $0 < \lambda(A) < 1$, we have

(7.4) $$\int_{A'} \chi_j d\lambda \neq 0 \qquad \text{for every } j.$$

We shall now prove that the measures λ_1 and λ_2 cannot have a common factor. If this is not true, then let μ be a common factor. Then there exist measures α_1, α_2 such that

(7.5) $$\lambda_1 = \alpha_1 * \mu, \qquad \lambda_2 = \alpha_2 * \mu.$$

From the definitions of λ_1 and λ_2 and (7.5) we have

(7.6) $$\lambda(A)\lambda_1 + \lambda(A')\lambda_2 = \lambda = (\lambda(A)\alpha_1 + \lambda(A')\alpha_2) * \mu.$$

Taking the characteristic functionals on both sides of (7.5) and (7.6) we get

(7.7)
$$\int \chi_j d\lambda_1 = \left(\int \chi_j d\alpha_1 \right)\left(\int \chi_j d\mu \right),$$
$$\int \chi_j d\lambda_2 = \left(\int \chi_j d\alpha_2 \right)\left(\int \chi_j d\mu \right),$$
$$\left(\lambda(A) \int \chi_j d\alpha_1 + \lambda(A') \int \chi_j d\alpha_2 \right)\left(\int \chi_j d\mu \right) = 0 \qquad \text{for } j \neq 0.$$

From condition (2) of the lemma and (7.7) we deduce that

$$\int \chi_j d\mu \neq 0 \qquad \text{for all } j.$$

Thus from (7.7) we have

$$\lambda(A) \int \chi_j d\alpha_1 + \lambda(A') \int \chi_j d\alpha_2 = 0 \qquad \text{for all } j \neq 0,$$

which is the same as saying

(7.8) $$\lambda(A)\alpha_1 + \lambda(A')\alpha_2 = \lambda.$$

From (7.5) and the definition of λ_1 and λ_2 we get

$$\int \alpha_1(A' - x)d\mu(x) = \lambda_1(A') = 0,$$
$$\int \alpha_2(A - x)d\mu(x) = \lambda_2(A) = 0.$$

Consequently

$$\alpha_1(A' - x) = 0 \ \ a \cdot e(\mu),$$
$$\alpha_2(A - x) = 0 \ \ a \cdot e(\mu).$$

Thus there exists a point x_0 such that

(7.9) $$\alpha_1(A' - x_0) = \alpha_2(A - x_0) = 0.$$

(7.8) and (7.9) imply that

$$\alpha_1(E) = \frac{\lambda(E \cap [A - x_0])}{\lambda(A)} = [\lambda_1 * (-x_0)](E),$$

$$\alpha_2(E) = \frac{\lambda(E \cap [A' - x_0])}{\lambda(A')} = [\lambda_2 * (-x_0)](E).$$

Thus from (7.5) and (7.10) we obtain

(7.11) $\qquad \lambda_1 = \lambda_1 * (-x_0) * \mu, \qquad \lambda_2 = \lambda_2 * (-x_0) * \mu.$

Taking characteristic functionals on both sides of (7.11) we have

$$\int \chi_j d((-x_0) * \mu) = 1 \qquad \qquad \text{for all } j.$$

Thus μ is degenerate at the point x_0. The proof of the lemma is complete.

LEMMA 7.4. *In any infinite compact group G there exists a set A possessing the properties (1) and (2) of Lemma 7.3.*

Proof. Let $S(\lambda)$ be the measure ring obtained by considering the space of Borel subsets of G modulo λ-null sets. This is a complete metric space with the distance $d(E, F) = \lambda(E \triangle F)$ where E and F belong to $S(\lambda)$ (cf. [7, pp. 165–169]). Let $\chi_0, \chi_1, \cdots$ be the characters of G, χ_0 being the identity. We consider the following mapping from $S(\lambda)$ to the complex plane. For any $E \in S(\lambda)$, we write

$$f_j(E) = \int_E \chi_j d\lambda.$$

The mapping f_j is obviously continuous. Hence the sets

$$V_j = \left[E : \int_E \chi_j d\lambda \neq 0 \right]$$

are open in $S(\lambda)$. We shall now prove that each V_j is dense in $S(\lambda)$. Let $A \in S(\lambda)$ and

$$\int_A \chi_j d\lambda = 0.$$

Let $\lambda(A) = c > 0$. Since λ is nonatomic, for any $0 < \epsilon < c$ there exists a set $B \subset A$ such that $\epsilon/2 < \lambda(B) < \epsilon$. Let C be any subset of B for which

$$\int_C \chi_j d\lambda \neq 0.$$

Such a C exists, for otherwise χ_j will vanish in B almost everywhere but at the same time $|\chi_j| = 1$. The set $A \cap C'$ has the property

$$d(A \cap C', A) = \lambda((A \cap C') \triangle A) = \lambda(C) < \epsilon.$$

Since this is true for any sufficiently small ϵ it is possible to get A as a limit of elements belonging to V_j. Since the class of sets A with $\lambda(A) > 0$ is dense in the ring $S(\lambda)$ it follows that the sets V_j are actually dense in $S(\lambda)$. By the Baire category theorem it follows that $\cap_{j=1}^{\infty} V_j$ is dense in $S(\lambda)$. Thus there exist Borel sets with the required properties.

LEMMA 7.5. *In any locally compact separable metric abelian group G there exist two absolutely continuous measures with compact supports which do not have a common factor.*

We shall prove this lemma in two steps. First of all let us assume that G is a finite dimensional vector space. Let A_1 and A_2 be two cubes in G such that the ratio of the lengths of their sides is irrational.

Then the uniform distributions μ_1 and μ_2, concentrated in A_1 and A_2 respectively, cannot have a common factor. For, if they have, then at least one of the one-dimensional marginal distributions of μ_1 must have a common factor with the corresponding marginal distribution of μ_2. Since the corresponding marginal distributions of μ_1 and μ_2 are rectangular distributions in the real line with the ratio of the lengths of their supports irrational, it follows from Theorem 2.5 that they cannot have a common factor. This proves the lemma in the case when G is a vector space.

If G is an infinite compact group, we have by Lemma 7.3 two absolutely continuous measures which do not have a common factor.

Now a result of Pontrjagin [15] states that for any general locally compact group G there exists an open subgroup H such that

$$H = V \oplus Z$$

where V is a vector group, Z a compact group and $\oplus$ denotes the direct sum. In V we take any two absolutely continuous measures μ_1 and μ_2 without any common factor. If Z is infinite we take two absolutely continuous measures ν_1 and ν_2 in Z without common factor. If Z is finite we take ν_1 and ν_2 to be any two degenerate measures. We form the product measures

$$\lambda_1 = \mu_1 \times \nu_1, \qquad \lambda_2 = \mu_2 \times \nu_2$$

in H. Since H is open λ_1 and λ_2 are absolutely continuous with respect to the Haar measure in G. Since none of the marginals of λ_1 and λ_2 have a common factor we conclude that λ_1 and λ_2 themselves cannot have a common factor. This completes the proof of the lemma.

THEOREM 7.1. *In any locally compact noncompact complete separable metric* *abelian group G the set of all absolutely continuous indecomposable distributions is a dense G_δ in $\alpha(G)$.*

That the set under consideration is a G_δ follows from the remarks made in §4. It remains to be proved that it is dense. It is clear that the set of all absolutely continuous measures with compact supports is everywhere dense. Thus it remains only to prove that any absolutely continuous measure μ with compact support is a limit of a sequence of absolutely continuous indecomposable measures. Let the support of μ be K_0. Let μ_1 and μ_2 be two absolutely continuous measures with compact supports K_1 and K_2 and having no common factor. Such measures exist because of Lemma 7.5. Let

$$K = K_0 \cup K_1 \cup K_2.$$

By using Lemma 7.3 we choose two points g, $h \in G$ such that K, $K+g$, $K+h$ satisfy conditions (1) and (2) of Lemma 7.1. We write

$$\alpha_1 = \mu, \qquad \alpha_2 = \mu_1 * g, \qquad \alpha_3 = \mu_2 * h$$

and

$$\mu_n = \left(1 - \frac{2}{n}\right)\alpha_1 + \frac{1}{n}\alpha_2 + \frac{1}{n}\alpha_3, \qquad n \geq 2.$$

From Lemma 7.1 it follows that μ_n is indecomposable. It is obvious that μ_n converges in norm to α_1 which is the same as μ. This completes the proof of the theorem.

REMARK. In the above theorem the assumption of noncompactness of G has played a crucial role. The question arises—is this assumption necessary? Or, more precisely, if G is an infinite compact group, is the collection of indecomposable distributions in $\alpha(G)$ dense in $\alpha(G)$? The answer is not known.

REFERENCES

1. A. I. Khinchine, *Contribution à l'arithmétique des lois de distribution*, Bull. Math. Univ. Moscou 1 (1937), 6–17.
2. H. Cramér, *Problems in probability theory*, Ann. Math. Statist. 18 (1948), 165–193.
3. P. Lévy, *Sur une classe de lois de probabilité indécomposables*, C. R. Acad. Sci. Paris 235 (1952), 489–492.
4. D. Dugué, *Sur certains exemples de décomposition en arithmétique des lois des probabilités*, Ann. Inst. H. Poincaré 12 (1951), 159–181.
5. D. Dugué et R. A. Fisher, *Un résultat assez inattendu d'arithmétique des lois des probabilités*, C. R. Acad. Sci. Paris 227 (1948), 1205–1207.
6. W. Rudin. *Representation of functions by convolutions*, J. Math. Mech. 7 (1958), 103–116.
7. P. R. Halmos, *Measure theory*, Van Nostrand, New York, 1950.
8. Yu. V. Prohorov, *Convergence of random processes and limit theorems in the theory of probability*, Teor. Veroyatnost. i Primenen. 1 (1956), 177–238. (Russian)
9. V. S. Varadarajan, *Weak convergence of measures on separable metric spaces*, Sankhyā 19 (1958), 15–22.

10. R. Ranga Rao, *Some problems in probability theory*, thesis submitted to Calcutta University, 1960.

11. B. V. Gnedenko and A. N. Kolmogorov, *Limit theorems for sums of independent random variables*, Addison-Wesley, Cambridge, Mass., 1954.

12. N. N. Vorobev, *Addition of independent random variables on finite abelian groups*, Mat. Sbornik **34** (1954), 89–126. (Russian)

13. R. Sikorski, *Boolean algebras*, Springer, Berlin, 1960.

14. F. Hausdorff, *Set theory*, Chelsea, New York, 1957.

15. L. S. Pontrjagin, *Topological groups*, Princeton Univ. Press, Princeton, N. J., 1939.

16. André Weil, *Sur les espaces à structure uniforme et sur la topologie générale*, Hermann, Paris, 1937.

INDIAN STATISTICAL INSTITUTE,
CALCUTTA, INDIA

Illinois J. Math. 7 (1963) 337-369

PROBABILITY DISTRIBUTIONS ON LOCALLY COMPACT ABELIAN GROUPS

BY

K. R. Parthasarathy, R. Ranga Rao, and S. R. S. Varadhan

1. Introduction

For probability distributions on the real line there are three main theorems on which the entire study of limit theorems for sums of independent random variables is based. These are (1) the Lévy-Khinchin representation of an infinitely divisible distribution, (2) the criteria for weak convergence of such distributions, and (3) Khinchin's theorem on sums of infinitesimal summands stating that these converge weakly if and only if certain associated infinitely divisible laws converge. For a precise statement of these results we refer to Kolmogorov and Gnedenko [3].

During the last two decades or so these results have been extended by many authors to varying degrees of generality. We mention in particular the works of Lévy [12], Kawada and Itô [17], Takano [9], Bochner [1], [2], Hunt [4], Urbanik [13], [14], Kloss [16]. In this paper we study probability distributions on a locally compact abelian (separable) group and obtain definitive extensions of all the three main results mentioned above.

The preliminaries are developed in Section 2. We mention in particular the concept of shift-compactness introduced therein and the important role that Theorem 2.1 plays in our study. A slightly modified notion of an infinitely divisible law is given in this paper to take into account the fact that the group may not be divisible.

The main results of the paper are the following. Weak limits of sums of uniformly infinitesimal random variables (with values in a group) are infinitely divisible. These limits can be obtained from certain accompanying infinitely divisible distributions if they have no idempotent factors. If μ is any infinitely divisible distribution without an idempotent factor, then its characteristic functional $\hat{\mu}(y)$, defined on the character gro .p, has the form

$$(x_0, y) \exp\left\{ \int [(x, y) - 1 - ig(x, y)] \, dF(x) - \phi(y) \right\}$$

where (x, y) is the value of the character y at x, x_0 is a fixed element of the group, $g(x, y)$ is a fixed function independent of μ, F is a σ-finite measure which integrates the function $(x, y) - 1 - ig(x, y)$ and has finite mass outside each neighborhood of the identity, and $\phi(y)$ is a nonnegative continuous function satisfying the equality

$$\phi(y_1 + y_2) + \phi(y_1 - y_2) = 2[\phi(y_1) + \phi(y_2)].$$

Received December 18, 1961.

Distributions μ with characteristic function of the form

$$(x_0, y) \exp(-\phi(y)),$$

where $\phi(y)$ has the above mentioned property, are precisely those characterized by the following algebraic property: $\mu = \alpha * \beta$, where α is a (generalized) Poisson distribution and β is infinitely divisible, implies that α is degenerate at the identity. We call them Gaussian since they are really so in the case of finite-dimensional Euclidean spaces.

One interesting feature of the representation of an infinitely divisible distribution without idempotent factors is its nonuniqueness. However, the Gaussian component is always unique. If F_1 and F_2 are two measures which occur in two different representations, their difference is always concentrated in a subgroup H of X characterized by the following property: H is the smallest closed subgroup containing all compact subgroups of X. Thus nonuniqueness is due to the prevalence of compact subgroups. If, however, we take a weakly continuous one-parameter convolution semigroup of measures $\{\mu_t\}$, then there exists a unique representation

$$\hat{\mu}_t(y) = (x_t, y) \exp\left[t\left(\int [(x, y) - 1 - ig(x, y)] \, dF - \phi(y)\right)\right].$$

In the last section we prove the following generalization of a result due to Khinchin [5] on the factorization of arbitrary probability distributions. Any distribution is the convolution of the normalized Haar measure of a compact subgroup, a finite or a countable number of indecomposable distributions, and an infinitely divisible distribution without indecomposable factors.

2. Preliminaries

2.1. All groups considered in this paper are locally compact abelian separable metric groups. Let X denote such a group, and let Y be its character group. For $x \in X$ and $y \in Y$, let (x, y) denote the value of the character y at x. By duality theory the relation between X and Y is perfectly symmetric, i.e., X is the character group of Y. Further, if G is a closed subgroup of X and H is the annihilator of G in Y, i.e., the set

$$H = [y : (x, y) = 1 \text{ for all } x \in G],$$

then G and Y/H are character groups of each other. These facts and some well-known results on the structure of locally compact abelian groups will be freely used in the sequel. For these details we refer to A. Weil [10], and Pontrjagin [18].

2.2. By a measure on X we shall mean a nonnegative completely additive set function defined on the Borel σ-field of subsets of X. We shall refer to probability measures as distributions. Let $\mathfrak{M}$ denote the class of all distributions. For λ, $\mu \in \mathfrak{M}$ we write

$$(\lambda * \mu)(E) = \int \mu(E - x) \, d\lambda$$

for any Borel set E. $\lambda * \mu$ is a distribution obtained by the convolution operation. With this operation $\mathfrak{M}$ becomes a commutative semigroup. If μ is the distribution degenerate at a point $x \in X$, then we write $\lambda * x$ for $\lambda * \mu$. We call $\lambda * x$ the shift or translate of λ by the element x. If λ_1, λ_2, $\cdots$, λ_n are some n distributions, then the distribution $\mu = \lambda_1 * \cdots * \lambda_n$ has an obvious meaning. We denote μ by $\prod_{i=1}^{n} \lambda_i$. If all the distributions λ_i are identically equal to a single distribution λ, we write $\mu = \lambda^n$. These definitions can obviously be extended to all measures, and we shall have occasion to use them.

For any measure μ we write $\bar{\mu}(A) = \mu(-A)$ where $-A$ is the set of all inverses of elements in A. Then $\bar{\mu}$ is also a measure. We denote by $| \mu |^2$ the measure $\mu * \bar{\mu}$.

We introduce in $\mathfrak{M}$ the weak topology, defined as follows: For μ_n, $\mu \in \mathfrak{M}$, μ_n is said to converge to μ (or $\mu_n \Rightarrow \mu$) if $\int f \, d\mu_n \to \int f \, d\mu$ for each bounded continuous function f on X. We assume some familiarity with this convergence and in particular the description of compact subsets in this topology.

2.3. For each $\mu \in \mathfrak{M}$ its characteristic function $\hat{\mu}(y)$ is a function on the character group Y, defined as follows:

$$\hat{\mu}(y) = \int_x (x, y) \, d\mu(x).$$

Some of the basic properties of the characteristic function $\hat{\mu}(y)$ are given below.

(1) $\hat{\mu}(y)$ is a uniformly continuous function of y,

(2) $\hat{\mu}(y)$ determines μ uniquely,

(3) $(\mu * \lambda)^{\wedge}(y) = \hat{\mu}(y)\hat{\lambda}(y)$ for all $y \in Y$, and μ, $\lambda \in \mathfrak{M}$,

(4) $\hat{\bar{\mu}}(y) = \overline{\hat{\mu}(y)}$,

(5) $\mu_n \Rightarrow \mu$ if and only if $\hat{\mu}_n(y) \to \hat{\mu}(y)$ uniformly over compact subsets of Y,

(6) if $\hat{\mu}_n(y)$ converges to a limit uniformly on each compact subset of Y, then there is a $\mu \in \mathfrak{M}$ such that

(i) $\hat{\mu}(y) = \lim_{n \to \infty} \hat{\mu}_n(y)$ and   (ii) $\mu_n \Rightarrow \mu$.

2.4. A distribution μ is said to be idempotent if $\mu^2 = \mu * x$ for some $x \in X$. If we write $\lambda = \mu * (-x)$, then it is clear that $\lambda^2 = \lambda$, so that $\hat{\lambda}(y) = 0$ or 1. From the inequality

(2.1) $1 - R(x, y_1 + y_2) \leq 2[(1 - R(x, y_1)) + (1 - R(x, y_2))]$

(R denoting the real part) it is clear that the set of all y for which $\hat{\lambda}(y) = 1$ is a both open and closed subgroup of Y. It is not difficult to see that the annihilator G of this subgroup is compact and λ is the normalized Haar measure of G. Thus μ is the translate of the Haar distribution of a compact subgroup.

2.5.   For α, $\mu \in \mathfrak{M}$, α is said to be a factor of μ ($\alpha \prec \mu$ in symbols) if there is a $\beta \in \mathfrak{M}$ such that $\mu = \alpha * \beta$. A distribution μ whose only factors are either degenerate distributions or translates of μ is said to be indecomposable. We denote by $F(\mu)$ the collection of all factors of μ.

Two distributions α, β are said to be equivalent ($\alpha \sim \beta$ in symbols) if each is a shift of the other. "$\sim$" is an equivalence relation. For each α, $\tilde{\alpha}$ denotes the equivalence class containing α. The collection $\tilde{\mathfrak{M}}$ of equivalence classes forms a semigroup. $\tilde{\mathfrak{M}}$ will be endowed with the quotient topology.

DEFINITION 2.1.   A subset $\mathfrak{N} \subset \mathfrak{M}$ is said to be *shift-compact* if its image $\tilde{\mathfrak{N}}$ in $\tilde{\mathfrak{M}}$ is conditionally compact.

The following theorem proved in [8] (see Section 3) plays a fundamental role in our study.

THEOREM 2.1.   *Let $\mathfrak{N}$ be a conditionally compact subset of $\tilde{\mathfrak{M}}$. Then $F(\mathfrak{N})$ consisting of the totality of all factors of elements belonging to $\mathfrak{N}$ is shift-compact.*

COROLLARY 1.   *A subset $\mathfrak{N} \subset \mathfrak{M}$ is shift-compact if and only if the set $|\mathfrak{N}|^2$ consisting of all elements $\alpha \in \mathfrak{M}$ of the form $|\mu|^2$ with $\mu \in \mathfrak{N}$, is conditionally compact.*

COROLLARY 2.   *For any distribution μ, $F(\mu)$ is shift-compact.*

COROLLARY 3.   *Suppose $\alpha_1 \prec \alpha_2 \prec \cdots \prec \alpha_n \prec \cdots$ and $\alpha_n \prec \mu$ for all n. Then there is a translate α'_n of α_n for every n such that α'_n is weakly convergent.*

Proof of Corollary 3. Since $\alpha_n \in F(\mu)$ for each n, $\{\alpha_n\}$ is shift-compact. Suppose that β and β' are any two limits of shifts of α_n. Then the fact that $\alpha_n \prec \alpha_{n+1}$ implies that β and β' are translates of each other. The corollary is an easy consequence of this fact.

The following corollary is proved in a similar manner.

COROLLARY 4.   *Suppose $\alpha_1 \succ \alpha_2 \cdots$. Then there exists a translate α'_n of α_n for each n such that α'_n weakly converges to a limit.*

Remark. Corollaries 2–4 have been obtained earlier by Itô [11] for distributions in the real line.

3. Two auxiliary lemmas

LEMMA 3.1.   *For each compact set $C \subset Y$ there are a neighborhood N_C of the identity in X and a finite set $E \subset C$ such that*

$$\sup_{y \in C} [1 - R(x, y)] \leqq M \cdot \sup_{y \in E} [1 - R(x, y)]$$

for all $x \, \epsilon \, N_C$, *where* M *is a finite constant depending on* C. *(Here* $R(x, y)$
denotes the real part of (x, y).)

Proof. From the inequality

$$1 - R(x_1 + x_2, y) \leq 2[(1 - R(x_1, y)) + (1 - R(x_2, y))]$$

it is clear that if the lemma is valid in two groups X_1 and X_2, it is valid for
their direct sum $X_1 \oplus X_2$. Let now Y' denote the closed subgroup generated by C, and Φ its annihilator in X. If τ denotes the canonical homomorphism from X onto $X' = X/\Phi$, it is obvious that $R(x, y) = R(\tau(x), y)$
for all $x \, \epsilon \, X$ and $y \, \epsilon \, Y'$. It is thus sufficient to prove the lemma when the
groups concerned are X' and Y' instead of X and Y. Since Y' is compactly
generated, it is of the form $V \oplus C \oplus I^r$ where V is a finite-dimensional vector
group, C is a compact group, and I^r is the product of r copies of the integer
group. Hence X' is of the form $V \oplus D \oplus K^r$ where D is a discrete group and
K^r is the product of r copies of the circle group. Since the lemma is trivially
valid in the case of the real line, discrete group, and compact group, the proof
of the lemma is complete.

LEMMA 3.2. *For any* $y \, \epsilon \, Y$ *there is a continuous function* $h_y(x)$ *on* X *with
the following properties*:
   (1) $| \, h_y(x) \, | \leq \pi$ *for all* $x \, \epsilon \, X$, *and* $h_y(-x) = -h_y(x)$,
   (2) $(x, y) = \exp(ih_y(x))$ *for all* $x \, \epsilon \, N_y$ *where*

$$N_y = [x : | \, (x, y) - 1 \, | \leq \tfrac{1}{2}].$$

Proof. Let $(x, y) = \exp(i\phi(x))$ where $-\pi \leq \phi(x) < \pi$. Then it is not
difficult to verify that $\phi(x)$ is a continuous function of x in the closed set
N_y. Now choose any continuous extension of $\phi(x)$ to X such that the first
condition is fulfilled. This will serve our purpose.

LEMMA 3.3. *There is a function* $g(x, y)$ *defined on the product space* $X \times Y$
possessing the following properties:
   (1) $g(x, y)$ *is a continuous function of both the variables* x *and* y.
   (2) $\sup_{x \epsilon X} \sup_{y \epsilon C} | \, g(x, y) \, | < \infty$ *for each compact set* $C \subset Y$.
   (3) $g(x, y_1 + y_2) = g(x, y_1) + g(x, y_2)$ *for each* $x \, \epsilon \, X$ *and* $y_1, y_2 \, \epsilon \, Y$, *and*
$g(-x, y) = -g(x, y)$.
   (4)   *If* C *is any compact subset of* Y, *then there is a neighborhood* N_C *of the
identity in* X *such that* $(x, y) = \exp[ig(x, y)]$ *for all* $x \, \epsilon \, N_C$ *and* $y \, \epsilon \, C$.
   (5)   *If* C *is any compact subset of* Y, *then* $g(x, y)$ *tends to zero uniformly in*
$y \, \epsilon \, C$ *as* x *tends to the identity of the group* X.

Proof. We shall reduce the proof of the proposition to the case of certain
simple groups by making use of the structure theory. Suppose that the
proposition is true for an open subgroup G of X. Let H and Y be the character groups of G and X respectively. Since H can be obtained as a quotient
group of Y by taking the quotient with respect to the annihilator of G in Y,

there is a canonical homomorphism τ from Y to H. Suppose $g(x, h)$ has been defined for $x \in G$ and $h \in H$ with the required properties. We extend the definition of g as follows. For $x \in G$ and $y \in Y$ we define

$$g(x, y) = g(x, \tau(y)).$$

For $x \notin G$, we define

$$g(x, y) = 0 \quad \text{for all } y \in Y.$$

Since an open subgroup is closed, the continuity of $g(x, y)$ follows immediately. The rest of the properties of $g(x, y)$ are immediate consequences of their validity in $G \times H$.

In the case of a general group X we take G to be the group generated by a compact neighborhood of the identity. This is both open and closed in X. This group G has the simple structure, $V \oplus C \oplus \Gamma^r$ where V is a finite-dimensional vector group, C is a compact group, and Γ^r is the product of the integer group taken r times. We now observe that if functions $g_1(x, y)$ and $g_2(u, v)$ with the properties mentioned in the lemma exist in groups X and U with character groups Y and V respectively, then a function $g(\xi, \eta)$ with the same properties exists for $\xi \in X \oplus U$ and $\eta \in Y \oplus V$. We have only to define

$$g(\xi, \eta) = g_1(x, y) + g_2(u, v)$$

where x and u are projections of ξ into X and U respectively, and y and v are projections of η into Y and V respectively. Thus it is enough to construct $g(x, y)$ in the case of a real line, a compact group, and the integer group. In the case of the integer group we can take $g(x, y)$ to be identically zero. In the case of the real line we can take $g(x, y) = \theta(x)y$ where

$$\begin{aligned} \theta(x) &= x, & x &\in [-1, 1], \\ &= 1, & x &> 1, \\ &= -1, & x &< -1. \end{aligned}$$

(Note that the character group of the real line is itself.) Thus, in order to complete the proof of the lemma it is enough to consider the case of a compact group X.

Let X be a compact group with Y as character group. Let X_0 be the component of identity in X, Y_1 the annihilator of X_0 in Y, $X_1 = X/X_0$, and $Y_0 = Y/Y_1$. Then Y_0 is the character group of X_0, and Y_1 is the character group of X_1. Since X_0 is connected and compact, Y_0 is a discrete torsion-free group. Let $\{d_\alpha\}$ be a maximal family of mutually independent elements in Y_0. Then for $d \in Y_0$ there exist elements $d_{\alpha_1}, \cdots, d_{\alpha_k}$ from the maximal family and integers $n, n_1, \cdots, n_k$ $(n > 0)$ such that

(3.1) $$nd = n_1 d_{\alpha_1} + \cdots + n_k d_{\alpha_k}.$$

This representation is unique except for multiplication by an integer on both sides.

Each element of Y_0 is a coset of Y_1 in Y. We take the coset d_α and pick out an element y_α of Y from this coset. We fix the elements y_α. We define

$$g(x, y_\alpha) = h_{y_\alpha}(x)$$

for every α where $h_{y_\alpha}(x)$ is as in Lemma 3.2. Let now $y \epsilon Y$ be arbitrary. Then y belongs to some coset of Y_1 which is an element of Y_0. If this element is denoted by d, then there exist integers $n, n_1, \cdots, n_k$ $(n > 0)$ and elements $d_{\alpha_1}, \cdots, d_{\alpha_k}$ from the collection $\{d_\alpha\}$ such that equation (3.1) is satisfied. We define

$$g(x, y) = (n_1/n)g(x, y_{\alpha_1}) + \cdots + (n_k/n)g(x, y_{\alpha_k}).$$

We shall now prove that the function $g(x, y)$ constructed in this way has all the required properties.

Since $g(x, y)$ is continuous in x for each fixed y and Y is discrete, the continuity of $g(x, y)$ in both the variables follows immediately. Properties (2) and (3) are obvious from the nature of the construction.

Since compact sets in Y are finite sets, it is enough to prove property (4) for each $y \epsilon Y$. For any $y \epsilon Y$, let $[y]$ denote the coset of Y_0 to which y belongs. Then $[y]$ is an element of Y_0. If we write $[y]$ for d and $[y_\alpha]$ for d_α, then equation (3.1) can be written as

$$(3.2) \qquad n[y] = n_1[y_{\alpha_1}] + \cdots + n_k[y_{\alpha_k}].$$

For any two elements $y_1, y_2 \epsilon [y]$ it is clear that $y_1 - y_2 \epsilon Y_1$. Since Y_1 is the character group of X/X_0 which is totally disconnected, every element of Y_1 is of finite order. Hence, for any $y \epsilon Y_1$ there exists a neighborhood of the identity in X where $(x, y) = 1$. Thus for any two elements $y_1, y_2 \epsilon [y]$ there exists a neighborhood of the identity in X where $(x, y_1) = (x, y_2)$.

Making use of the remarks made in the previous paragraph we shall complete the proof of the lemma. From the construction of $g(x, y)$ and Lemma 3.2 it is clear that, for each y_α, there exists a neighborhood of the identity in X where $e^{ig(x, y_\alpha)} = (x, y_\alpha)$. Let now $y \epsilon Y$ be arbitrary. From (3.2) it is clear that there exist elements $y_{\alpha_j 1}, \cdots y_{\alpha_j n_j}$ in $[y_{\alpha_j}]$, for $j = 1, 2, \cdots k$, such that

$$(3.3) \qquad ny = \sum_{j=1}^{k} (y_{\alpha_j 1} + \cdots + y_{\alpha_j n_j}).$$

From the remarks made in the previous paragraph it follows that there exists a neighborhood of the identity in X (depending on $y_{\alpha_j r}$ and y_{α_j}) where

$$(x, y_{\alpha_j r}) = (x, y_{\alpha_j}).$$

Denoting by N the intersection of all the neighborhoods corresponding to $y_{\alpha_j r}$ $(r = 1, 2, \cdots, n_j$ and $j = 1, 2, \cdots, k)$ we have

$$(3.4) \quad (x, y_{\alpha_j r}) = (x, y_{\alpha_j}) \qquad \text{for } x \epsilon N, r = 1, 2, \cdots, n_j, j = 1, 2, \cdots, k.$$

From (3.3) and (3.4) we have

$$(x, y)^n = (x, ny) = \prod_{j=1}^{k} (x, y_{\alpha_j})^{n_j} \qquad \text{for } x \epsilon N.$$

Since there are neighborhoods of the identity where $(x, y_{\alpha_j}) = \exp\{ig(x, y_{\alpha_j})\}$, it follows that there exists a neighborhood of the identity where

$$(x, y)^n = e^{ing(x,y)}.$$

Since (x, y) and $e^{ig(x,y)}$ are continuous functions nonvanishing at the identity of X, there exists a neighborhood of the identity where

$$(x, y) = e^{ig(x,y)}.$$

Property (5) is obvious from property (1) and the fact that $g(x, y)$ vanishes when either x or y is the identity of the corresponding group. This completes the proof of the lemma.

In the following paragraphs we give examples of the function $g(x, y)$ for some particular groups.

Example 1. Let $X = Y = R^n$. If $x = (x_1, \cdots, x_n) \in X$ and $y = (y_1, y_2, \cdots, y_n) \in Y$, then

$$g(x, y) = \sum_{i=1}^{n} \phi_i(x_i) y_i$$

where $\phi_i(t)$ $(i = 1, 2, \cdots, n)$ are bounded continuous functions on the real line such that $\phi_i(t) = t$ in a neighborhood of $t = 0$ and $\phi_i(-t) = -\phi_i(t)$.

Example 2. Let K denote the circle group, and let $X = K^n$ and $Y = I^n$, I being the integer group. Let

$$X = [(x_1, \cdots, x_n): -1 < x_i \leq 1 \text{ for } i = 1, 2, \cdots, n]$$

addition being taken modulo 2. Then if $y = (y_1, \cdots, y_n) \in I^n$,

$$g(x, y) = \sum_{i=1}^{n} \phi_i(x_i) y_i$$

where the functions $\phi_i(t)$ are as in Example 1.

Example 3. $Y = $ additive group of rationals, and X is its character group. Let $\phi(x)$ be a bounded continuous function on X such that $\exp(i\phi(x)) = (x, y_0)$ for x in a neighborhood of the identity in X, where y_0 is a fixed element of Y, other than the identity. Then

$$g(x, y) = \phi(x)y/y_0.$$

Example 4. If X is totally disconnected, then every homomorphism of Y into the real line is trivial, so that in this case $g(x, y) \equiv 0$ for all $x \in X$ and $y \in Y$.

4. Infinitely divisible distributions

We shall now introduce the definition of infinitely divisible distributions and study some of their elementary properties.

DEFINITION 4.1. A distribution μ is said to be *infinitely divisible* if, for each n, there are elements $x_n \in X$ and $\lambda_n \in \mathfrak{M}$ such that $\mu = \lambda_n^n * x_n$.

We remark that the definition given here is slightly different from the classical definition in the case of the real line. Such a modification is necessary if we want to avoid the role of divisibility of elements from the group. As yet it is not clear whether there exists a single element x of the group X with the property $\mu = \lambda_n^n * x$ for every n. That such is the case for every infinitely divisible distribution will be obvious from the representation which we shall give later in Section 7.

THEOREM 4.1. *The infinitely divisible distributions form a closed subsemigroup of $\mathfrak{M}$.*

Proof. If λ and μ are infinitely divisible, it is obvious from the definition that $\lambda * \mu$ is also infinitely divisible. Let now μ_k, $k = 1, 2, \cdots$, be a sequence of infinitely divisible distributions weakly converging to μ. For any fixed integer n, let

(4.1) $$\mu_k = \lambda_{kn}^n * x_{kn}.$$

From Theorem 2.1 it is clear that there exists a subsequence of λ_{kn} which after a suitable shift converges to a distribution λ_n. Since $\mu_k \Rightarrow \mu$, it is obvious from (4.1) that there exists an element x_n such that $\mu = \lambda_n^n * x_n$. This completes the proof.

The normalized Haar measure of a compact subgroup is an example of an infinitely divisible distribution. We shall now prove a result concerning the absence of zeros for the characteristic function of an infinitely divisible distribution without idempotent factors.

THEOREM 4.2. *Let $\hat{\mu}(y)$ be the characteristic function of an infinitely divisible distribution μ. If $\hat{\mu}(y_0) = 0$ for some character y_0, then μ has an idempotent factor.*

Proof. From the definition of infinite divisibility it follows that, for each n, there exist an element $x_n \in X$ and a distribution λ_n such that $\mu = \lambda_n^n * x_n$. Since $\hat{\mu}(y_0) = 0$, $\hat{\lambda}_n(y_0)$ also vanishes for every n. By Theorem 2.1, λ_n is shift-compact. Let λ be a limit of shifts of λ_n. Then $\hat{\lambda}(y_0) = 0$ and is hence a nondegenerate distribution. It is also clear that every power of λ is a factor of μ. Thus the sequence λ^n is shift-compact, and any limit of shifts is a nondegenerate idempotent factor of μ. This completes the proof.

DEFINITION 4.2. If F is any totally finite measure on X the *distribution $e(F)$ associated with F* is defined as follows:

$$e(F) = e^{-F(X)}[1 + F + F^2/2! + \cdots + F^n/n! + \cdots]$$

where 1 is used to denote the measure with unit mass and degenerate at the identity.

$e(F)$ is obviously an infinitely divisible distribution since $e(F) = [e(F/n)]^n$.

Its characteristic function is given by

$$(e(F))^{\wedge}(y) = \exp\left[\int ((x, y) - 1) \, dF(x)\right].$$

Suppose F_n is a sequence of totally finite measures and we form the sequence $e(F_n)$. We shall now obtain a necessary condition (which will be shown to be sufficient in Section 9) for the shift-compactness of $e(F_n)$.

THEOREM 4.3.   *Let* $\mu_n = e(F_n)$ *where* F_n *is a sequence of totally finite measures. Then, in order that*
(a) μ_n *be shift-compact,*
(b)   *if* μ *is any limit of shifts of* $\{\mu_n\}$, *then* μ *have no idempotent factor, the following conditions are necessary*:
   (i)   *For each neighborhood* N *of the identity the family* $\{F_n\}$ *restricted to* $X - N$ *is weakly conditionally compact.*
   (ii)   *For each* $y \in Y$

$$\sup_n \int [1 - R(x, y)] \, dF_n < \infty,$$

$R(x, y)$ *denoting the real part of* (x, y).

Before proceeding to the proof of this theorem we shall prove the following.

LEMMA 4.1.   *Let* F_n *be as in Theorem 4.3, and suppose* $\sup_n F_n(N') \leq k$ *where* N' *denotes the complement of a symmetric neighborhood* N *of the identity. If the sequence* $e(F_n)$ *is shift-compact, then the sequence of measures* F_n *is tight when restricted to* N'.

Proof.   Let G_n denote the restriction of F_n to N'. Then $e(G_n)$ is a factor of $e(F_n)$. Since $\{e(F_n)\}$ is shift-compact, so is the sequence $\{e(G_n)\}$ by Theorem 2.1. Let $H_n = G_n + \bar{G}_n$. Then, by Corollary 1 of Theorem 2.1, the sequence $\{e(H_n)\}$ is compact. Hence, for any $\varepsilon > 0$ there exists a compact set C such that $e(H_n)(C') < \varepsilon$. Since $e(H_n) = e^{-H_n(X)}[\sum_{r=0}^{\infty} H_n^r / r!]$, we have

$$\varepsilon > e(H_n)(C') \geq e^{-H_n(X)} H_n(C') \geq e^{-2k} H_n(C')$$
$$\geq e^{-2k} G_n(C')$$

for all n. Since k is a constant not depending on n and ε is arbitrary, it follows that the family $\{G_n\}$ is tight.

Proof of Theorem 4.3.   Since any neighborhood of the identity contains a symmetric neighborhood, we can assume that N is symmetric. Suppose (a) and (b) are valid. Let, if possible, $\sup_n F_n(N') = \infty$. We can then choose a subsequence for which

(4.2) $F_{n_k}(N') \geq 2k$ for $k = 1, 2, \cdots$.

Let L_k, $k = 1, 2, \cdots$, be measures such that

$$L_k(A) \leqq (1/k)F_{n_k}(A) \qquad \text{for every Borel set } A,$$

(4.3) $$L_k(N) = 0,$$

$$L_k(N') = 1.$$

The distribution $\lambda_k = e(L_k)$ is a factor of $e(F_{n_k})$, and the shift-compactness of $e(F_n)$ implies the shift-compactness of $\{\lambda_k\}$ by Theorem 2.1. Let λ be any limit of shifts of λ_k. From (4.2) and (4.3) it follows that any power of λ is a factor of μ. Thus the sequence λ^n is shift-compact, and any limit of shifts of λ^n is a factor of μ. Since these limits will be idempotent and μ has no idempotent factors, it follows that any such limit must be degenerate. Since λ^n is an increasing sequence (in the order $\prec$), it follows that λ itself must be degenerate. Thus the sequence $|\lambda_k|^2$ converges to the distribution degenerate at the identity. Hence

$$e(L_k + \bar{L}_k)(N') \to 0 \quad \text{as} \quad k \to \infty.$$

But

$$e(L_k + \bar{L}_k)(N') = e^{-(L_k+\bar{L}_k)(X)} \sum \frac{(L_k + \bar{L}_k)^r}{r!} (N') \geqq e^{-2} L_k(N') = e^{-2},$$

which is a contradiction. Thus we have $\sup_n F_n(N') < \infty$. Now an application of Lemma 4.1 shows that condition (i) of the theorem is necessary.

In order to prove the necessity of (ii) we observe that $e(F_n + \bar{F}_n) = |e(F_n)|^2$ is a compact sequence, and an application of Theorem 4.2 shows that any limit of $|e(F_n)|^2$ has a nonvanishing characteristic function. Thus, for any y,

$$\lim_n \exp\left\{\int [R(x, y) - 1] d(F_n + \bar{F}_n)\right\} \neq 0,$$

which implies condition (ii). The proof is thus complete.

5. General limit theorems for sums of infinitesimal summands

In the case of the real line a well-known result due to Bawly and Khinchin (see [3, Chapter 4]) asserts that the limit of sums of infinitesimal random variables is infinitely divisible, and it can be obtained as the limit of a certain accompanying sequence of infinitely divisible distributions. The purpose of this section is to introduce the notion of infinitesimal distributions in a group and prove a generalized version of the above-mentioned result in the case when the limiting distribution has no idempotent factor.

DEFINITION 5.1.   A triangular sequence $\{\alpha_{nj}\}$, $j = 1, 2, \cdots$, k_n of distributions is said to be *uniformly infinitesimal* if

$$\lim_{n \to \infty} \sup_{1 \leq j \leq k} \sup_{y \in K} |\hat{\alpha}_{nj}(y) - 1| = 0$$

for each compact set $K \subset Y$.

Before going to the statement of the main result of this section we shall prove a lemma which will be often used in the sequel.

LEMMA 5.1. *Let* $\mu_n = \prod_{j=1}^{k_n} \alpha_{nj}$ *where the sequence* $\{\alpha_{nj}\}$ *is uniformly infinitesimal. If the distribution* μ *is a limit of shifts of* μ_n, *then the set of characters* $[y : \hat{\mu}(y) \neq 0]$ *is an open subgroup of* Y, *and consequently the normalized Haar measure of the annihilator of this subgroup in* X *is a factor of* μ.

Proof. Since μ is a limit of shifts of μ_n, it is clear that for a subsequence (which we shall denote by μ_n itself) $| \mu_n |^2 \Rightarrow | \mu |^2$, and hence

(5.1) $\lim_{n \to \infty} \prod_{j=1}^{k_n} | \hat{\alpha}_{nj}(y) |^2 = | \hat{\mu}(y) |^2.$

If $\hat{\mu}(y) \neq 0$, it is obvious that $\hat{\mu}(-y) \neq 0$. (5.1) implies that a necessary and sufficient condition that $\hat{\mu}(y) \neq 0$ is that

(5.2) $\sup_n \sum_{j=1}^{k_n} (1 - | \hat{\alpha}_{nj}(y) |^2) < \infty.$

Thus if we make use of the inequality

$$1 - \phi(y_1 + y_2) \leq 2[(1 - \phi(y_1)) + (1 - \phi(y_2))]$$

for any real-valued characteristic function ϕ, it is clear that the validity of (5.2) for y_1 and y_2 implies its validity for $y_1 + y_2$. The continuity of $\hat{\mu}(y)$ now implies that the set $[y : \hat{\mu}(y) \neq 0]$ is an open subgroup of Y.

We choose and fix a function $g(x, y)$ defined on $X \times Y$ and satisfying all the properties mentioned in Lemma 3.3. The main theorem of this section can now be stated as follows:

THEOREM 5.1. *Let* $\{\alpha_{nj}\}$ *be a uniformly infinitesimal sequence of distributions, and let*

$$\mu_n = \prod_{j=1}^{k_n} \alpha_{nj}.$$

Suppose that $\{\mu_n\}$ *is shift-compact such that no limit of shifts of* μ_n *has an idempotent factor. Let*

$$\beta_{nj} = e(\alpha_{nj} * x_{nj})$$

where x_{nj} *is that element of the group* X *defined by the equality*

$$(x_{nj}, y) = \exp\left[-i \int g(x, y) \, d\alpha_{nj} \right].$$

If $\lambda_n = (\prod_{j=1}^{k_n} \beta_{nj}) * x_n$, *where* $x_n = -\sum_j x_{nj}$, *then*

$$\lim_{n \to \infty} \sup_{y \in K} | \hat{\lambda}_n(y) - \hat{\mu}_n(y) | = 0$$

for each compact set K *of* Y.

Proof. During the course of the proof of the theorem we shall adopt the following conventions: We denote by $C_1, C_2, \cdots$ constants depending only on the compact set K (and not on n). All the statements that we make are for sufficiently large n. By N we denote any arbitrarily small neighborhood of the identity in X.

Turning to the proof of the theorem, we observe that the elements x_{nj} are well defined, since, from the properties of $g(x, y)$, it follows that $\exp\left[-i \int g(x, y)\, d\alpha\right]$ is a character on Y for any distribution α. Further, for any neighborhood N of the identity all the points x_{nj} are in N for sufficiently large n. Therefore the uniform infinitesimality of $\{\alpha_{nj}\}$ implies the uniform infinitesimality of the sequence $\{\beta_{nj}\}$. Thus $\hat{\lambda}_n(y)$ and $\hat{\mu}_n(y)$ are nonvanishing in K, and hence we use the logarithmic notation freely. Since no limit of shifts of μ_n has an idempotent factor, it follows from Lemma 5.1 that the sequence $\hat{\mu}_n(y)$ is uniformly bounded away from zero for all $y \in K$. Thus it is enough to prove that

$$\lim_{n \to \infty} \sup_{y \in K} |\log \hat{\lambda}_n(y) - \log \hat{\mu}_n(y)| = 0.$$

We have

$$\log \hat{\lambda}_n(y) = \sum_j \log \hat{\beta}_{nj}(y) - \sum_j \log (x_{nj}, y)$$

$$= \sum_j \log \hat{\beta}_{nj}(y) + i \sum_j \int g(x, y)\, d\alpha_{nj}(x)$$

$$= \sum_j [(\alpha_{nj} * x_{nj})\hat{\ }(y) - 1] + i \sum_j \int g(x, y)\, d\alpha_{nj}(x),$$

and

$$\log \hat{\mu}_n(y) = \sum_j \log \hat{\alpha}_{nj}(y).$$

Writing $\theta_{nj} = \alpha_{nj} * x_{nj}$, we obtain

$$|\log \hat{\lambda}_n(y) - \log \hat{\mu}_n(y)| = \left| \sum_j (\hat{\theta}_{nj}(y) - 1) + i \sum_j \int g(x, y)\, d\alpha_{nj} \right.$$

$$\left. - \sum_j \log \hat{\theta}_{nj}(y) + \sum_j \log (x_{nj}, y) \right|$$

$$= \left| \sum_j (\hat{\theta}_{nj}(y) - 1) - \sum_j \log \hat{\theta}_{nj}(y) \right|$$

$$\leq C_1 (\sum_j |1 - \hat{\theta}_{nj}(y)|) \sup_j |1 - \hat{\theta}_{nj}(y)|.$$

Since $\{\theta_{nj}\}$ is uniformly infinitesimal, it is clear from the above inequality that it is enough to prove that

(5.3) $$\sup_n \sup_{y \in K} \left[\sum |1 - \hat{\theta}_{nj}(y)| \right] < \infty.$$

We have, for any neighborhood N of the identity in X

(5.4)
$$|1 - \hat{\theta}_{nj}(y)| \leq \left| \int_N (1 - (x, y))\, d\theta_{nj} \right| + \left| \int_{N'} (1 - (x, y))\, d\theta_{nj} \right|$$

$$\leq \left| \int_N (1 - (x, y))\, d\theta_{nj} \right| + 2\theta_{nj}(N').$$

From property (4) of $g(x, y)$ in Lemma 3.3 it follows that there exists a neighborhood of the identity in X where

$$(x, y) = e^{ig(x,y)} \qquad \text{for } y \, \epsilon \, K.$$

In such a neighborhood we have, for $y \, \epsilon \, K$,

(5.5) $$| \, 1 - (x, y) + ig(x, y)| \leqq C_2 \, g^2(x, y).$$

(5.4) and (5.5) imply

(5.6) $$| \, 1 - \hat{\theta}_{nj}(y) \, | \leqq \left| \int_N g(x, y) \, d\theta_{nj} \right| + C_2 \int_N g^2(x, y) \, d\theta_{nj} + 2\theta_{nj}(N')$$

for all $y \, \epsilon \, K$.   By property (2) of $g(x, y)$ in Lemma 3.3

(5.7)
$$\left| \int_X g(x, y) \, d\theta_{nj} \right| = \left| \int_X g(x + x_{nj}, y) \, d\alpha_{nj}(x) \right|$$

$$\leqq \left| \int_N g(x + x_{nj}, y) \, d\alpha_{nj}(x) \right| + C_3 \, \alpha_{nj}(N').$$

Since all the x_{nj} will be in any small neighborhood of the identity after a certain stage, and since $e^{ig(x,y)} = (x, y)$ for $x \, \epsilon \, N$ and $y \, \epsilon \, K$, we conclude, by making use of property (5) of $g(x, y)$ in Lemma 3.3, that

(5.8) $$g(x + x_{nj}, y) = g(x, y) + g(x_{nj}, y)$$

for $x \, \epsilon \, N$ and $y \, \epsilon \, K$.   Further

$$e^{ig(x_{nj},y)} = (x_{nj}, y) = \exp\left\{ - i \int g(x, y) \, d\alpha_{nj} \right\}$$

for all $y \, \epsilon \, K$ and sufficiently large n.   By property (5) of $g(x, y)$ in Lemma 3.3, we get

(5.9) $$g(x_{nj}, y) = - \int g(x, y) \, d\alpha_{nj} .$$

(5.8) and (5.9) imply

$$\left| \int_N g(x + x_{nj}, y) \, d\alpha_{nj} \right| = \left| \int_N (g(x, y) + g(x_{nj}, y)) \, d\alpha_{nj} \right|$$

$$= \left| \int_N g(x, y) \, d\alpha_{nj} - \alpha_{nj}(N) \int g(x, y) \, d\alpha_{nj} \right|$$

$$= \left| \alpha_{nj}(N') \int_N g(x, y) \, d\alpha_{nj} - \alpha_{nj}(N) \int_{N'} g(x, y) \, d\alpha_{nj} \right|$$

$$\leqq C_4 \, \alpha_{nj}(N').$$

The above inequality together with (5.7) implies that

$$(5.10) \qquad \left| \int g(x, y) \, d\theta_{nj} \right| \leqq C_5 \, \alpha_{nj}(N') \qquad \text{for } y \, \epsilon \, K.$$

(5.6), (5.10), and property (2) of Lemma 3.3 give

$$| 1 - \hat{\theta}_{nj}(y)| \leqq C_2 \int g^2(x, y) \, d\theta_{nj} + C_6 \, \theta_{nj}(N') + C_7 \, \alpha_{nj}(N')$$

for $y \, \epsilon \, K$. Thus, in order to complete the proof of the theorem we have only to show that

$$(5.11) \qquad \lim \sup_n \sum_j \theta_{nj}(N') < \infty,$$

$$(5.12) \qquad \lim \sup_n \sum_j \alpha_{nj}(N') < \infty,$$

$$(5.13) \qquad \lim \sup_n \sup_{y \epsilon K} \sum_j \int_N g^2(x, y) \, d\theta_{nj} < \infty.$$

To this end we consider the distribution

$$| \mu_n |^2 = \prod_{j=1}^{k_n} | \alpha_{nj} |^2.$$

Since $| \mu_n |^2$ is compact and no limit of $| \mu_n |^2$ has an idempotent factor, according to Lemma 5.1, $| \hat{\mu}_n(y) |^2$ is bounded away from zero uniformly for $y \, \epsilon \, K$ and in n. Thus

$$\lim \sup_n \sup_{y \epsilon K} \sum_j (1 - | \hat{\alpha}_{nj}(y)|^2) < \infty.$$

This is the same as (5.3) with $| \alpha_{nj} |^2$ replacing θ_{nj}, and hence

$$\lim_{n \to \infty} \sup_{y \epsilon K} [\exp \left(\sum_j (| \hat{\alpha}_{nj}(y)|^2 - 1) \right) - | \hat{\mu}_n(y)|^2] = 0.$$

Thus the sequence $e(\sum_j | \alpha_{nj} |^2)$ is compact. We now appeal to Theorem 4.3. Then

$$(5.14) \qquad \lim \sup_n \sum_j | \alpha_{nj} |^2(N') < \infty,$$

$$(5.15) \qquad \lim \sup_n \sum_j \int (1 - R(x, y)) \, d | \alpha_{nj} |^2 < \infty.$$

We now choose a neighborhood V of the identity such that $V + V \subset N$. Then

$$\sum_j \alpha_{nj}(N') \leqq \sum_j \alpha_{nj}((V + V)')$$
$$\leqq \sum_j \inf_{x \epsilon V} \alpha_{nj}((V + x)') = \sum_j \inf_{x \epsilon V} \alpha_{nj}(V' + x)$$
$$\leqq \sum_j [\alpha_{nj}(V)]^{-1} \int_V \alpha_{nj}(V' + x) \, d\alpha_{nj}$$
$$\leqq \sum_j [\alpha_{nj}(V)]^{-1} \int \alpha_{nj}(V' + x) \, d\alpha_{nj}$$
$$\leqq (\sup_j [\alpha_{nj}(V)]^{-1}) \sum_j | \alpha_{nj} |^2(V').$$

Since $\{\alpha_{nj}\}$ is uniformly infinitesimal, $\sup_j [\alpha_{nj}(V)]^{-1} < 1 + \varepsilon$, for any given $\varepsilon > 0$ and all sufficiently large n depending on ε. The above inequality and the validity of (5.14) for any neighborhood N of the identity imply (5.12). Since $|\alpha_{nj}|^2 = |\theta_{nj}|^2$ and $\{\theta_{nj}\}$ is uniformly infinitesimal, the same argument leads to (5.11).

From (5.15) we have, for any neighborhood V of the identity in X,

$$(5.16) \qquad \limsup_n \sum_j \iint_{V \times V} [1 - R(x_1 - x_2, y)] \, d\theta_{nj}(x_1) \, d\theta_{nj}(x_2) < \infty.$$

We now choose V such that $V - V \subset N$. Then

$$R(x_1 - x_2, y) = \cos g(x_1 - x_2, y).$$

Since $1 - \cos \theta > \theta^2/4$ for sufficiently small θ, we have from property (5) of $g(x, y)$ in Lemma 3.3,

$$1 - R(x_1 - x_2, y) \geqq \tfrac{1}{4} g^2(x_1 - x_2, y)$$

for $y \, \epsilon \, K$. Since $e^{ig(x,y)} = (x, y)$ for $x \, \epsilon \, N$ and $y \, \epsilon \, K$, the same property of $g(x, y)$ gives

$$g(x_1 - x_2, y) = g(x_1, y) - g(x_2, y), \qquad x_1, x_2 \, \epsilon \, V, \quad y \, \epsilon \, K.$$

Thus, for $x_1, x_2 \, \epsilon \, V$, $y \, \epsilon \, K$,

$$(5.17) \quad 1 - R(x_1 - x_2, y) \geqq \tfrac{1}{4}[g^2(x_1, y) + g^2(x_2, y) - 2g(x_1, y)g(x_2, y)].$$

(5.16) and (5.17) imply

$$(5.18) \quad \limsup_n \sup_{y \epsilon K} \left\{ \sum_j \int_V g^2(x, y) \, d\theta_{nj}(x) - \left(\int_V g(x, y) \, d\theta_{nj} \right)^2 \right\} < \infty.$$

(5.10), (5.12), and (5.18) imply (5.13). This completes the proof of the theorem.

THEOREM 5.2. *If $\{\alpha_{nj}\}$ is uniformly infinitesimal, $\mu_n = \prod_j \alpha_{nj}$, and $\mu_n \Rightarrow \mu$, then μ is infinitely divisible.*

Proof. If μ has no idempotent factor, then it is also a limit of the sequence λ_n where λ_n is constructed as in Theorem 5.1. Since λ_n is infinitely divisible for each n, μ is also infinitely divisible.

Now let us consider the general case. By Lemma 5.1 the set $[y : \hat{\mu}(y) \neq 0]$ is an open subgroup H. If G is the annihilator of this subgroup in X, then the normalized Haar measure of the compact group G is a factor of μ. Let τ be the canonical homomorphism from X to X/G. Then the sequence $\{\alpha_{nj} \tau^{-1}\}$ is uniformly infinitesimal in X/G, and $\mu_n \tau^{-1} \Rightarrow \mu \tau^{-1}$. $\mu \tau^{-1}$ has no idempotent factor and hence is infinitely divisible. For $y \, \epsilon \, H$, $\mu(y) = \hat{\mu}\tau^{-1}(y)$ and $\hat{\mu}(y) = 0$ for $y \, \epsilon \, H$. Thus μ itself is infinitely divisible.

Remark. In the statement of Theorem 5.2 we have assumed that $\{\alpha_{nj}\}$ is uniformly infinitesimal. However, it is enough to assume the existence of a sequence $\{x_{nj}\}$ of elements from the group with the property that $\{\alpha_{nj} * x_{nj}\}$ is uniformly infinitesimal. This is equivalent to the statement that any limit of shifts of α_{nj} is degenerate.

6. Gaussian distributions

In this section we give an algebraic definition of a Gaussian distribution and obtain its representation. This definition is also consistent with the classical definition of Gaussian laws in the case of the finite-dimensional vector spaces. Another definition of a Gaussian law is due to Urbanik [14] who also discussed some of its properties.

From the point of view of limit theorems, the Gaussian laws arise very naturally as follows. Suppose F_n is a sequence of finite measures on the group X such that (1) outside each neighborhood of the identity $F_n \rightarrow 0$ as $n \rightarrow \infty$, and (2) $e(F_n)$ converges to a limit after a suitable shift. If the total mass of F_n is not uniformly bounded, $e(F_n)$ may actually converge to a non-degenerate distribution. These are precisely the Gaussian laws.

DEFINITION 6.1. A distribution μ is said to be *Gaussian* if it has the following properties: (i) μ is infinitely divisible, and (ii) if $\mu = e(F) * \alpha$ where α is infinitely divisible, then F is degenerate at the identity.

THEOREM 6.1. *A function on Y is the characteristic function of a Gaussian distribution on X if and only if it has the form*

$$(x, y) \exp [-\phi(y)]$$

where x is a fixed point of X and $\phi(y)$ is a continuous, nonnegative function on Y satisfying the equality

$$(6.1) \qquad \phi(y_1 + y_2) + \phi(y_1 - y_2) = 2[\phi(y_1) + \phi(y_2)]$$

for all y_1, y_2 in Y.

Proof. Let μ be Gaussian. Then μ cannot have a nondegenerate idempotent factor. For, otherwise, the Haar measure of some compact subgroup will be a factor of μ, and hence if F is any measure concentrated in that subgroup, then $\mu = e(F) * \mu$. This contradicts property (ii) of Definition 6.1. From the definition of infinite divisibility it follows that, for each n, there exist a distribution α_n and an element g_n of X such that $\mu = \alpha_n^n * g_n$.

Since μ has no idempotent factors, any limit of shifts of α_n is degenerate, and hence α_n's can be shifted so as to converge to the distribution degenerate at the identity. All these shifts may be absorbed into g_n, so that α_n itself can be assumed to converge to the distribution degenerate at the identity. We now write (as in the proof of Theorem 5.1)

$$\theta_n = \alpha_n * x_n, \qquad \beta_n = e(\theta_n), \qquad \lambda_n = \beta_n^n * (-nx_n) * g_n,$$

where the element x_n is determined by the identity

$$(x_n, y) = \exp\left[-i \int g(x, y)\, d\alpha_n\right].$$

The absence of idempotent factors for μ implies

$$\lim \sup_n \sup_{y \in K} |\hat{\lambda}_n(y) - \hat{\mu}(y)| = 0$$

by Theorem 5.1.

Thus

$$|\hat{\mu}(y)| = \lim_{n \to \infty} \exp\left(n \int [R(x, y) - 1]\, d\theta_n\right).$$

We shall first show that the function

$$\phi(y) = \lim_{n \to \infty} \left(n \int [1 - R(x, y)]\, d\theta_n\right)$$

satisfies (6.1). We write $P_n = n\theta_n$. Then $e(P_n)$ is a shift of λ_n, and hence $e(P_n)$ is shift-compact. Now Theorem 4.3 implies that P_n restricted to N' is tight for every neighborhood N of the identity. But any limit P of P_n restricted to N' will be such that $\mu = e(P) * \alpha$ where α is also infinitely divisible. From condition (ii) of Definition 6.1 it follows that the mass of P_n outside every neighborhood of the identity tends to zero. Thus

$$(6.2) \qquad \phi(y) = \lim_{n \to \infty} \int_N [1 - R(x, y)]\, dP_n$$

for every neighborhood N of the identity. (6.2) and the following identity[1]

$$\lim_{x \to e} \frac{[1 - R(x, y_1 + y_2)] + [1 - R(x, y_1 - y_2)]}{2[1 - R(x, y_1)] + 2[1 - R(x, y_2)]} = 1$$

(e denoting the identity element of the group X) imply

$$\phi(y_1 + y_2) + \phi(y_1 - y_2) = 2[\phi(y_1) + \phi(y_2)].$$

Thus, in order to complete the proof of the theorem it suffices to show that $\hat{\mu}(y)/|\hat{\mu}(y)|$ is a character on Y. Let us denote this by $\chi(y)$. It is not difficult to verify that, for every neighborhood N of the identity,

$$(6.3) \qquad \begin{aligned} &\chi(y_1 + y_2)[\chi(y_1)\chi(y_2)]^{-1} \\ &= \exp\left(\lim_{n \to \infty} \int [I(x, y_1 + y_2) - I(x, y_1) - I(x, y_2)]\, dP_n\right) \end{aligned}$$

where $I(x, y)$ denotes the imaginary part of (x, y). For any given $\varepsilon > 0$, we

---

[1] Since both the numerator and denominator can vanish, this limiting relation is to be interpreted as follows: the numerator lies between $(1 - \varepsilon)$ and $(1 + \varepsilon)$ times the denominator, if x is near enough to e.

choose a neighborhood N of the identity such that

$$|I(x, y_1)| < \varepsilon, \qquad |I(x, y_2)| < \varepsilon$$

for $x \in N$. Since

$$|I(x, y_1 + y_2) - I(x, y_1) - I(x, y_2)|$$
$$\leq |I(x, y_1)| |1 - R(x, y_2)| + |I(x, y_2)| |1 - R(x, y_1)|$$

and

$$\limsup \int [1 - R(x, y)] \, dP_n < \infty$$

(by Theorem 4.3), we have

$$\left| \int_N [I(x, y_1 + y_2) - I(x, y_1) - I(x, y_2)] \, dP_n \right| < C \cdot \varepsilon$$

where C is a constant depending only on y_1 and y_2. Since ε is arbitrary, the right side of (6.3) is equal to unity. The continuity of $\chi(y)$ is obvious. This shows that χ is a character on Y. Thus there exists an element $x \in X$ such that

$$\hat{\mu}(y) = (x, y) \exp(-\phi(y)).$$

This proves the necessity part.

Conversely, let $\hat{\mu}(y) = (x, y) \exp(-\phi(y))$ where $\phi(y)$ is a nonnegative continuous function of y satisfying (6.1). Let $y_1, \cdots, y_k$ be some k characters. Then it is easily verified that $\exp[-\phi(n_1 y_1 + \cdots + n_k y_k)]$, considered as a function of integers $n_1, \cdots, n_k$, is positive-definite in the product of integer group taken k times. This implies the positive-definiteness of $\exp(-\phi(y))$, and hence this function is the characteristic function of a measure. Since $\hat{\mu}(y) = 1$ at the identity of Y, the measure is a distribution. The infinite divisibility of μ is obvious. We shall now prove property (ii) of Definition 6.1. Let, if possible, $\hat{\mu}(y) = \hat{\mu}_1(y) \hat{\mu}_2(y)$, where $\mu_1 = e(F)$ and μ_2 is infinitely divisible. Since $\hat{\mu}(y)$ does not vanish for any y, $\hat{\mu}_2(y)$ also does not vanish, and hence by Theorem 5.1, μ_2 is a limit of distributions of the type $e(H)$. From (2.1) it is clear that for any finite measure H

$$(6.4) \qquad \begin{aligned} -\log |(e(H))^{\wedge}(y_1 + y_2)| &- \log |(e(H))^{\wedge}(y_1 - y_2)| \\ &\leq 2[-\log |(e(H))^{\wedge}(y_1)| - \log |(e(H))^{\wedge}(y_2)|]. \end{aligned}$$

Thus (6.4) is also valid when $e(H)$ is replaced by either μ_1 or μ_2. Substituting μ_1 and μ_2 for $e(H)$ in (6.4) and adding, we get $\phi(y_1 + y_2) + \phi(y_1 - y_2) \leq 2[\phi(y_1) + \phi(y_2)]$. Since equality holds good in this case, we must have

$$\int ([1 - R(x, y_1 + y_2)] + [1 - R(x, y_1 - y_2)]) \, dF$$
$$= 2 \int ([1 - R(x, y_1)] + [1 - R(x, y_2)]) \, dF$$

for each y_1, $y_2 \in Y$, i.e.,

$$\int [1 - R\,(x, y_1)]\,[1 - R\,(x, y_2)]\,dF = 0.$$

Since F is a measure, this implies that F must be degenerate at the origin. This completes the proof.

Remark. 1. Consider real-valued continuous functions $\psi(y_1, y_2)$ defined for y_1, $y_2 \in Y$ possessing the following properties:

   (i) $\psi(y_1, y_2) = \psi(y_2, y_1)$,
   (ii) $\psi(y_1 + y_2, y_3) = \psi(y_1, y_3) + \psi(y_2, y_3)$,
   (iii) $\psi(y, y) \geqq 0$.

Clearly for any such function ψ, $\phi(y) = \psi(y, y)$ satisfies the identity (6.1). Conversely, any nonnegative continuous function satisfying (6.1) can be obtained in this way. In fact, $\psi(y_1, y_2)$ can be recovered from $\phi(y)$ by the relation

$$\psi(y_1, y_2) = \tfrac{1}{2}[\phi(y_1 + y_2) - \phi(y_1) - \phi(y_2)].$$

Remark 2. If X_0 is the component of the identity in X, then its annihilator Y_0 is the smallest closed subgroup containing all compact subgroups of Y. Consequently any $\psi(y_1, y_2)$ with the properties stated above vanishes identically in Y_0, i.e., $\psi(y_1, y_2) = 0$ if y_1 or $y_2 \in Y_0$. Thus if μ is a symmetric Gaussian distribution on X, then $\hat{\mu}(y) = 1$ for $y \in Y_0$. In other words μ is necessarily concentrated in X_0. Theorem 6.1 and Remark 1 above also show that every connected locally compact group has nontrivial Gaussian measures defined on it.

7. Representation of infinitely divisible distributions

As we have mentioned in the introduction, the characteristic function of any infinitely divisible distribution on the real line possesses the famous Lévy-Khinchin representation [3]. The purpose of this section is to obtain such a canonical representation in the case of a general locally compact abelian group.

DEFINITION 7.1. An infinitely divisible distribution λ is said to be a *proper factor* of another infinitely divisible distribution μ, if $\mu = \lambda * \alpha$ and α is infinitely divisible.

LEMMA 7.1. *The set of proper factors of an infinitely divisible distribution is closed.*

LEMMA 7.2. *If $e(F_n)$ converges to the distribution degenerate at the identity, then $F_n(N') \to 0$ as $n \to \infty$ for every neighborhood N of the identity.*

The proof of both the lemmas is quite elementary and is left to the reader.

THEOREM 7.1. *If μ is an infinitely divisible distribution without idempotent factors, then $\hat{\mu}(y)$ has a representation*

$$(7.0) \qquad \hat{\mu}(y) = (x_0, y) \exp\left[\int [(x, y) - 1 - ig(x, y)] \, dF(x) - \phi(y) \right]$$

where x_0 is a fixed element of X, $g(x, y)$ is a function on $X \times Y$ which is independent of μ and has the properties mentioned in Lemma 3.3, F is a σ-finite measure with finite mass outside every neighborhood of the identity in X which satisfies

$$\int [1 - R\,(x, y)] \, dF < \infty \qquad\qquad \text{for every } y,$$

and $\phi(y)$ is a nonnegative continuous function satisfying

$$\phi(y_1 + y_2) + \phi(y_1 - y_2) = 2[\phi(y_1) + \phi(y_2)]$$

for each y_1, $y_2 \in Y$. Conversely, any function of the type (7.0) is the characteristic function of an infinitely divisible distribution.

Proof. Let μ be any infinitely divisible distribution without an idempotent factor. Choose and fix a sequence $\{N_k\}$ of neighborhoods of the identity in X descending to the identity. Let μ_1 be that proper factor of μ which is of the type $e(F)$ and for which $F(N_1) = 0$ and $F(N_1') = 0$ and $F(N_1')$ is maximum. Such an $e(F)$ exists because of Theorems 2.1 and 4.3 and Lemma 7.1. Let the F at which the maximum is attained be F_1, and let $\mu = \mu_1 * \lambda_1$ and $\mu_1 = e(F_1)$. Since λ_1 is infinitely divisible and without idempotent factors, the same argument can be applied to λ_1 and the neighborhood N_2. Thus there exists a measure F_2 for which $F_2(N_2) = 0$, $F_2(N_2')$ is a maximum, $\mu_2 = e(F_2)$, $\lambda_1 = \mu_2 * \lambda_2$, and λ_2 is infinitely divisible. Repeating this procedure we can write

$$(7.1) \qquad\qquad \mu = \mu_1 * \mu_2 * \cdots * \mu_n * \lambda_n,$$

$$(7.2) \qquad\qquad \lambda_{n-1} = \mu_n * \lambda_n,$$

$$(7.3) \qquad\qquad \mu_n = e(F_n), \qquad F_n(N_n) = 0.$$

λ_n is infinitely divisible, and $F_n(N_n')$ is a maximum in the sense explained earlier. Thus by Theorem 2.1 there exist shifts of $\mu_1 * \cdots * \mu_n$ and λ_n converging to ν and λ respectively, and $\mu = \nu * \lambda$. We now assert that λ cannot have a proper factor of the type $e(F)$. Suppose, on the contrary, $e(F)$ is a proper factor of λ. Then it will have a positive mass outside some N_k. Further, since the sequence λ_n is descending (in the order $\prec$), $e(F)$ is a proper factor of λ_k. $\lambda_k = e(F) * \theta$ where θ is infinitely divisible. If F' is the restriction of F to N_k', then (7.2) and (7.3) imply that

$$\lambda_{k-1} = e(F_k + F') * e(F - F') * \theta.$$

This is a contradiction since the total mass of $F_k + F'$ exceeds that of F_k. Thus λ has no proper factor of the type $e(F)$ and is therefore a Gaussian distribution. An application of Theorem 6.1 leads to the existence of a function $\phi(y)$ and an element $x \in X$ for which

$$(7.4) \qquad \hat{\lambda}(y) = (x, y) \exp [-\phi(y)]$$

and

$$\phi(y_1 + y_2) + \phi(y_1 - y_2) = 2[\phi(y_1) + \phi(y_2)], \qquad y_1, y_2 \in Y.$$

Now we write $H_n = F_1 + F_2 + \cdots + F_n$. From the construction of the distributions $\nu_n = \mu_1 * \cdots * \mu_n$, it is clear that

$$\hat{\nu}_n(y) = (x_n, y) \exp \left[\int [(x, y) - 1] dH_n(x) \right]$$

for some element $x_n \in X$. Since $\exp [i \int g(x, y) dH_n(x)]$ is a character on Y, it can be considered as an element of X. Thus there exists an element $z_n \in X$ such that

$$(7.5) \qquad \hat{\nu}_n(y) = (z_n, y) \exp \left[\int [(x, y) - 1 - ig(x, y)] dH_n(x) \right].$$

Since $e(H_n)$ is a factor of μ and H_n increases as $n \to \infty$, it follows from the shift-compactness of $e(H_n)$, Theorem 4.3, and Lemma 2.1 that H_n increases to a σ-finite measure H for which $H(N') < \infty$ for every neighborhood N of the identity and

$$\int \sup_{y \in K} [1 - R(x, y)] dH < \infty,$$

for every compact $K \subset Y$. Since $[(x, y) - 1 - ig(x, y)]$ is bounded uniformly in $y \in K$, by property (2) of $g(x, y)$ in Lemma 3.3, we have, for every neighborhood N of the identity,

$$\lim_{n \to \infty} \int_{N'} [(x, y) - 1 - ig(x, y)] dH_n = \int_{N'} [(x, y) - 1 - ig(x, y)] dH,$$

uniformly in $y \in K$. When N is sufficiently small, we have, by the properties (4) and (5) of $g(x, y)$ in Lemma 3.3,

$$(x, y) = e^{ig(x,y)} \qquad \qquad \text{for} \quad x \in N, \quad y \in K,$$

$$g^2(x, y) \leqq C_1[1 - R(x, y)], \qquad \qquad x \in N, \quad y \in K,$$

where C_1 is a constant depending on K only. Thus

$$(7.6) \qquad |(x, y) - 1 - ig(x, y)| \leqq C_2[1 - R(x, y)], \quad x \in N, \quad y \in K,$$

where C_2 is a constant depending on K only. Thus

$$\int_N \sup_{y \in K} |(x, y) - 1 - ig(x, y)| \, dH_n \leqq C_2 \int_N \sup_{y \in K} [1 - R(x, y)] \, dH.$$

The above inequality implies the convergence of

$$\int [(x, y) - 1 - ig(x, y)] \, dH_n \quad \text{to} \quad \int [(x, y) - 1 - ig(x, y)] \, dH$$

uniformly in $y \in K$. Now (7.5) implies that $\hat{\rho}_n(y)$, after a suitable shift, converges uniformly over compact sets to $\exp \int [(x, y) - 1 - ig(x, y)] \, dH$. This completes the proof of the first part since ν_n converges to ν after a suitable shift, $\mu = \nu * \lambda$, and λ satisfies (7.4).

To prove the converse, we first observe that if F is a totally finite measure, then $\exp[i \int g(x, y) \, dF]$ is a character, and hence $\hat{\mu}(y)$ given by (7.0) is the characteristic function of an infinitely divisible distribution. In the general case we consider a sequence F_n of totally finite measures increasing to F. If K is any compact subset of Y, then, according to Lemma 2.1, $1 - R(x, y)$ is uniformly integrable with respect to F for $y \in K$. (7.6) implies the uniform integrability of $[(x, y) - 1 - ig(x, y)]$ for $y \in K$. This shows that the function

$$(x_0, y) \exp \left(\int [(x, y) - 1 - ig(x, y)] \, dF_n - \phi(y) \right)$$

converges uniformly over compact sets to $\hat{\mu}(y)$. Thus, by Theorem 5.2' $\hat{\mu}(y)$ is infinitely divisible.

Remark. If the group X is totally disconnected, then the representation takes a simpler form. For in such groups $\phi(y) = 0$ (see Remark 2 following Theorem 6.1), and $g(x, y) = 0$ (see Example 4 at the end of Section 3). Thus every infinitely divisible distribution without idempotent factors has the representation

$$\hat{\mu}(y) = (x_0, y) \exp \left[\int [(x, y) - 1] \, dF \right]$$

where F is a σ-finite measure which has finite mass outside each neighborhood of the identity.

8. Uniqueness of the representation

In the case of the real line it is well known that the canonical representation of an infinitely divisible distribution is unique. However, this is not true in the case of a general group. We shall show that the nonuniqueness is essentially due to the presence of compact subgroups in the original group or equivalently due to the disconnectedness of the character group.

Before proceeding to the statement of the main result of this section we shall explain a few conventions and prove an elementary lemma. If μ is any infinitely divisible distribution without idempotent factors, we say that μ has the representation (x_0, F, ϕ) where x_0, F, and ϕ are as in Theorem 7.1. If F is any signed measure, we denote by F_y the measure given by

$$F_y(A) = \int_A [1 - R(x, y)] \, dF.$$

LEMMA 8.1. *Let μ be a totally finite signed measure. If $\hat{\mu}(y)$ is constant on the cosets of a closed subgroup Y_0 of Y, then μ vanishes identically on the complement of the annihilator of Y_0 in X.*

Proof. Let $Y_1 = Y/Y_0$, and let X_1 be the annihilator of Y_0 in X. $\hat{\mu}(y)$, being constant on cosets of Y_0, can be considered as a function on Y_1. Then $\hat{\mu}(y_1)$, $y_1 \epsilon Y_1$ is the characteristic function of a signed measure on X_1. Since, for $x \epsilon X$, (x, y) remains constant on cosets of Y_1 and $\hat{\mu}(y)$ has the same property, we can write

$$\hat{\mu}(y) = \int_{X_1} (x, y_1)\, d\lambda = \int_{X_1} (x, y)\, d\lambda$$

where y_1 denotes that coset of Y_1 to which y belongs. This shows that the signed measures μ and λ are identical, and hence μ vanishes identically on the complement of X_1.

THEOREM 8.1. *If (x_1, F_1, ϕ_1) and (x_2, F_2, ϕ_2) are two representations of the same infinitely divisible distribution without idempotent factors, then* (i) $\phi_1 = \phi_2$, *and* (ii) *the signed measure $F_1 - F_2$ vanishes identically on the complement of the annihilator of the component of identity of the character group Y.*

Proof. Writing $F = F_1 - F_2$, $\phi = \phi_1 - \phi_2$, and $x_0 = x_2 - x_1$, we have

(8.1) $$\exp \int [(x, y) - 1 - ig(x, y)]\, dF = (x_0, y)\, \exp \phi(y),$$

(8.2) $$\phi(y_1 + y_2) + \phi(y_1 - y_2) - 2[\phi(y_1) + \phi(y_2)] = 0, \quad y_1, y_2 \epsilon Y.$$

Equating the logarithm of the absolute value on both sides of (8.1) we obtain

(8.3) $$\phi(y) = \int [R(x, y) - 1]\, dF.$$

Substituting the values of the above expression at $y_1 + y_2$, $y_1 - y_2$, y_1, and y_2 in (8.2), we get

(8.4) $$\int (1 - R(x, y_1))(1 - R(x, y_2))\, dF = 0.$$

(8.4) can be rewritten as

$$\int [1 - R(x, y_1)]\, dF_{y_2} = 0, \qquad \text{for} \quad y_1, y_2 \epsilon Y.$$

Since F_{y_2} is totally finite, we have

(8.5) $$\int (x, y)\, d(F_{y_2} + \bar{F}_{y_2}) = 2F_{y_2}(X).$$

Since the right side of (8.5) is constant when y_2 is fixed, we conclude that the signed measure $F_{y_2} + \bar{F}_{y_2}$ is degenerate at the identity. But the mass of F_{y_2} at the identity is zero. Thus $F_{y_2} + \bar{F}_{y_2} = 0$. In particular $F_{y_2}(X) = 0$, i.e.,

$$(8.6) \qquad \int [1 - R(x, y_2)] \, dF = 0 \qquad \text{for every } y_2.$$

(8.3) and (8.6) imply the equality of ϕ_1 and ϕ_2.

In order to prove the second part of the theorem we make use of the equality of ϕ_1 and ϕ_2 and rewrite (8.1) as

$$(8.7) \qquad \exp \int [(x, y) - 1 - ig(x, y)] \, dF = (x_0, y), \qquad y \in Y.$$

Substituting $y = y_1 + y_2$, $y_1 - y_2$, and y_1 successively in (8.7) and dividing the product of the first two by the square of the third, we obtain

$$(8.8) \qquad \exp \int (x, y_1)[1 - R(x, y_2)] \, dF = 1, \qquad y_1, y_2 \in Y,$$

or equivalently

$$\int (x, y_1)[1 - R(x, y_2)] \, dF = 2\pi n(y_1, y_2)$$

where $n(y_1, y_2)$ is an integer-valued continuous function of y_1 and y_2. We fix y_2 for the present. Then $n(y_1, y_2)$ remains constant on every connected subset of Y and, in particular, on the cosets of the component of identity in Y. This implies, by Lemma 8.1, that the signed measure F_{y_2} vanishes identically on the complement of the annihilator in X of the component of the identity of Y. Since this is true for each y_2, it follows that F itself vanishes identically outside this annihilator. This completes the proof of the theorem.

Remark 1. It is not difficult to show that the annihilator of the component of the identity of Y is the smallest closed subgroup containing all compact subgroups of X. This reflects the role of compact subgroups in making the representation nonunique. In particular, if the group X has no compact subgroups, then the representation is unique.

Remark 2. It was shown in the course of the proof of Theorem 8.1 that the measure F_y is antisymmetric for each character y, i.e., $F_y(A) = -F_y(-A)$ for every Borel set. But if every element in the group were of order two, then such a measure would be identically zero. Coupling this with Remark 1 we can say that if the group X is such that every compact subgroup of X consists only of elements of order two, then the representation is unique.

Remark 3. Conversely, if X is a compact group such that not all elements of X are of order two, then the representation is not unique as can be seen from the following example. We take an element y_0 in the character group which is not of order two and consider the function

$$f(x) = 2\pi i[\overline{(x, y_0)} - (x, y_0)].$$

$f(x)$ is real and not identically zero. If h denotes the normalized Haar measure of X, we have

$$\int (x, y)f(x) \, dh(x) = 2\pi i \qquad \text{if} \quad y = y_0,$$

$$= -2\pi i \qquad \text{if} \quad y = -y_0,$$

$$= 0 \qquad \text{otherwise.}$$

Writing f^+ and f^- to denote the positive and negative parts of $f(x)$, we define the two measures

$$F_1(A) = \int_A f^+(x) \, dh, \qquad F_2(A) = \int_A f^-(x) \, dh.$$

Then $F_1 \neq F_2$, but

$$\exp \int [(x, y) - 1 - ig(x, y)] \, dF_1 = (x_0, y) \exp \int [(x, y) - 1 - ig(x, y)] \, dF_2$$

where $(x_0, y) = \exp [i \int g(x, y) \, d(F_2 - F_1)]$ is a character on Y and hence an element of X. Thus $(e, F_1, 0)$ and $(x_0, F_2, 0)$ are two representations of the same infinitely divisible distribution.

9. Compactness criteria

When the group is the real line, necessary and sufficient conditions that a sequence of infinitely divisible distributions may converge to a given infinitely divisible distribution have been obtained in terms of their representations (cf. Gnedenko and Kolmogorov [3]). Such a result fails to be valid in the general case because of the nonuniqueness of the representations. However, it is possible to obtain conditions for the compactness of a family of infinitely divisible distributions in terms of their representations.

Before proceeding to state the main result of this section we shall investigate what happens to the representation when we pass over from a group to its quotient group. Let $G \subset X$ be some closed subgroup of X, and $X' = X/G$ the quotient group. Let τ denote the canonical homomorphism from X to X'. If Y' is the character group of X', we choose and fix a function $g'(x', y')$ defined on $X' \times Y'$ and satisfying all the properties of Lemma 3.3. We observe that Y' is the annihilator of G in Y and hence a subgroup of Y. Any infinitely divisible distribution μ' on X' without idempotent factors has a representation (x', F', ϕ') (with g replaced by g') according to Theorem 7.1.

LEMMA 9.1. *Let μ be an infinitely divisible distribution on X with a representation (x, F, ϕ). If $\mu' = \mu\tau^{-1}$, $x' = \tau x$, $F' = F\tau^{-1}$, and ϕ' is the restriction of ϕ to Y', then μ' is an infinitely divisible distribution on X' and is a shift of the distribution represented by (x', F', ϕ').*

Proof. Let $\mu = \mu_1 * \mu_2$ where μ_2 is the unique Gaussian component of μ. Clearly $\mu' = \mu\tau^{-1} = \mu_1 \tau^{-1} * \mu_2 \tau^{-1}$. Since $e(\alpha) \tau^{-1} = e(\alpha\tau^{-1})$ for every finite measure α, it follows easily that $\mu_1 \tau^{-1}$ and the distribution represented by $(x', F', 0)$ are shifts of each other. Thus $\mu_2 \tau^{-1}$ is the Gaussian component of μ'. This completes the proof.

We shall now prove the following.

THEOREM 9.1. *Let $\{\mu_\alpha\}$ be a family of infinitely divisible distributions without idempotent factors and with representations $\{(x_\alpha, F_\alpha, \phi_\alpha)\}$. Necessary and sufficient conditions that $\{\mu_\alpha\}$ be shift-compact and any limit of shifts of $\{\mu_\alpha\}$ be devoid of idempotent factors, are*

(1) *The family $\{F_\alpha\}$ of measures is compact when restricted to N' for every neighborhood N of the identity.*

$$(2) \quad \sup_\alpha \left[\int [1 - R(x, y)] \, dF_\alpha + \phi_\alpha(y) \right] < \infty \quad \textit{for all } y.$$

Proof. The necessity of the above two conditions is obvious in view of Theorem 4.3. Regarding sufficiency we first observe that if $\{\mu_\alpha\}$ is a shift-compact family, then condition (2) is sufficient to ensure the absence of idempotent factors in any limit of shifts of μ_α. We shall now prove shift-compactness.

We have only to prove the compactness of the family $\{|\mu_\alpha|^2\}$. We now observe that if (x, F, ϕ) is a representation of an infinitely divisible distribution μ in a group X and τ is a continuous homomorphism of X onto another group X', then $\mu\tau^{-1}$ is a shift of the distribution represented by $(\tau x, F\tau^{-1}, \phi')$ where ϕ' is the restriction of ϕ to the character group of X' (which is a subgroup of Y). Further if $\{(x_\alpha, F_\alpha, \phi_\alpha)\}$ satisfies conditions (1) and (2), so does the family $\{(\tau x_\alpha, F_\alpha \tau^{-1}, \phi_\alpha')\}$. Making use of these remarks we shall reduce the proof of the general case to that of certain simple groups.

In order that a family of measures be compact, it is necessary and sufficient that the family be tight. If C is a compact subgroup of X, and if the family of measures induced by the canonical homomorphism on X/C is tight, then the original family itself is tight. We now choose the group C in such a manner that X/C has the structure $V \oplus D \oplus K^r$ where V is a vector group, D a discrete group, and K^r the r-dimensional torus. The existence of such a compact subgroup is well known. But a family of measures in the product of two topological spaces is tight as soon as the two marginal families are so. Thus it is enough to prove the sufficiency of (1) and (2) in the case of the real line, discrete groups, and compact groups. In the case of the real line the boundedness of $\int [1 - R(x, y)] \, dF_\alpha$ implies the boundedness of $\int_{|x| < \varepsilon} x^2 \, dF_\alpha$ for a suitable ε, which together with condition (1) implies the equicontinuity of the family of functions $\exp\{-\int (1 - R(x, y)) \, dF_\alpha\}$. Since $\phi_\alpha(y)$ assumes the form $\sigma_\alpha^2 y^2$ and hence σ_α^2 is bounded, it is clear that $\{|\mu_\alpha(y)|^2\}$ is equicontinuous. But equicontinuity implies compactness. In the case of a discrete group, identity itself is an open set, and hence the family $\{F_\alpha\}$ is compact outside the identity. This, together with the fact that every infinitely divisible distribution without idempotent factors is a shift of $e(F)$ for some F with zero mass at the identity, implies the required result. In the case of compact groups any family of measures is compact. This completes the proof of the theorem.

COROLLARY 9.1. *In addition to the conditions* (1) *and* (2) *of Theorem* 9.1 *the condition that* $\{x_\alpha\}$ *be a conditionally compact set is necessary and sufficient to ensure the conditional compactness of* $\{\mu_\alpha\}$ *with representations*

$$(x_\alpha, F_\alpha, \phi_\alpha).$$

Proof. From (7.6) we have

$$|(x, y) - 1 - ig(x, y)| \leqq C[1 - R(x, y)]$$

for all $x \in N$, $y \in K$ where C is a constant depending on K only, N is a sufficiently small neighborhood of the identity in X, and K is a compact subset of Y. This implies the equicontinuity of the family of functions

$$\exp\left[\int [(x, y) - 1 - ig(x, y)]\, dF_\alpha - \phi_\alpha(y)\right].$$

Hence $\{x_\alpha\}$ is conditionally compact if and only if $\{\mu_\alpha\}$ is so.

For any σ-finite measure F which has finite mass outside every neighborhood of the identity and integrates the function $1 - R(x, y)$ for every y, we write

(9.1) $$(E(F))^\wedge(y) = \exp \int [(x, y) - 1 - ig(x, y)]\, dF.$$

Theorem 7.1 implies that $(E(F))^\wedge(y)$ is the characteristic function of an infinitely divisible distribution. By proceeding along the same lines as in the proof of Theorem 9.1 it is possible to prove the following.

THEOREM 9.2. *Let* $\{\mu_n\}$ *be a sequence of infinitely divisible distributions with representations* (x_n, F_n, ϕ_n). *Let* μ_n *converge to* μ *after a suitable shift, and* F_n *to* F *outside each neighborhood of the identity. Then* μ *has a representation* (x, F, ϕ) *for a suitable choice of* x *and* ϕ.

10. Representation of convolution semigroups

We have observed earlier that the representation of an infinitely divisible distribution is not unique. We shall now consider the representation problem for a one-parameter convolution semigroup of distributions. By such a semigroup we mean a family $\{\mu_t\}$ of distributions indexed by $t \geqq 0$ such that $\mu_t * \mu_s = \mu_{t+s}$. We shall further assume that μ_t converges weakly to the distribution degenerate at the identity when $t \to 0$. Obviously, for such semi-groups $\hat{\mu}_t(y) \neq 0$ for any $t > 0$ and $y \in Y$. That such a semigroup has a unique canonical representation is the content of the following.

THEOREM 10.1. *Let $\{\mu_t\}$ be a one-parameter convolution semigroup of distributions such that μ_t converges weakly to the distribution degenerate at the identity as $t \to 0$. Then $\hat{\mu}_t(y)$ has the canonical representation*

$$\hat{\mu}_t(y) = (x_t, y) \exp \left(t \int [(x, y) - 1 - ig(x, y)] \, dF - t\phi(y) \right)$$

where F and ϕ are as in Theorem 7.1 and $\{x_t\}$ is a continuous one-parameter semigroup in X. Moreover, $\{x_t\}$, F, and ϕ are uniquely determined by $\{\mu_t\}$.

Proof. Since $\mu_t = (\mu_{t/n})^n$, μ_t is infinitely divisible. As remarked at the beginning, $\hat{\mu}_t(y)$ is nonvanishing at any point and hence has no idempotent factor. Thus by Theorem 7.1, μ_t has a representation (z_t, F_t, ϕ_t). The uniqueness of ϕ_t implies that $\phi_t = t\phi_1$. We write $\phi = \phi_1$ and $\hat{\lambda}_t(y) = \hat{\mu}_t(y)e^{t\phi}$. Then $\{\lambda_t\}$ is a weakly continuous convolution semigroup, and λ_t has neither idempotent nor Gaussian factors. For any σ-finite measure F which has finite mass outside every neighborhood of the identity and integrates the function $1 - R(x, y)$, we define the distribution $E(F)$ as in (9.1). We now observe that the distribution $E(n!F_{1/n}!)$ is a shift of λ_1 for every n. By Theorem 4.3 the sequence of measures $n!F_{1/n}!$ is compact outside every neighborhood of the identity. Thus we can choose a subsequence $n_k!F_{1/n_k}!$ which converges to a measure F outside every neighborhood of the identity. Thus by Theorem 9.2, λ_1 has a representation $(z, F, 0)$ for some $z \in X$. We now define

$$\hat{P}_t(y) = \exp \left(t \int [(x, y) - 1 - ig(x, y)] \, dF \right).$$

If $t = p/q$ is rational, then $E((p/q)n!F_{1/n}!)$ is a shift of $\lambda_{p/q}$ for all sufficiently large n. Since $(p/q)n_k!F_{1/n_k}!$ converges to $(p/q)F$ outside every neighborhood of the identity, another application of Theorem 9.2 shows that $\lambda_{p/q}$ is a shift of $P_{p/q}$. By the continuity of the semigroups it is clear that, for every t, λ_t is a shift of P_t. Thus $\hat{\lambda}_t(y)$ can be written as $(x_t, y)\hat{P}_t(y)$. Then x_t automatically becomes a continuous one-parameter semigroup in X.

If now $(x_t, tF, t\phi)$ and $(x_t', tF', t\phi)$ are two representations of the semigroup $\{\mu_t\}$, then by proceeding in the same way as in the proof Theorem 8.1 we obtain

$$\exp \left[t \int (x, y_1)[1 - R(x, y)] \, d(F - F') \right] = 1$$

for every t. But this can happen only if $F - F' = 0$. Thus the representation of the semigroup is always unique.

COROLLARY. *If μ_t is symmetric for each t, i.e., $\mu_t(A) = \mu_t(-A)$ for each Borel set A, then the representation stated in Theorem 10.1 takes the simpler form*

$$\hat{\mu}_t(y) = \exp\left[t \int \left[(x, y) - 1 \right] dF - t\phi(y) \right]$$

where F is a symmetric measure which integrates $[1 - (x, y)]$ for each y, and $\phi(y)$ is continuous, nonnegative, and satisfies (6.1).

Remark. The above corollary could be used to give a representation for the "negative definite" functions of Herz [15, p. 198]. (The problem of obtaining such a representation for "negative definite" functions is raised by Herz on page 207 of [15].) A function $\nu(y)$ on Y is said to be "negative definite" if

(i) $\nu(y)$ is continuous, $\nu(y) = \nu(-y)$, and $\nu(y) \geq \nu(0) = 0$,   and

(ii)   if $y_1, \cdots, y_n$ are arbitrary elements of Y and $c_1, c_2, \cdots, c_n$ are real numbers such that $\sum c_j = 0$, then

$$\sum \sum c_i c_j \nu(y_i - y_j) \leq 0.$$

From the corollary above and Theorem 3.1 of [15, p. 198], it follows that $\nu(y)$ is negative definite if and only if it is of the form

$$\nu(y) = \int_X [1 - (x, y)] dF + \phi(y)$$

where F and ϕ have properties stated in the above corollary.[2]

11. A decomposition theorem

According to a theorem of Khinchin [5] any distribution on the real line can be written as the convolution of two distributions one of which is the convolution of a finite or countable number of indecomposable distributions, and the other of which is a distribution without indecomposable factors. Further, any distribution which is not infinitely divisible has an indecomposable factor. The object of this section is to extend this result to the general case with a slight modification to counteract the existence of idempotent factors.

THEOREM 11.1. *Let μ be any distribution on X. Then it can be written as $\lambda_G * \lambda$ where λ is a distribution without any idempotent factor and λ_G is the maximal idempotent factor of μ.*

Proof. Let H be the group generated by the set of all characters y at which $\hat{\mu}(y) \neq 0$. H is open, and hence its annihilator G in X is compact. The normalized Haar measure λ_G of G is the required maximal idempotent factor. If we denote by τ the canonical homomorphism from X to X/G, then the distribution $\mu\tau^{-1}$ in X/G has no idempotent factors. We now choose a Borel

---

[2] The authors wish to thank the referee for pointing out that this representation could be derived from the results of the paper.

set $A \subset X$ such that τ restricted to A maps A onto X/G in a one-to-one manner. The existence of such a Borel set follows from a result due to Mackey [7, Lemma 1.1, p. 102]. By a result of Kuratowski [6, p. 251] it follows that the inverse of τ from X/G to A is a measurable map. Hence $\mu\tau^{-1}$ induces a measure λ on A. This λ satisfies the requirements of the theorem.

We shall now introduce a function $\theta(\alpha)$ defined for all factors of a distribution μ and similar to the function introduced by Khinchin. Let μ be a distribution without any idempotent factor. Then there is a sequence $y_1, y_2, \cdots$ of characters in Y such that $\hat\mu(y_i) \neq 0$ for $i = 1, 2, \cdots$ and the smallest closed subgroup generated by this sequence is Y. Since $-\log |\hat\mu(y_i)|$ is well defined, we can choose a sequence $\varepsilon_n > 0$ such that

$$-\sum_{n=1}^{\infty} \varepsilon_n \log |\hat\mu(y_n)| < \infty.$$

This implies at once that, for every factor α of μ, the function

$$\theta(\alpha) = -\sum_{n=1}^{\infty} \varepsilon_n \log |\hat\alpha(y_n)|$$

is well defined and has the following obvious properties:

(11.1)

    (i) $\theta(\alpha) \geqq 0$.

    (ii) $\theta(\alpha) = 0$ if and only if α is degenerate.

    (iii) $\theta(\alpha_1 * \alpha_2) = \theta(\alpha_1) + \theta(\alpha_2)$.

    (iv) If $\alpha_n \Rightarrow \alpha$, then $\theta(\alpha_n) \to \theta(\alpha)$.

    (v) $\theta(\alpha) = \theta(\beta)$ if α is a shift of β.

THEOREM 11.2. *Let μ be a distribution without any indecomposable or idempotent factors. Then μ is infinitely divisible.*

Proof. In view of the remark made under Theorem 5.2 it is enough to factorize μ in the form

(11.2) $$\mu = \alpha_{n1} * \cdots * \alpha_{n2^n}$$

where $\{\alpha_{nj}\}$ is such that any limit of their shifts is degenerate. From the properties (i)–(v) of the θ function it is clear that it is sufficient to factorize μ in the form (11.2) with $\theta(\alpha_{nj}) = 2^{-n}\theta(\mu)$. If μ satisfies the conditions stated in the theorem, then any factor of μ also satisfies them. Thus it suffices to prove that μ can be written as $\mu_1 * \mu_2$ with $\theta(\mu_1) = \theta(\mu_2) = \frac{1}{2}\theta(\mu)$. A repetition of this argument will then complete the proof. In order to do this we first observe that

$$\inf_{\alpha \in F(\mu), \theta(\alpha) \neq 0} \theta(\alpha) = 0$$

where $F(\mu)$ is the class of all factors of μ. For, otherwise, the class $F(\mu)$ being shift-compact, the infimum would be attained at an indecomposable distribution. But this is a contradiction. Thus there are factors of μ with arbitrarily small θ-values. We now take two distributions μ_1, μ_2 for which $\mu =$

$\mu_1 * \mu_2$ and $|\theta(\mu_1) - \theta(\mu_2)|$ is minimum. From the shift-compactness of $F(\mu)$ it follows that the minimum is attained. This minimum has to be zero. For, otherwise, by transferring a factor of μ_1 or μ_2 with an arbitrarily small θ-value to μ_2 or μ_1 we can make $|\theta(\mu_1) - \theta(\mu_2)|$ smaller. This completes the proof of the theorem.

THEOREM 11.3. *Any distribution μ on X can be written as $\lambda_H * \lambda_1 * \lambda_2$ where λ_H is the maximal idempotent factor of μ, λ_1 is the convolution of a finite or a countable number of indecomposable distributions, and μ_2 is an infinitely divisible distribution without indecomposable or idempotent factors.*

Proof. An application of Theorem 11.1 shows that μ can be written as $\lambda_H * \lambda$ where λ_H is the maximal idempotent factor and λ has no idempotent factors. Thus we can define a θ function on $F(\lambda)$ satisfying the properties (i)–(v) of (11.1). Let now δ_1 be the maximum of $\theta(\alpha)$ as α varies over the indecomposable factors of λ. If δ_1 is greater than zero, we write $\lambda = \alpha_1 * P_1$ where α_1 is indecomposable and $\theta(\alpha_1) \geqq \delta_1/2$. We now denote by δ_2 the maximum of $\theta(\alpha)$ as α varies over the indecomposable factors of P_1. If $\delta_2 > 0$, then we write $P_1 = \alpha_2 * P_2$ where α_2 is indecomposable and $\theta(\alpha_2) \geqq \delta_2/2$. We repeat this argument. If the process terminates at the nth stage, then $\lambda = \alpha_1 * \cdots * P_n$ and $\delta_{n+1} = 0$, which means that P_n has no indecomposable factors. Otherwise the process continues ad infinitum. Since $\sum_1^\infty \theta(\alpha_i)$ is convergent, $\theta(\alpha_n) \to 0$ as $n \to \infty$. The sequence $\alpha_1 * \cdots * \alpha_n$ being monotonic (in the order $\prec$) converges after a suitable shift. Absorbing this shift in α_n we can assume that $\alpha_1 * \cdots * \alpha_n$ converges to a distribution λ_1. Automatically P_n will converge to a distribution λ_2. If λ_2 has an indecomposable factor α, then it is a factor of P_n. Thus $\theta(\alpha) \leqq \delta_n$ for each n. But $\delta_n \leqq 2\theta(\alpha_n)$ and hence tends to zero. Therefore $\theta(\alpha) = 0$, or equivalently λ_2 has no indecomposable factors.

12. Concluding remarks

In obtaining the representation for an infinitely divisible distribution as well as in the proof of the theorem on accompanying laws (Theorems 5.1 and 7.1), we have assumed that the distributions under consideration do not have any idempotent factors. If a distribution μ has an idempotent factor, then, as in the proof of Theorem 11.1, we can construct a compact subgroup G such that the measure induced by the canonical homomorphism onto X/G has no idempotent factors. Since uniform infinitesimality and infinite divisibility are preserved by this map, these results can now be discussed in the quotient group X/G.

Another assumption we have made in the paper is that the group is separable (i.e., satisfies the second axiom of countability). All the results of Sections 3–10 are easily extended, with little essential modification in the proofs, if in place of this restriction on the group, we suppose that the measures under consideration have supports contained in a σ-compact subset.

REFERENCES

1. S. BOCHNER, *Harmonic analysis and the theory of probability*, Berkeley, University of California Press, 1955.
2. ———, *General analytical setting for the central limit theory of probability*, Calcutta Mathematical Society Golden Jubilee Commemoration Volume (1958), Part I, pp. 111–128.
3. B. V. GNEDENKO AND A. N. KOLMOGOROV, *Limit distributions for sums of independent random variables*, trans. from the Russian, Cambridge, Mass., Addison-Wesley Pub. Co., 1954.
4. G. A. HUNT, *Semi-groups of measures on Lie groups*, Trans. Amer. Math. Soc., vol. 81 (1956), pp. 264–293.
5. A. KHINCHIN, *Contribution à l'arithmétique des lois de distribution*, Bull. Univ. État Moscou, Sér. Int., Sect. A: Math. et Mécan., vol. 1 (1937), pp. 6–17.
6. C. KURATOWSKI, *Topologie I*, 2nd ed., Warsaw, 1948.
7. G. W. MACKEY, *Induced representations of locally compact groups. I*, Ann. of Math. (2), vol. 55 (1952), pp. 101–139.
8. K. R. PARTHASARATHY, R. RANGA RAO, AND S. R. S. VARADHAN, *On the category of indecomposable distributions on topological groups*, Trans. Amer. Math. Soc., vol. 102 (1962), pp. 200–217.
9. K. TAKANO, *Central convergence criterion in the multidimensional case*, Ann. Inst. Statist. Math., Tokyo, vol. 7 (1956), pp. 95–102.
10. A. WEIL, *L'intégration dans les groupes topologiques et ses applications*, 2 éd., Paris, Hermann, 1951.
11. K. ITÔ, *On stochastic processes (I)*, Japanese J. Math., vol. 18 (1942), pp. 261–301.
12. P. LÉVY, *L'addition des variables aléatoires définies sur une circonférence*, Bull. Soc. Math. France, vol. 67 (1939), pp. 1–41.
13. K. URBANIK, *Poisson distributions on compact Abelian topological groups*, Colloq. Math., vol. 6 (1958), pp. 13–24.
14. ———, *Gaussian measures on locally compact Abelian topological groups*, Studia Math., vol. 19 (1960), pp. 77–88.
15. C. HERZ, *The spectral theory of bounded functions*, Trans. Amer. Math. Soc., vol. 94 (1960), pp. 181–232.
16. B. M. KLOSS, *Limiting distributions on bicompact Abelian groups*, Teor. Veroyatnost. i Primenen., vol. 6 (1961), pp. 392–421 (in Russian with English summary).
17. Y. KAWADA AND K. ITÔ, *On the probability distribution on a compact group. I*, Proc. Phys.-Math. Soc. Japan (3), vol. 22 (1940), pp. 977–998.
18. L. PONTRJAGIN, *Topological groups*, trans. from the Russian, Princeton University Press, 1946.

INDIAN STATISTICAL INSTITUTE
CALCUTTA, INDIA

ТЕОРИЯ ВЕРОЯТНОСТЕЙ

Том IX   И ЕЕ ПРИМЕНЕНИЯ   *Выпуск 1*

1964

EXTENSION OF STATIONARY STOCHASTIC PROCESSES

K. R. PARTHASARATHY, S. R. S. VARADHAN

1. Introduction. In this paper the following problem is solved. Suppose x_t is a stochastically continuous and relatively stationary process defined in the unit interval $[0, 1]$. Does there exist a stationary process X_t defined on the entire real line with the property that the distributions of X_t and x_t coincide when $t \in [0, 1]$? The answer to this question is always in the affirmative. However we do not know when the extension is unique and when does there exist an ergodic extension.

As a particular case we deduce the well known result of Krein [1] that a continuous positive definite function defined on the interval $[-1,1]$ has a positive definite extension to the entire real line.

2. A preliminary lemma on the extension of invariant linear functionals. Let X be any topological space and $C(X)$ the space of all real valued bounded continuous functions on X. $C(X)$ becomes a Banach lattice under the usual supremum norm and ordering. Let $B_0 \subset C(X)$ be a subspace of $C(X)$ containing constants. Let further T_g, $g \in G$ be a group of homeomorphisms of X into itself. Each T_g, $g \in G$ induces an automorphism U_g of the Banach lattice $C(X)$ in a natural manner. We are interested in the problem of extending a nonnegative invariant linear functional on B_0 to the entire space X. For our purpose it is enough to consider the special case when G is the group of rationals under addition. In what follows we tacitly assume that G is the group of rationals under addition. However, we may remark that the problem mentioned above is meaningful when $C(X)$ is replaced by an arbitrary Banach lattice and U_g, $g \in G$ is an arbitrary group of order-preserving automorphisms of the Banach lattice into itself.

Lemma 2.1. *Let Λ_0 be a non-negative invariant linear functional defined on B_0 and invariant under U_g for all $g \in G$. Then there exists an extension Λ of Λ_0 to the whole of $C(X)$ which is non-negative and invariant.*

P r o o f. Let f be any fixed element outside B_0. We shall then show that there is an extension Λ_1 of Λ_0 to the smallest closed invariant subspace B_1 containing B_0 and f. A routine zornification will then complete the proof.

To this end we define

$$\theta = \sup_{l} \sup_{\substack{a_1+a_2+\dots+a_l=1 \\ g_1, g_2, \dots, g_l \in G}} \sup_{\substack{\varphi \in B_0 \\ \varphi \leqslant a_1 U_{g_1} f + \dots + a_l U_{g_l} f}} \Lambda_0(\varphi), \qquad (2.1)$$

$$\theta' = \inf_{\substack{m \ b_1+b_2+\ldots+b_m=1 \\ h_1, \ h_2, \ldots, h_m \in G}} \inf_{\substack{\psi \in B_0 \\ \psi \geqslant b_1 U_{h_1}f+\ldots+b_m U_{h_m}f}} \Lambda_0(\psi). \tag{2.2}$$

Suppose now that φ and ψ are any two functions in B_0, a_1, a_2, ..., a_l; b_1, b_2, ..., b_m are real numbers and g_1, g_2, ..., g_l; h_1, h_2, ..., h_m are elements of G (i. e. rational numbers) with the following properties:

(I)	$a_1 + a_2 + \ldots + a_l = 1,$
(II)	$b_1 + b_2 + \ldots + b_m = 1,$
(III)	$\varphi \leqslant a_1 U_{g_1}f + \ldots + a_l U_{g_l}f,$
(IV)	$\psi \geqslant b_1 U_{h_1}f + \ldots + b_m U_{h_m}f.$

$$\tag{2.3}$$

We shall now show that $\Lambda_0(\varphi) \leqslant \Lambda_0(\psi)$. First of all, we observe that there exists a rational α such that g_1, g_2, ..., g_l, h_1, ..., h_m are all integral multiples of α.

We define the operator A_n in $C(X)$ as follows:

$$A_n = \frac{U_\alpha + U_{2\alpha} + \ldots + U_{n\alpha}}{n}. \tag{2.4}$$

Obviously, for any integer k, $U_{k\alpha} = U_\alpha^k$. It is not difficult to verify that

$$\lim_{n \to \infty} \| A_n f - A_n U_g f \| = 0$$

whenever g is any one of the elements g_1, g_2, ..., g_l, h_1, h_2, ..., h_m. Further A_n is an order-preserving operator for every n. Thus properties (III) and (IV) of (2.3) imply

$$A_n\varphi \leqslant A_n\left[\sum_{r=1}^{l} a_r U_{g_r}f\right], \tag{2.5}$$

$$A_n\psi \geqslant A_n\left[\sum_{r=1}^{m} b_r U_{h_r}f\right]. \tag{2.6}$$

From (2.5), (2.6) and properties (I), (II) of (2.3) we get

$$A_n\varphi - A_nf \leqslant \left(\sum_1^l |a_r|\right) \sup_{1 \leqslant r \leqslant l} \| A_nf - A_n U_{g_r}f \|,$$

$$A_n\psi - A_nf \geqslant -\left(\sum_1^m |b_r|\right) \sup_{1 \leqslant r \leqslant m} \| A_nf - A_n U_{h_r}f \|.$$

Thus, subtracting both the sides of the second inequality from the corresponding sides of the first above we obtain

$$A_n\varphi - A_n\psi \leqslant \varepsilon_n$$

where ε_n is a sequence of positive numbers tending to zero as $n \to \infty$. Since φ, $\psi \in B_0$ we have $\Lambda_0(A_n\varphi) = \Lambda_0(\varphi)$, $\Lambda_0(A_n\psi) = \Lambda_0(\psi)$. The fact that constants belong to B_0 and Λ_0 is non-negative on B_0 implies that

$$\Lambda_0(\varphi) \leqslant \Lambda_0(\psi). \tag{2.7}$$

The definition of θ and θ' together with (2.7) implies

$$\theta \leqslant \theta'. \tag{2.8}$$

Now choose and fix a t such that

$$\theta \leqslant t \leqslant \theta'.$$

Let now $\varphi \in B_0$ and $a_1, a_2, \ldots, a_l$ be constants such that

$$\varphi + a_1 U_{g_1} f + \ldots + a_l U_{g_l} f \geqslant 0, \tag{2.9}$$

$$a_1 + a_2 + \ldots + a_l \neq 0. \tag{2.10}$$

We shall now show that

$$\Lambda_0(\varphi) + (a_1 + a_2 + \ldots + a_l) t \geqslant 0. \tag{2.11}$$

Without loss of generality we can assume that $a_1 + a_2 + \ldots + a_l > 0$. As before we choose an $\alpha \in G$ such that $g_1, g_2, \ldots, g_l$ are integral multiples of α and define A_n as in (2.4). From (2.9) we have

$$A_n\varphi + a_1 A_n U_{g_1} f + \ldots + a_l A_n U_{g_l} f \geqslant 0.$$

Thus

$$A_n\varphi + (a_1 + \ldots + a_l) A_n f \geqslant \varepsilon_n \tag{2.12}$$

where

$$\varepsilon_n = -\left(\sum_i |a_i|\right) \sup_{1 \leqslant i \leqslant l} \| A_n f - A_n U_{g_i} f \|. \tag{2.13}$$

Inequality (2.12) can be rewritten as

$$A_n f \geqslant \frac{\varepsilon_n}{a_1 + \ldots + a_l} - A_n\left(\frac{\varphi}{a_1 + \ldots + a_l}\right). \tag{2.14}$$

The function on the right side of (2.14) is an element of B_0. Thus by the definition of θ we have

$$\Lambda_0\left(\frac{\varepsilon_n}{a_1 + \ldots + a_l}\right) - \frac{\Lambda_0(\varphi)}{a_1 + \ldots + a_l} \leqslant \theta \leqslant t.$$

Since $\varepsilon_n \to 0$ as $n \to \infty$ we have, in the limit,

$$\Lambda_0(\varphi) + t(a_1 + \ldots + a_l) \geqslant 0.$$

A similar argument can be given when $a_1 + \ldots + a_l < 0$. In that case we have to use the inequality $t \leqslant \theta'$. If $a_1 + a_2 + \ldots + a_l = 0$, we have by the same analysis as in the derivation of (2.12),

$$A_n \varphi \geqslant \varepsilon_n$$

where ε_n is given by (2.13). Applying Λ_0 we have

$$\Lambda_0(\varphi) = \Lambda_0(A_n \varphi) \geqslant \Lambda_0(\varepsilon_n).$$

Since $\varepsilon_n \to 0$ as $n \to \infty$ we have in the limit

$$\Lambda_0(\varphi) \geqslant 0$$

which can be rewritten as

$$\Lambda_0(\varphi) + (a_1 + \ldots + a_l) t \geqslant 0. \tag{2.15}$$

Thus (2.15) holds good whenever

$$\varphi + a_1 U_{g_1} f + \ldots + a_l U_{g_l} f \geqslant 0$$

for some $g_1, g_2, \ldots, g_l \in G$.

Now we define, for any $\varphi \in B_0$, $g_1, \ldots, g_l \in G$ and constants $a_1, \ldots, a_l$

$$\Lambda_1(\varphi + a_1 U_{g_1} f + \ldots + a_l U_{g_l} f) = \Lambda_0(\varphi) + (a_1 + \ldots + a_l) t.$$

If

$$\varphi + a_1 U_{g_1} f + \ldots + a_l U_{g_l} f = \varphi' + b_1 U_{h_1} f + \ldots + b_m U_{h_m} f$$

then it follows from the non-negativity of Λ_1 that

$$\Lambda_1(\varphi + a_1 U_{g_1} f + \ldots + a_l U_{g_l} f) = \Lambda_1(\varphi' + b_1 U_{h_1} f + \ldots + b_m U_{h_m} f).$$

Thus, Λ_1 is unambiguously defined on the smallest invariant linear manifold generated by B_0 and f. Non-negativity implies boundedness. Thus Λ_1 can be extended to the closure of this linear manifold without the destruction of the properties of invariance and non-negativity. This completes the proof of the lemma.

3. **Extension of stationary stochastic processes.** Suppose x_t is a stochastic process with the following properties: a) x_t is defined for $t \in [0, 1]$; b) x_t takes values in the real line; c) x_t is stochastically continuous; d) x_t is relatively stationary in $[0, 1]$, i. e., whenever $t_1, t_2, \ldots, t_n$; $t_1 + h, t_2 + h, \ldots$ $\ldots, t_n + h$ belong to $[0, 1]$ the joint distributions of the random vectors $(x_{t_1}, x_{t_2}, \ldots, x_{t_n})$ and $(x_{t_1+h}, \ldots, x_{t_n+h})$ are same. If the above conditions are satisfied then the question arises as to whether there exists a stationary process x_t, $t \in R$ (real line) whose restriction to the unit interval is the given process. That such an extension always exists is the content of the following

Theorem 3.1. *Let x_t be a real valued, stochastically continuous and relatively stationary process defined in the interval $t \in [0, 1]$. Then there exists a stochastically continuous stationary process X_t, $t \in R$ (real line) whose restriction to the unit interval is x_t.*

Remark. x_t is assumed to be real valued only for convenience. The theorem is valid even for complex valued processes.

Proof of theorem 3.1. Let $G \subset R$ be the set of all rationals and $G_1 \subset [0, 1]$ be the set of rationals in the unit interval. We consider the process x_t, $t \in G_1$. Suppose there exists an extension X_t, $t \in G$. Then we can get the extension to the entire real line by using stationarity and stochastic continuity. Thus it is enough to get the extension X_t for $t \in G$ with the following properties: if $t_1, \ldots, t_n \in G$, $t_1 + r, \ldots, t_n + r \in G$ then the joint distributions of $(X_{t_1}, \ldots, X_{t_n})$ and $(X_{t_1+r}, \ldots, X_{t_n+r})$ are same; X_t coincides with x_t when $t \in G_1$.

Further it is enough to consider the case when x_t takes values in a bounded interval, say $[0, 1]$. For, if not, we can compactify the real line and consider the process x_t in this space. We can make use of the following fact to ultimately get the process in the real line: if (X, Y) is a random vector such that the marginal distributions of X and Y have no mass at infinity then the joint distribution of (X, Y) also has no mass at infinity.

Thus we shall now consider a process x_t, $t \in G_1$, which is relatively stationary in G_1. We shall get a stationary extension for $t \in G$. Let $\mathfrak{X} = I^G$ be the space of all real valued functions on G taking values in the unit interval $I = [0, 1]$. $\mathfrak{X}$ is a compact topological space under the product topology. A point $z \in \mathfrak{X}$ can be written as $\{z_t, t \in G\}$ where z_t is a real number in the interval $[0, 1]$ corresponding to each t. Suppose $E \subset G$ is some subset and the value of an element $f \in C(\mathfrak{X})$ (the space of continuous functions on $\mathfrak{X}$) is completely determined by the coordinates z_t, $t \in E$. Then we shall write $f(E)$ for f.

Let C_0 be the subspace of all those bounded continuous functions which are completely determined by the coordinates z_t, $t \in G_1$. The stochastic process x_t, $t \in G_1$ can be viewed upon as a nonnegative bounded linear functional Λ_0 on the subspace C_0.

For any $t \in G$ let T_t denote the shift by t in the space $\mathfrak{X}$. If $z = \{z_t, t \in G\}$ and $y = T_{t_0} z = \{y_t, t \in G\}$ then $y_t = z_{t+t_0}$. The relative-stationarity of the given process x_t can then be described as follows: if the functions $f(z)$ and $g(z) = f(T_t z)$ belong to C_0 then $\Lambda_0(f) = \Lambda_0(g)$. It is clear that T_t induces an automorphism U_t in $C(\mathfrak{X})$ in a natural way. C_0 is not invariant under T_t. If Λ_0 is extended to the smallest closed invariant subspace generated by C_0 then an application of lemma 2.1 and the famous Riesz representation theorem will complete the proof. Thus we shall now concentrate our attention on the extension of Λ_0 to the smallest closed invariant subspace generated by C_0.

For any $f_1, f_2, \ldots, f_n \in C_0$ and $t_1, t_2, \ldots, t_n \in G$ define

$$\Lambda(U_{t_1}f_1 + U_{t_2}f_2 + \ldots + U_{t_n}f_n) = \Lambda_0(f_1) + \Lambda_0(f_2) + \ldots + \Lambda_0(f_n).$$

If we show the non-negativity of Λ then this will establish the unambiguity of the definition of Λ on the smallest invariant linear manifold generated by C_0. In order to establish non-negativity of Λ we have to only show that

$$\Lambda_0(f_1) + \Lambda_0(f_2) + \ldots + \Lambda_0(f_n) \geqslant 0 \tag{3.1}$$

whenever
$$U_{t_1}f_1 + U_{t_2}f_2 + \ldots + U_{t_n}f_n \geqslant 0 \qquad (3.2)$$

for $f_1, f_2, \ldots, f_n \in C_0$, $t_1 < t_2 < \ldots < t_n \in G$. We shall do [this by induction. Suppose (3.1) is true whenever (3.2) is true for n. We shall show that the same is true for $n + 1$. Let

$$f = U_{t_1}g_1 + U_{t_2}g_2 + \ldots + U_{t_{n+1}}g_{n+1} \geqslant 0, \qquad (3.3)$$

$$t_1 < t_2 < \ldots < t_{n+1} \in G, \qquad (3.4)$$

$$g_1, g_2, \ldots, g_{n+1} \in C_0. \qquad (3.5)$$

Let $E_r = G_1 + t_r$ and $h_r = U_{t_r}g_r$. Then the value of h_r is completely deter-. mined by the coordinates corresponding to E_r. Thus we can write $h_r = h_r(E_r)$. Let Δ_{n+1} be the set of all points in E_{n+1} but not in E_n. Then Δ_{n+1} is disjoint from $E_1, E_2, \ldots, E_n$. The inequality (3.3) can be written as

$$h_1(E_1) + h_2(E_2) + \ldots + h_{n+1}(E_{n+1}) \geqslant 0.$$

Thus
$$\min_{\Delta_{n+1}} \{h_1(E_1) + h_2(E_2) + \ldots + h_{n+1}(E_{n+1})\} \geqslant 0 \qquad (3.6)$$

where the minimum is taken over all coordinates corresponding to Δ_{n+1}. Since Δ_{n+1} is disjoint from $E_1 \cup E_2 \cup \ldots \cup E_n$, (3.6) can be written as

$$h_1(E_1) + h_2(E_2) + \ldots + h_n(E_n) + \min_{\Delta_{n+1}} h_{n+1}(E_{n+1}) \geqslant 0. \qquad (3.7)$$

Let
$$u_n(E_n) = \min_{\Delta_{n+1}} h_{n+1}(E_{n+1}), \quad v_n(E_n) = h_n(E_n) + u_n(E_n).$$

Obviously u_n and v_n are determined completely by the coordinates corresponding to E_n. From the compactness of the product space it also follows that u_n and v_n are continuous functions. Thus (3.7) can be written as

$$h_1(E_1) + h_2(E_2) + \ldots + h_{n-1}(E_{n-1}) + v_n(E_n) \geqslant 0.$$

From the validity of non-negativity for n, we obtain

$$\Lambda_0(g_1) + \Lambda_0(g_2) + \ldots + \Lambda_0(g_{n-1}) + \Lambda_0(U_{-t_n}v_n) \geqslant 0. \qquad (3.8)$$

But
$$\Lambda_0(U_{-t_n}v_n) = \Lambda_0(g_n) + \Lambda_0(U_{-t_n}u_n). \qquad (3.9)$$

Since $U_{-t_{n+1}}u_n$ and $U_{-t_n}u_n$ are in C_0 it follows from relative stationarity that

$$\Lambda_0(U_{-t_{n+1}}u_n) = \Lambda_0(U_{-t_n}u_n).$$

Since
$$h_{n+1}(E_{n+1}) \geqslant u_n(E_n)$$

it follows that

$$g_{n+1} \geqslant U_{-t_{n+1}} u_n. \tag{3.10}$$

Both the sides of (3.10) are in C_0. From the non-negativity of Λ_0 in C_0 it follows that

$$\Lambda_0(g_{n+1}) \geqslant \Lambda_0(U_{-t_{n+1}} u_n) = \Lambda_0(U_{-t_n} u_n). \tag{3.11}$$

Thus (3.8), (3.9) and (3.11) imply that

$$\Lambda_0(g_1) + \Lambda_0(g_2) + \ldots + \Lambda_0(g_{n+1}) \geqslant 0.$$

Since Λ_0 is non-negative the statement is valid for $n = 0$. This completes the proof of non-negativity and unambiguity of the functional Λ. Now Λ can be extended to the smallest closed invariant subspace containing C_0 by limiting process. By the very definition Λ is invariant under U_t, $t \in G$. This completes the proof of theorem 3.1.

Remark 1. As a particular case of theorem 3.1 it follows that if $\varphi(t)$ nis a continuous positive definite function in the interval $[-1, 1]$ then it can be extended as continuous positive definite function on the real line. To do this we have to consider the Gaussian process in $[-1/2, 1/2]$ with $\varphi(t)$ as the covariance function, extend the process to $(-\infty, +\infty)$ and take the covariance function of the extended process as the required extension.

Remark 2. The following questions arise in this connection. When is the extension of the process unique? When is an ergodic extension possible?

Remark 3. It is clear from the proofs that theorem 3.1 is valid in the case when the process x_t takes values in any locally compact separable metric space. In particular, if $x_t (0 \leqslant t \leqslant 1)$ is a relatively stationary finite-dimensional Gaussian process then there exists a stationary Gaussian extension in $-\infty < t < \infty$.

Indian Statistical Institute, Calcutta.

Поступила в редакцию
10.12.62

REFERENCES

[1] М. Г. Крейн, О проблеме продолжения эрмитово положительных непрерывных функций, ДАН СССР, **XXVI**, 1 (1940), 17—21.

Math. Student 32 (1964) 17-21

LIMIT THEOREMS IN PROBABILITY

by

S. R. S. Varadhan

The collection of all probability distributions on the real line admits of two important structures one algebraic and the other topological. The interplay between these two structures has been studied in some detail.

Let $\mathcal{M}$ denote the collection of all probability distributions on the real line. For any two elements α, β of $\mathcal{M}$ their convolution $\alpha*\beta$ is defined by the relation

$$\alpha*\beta(A) = \int_R \alpha(A-x)d\beta(x)$$

Under this operation $\mathcal{M}$ becomes a commutative semi-group. One can also define a notion of convergence in $\mathcal{M}$: A sequence α_n in $\mathcal{M}$ converges to α in $\mathcal{M}$ if and only if

$$\int f(x)d\alpha_n(x) \to \int f(x)d\alpha(x)$$

for every bounded continuous function $f(x)$ on R. One can verify that this convergence is actually induced by a metric on $\mathcal{M}$. The topology on $\mathcal{M}$ is compatible with the semi-group operation and $\mathcal{M}$ is actually a topological semi-group.

The main tool in the study of this semi-group is the notion of Fourier transform. For any $\alpha\varepsilon\mathcal{M}$ its 'characteristic function' is defined by the relation

$$\alpha(t) = \int e^{itx}d\alpha(x).$$

It is well known that $\alpha(t)$ determines α uniquely, $\alpha*\beta(t) = \alpha(t).\beta(t)$ and $\alpha_n \to \alpha$ if and only if $\alpha_n(t) \to \alpha(t)$ uniformly over all t in every bounded interval.

The theory revolves around the following main problem : Let, for each n, there be given k_n elements of $\mathcal{M}$, $\{\alpha_{nj}\}$ $j=1, 2,...k_n$. Form the convolution

$$\mu_n = \alpha_{n_1}*\alpha_{n_2}...*\alpha_{nk_n}.$$

Under the assumption that for sufficiently large n the various α_{nj} are nearly the unit element of $\mathcal{M}$ (the distribution giving mass 1 to the point 0) what can one say about the asymptotic behaviour of μ_n? The assumption known as uniform infinitesimality is formulated in precise terms as follows :

$$\lim_{n\to\infty} \sup_{|t|\leqslant T} \sup_{1\leqslant j\leqslant k_n} |\alpha_{nj}(t)-1| = 0$$

for every finite T.

It turns out that every limit (along any subsequence) of μ_n is infinitely divisible in the sense that if μ is the limit it can be put in the form λ_n^n for every n. One then obtains a complete characterization of all infinitely divisible distributions and the conditions for the convergence of μ_n to a specified infinitely divisible distribution μ.

We shall examine now what happens when the real line is replaced by a more general object. Let X be a locally compact abelian group which is separable, Y its character group. Then the class $\mathcal{M}$ of distribution on X becomes in a natural manner a topological semigroup. The convolution operation is

$$\alpha*\beta(A) = \int_X \alpha(A-x)d\beta(x)$$

and the convergence notion is

$$\int_X f(x)d\alpha_n(x) \to \int_X f(x)d\alpha(x)$$

for every bounded continuous function $f(x)$ on X. For every $\alpha\varepsilon\mathcal{M}$ the characteristic function is defined on Y by the formula

$$\alpha(y) = \int_X (x,y)d\alpha(x)$$

where (x, y) denotes the value of the character y at the point x. The relations between α and $\alpha(y)$ valid for the real line continue to be valid even in this set up.

We shall now state the various results known concerning the semigroup. A sequence μ_n in $\mathcal{M}$ is called shift compact if there are elements x_n in such that μ_n translated by x_n is compact. We shall denote this translation by μ_n*x_n. If $\mu = \alpha*\beta$ for some β then α is called a factor of μ. α is idempotent if $\alpha^2 = \alpha$. Idempotent distributions are precisely normalised Haar measures of compact sub-groups.

Theorem 1. Let $\mu_n = \alpha_n*\beta_n$ for each n. If μ_n is compact then α_n and β_n are shift compact.

Definition. A distribution μ is said to be infinitely divisible if for each n there are elements x_n in X and $\lambda_n\varepsilon\mathcal{M}$ such that

$$\mu = \lambda_n^n *x_n.$$

Remark. This modification of the classical definition is intended to avoid the consequence of the presence of elements that are not divisible. In a divisible group this definition is equivalent to the classical definition as can be easily seen.

Theorem 2. The totality of infinitely divisible distributions is a closed sub-semigroup of $\mathcal{M}$.

Definition. If F is any totally finite measure on X the distribution $e(F)$ associated with it is defined as follows.

$$e(F) = e^{-F(X)}[1 + F + \frac{F^2}{2!} + \dots].$$

For $\mu \varepsilon \mathcal{M}$ we denote by $\mu(y)$ its characteristic function. We then have

$$e(F)(y) = [\int [(x, y) - 1]dF].$$

Theorem 3. If μ is infinitely divisible and if its characteristic function vanishes at some point then μ has an idempotent factor.

Theorem 4. Let $\mu_n = e(F_n)$. Then the necessary and sufficient conditions that

(a) μ_n is shift compact,

(b) if μ is any limit of shifts of μ_n then μ has no idempotent factors,

are

(i) for each neighbourhood N of identity ε the family F_n restricted to $X - N$ is weakly conditionally compact,

(ii) for each $y \varepsilon Y$

$$\sup_n \int [1 - \text{Real } (x, y)]dF_n < \infty .$$

Definition. A double sequence $\alpha_{nj} : j = 1, 2, \dots, k_n ; n = 1, 2, \dots$ of distributions is said to be uniformly infinitesimal if

$$\lim_{n \to \infty} \sup_{1 \leqslant j \leqslant k_n} \sup_{y \varepsilon K} \mid \alpha_{nj}(y) - 1 \mid = 0$$

for every compact subset $K \subset Y$.

Lemma. There exists a real valued function $g(x, y)$ on $X \times Y$ which is continuous in both the variables x and y and such that

(i) $g(x, y_1 + y_2) = g(x, y_2)$ for each $x \varepsilon X$ and $y_1, y_2 \varepsilon Y$.

(ii) for any compact subset $K \subset Y$.

$$\sup_{x \varepsilon X} \sup_{y \varepsilon K} \mid g(x, y) \mid < +\infty.$$

(iii) for each compact subset $K \subset Y$ there exists a neighbourhood N_K of the identity in X such that

$$(x, y) = e^{ig(x, y)}$$

holds for $x \varepsilon N_K$ and $y \varepsilon K$,

155

(iv) $g(x, y) = 0$ whenever $x = e$ for any y.

Theorem 5. Let α_{nj} be a uniformly infinitesimal sequence and let

$$\mu_n = \prod_{j=1}^{k_n} \alpha_{nj}.$$

Suppose $\beta_{nj} = e[\alpha_{nj} * g_{nj}]$ where g_{nj} is that element of the group X defined by the equality

$$(g_{nj}, y) = \exp - i \int g(x, y) d\alpha_{nj}.$$

If

$$\lambda_n = \prod_{j=1}^{k_n} \beta_{nj} * g_n \text{ where } g_n = -\sum_j g_{nj},$$

then

$$\lim_{n \to \infty} \sup_{y \varepsilon K} |\lambda_n(y) - \mu_n(y)| = 0$$

for every compact set K of Y.

Corollary. Limit distribution of sums of uniformly infinitesimal independent random variables is infinitely divisible.

Definition. A distribution μ is said to be Gaussian if it has the following properties :

(i) μ is infinitely divisible,

(ii) $\mu = e(F) * \alpha$, where α is infinitely divisible implies that F vanishes outside the identity.

Theorem 6. A distribution μ is Gaussian if and only if $\mu(y)$ is of the form

$$\mu(y) = (x. y) \exp [-\phi(y)]$$

where x is a fixed element of X and $\phi(y)$ a continuous non-negative function of y satisfying the equation

$$\phi(y_1 + y_2) + \phi(y_1 - y_2) = 2[\phi(y_1) + \phi(y_2)]$$

for every pair y_1, y_2 in Y.

Theorem 7. A distribution μ on X is infinitely divisible without an idempotent factor if and only if $\mu(y)$ is of the form

$$\mu(y) = (x_o, y) \exp [\int [(x, y) - 1 - ig(x, y)] dM - \phi(y)]$$

where

(i) x_o is a fixed element of X,

(ii) $\phi(y)$ is a function as in Theorem 6,

(iii) μ is a σ-finite measure giving finite mass outside every neighbourhood of the identity,

(iv) $\int [1 - \text{Real} (x, y)] d\mu < +\infty$ for each $y \varepsilon Y$.

Remark. This representation is not in general unique. If the group X has no compact sub-groups or more generally if every compact sub-group consists only of elements of order two then the representation is unique. In the contrary case examples can be constructed to exhibit non-uniqueness.

Corollary. Every one parametric weakly continuous semi-group μ_t has a unique representation

$$\mu_t(y) = (x_t, y) \exp\left[t \int [(x, y) - 1 - ig(x, y)]dM - t\phi(y)\right]$$

where x_t is a continuous semi-group in X and M and ϕ are as in Theorem 7.

REFERENCES

1. B, V, GNEDENKO and A, N. KOLMOGOROV, Limit distributions for sums of independent random variables, trans., from the Russian, Cambridge, Mass., Addison Wesley Pub. Co., 1954.

2. K. R. PARTHASARATHY, R. RANGA RAO, and S.R.S. VASADHAN, Probability distributions on locally compact abelian groups, Illinois, J. Math. Vol. 7, No. 2, 1963, p.p. 337—369.

Indian Statistical Institute,
Calcutta.

COMMUNICATIONS ON PURE AND APPLIED MATHEMATICS, VOL. XXVI, 497–524 (1973)

A Limit Theorem with Strong Mixing in Banach Space and Two Applications to Stochastic Differential Equations*

G. C. PAPANICOLAOU AND S. R. S. VARADHAN

1. Introduction

In [1] the asymptotic behavior of the expected value of the solution of an abstract stochastic equation was investigated. This work was motivated by the results of Khasminskii [2] and Stratonovich [3] for stochastic ordinary differential equations and other works (cf. [4], [5], [6]) concerned with operator equations. The results obtained in [1] were limited by severe restrictions on the allowed from of the stochastic perturbation.

Recently, Cogburn and Hersh [7] have generalized the results of [1] considerably by allowing a much broader class of stochastic perturbations and requiring only a strong mixing condition. Our aim here is to improve the results of [7] by giving an estimate for the error committed in the asymptotic approximation. If ε denotes the small parameter of the problem we show that the error is $O(\varepsilon)$ as $\varepsilon \to 0$. The results in [1], [2], [3] and [7] show only that the error is $o(1)$. Our estimate is best possible since it is achieved for the classical central limit theorem which is a special case of our Theorem 2.

In Section 2 we state and prove an abstract limit theorem, Theorem 1. In Section 3 we apply the theorem of Section 2 to obtain a result similar to that of Khasminskii in [2] with error estimate $O(\varepsilon)$, Theorem 2. In Section 4 we apply Theorem 1 to obtain a limit theorem for random motion in a Lie group. This is the content of Theorem 3. Theorem 4 contains a result which is needed in Theorem 3 but it is stated as a theorem since it concerns the behavior of the solution of a diffusion equation on the group and is of independent interest. Theorem 3 has a number of interesting applications (cf. [8], [9]).

The nature of the hypotheses of the abstract theorem, Theorem 1, can best be understood by the form they take in the concrete applications of Theorems 2 and 3. In fact, Theorem 1 is a simple abstraction of the situation encountered in the other theorems. It has, however, other interesting applications in connection with random evolution equations (cf. [7], [10], [11], [12], [13]).

*Research performed at the Courant Institute of Mathematical Sciences and supported by NSF Grant #GP-32996X2 and Air Force Grant #AF-AFOSR-72-2307. Reproduction in whole or in part is permitted for any purpose of the United States Government.

The first named author wishes to thank R. Hersh, D. Stroock and E. B. Davies for several enlightening conversations concerning the present problem.

2. A Limit Theorem in Banach Space

We begin with some definitions. Then we state the problem under consideration and the main result.

Let $(\Omega, \mathscr{F}, P)$ be a probability space and $\mathscr{F}_s^t \subset \mathscr{F}$, $0 \leq s \leq t \leq \infty$, a family of σ-algebras such that $\mathscr{F}_{s_1}^{t_1} \subset \mathscr{F}_s^t$, $0 \leq s \leq s_1 \leq t_1 \leq t \leq \infty$. Let the conditional probabilities $P(A \mid \mathscr{F}_0^s)$ have a regular version (cf. [15], p. 354) i.e., there exists a function $P_s(A \mid \omega)$ on $\mathscr{F} \times \Omega$ such that

   (i) $P_s(A \mid \cdot)$ is $\mathscr{F}_0^s$ measurable for every $A \in \mathscr{F}$,

and

   (ii) $P_s(\cdot \mid \omega)$ is a probability measure on $\mathscr{F}$ for every $\omega \in \Omega$.

We assume that the strong mixing property holds as follows:

$$(2.1) \quad \sup_{\substack{s \geq 0 \\ B \in \mathscr{F}_0^s}} \sup_{A \in \mathscr{F}_{t+s}^\infty} |P(A \mid B) - P(A)| = \sup_{\substack{s \geq 0 \\ A \in \mathscr{F}_{t+s}^\infty}} \text{ess.} \sup_{\omega \in \Omega} |P_s(A \mid \omega) - P(A)|$$

$$= \rho(t) \downarrow 0, \qquad\qquad t \to \infty.$$

Let L_0 be a separable Banach space, its norm being denoted by $\| \; \|_0$. Let L_k, $1 \leq k \leq 4$, be dense linear subspaces of L_0 such that $L_{k-1} \supset L_k$. Let $\| \; \|_k$ be a norm on L_k, $1 \leq k \leq 4$, so that the L_k spaces normed this way are also separable Banach spaces. We assume that $\|f\|_{k-1} \leq \|f\|_k$, $f \in L_k$, $1 \leq k \leq 4$.

For each $t \in [0, \infty)$ and $\omega \in \Omega$, let $V(t, \omega)$ be a bounded linear operator from L_k into L_{k-1}, $1 \leq k \leq 4$. We shall assume that the operators $V(t, \omega)$ are strongly measurable (cf. [14], p. 72) with respect to the σ-algebra generated by $B \times \mathscr{F}$, where B denotes the Borel sets on $[0, \infty)$. We assume further that, for $f \in L_k$, $1 \leq k \leq 4$, $V(t)f = V(t, \omega)f \in L_{k-1}$ is strongly $\mathscr{F}_t^t$ measurable as a function of $\omega \in \Omega$, t fixed.

Let $\|V(t, \omega)\|_k$ denote the norm of V as an operator on $L_k \to L_{k-1}$, $1 \leq k \leq 4$, i.e.,

$$(2.2) \qquad\qquad \|V(t, \omega)\|_k = \sup_{f \in L_k, f \neq 0} \frac{\|Vf\|_{k-1}}{\|f\|_k}.$$

Since the spaces L_k are separable, $\|V(t, \omega)\|_k$ is an $\mathscr{F}_t^t$ measurable function of ω and jointly measurable in t and ω. We shall assume that $\|V(t, \omega)\|_k$ satisfies one of the following two hypotheses:

(V.i)   There exist constants c_k such that, for almost all $\omega \in \Omega$,

$$(2.3) \qquad\qquad \|V(t, \omega)\|_k \leqq c_k , \qquad\qquad 1 \leqq k \leqq 4, t \geqq 0 .$$

(V.ii)   There exist constants $\tilde{c}_k$ such that

$$(2.4) \qquad\qquad E^{1/5}\{\|V(t, \omega)\|_k^5\} \leqq \tilde{c}_k , \qquad\qquad 1 \leqq k \leqq 4 , t \geqq 0 .$$

Here and below $E\{\cdot\}$ stands for integration over Ω relative to P. Note also that

$$(2.5) \qquad\qquad \|Vf\|_{k-1} \leqq \|V\|_k \|f\|_k ,$$

where we omit the $\omega \in \Omega$ for convenience.

The object of our study are the asymptotic properties of the expected value of the solution of the Cauchy problem

$$(2.6) \qquad\qquad \frac{dy(t)}{dt} = \varepsilon V(t) y(t) , \qquad\qquad t \geqq 0 , y(0) = f .$$

Here ε is a small real parameter. For the asymptotic analysis it is best to rescale the problem (2.6) by defining

$$(2.7) \qquad\qquad \tau = \varepsilon^2 t , \qquad y^{(\varepsilon)}(\tau) = y(\tau/\varepsilon^2) ,$$

so that

$$(2.8) \qquad\qquad \frac{dy^{(\varepsilon)}(\tau)}{d\tau} = \frac{1}{\varepsilon} V(\tau/\varepsilon^2) y^{(\varepsilon)}(\tau) , \qquad y^{(\varepsilon)}(0) = f .$$

In order to make the sense in which (2.8) is to be taken precise, we introduce the solution operators $U^{(\varepsilon)}(\tau, \sigma)$ by

$$(2.9) \qquad\qquad y^{(\varepsilon)}(\tau) = U^{(\varepsilon)}(\tau, \sigma) y^{(\varepsilon)}(\sigma) , \qquad\qquad \tau \geqq \sigma \geqq 0 .$$

We call the operators $U^{(\varepsilon)}(\tau, \sigma)$ propagators and assume that the following hypotheses hold.

(P.i) $U^{(\varepsilon)}(\tau, \sigma)$ are contraction operators on $L_0 \to L_0$ and $U^{(\varepsilon)}(\tau, \sigma) f$, $f \in L_0$, is strongly $\mathscr{F}_{\sigma/\varepsilon^2}^{\tau/\varepsilon^2}$ measurable. The $U^{(\varepsilon)}(\tau, \sigma)$ are bounded linear operators on $L_k \to L_k$, $k = 1, \cdots, 4$.

(P.ii)   The finite propagator property holds:

$$U^{(\varepsilon)}(\tau, \sigma) U^{(\varepsilon)}(\sigma, s) = U^{(\varepsilon)}(\tau, s) , \qquad 0 \leqq s \leqq \sigma \leqq \tau$$

(2.10)

$$U^{(\varepsilon)}(\tau, \tau) = I .$$

(P.iii)   The infinitesimal forward and backward propagator properties hold:

$$U^{(\varepsilon)}(\tau, \sigma) f = f + \frac{1}{\varepsilon} \int_\sigma^\tau V(s/\varepsilon^2) U^{(\varepsilon)}(s, \sigma) f \, ds ,$$

(2.11) $f \in L_1 ,$

$$U^{(\varepsilon)}(\tau, \sigma) f = f + \frac{1}{\varepsilon} \int_\sigma^\tau U^{(\varepsilon)}(\tau, s) V(s/\varepsilon^2) f \, ds .$$

Integrals of L_k-valued functions are in Bochner's sense and for their existence it is sufficient that the integrand be strongly measurable and its norm integrable (cf. [14], p. 80). Other properties required in the calculations below, such as Fubini's theorem for vector-valued functions, are examined in detail in [14]. The use of such properties is not stated explicitly in what follows.

In the applications of Sections 3 and 4 the role of τ and σ in (P.i)–(P.iii) is reversed. However, we adhere to the present format because this is the way results are frequently presented in the physical literature (cf. [4]).

We adopt the convention that C denotes a constant independent of ε, t, τ, σ, etc. In particular situations we distinguish different constants by using small c's with primes etc. in order to make an argument more explicit.

We now state the main result of this section.

THEOREM 1.   Let $U^{(\varepsilon)}(\tau, \sigma)$, $0 \leqq \sigma \leqq \tau$, be defined as above and have properties (P.i)–(P.iii). Let the following conditions hold:

(1.i)   For $f \in L_1$,

(2.12) $E\{V(t) f\} = 0 .$

(1.ii)   For $f \in L_2$, the strong limit in L_0,

(2.13) $\displaystyle \lim_{t \to \infty} \frac{1}{t} \int_{t_0}^{t_0+t} \int_{t_0}^s E\{V(s)V(\sigma)f\} \, d\sigma \, ds ,$

exists uniformly in t_0 and defines a bounded operator $\bar{V}$ on $L_2 \to L_0$. We assume specifically that

(2.14) $\displaystyle \sup_{t_0 \geqq 0} \left\| \bar{V}f - \frac{1}{t} \int_{t_0}^{t_0+t} \int_{t_0}^s E\{V(s)V(\sigma)f\} \, d\sigma \, ds \right\| \leqq C \frac{\|f\|_2}{t} , \qquad t \to \infty ,$

and that $\bar{V}$ has a closed extension in L_0 with domain $\mathscr{D}_{\bar{V}} \supset L_2$ which generates a strongly continuous contraction semigroup on L_0 denoted by $e^{\tau \bar{V}}$, $\tau \geqq 0$. The set $L_4 \cap \mathscr{D}_{\bar{V}}^2$ is assumed dense in L_0. ($\mathscr{D}_{\bar{V}}^2$ is the domain of $\bar{V}^2$).

   (1.iii) $V(t)$, $t \geqq 0$, satisfies (V.1).
   (1.iv)   The following inequalities are satisfied:

$$(2.15) \qquad \| U^{(\varepsilon)}(\tau, \sigma) f \|_1 \leqq c_1' \, e^{\alpha(\tau-\sigma)/\varepsilon} \, \| f \|_1 \qquad a.s.,$$

$$(2.16) \qquad \| U^{(\varepsilon)}(\tau, \sigma) f \|_2 \leqq c_2' \left(1 + \frac{\tau - \sigma}{\varepsilon} \right) e^{\alpha(\tau-\sigma)/\varepsilon} \| f \|_2 \qquad a.s.,$$

where α, c_1' and c_2' are positive constants.

   (1.v)   The inequality

$$(2.17) \qquad \sup_{0 \leqq \tau \leqq \tau_0} \| e^{\tau \bar{V}} f \|_k \leqq c_k'' \, \| f \|_k , \qquad\qquad 1 \leqq k \leqq 4 ,$$

is satisfied, where τ_0 is a fixed point in $[0, \infty)$ and c_k'', $k = 1, 2, \cdots$, are positive constants depending on τ_0.

   (1.vi)   In addition,

$$(2.18) \qquad \int_0^\infty \rho^{1/2}(s) \, ds < \infty .$$

   Under these hypotheses,

$$(2.19) \qquad \sup_{0 \leqq \tau \leqq \tau_0} \| E \{ U^{(\varepsilon)}(\tau, 0) f \} - e^{\tau \bar{V}} f \|_0 \leqq C(f; \tau_0) \varepsilon , \qquad f \in L_4 \cap \mathscr{D}_{\bar{V}}^2 .$$

$C(f; \tau_0)$ denotes a positive constant that depends on f and τ_0. Thus $E\{U^{(\varepsilon)}(\tau, 0) f\}$ converges strongly in L_0 to $e^{\tau \bar{V}} f$ as $\varepsilon \to 0$, $0 \leqq \tau \leqq \tau_0$, and the error in the approximation is $O(\varepsilon)$.

   Before proceeding with the proof we insert a lemma which we shall use often. This lemma is a generalization of a result in [16], p. 222, to L_k-valued random variables; it is also used in [7].

   LEMMA 1.   Let $\mathscr{L}$, $\mathscr{M}$, $\mathscr{N}$ be Banach spaces and let $W(\omega)$ and $U(\omega)$, $\omega \in \Omega$, be bounded operators from $\mathscr{N} \to \mathscr{M}$ and $\mathscr{M} \to \mathscr{L}$, respectively. Let U be strongly $\mathscr{F}_{t+s}^\infty$ measurable and W strongly $\mathscr{F}_0^s$ measurable. Then, for any $f \in \mathscr{N}$,

$$(2.20) \qquad \| E \{ U W f \} - E \{ U E \{ W f \} \} \|_{\mathscr{L}} \leqq 2\rho(t) \sup_{\omega, \omega' \in \Omega} \| U(\omega) W(\omega') f \|_{\mathscr{L}} .$$

Proof of Lemma 1: Set $\mu_s(A \mid \omega) = P_s(A \mid \omega) - P(A)$. For fixed ω and s, $\mu_s(A \mid \omega)$ is a signed measure. Let $|\mu_s| (A \mid \omega)$ denote the variation of μ_s on A. From the Hahn decomposition theorem we know that

$$(2.21) \qquad \sup_{A \in \mathscr{F}_{t+s}^\infty} |\mu_s| (A \mid \omega) \leqq 2\rho(t) ,$$

where we have used (2.1). On using Fubini's theorem for vector-valued functions, we obtain

$$\|E\{UWf\} - E\{UE\{Wf\}\}\|_{\mathscr{L}} = \|E\{UE\{Wf \mid \mathscr{F}_0^s\}\} - E\{UE\{Wf\}\}\|_{\mathscr{L}}$$

$$= \left\| \int\!\!\int U(\omega') W(\omega) f \mu_s(d\omega \mid \omega') P(d\omega') \right\|_{\mathscr{L}}$$

$$\leqq \int\!\!\int \|U(\omega') W(\omega) f\|_{\mathscr{L}} |\mu_s| (d\omega \mid \omega') P(d\omega')$$

$$\leqq \int \sup_\omega \|U(\omega') W(\omega) f\|_{\mathscr{L}} \sup_{A \in \mathscr{F}_{t+s}^\infty} |\mu_s| (A \mid \omega') P(d\omega')$$

$$\leqq 2\rho(t) \sup_{\omega, \omega' \in \Omega} \|U(\omega') W(\omega) f\|_{\mathscr{L}} .$$

Since we omit the ω's throughout, we shall also omit the sup in (2.20) when we apply it below. It is, of course, understood that the supremum is taken.

Proof of Theorem 1: Fix $\tau \in [0, \tau_0]$ and let ε go to zero through values that render $m = \tau/\varepsilon$ an integer going to infinity.

We use the finite propagator property (P.ii) to obtain the decomposition

$$(2.22) \quad \begin{aligned} &E\{(U^{(\varepsilon)}(\tau, 0) - e^{\tau \mathit{V}})f\} \\ &\qquad = \sum_{k=0}^{m-1} E\{U^{(\varepsilon)}(\tau, (k+1)\varepsilon)[U^{(\varepsilon)}((k+1)\varepsilon, k\varepsilon) - e^{\varepsilon \mathit{V}}]e^{k\varepsilon \mathit{V}} f\} . \end{aligned}$$

We decompose the right side of (2.22) into two parts and use property (P.i) to obtain the estimate

$$(2.23) \quad \begin{aligned} &\|E\{U^{(\varepsilon)}(\tau, 0)f\} - e^{\tau \mathit{V}} f\|_0 \\ &\qquad \leqq \sum_{k=0}^{m-1} \|[E\{U^{(\varepsilon)}((k+1)\varepsilon, k\varepsilon)\} - e^{\varepsilon \mathit{V}}]e^{k\varepsilon \mathit{V}} f\|_0 \\ &\qquad\quad + \sum_{k=0}^{m-1} \|E\{U^{(\varepsilon)}(\tau, (k+1)\varepsilon) U^{(\varepsilon)}((k+1)\varepsilon, k\varepsilon)e^{k\varepsilon \mathit{V}} f\} \\ &\qquad\qquad - E\{U^{(\varepsilon)}(\tau, (k+1)\varepsilon)E\{U^{(\varepsilon)}((k+1)\varepsilon, k\varepsilon)e^{k\varepsilon \mathit{V}} f\}\}\|_0 \\ &\qquad = I_1 + I_2 . \end{aligned}$$

Let us estimate I_2 first. Set $g_k = e^{k \varepsilon V} f$. Using (2.11), we have

$$
\begin{aligned}
(2.24) \quad I_2^{(k)} &\equiv \| E\{U^{(\varepsilon)}(\tau, (k+1)\varepsilon) U^{(\varepsilon)}((k+1)\varepsilon, k\varepsilon) g_k\} \\
&\qquad - E\{U^{(\varepsilon)}(\tau, (k+1)\varepsilon) E\{U^{(\varepsilon)}((k+1)\varepsilon, k\varepsilon) g_k\}\} \|_0 \\
&= \frac{1}{\varepsilon^2} \left\| E\left\{ \int_{(k+1)\varepsilon}^{\tau} U^{(\varepsilon)}(\tau, s) V(s/\varepsilon^2)\, ds \int_{k\varepsilon}^{(k+1)\varepsilon} V(\sigma/\varepsilon^2) U^{(\varepsilon)}(\sigma, k\varepsilon) g_k\, d\sigma \right\} \right. \\
&\qquad \left. - E\left\{ \int_{(k+1)\varepsilon}^{\tau} U^{(\varepsilon)}(\tau, s) V(s/\varepsilon^2)\, ds\, E\left\{ \int_{k\varepsilon}^{(k+1)\varepsilon} V(\sigma/\varepsilon^2) U^{(\varepsilon)}(\sigma, k\varepsilon) g_k\, d\sigma \right\} \right\} \right\|_0 \\
&= \varepsilon^2 \left\| \int_{(k+1)/\varepsilon}^{\tau/\varepsilon^2} \int_{k/\varepsilon}^{(k+1)/\varepsilon} [E\{U^{(\varepsilon)}(\tau, \varepsilon^2 s) V(s) V(\sigma) U^{(\varepsilon)}(\varepsilon^2 \sigma, k\varepsilon) g_k\} \right. \\
&\qquad \left. - E\{U^{(\varepsilon)}(\tau, \varepsilon^2 s) V(s) E\{V(\sigma) U^{(\varepsilon)}(\varepsilon^2 \sigma, k\varepsilon) g_k\}\}]\, d\sigma\, ds \right\|_0 .
\end{aligned}
$$

Note that in the first equality above the identity terms of (2.11) cancel and that in the second equality a simple change of variables and a rearrangement have been made. We may now use Lemma 1 in the last expression in (2.24), since of the four factors $U^{(\varepsilon)}, V(s), V(\sigma), U^{(\varepsilon)}$ the first two are $\mathscr{F}_s^\infty$ measurable and the last two $\mathscr{F}_0^\sigma$ measurable. Thus,

$$
\begin{aligned}
(2.25) \quad I_2^{(k)} &\leq \varepsilon^2 \int_{(k+1)/\varepsilon}^{\tau/\varepsilon^2} \int_{k/\varepsilon}^{(k+1)/\varepsilon} \rho(s - \sigma)\, d\sigma\, ds \\
&\qquad \times \sup_{\substack{s \geq 0 \\ k/\varepsilon \leq \sigma \leq (k+1)/\varepsilon}} \| U^{(\varepsilon)}(\tau, \varepsilon^2 s) V(s) V(\sigma) U^{(\varepsilon)}(\varepsilon^2 \sigma, k\varepsilon) g_k \|_0 .
\end{aligned}
$$

We use hypotheses (P.i), (1.iii) twice, (1.iv) once, and (1.v) once to obtain

$$
\begin{aligned}
(2.26) \quad \sup_{\substack{s \geq 0 \\ k/\varepsilon \leq \sigma \leq (k+1)/\varepsilon}} &\| U^{(\varepsilon)}(\tau, \varepsilon^2 s) V(s) V(\sigma) U^{(\varepsilon)}(\varepsilon^2 \sigma, k\varepsilon) g_k \|_0 \\
&\leq \sup_{\substack{s \geq 0 \\ k/\varepsilon \leq \sigma \leq (k+1)/\varepsilon}} \| V(s) V(\sigma) U^{(\varepsilon)}(\varepsilon^2 \sigma, k\varepsilon) g_k \|_0 \\
&\leq c_1 \sup_{k/\varepsilon \leq \sigma \leq (k+1)/\varepsilon} \| V(\sigma) U^{(\varepsilon)}(\varepsilon^2 \sigma, k\varepsilon) g_k \|_1 \\
&\leq c_1 c_2 \sup_{k/\varepsilon \leq \sigma \leq (k+1)/\varepsilon} \| U^{(\varepsilon)}(\varepsilon^2 \sigma, k\varepsilon) g_k \|_2 \\
&\leq 2 c_1 c_2 c_2' e^\alpha \| g_k \|_2 \leq c_1 c_2 c_2 2 e^\alpha C_2'' \| f \|_2 .
\end{aligned}
$$

Thus

$$(2.27) \qquad I_2^{(k)} \leq C\varepsilon^2 \, \|f\|_2 \int_{(k+1)/\varepsilon}^{\tau/\varepsilon^2} \int_{k/\varepsilon}^{(k+1)/\varepsilon} \rho(s - \sigma) \, d\sigma \, ds \, .$$

We show next that the double integral in (2.27) is bounded independently of k and $\varepsilon \in [0, 1]$, say. We have

$$
\begin{aligned}
(2.28) \qquad \int_{(k+1)/\varepsilon}^{\tau/\varepsilon^2} \int_{k/\varepsilon}^{(k+1)/\varepsilon} \rho(s - \sigma) \, d\sigma \, ds &\leq \int_{(k+1)/\varepsilon}^{\infty} \int_{k/\varepsilon}^{(k+1)/\varepsilon} \rho(s - \sigma) \, d\sigma \, ds \\
&= \int_0^{1/\varepsilon} \sigma\rho(\sigma) \, d\sigma + \frac{1}{\varepsilon} \int_{1/\varepsilon}^{\infty} \rho(\sigma) \, d\sigma \, .
\end{aligned}
$$

By the monotonicity of $\rho(s)$ and (1.vi) we have further

$$
\begin{aligned}
(2.29) \qquad \int_0^{1/\varepsilon} \sigma\rho(\sigma) \, d\sigma = \int_0^{1/\varepsilon} \rho(\sigma) \int_0^{\sigma} ds \, d\sigma &\leq \int_0^{1/\varepsilon} \rho^{1/2}(\sigma) \, d\sigma \int_0^{\sigma} \rho^{1/2}(s) \, ds \\
&\leq \left[\int_0^{\infty} \rho^{1/2}(\sigma) \, d\sigma \right]^2 .
\end{aligned}
$$

Similarly,

$$(2.30) \qquad \frac{1}{\varepsilon} \int_{1/\varepsilon}^{\infty} \rho(\sigma) \, d\sigma \leq \frac{1}{\varepsilon} \, \rho^{1/2}(1/\varepsilon) \int_0^{\infty} \rho^{1/2}(\sigma) \, d\sigma \, .$$

Since ρ decreases to zero and (1.vi) holds, it follows that $(1/\varepsilon)\rho^{1/2}(1/\varepsilon)$ is uniformly bounded for $\varepsilon \in [0, 1]$. Consequently, (2.30), (2.29), and (2.27) yield

$$(2.31) \qquad I_2^{(k)} \leq C\varepsilon^2 \, \|f\|_2 \, .$$

Using this last estimate in (2.23), we get finally the desired estimate

$$(2.32) \qquad I_2 \leq mC\varepsilon^2 \, \|f\|_2 \leq \tau_0 C \, \|f\|_2 \cdot \varepsilon \, .$$

The next step in the proof is the estimation of I_1. For this purpose we proceed as follows. We use the backward and forward infinitesimal propagator property (2.11) iterated four times to obtain the expansion

$$
\begin{aligned}
U^{(\varepsilon)}&((k+1)\varepsilon, k\varepsilon)f \\
&= f + \frac{1}{\varepsilon} \int_{k\varepsilon}^{(k+1)\varepsilon} V(s/\varepsilon^2) f\, ds \\
&\quad + \frac{1}{\varepsilon^2} \int_{k\varepsilon}^{(k+1)\varepsilon} \int_{s}^{(k+1)\varepsilon} V(\sigma/\varepsilon^2) V(s/\varepsilon^2) f\, d\sigma\, ds \\
&\quad + \frac{1}{\varepsilon^3} \int_{k\varepsilon}^{(k+1)\varepsilon} \int_{s}^{(k+1)\varepsilon} \int_{\sigma}^{(k+1)\varepsilon} V(p/\varepsilon^2) V(\sigma/\varepsilon^2) V(s/\varepsilon^2) f\, dp\, d\sigma\, ds \\
&\quad + \frac{1}{\varepsilon^4} \int_{k\varepsilon}^{(k+1)\varepsilon} \int_{s}^{(k+1)\varepsilon} \int_{\sigma}^{(k+1)\varepsilon} V(p/\varepsilon^2) \int_{\sigma}^{p} V(q/\varepsilon^2) U^{(\varepsilon)}(q, \sigma)\, dq \\
&\qquad \times\, V(\sigma/\varepsilon^2) V(s/\varepsilon^2) f\, dp\, d\sigma\, ds\,, \qquad\qquad f \in L_4\,.
\end{aligned}
$$

(2.33)

We take expected values in (2.33), rearrange and take norms. This yields

$$
\begin{aligned}
\| E\{U^{(\varepsilon)}&((k+1)\varepsilon, k\varepsilon)f\} - f - \varepsilon\bar{V}f\}\|_0 \\
&\leq \varepsilon \left\| \bar{V}f - \varepsilon \int_{k/\varepsilon}^{(k+1)/\varepsilon} \int_{k/\varepsilon}^{\sigma} E\{V(\sigma) V(s) f\}\, d\sigma\, ds \right\|_0 \\
&\quad + \varepsilon^3 \left\| \int_{k/\varepsilon}^{(k+1)/\varepsilon} \int_{s}^{(k+1)/\varepsilon} \int_{\sigma}^{(k+1)/\varepsilon} E\{V(p) V(\sigma) V(s) f\}\, dp\, d\sigma\, ds \right\|_0 \\
&\quad + \varepsilon^4 \left\| \int_{k/\varepsilon}^{(k+1)/\varepsilon} \int_{s}^{(k+1)/\varepsilon} \int_{\sigma}^{(k+1)/\varepsilon} \int_{\sigma}^{p} E\{V(p) V(q) U^{(\varepsilon)}(\varepsilon^2 q, \varepsilon^2 \sigma) \right. \\
&\qquad\qquad\qquad \left. \times\, V(\sigma) V(s) f\}\, dq\, dp\, d\sigma\, ds \right\|_0\,.
\end{aligned}
$$

(2.34)

The first term on the right side of the inequality (2.34) is less than or equal to a constant times $\varepsilon^2 \|f\|_2$ independently of k. This follows from (2.14). The other two terms are estimated in Lemmas 2 and 3 which are stated and proved at the end of this proof. On using (2.38) and (2.42) and the above observation, we have the estimate

(2.35) $$ \| E\{U^{(\varepsilon)}((k+1)\varepsilon, k\varepsilon)f\} - f - \varepsilon\bar{V}f\|_0 \leq C\varepsilon^2 \|f\|_4\,. $$

We recall here the convention about the constants C. By elementary considerations and the contractiveness of $e^{t\bar{V}}$, we find that

(2.36) $$ \| e^{\varepsilon\bar{V}} f - f - \varepsilon\bar{V}f\|_0 \leq \tfrac{1}{2}\varepsilon^2 \|\bar{V}^2 f\|_0\,, \qquad\qquad f \in \mathscr{D}_{\bar{V}}^2 $$

When we combine (2.35) and (2.36) we obtain the desired estimate of I_1 :

$$I_1 = \sum_{k=0}^{m-1} \| E\{U^{(\varepsilon)}((k+1)\varepsilon, k\varepsilon) e^{k\varepsilon V} f - e^{\varepsilon V} \cdot e^{k\varepsilon V} f \|_0$$

(2.37)
$$\leq \sum_{k=0}^{m-1} \varepsilon^2 [C \| e^{k\varepsilon V} f \|_4 + \tfrac{1}{2} \| V^2 e^{k\varepsilon V} f \|_0]$$

$$\leq \varepsilon \tau_0 (C c_4'' \| f \|_4 + \tfrac{1}{2} \| V^2 f \|_0] , \qquad\qquad f \in L_4 \cap \mathscr{D}_V^2 .$$

In the last inequality in (2.37) we have used the hypothesis (1.v) and the fact that V^2 and e^{sV} commute on $\mathscr{D}_V^2$. The estimates (2.37) and (2.32) imply that (2.19) holds. Since $U^{(\varepsilon)}$ and e^{tV} are contractions, and $L_4 \cap \mathscr{D}_V^2$ is dense in L_0, we also have the strong convergence in L_0. This completes the proof of the theorem.

LEMMA 2.   For $f \in L_3$, $k = 0, 1, 2, \cdots, 0 \leq \varepsilon \leq 1$,

$$(2.38) \quad J = \left\| \int_{k/\varepsilon}^{(k+1)/\varepsilon} \int_s^{(k+1)/\varepsilon} \int_\sigma^{(k+1)/\varepsilon} E\{V(p)V(\sigma)V(s)f\} \, dp \, d\sigma \, ds \right\|_0 \leq \frac{C}{\varepsilon} \| f \|_3 .$$

Proof: By a simple change of variables and the triangle inequality, we have

(2.39)
$$J \leq \tilde{J} \equiv \int_0^{1/\varepsilon} \int_0^{(1/\varepsilon)-s} \int_0^{(1/\varepsilon)-s-\sigma} \left\| E\left\{ V\left(s + \frac{k}{\varepsilon} + \sigma + p\right) V\left(s + \frac{k}{\varepsilon} + \sigma\right) \right.\right.$$
$$\left.\left. \times V\left(s + \frac{k}{\varepsilon}\right) f \right\} \right\|_0 dp \, d\sigma \, ds .$$

In order to estimate the integral in (2.39) by using Lemma 1 and hypothesis (1.i), we first rewrite it as follows:

$$\tilde{J} \leq \int_0^{1/\varepsilon} \left[\left[\int_0^{1/2\varepsilon} \int_0^p + \int_{1/2\varepsilon}^{1/\varepsilon} \int_0^{(1/\varepsilon)-p} \right] \left\| E\left\{ V\left(s + \frac{k}{\varepsilon} + \sigma + p\right) V\left(s + \frac{k}{\varepsilon} + \sigma\right) \right.\right.\right.$$
$$\left.\left. \times V\left(s + \frac{k}{\varepsilon}\right) f \right\} \right\|_0 d\sigma \, dp$$

(2.40)
$$+ \int_0^{1/2\varepsilon} \int_0^\sigma + \int_{1/2\varepsilon}^{1/\varepsilon} \int_0^{(1/\varepsilon)-\sigma} \left\| E\left\{ V\left(s + \frac{k}{\varepsilon} + \sigma + p\right) V\left(s + \frac{k}{\varepsilon} + \sigma\right) \right.\right.$$
$$\left.\left. \times V\left(s + \frac{k}{\varepsilon}\right) f \right\} \right\|_0 dp \, d\sigma \right] ds .$$

Now we use Lemma 1, (1.i), (1.iii) (three times) and obtain

(2.41)

$$
\begin{aligned}
J &\leq C \, \|f\|_3 \int_0^{1/\varepsilon} \left[\int_0^{1/2\varepsilon} \int_0^p + \int_{1/2\varepsilon}^{1/\varepsilon} \int_0^{(1/\varepsilon)-p} \rho(p) \, d\sigma \, dp \right. \\
&\qquad\qquad\qquad\qquad \left. + \int_0^{1/2\varepsilon} \int_0^\sigma + \int_{1/2\varepsilon}^{1/\varepsilon} \int_0^{(1/\varepsilon)-\sigma} \rho(\sigma) \, dp \, d\sigma \right] ds \\
&\leq 2C \, \|f\|_3 \frac{1}{\varepsilon} \left[\int_0^{1/2\varepsilon} \sigma\rho(\sigma) \, d\sigma + \int_{1/2\varepsilon}^{1/\varepsilon} \rho(\sigma)\left(\frac{1}{\varepsilon} - \sigma\right) d\sigma \right] \\
&\leq 2C \, \|f\|_3 \frac{1}{\varepsilon} \left[\left(\int_0^\infty \rho^{1/2}(\sigma) \, d\sigma \right)^2 + \frac{1}{\varepsilon} \, \rho^{1/2}\!\left(\frac{1}{2\varepsilon}\right) \int_0^\infty \rho^{1/2}(\sigma) \, d\sigma \right].
\end{aligned}
$$

Remarks identical to those following (2.30) show that the quantity in the square brackets in the last inequality in (2.41) is bounded independently of $\varepsilon \in [0, 1]$. This then completes the proof of the lemma.

LEMMA 3.   *For* $f \in L_4$, $k = 0, 1, 2, \cdots, 0 \leq \varepsilon \leq 1$,

(2.42)

$$
\begin{aligned}
\left\| \int_{k/\varepsilon}^{(k+1)/\varepsilon} \int_s^{(k+1)/\varepsilon} \int_\sigma^{(k+1)/\varepsilon} \int_\sigma^p \right. & E\{V(p)V(q)U^{(\varepsilon)}(\varepsilon^2 q, \varepsilon^2 \sigma) \\
& \left. \times \, V(\sigma)V(s)f\} \, dq \, dp \, d\sigma \, ds \right\|_0 \\
& \leq \frac{C \, \|f\|_4}{\varepsilon^2}.
\end{aligned}
$$

Proof: The proof of this lemma is very similar to the previous one.   However, now in addition to the hypotheses used above, we need (P.i) concerning the measurability of $U^{(\varepsilon)}$ and also the inequality (2.16) concerning the growth of the $\| \ \|_2$ norm of $U^{(\varepsilon)}$. In the decomposition analogous to (2.40), the additional integral is simply carried along. We shall omit the details of the derivation.

THEOREM 1′.   *Let all hypotheses of Theorem 1 hold with the following changes:*
(1.iii)′ $V(t)$ *satisfies* (V.ii).
(1.iv)′ $U^{(\varepsilon)}(\tau, \sigma)$, $\varepsilon \geq 0$, $0 \leq \sigma \leq \tau$, *satisfies*

(2.43) $E^{1/5}\{\| U^{(\varepsilon)}(\tau, \sigma)f\|_1^5\} \leq c_1' \, e^{\alpha(\tau-\sigma)/\varepsilon} \, \|f\|_1 ,$

(2.44) $E^{1/5}\{\| U^{(\varepsilon)}(\tau, \sigma)f\|_2^5\} \leq c_2'\!\left(1 + \frac{\tau - \sigma}{\varepsilon}\right) e^{\alpha(\tau-\sigma)/\varepsilon} \, \|f\|_2 .$

(1.vi)′   *Moreover, assume*

$$(2.45) \qquad \int_0^\infty \rho^{1/10}(s)\, ds < \infty .$$

Then the conclusion of Theorem 1 is again valid.

The proof of Theorem 1′ is very similar to that of Theorem 1. We require now the following analogue of Lemma 1 whose proof we omit since it is entirely analogous to the one for random variables given in [16], p. 222 (cf. also [7]).

LEMMA 1′. *Let* $\mathscr{L}$, $\mathscr{M}$, $\mathscr{N}$ *be separable Banach spaces and let* $W(\omega)$ *and* $U(\omega)$, $\omega \in \Omega$, *be bounded operators from* $\mathscr{N} \to \mathscr{M}$ *and* $\mathscr{M} \to \mathscr{L}$, *respectively. Let* U *be strongly* $\mathscr{F}_{t+s}^\infty$ *measurable and* W *strongly* $\mathscr{F}_0^s$ *measurable. Set*

$$(2.46) \qquad \begin{aligned} \|U(\omega)\|_{\mathscr{M},\mathscr{L}} &= \sup_{\substack{g \neq 0 \\ g \in \mathscr{M}}} \frac{\|U(\omega)g\|_{\mathscr{L}}}{\|g\|_{\mathscr{M}}}, \\[2mm] \|W(\omega)\|_{\mathscr{N},\mathscr{M}} &= \sup_{\substack{g \neq 0 \\ g \in \mathscr{N}}} \frac{\|W(\omega)g\|_{\mathscr{M}}}{\|g\|_{\mathscr{N}}}, \end{aligned}$$

and suppose that

$$(2.47) \quad E\{\|U(\omega)\|_{\mathscr{M},\mathscr{L}}^p\} < \infty , \qquad E\{\|W(\omega)\|_{\mathscr{N},\mathscr{M}}^q\} < \infty , \qquad \frac{1}{p} + \frac{1}{q} = 1 .$$

Then

$$(2.48) \qquad \begin{aligned} &\|E\{UWf\} - E\{UE\{Wf\}\}\|_{\mathscr{L}} \\ &\qquad \leqq 2\rho^{1/p}(t) E^{1/p}\{\|U\|_{\mathscr{M},\mathscr{L}}^p\} E^{1/q}\{\|W\|_{\mathscr{N},\mathscr{M}}^q\} \|f\|_{\mathscr{N}} . \end{aligned}$$

Proof of Theorem 1′: The proof is quite similar to the one for Theorem 1. We now use Lemma 1′ in place of Lemma 1. Specifically, we use Lemma 1′ on (2.24) with $p = q = 2$ so that in (2.25) ρ is replaced by $\rho^{1/2}$. Thus (2.45) is not used fully at this point. Similarly, hypotheses (1.iii)′ and (1.iv)′ are used to obtain the analogue of (2.26) (but only fourth moments are needed at this point). Lemma 1′ is used again to obtain the analogues of Lemmas 2 and 3. In Lemma 2 the new hypotheses are not used fully but in Lemma 3 they are. For Lemma 3 we first use Lemma 1′ with $p = 5$, $q = \frac{5}{4}$, so that ρ is replaced by $\rho^{1/5}$ and then use Hölder's inequality repeatedly. This brings about the appearance of fifth moments.

3. Application to Systems of Nonlinear Stochastic Ordinary Differential Equations

In this section we shall apply Theorem 1 to obtain a limit theorem for the solution of a system of stochastic equations.

Let $z(t) = z(t, \omega) \in R^n$, $t \geqq 0$, $\omega \in \Omega$, be defined as the solution of the equations

$$(3.1) \qquad \frac{dz(t)}{d} = \varepsilon F(z(t), t, \omega) , \qquad\qquad z(0) = z \in R^n .$$

Here ε is a real parameter and $F(z, t, \omega)$ is a mapping from $R^n \times [0, \infty) \times \Omega$ into R^n which has bounded and continuous derivatives in z up to order four for all $t \geqq 0$, $\omega \in \Omega$. For fixed $z \in R^n$, F is measurable in t and ω. Thus the solution of (3.1) exists and is unique and the process $z(t, \omega)$ is well defined. To study the process we let

$$(3.2) \qquad \tau = \varepsilon^2 t , \qquad z^{(\varepsilon)}(\tau) = z(\tau/\varepsilon^2) ,$$

so that

$$(3.3) \qquad \frac{dz^{(\varepsilon)}(\tau)}{d\tau} = \frac{1}{\varepsilon} F(z^{(\varepsilon)}(\tau), \tau/\varepsilon^2, \omega) , \qquad\qquad z^{(\varepsilon)}(0) = z .$$

In order to apply Theorem 1 we must introduce appropriate function spaces and operators on them. The space $C(R^n)$ of bounded continuous functions on R^n is a natural candidate for L_0 of Section 2. However, this space is not separable and the measurability hypotheses we need are conveniently verified in a separable space. We introduce therefore the one-point compactification of R^n, denoted by R_c^n, and $C(R_c^n) = L_0$, the space of bounded continuous functions on R_c^n with $\| \ \|_0$ the maximum norm. This space is a separable Banach space. The space $C(R_c^n)$ is sufficiently rich from the point of view of weak convergence of measures on R^n. On the other hand, because of this choice of L_0 (and $L_1, \cdots, L_4$) certain other hypotheses are unnecessarily restrictive (compare with the results in [2]). The example of this section gives a good indication of some of the disadvantages of formulating the limit theorem associated with (3.3) in terms of spaces and operators rather than following a more probabilistic approach.

Assume now that the hypotheses concerning F above hold with $z \in R_c^n$ instead of R^n. We have the following result.

THEOREM 2. *Let* $z^{(\varepsilon)}(\tau)$, $0 \leqq \tau \leqq \tau_0$, *be the process defined by* (3.3) *and assume the following conditions hold:*

(2.i) *For all* $z \in R_c^n$, $t \geqq 0$, $\omega \in \Omega$,

$$|F_i(z, t, \omega)| \leqq C, \qquad \left|\frac{\partial F_i}{\partial z_j}(z, t, \omega)\right| \leqq C,$$

(3.4)
$$i, j, k, l = 1, \cdots, n,$$

$$\left|\frac{\partial^2 F_i}{\partial z_j \partial z_k}(z, t, \omega)\right| \leqq C, \qquad \left|\frac{\partial^3 F_i}{\partial z_j \partial z_k \partial z_l}(z, t, \omega)\right| \leqq C.$$

(2.ii) *For each fixed* $z \in R_c^n$ *and* $t \geqq 0$,

(3.5)
$$E\{F_i(z, t, \omega)\} = 0, \qquad i = 1, \cdots, n.$$

(2.iii) $F(z, t, \omega)$ *as a function of* ω *is* $\mathscr{F}_t^t$ *measurable for fixed* t *and all* $z \in R_c^n$ *and the* σ-*algebras* $\mathscr{F}_s^t$ *satisfy the conditions stated above* (2.1). *In addition,*

$$\int_0^\infty \rho^{1/2}(s)\, ds < \infty.$$

(2.iv) *The limits*

(3.6)
$$a_{ij}(z) = \lim_{t \to \infty} \frac{1}{t} \int_{t_0}^{t_0+t} \int_{t_0}^s E\{F_i(z, \sigma, \omega) F_j(z, s, \omega)\}\, d\sigma\, ds,$$

$$i, j = 1, 2, \cdots, n,$$

(3.7)
$$b_j(z) = \lim_{t \to \infty} \frac{1}{t} \int_{t_0}^{t_0+t} \int_{t_0}^s \sum_{i=1}^n E\left\{F_i(z, \sigma, \omega) \frac{\partial F_j}{\partial z_i}(z, s, \omega)\right\} d\sigma\, ds,$$

exist uniformly in $z \in R_c^n$ *and* $t_0 \geqq 0$ *and the rate of approach of the double integrals to* a_{ij}, b_j *is not slower than* $1/t$, $t \to \infty$. *Assume further that* $a_{ij}(z)$ *and* $b_j(z)$, $i, j = 1, \cdots, n$, *have bounded continuous derivatives in* R_c^n *up to order* 4.

    Under these hypotheses and with $f \in C^4(R_c^n)$ *and* $w^{(\varepsilon)}(\tau, z) = E\{f(z^{(\varepsilon)}(\tau))\}$, *we have*

(3.8)
$$\sup_{0 \leq \tau \leq \tau_0} \sup_{z \in R_c^n} |w^{(\varepsilon)}(\tau, z) - w^{(0)}(\tau, z)| \leqq C(f; \tau_0) \cdot \varepsilon.$$

Here $C(f; \tau_0)$ *denotes a constant depending on* f *and* τ_0 *and* $w^{(0)}$ *is the solution of the Cauchy problem*

(3.9)
$$\frac{\partial w^{(0)}}{\partial \tau} = \sum_{i,j=1}^n a_{ij} \frac{\partial^2 w^{(0)}}{\partial z_i \partial z_j} + \sum_{j=1}^n b_j \frac{\partial w^{(0)}}{\partial z_j} = \Gamma w^{(0)}, \qquad \tau > 0,$$

$$w^{(0)}(0, z) = f(z).$$

Proof: We prove this theorem by reducing it to a special case of Theorem 1.

The Banach spaces L_k are $C^k(R_c^n)$, $k = 0, 1, \cdots$, and $\| \ \|_k$ is the sum of the maximum norm of the function and its derivatives up to order k in the usual manner. The spaces L_k are separable and the L_k, $k \geqq 1$, are dense in L_0. The operator $V(t)$ is defined on L_1 by

$$(3.10) \qquad V(t, \omega) = \sum_{j=1}^{n} F_j(z, t, \omega) \frac{\partial}{\partial z_j} .$$

Since the dual space L_0^* of L_0 is the space of finite signed Borel measures on R_c^n, the strong measurability of $V(t)$ is implied by the $\mathscr{F}_t^i$ measurability of

$$\int (V(t, \omega)f)(z)\mu(dz) , \qquad\qquad f \in L_1 , \quad \mu \in L_0^* .$$

The latter follows from our hypotheses on F.

Let $T_\sigma^\tau(\omega)z = T_\sigma^\tau z$ denote the solution of (3.3) for $\tau \geqq \sigma \geqq 0$ with $T_\sigma^\sigma z = z$. From our hypotheses it follows that $T_\sigma^\tau(\omega)z$ is a differentiable transformation of R^n into itself for fixed σ, τ, and ω, and $\mathscr{F}_\sigma$ measurable as a function of ω for all z (if z is the point at infinity, then $T_\sigma^\tau(\omega)z$ is the point at infinity for all $\tau \geqq \sigma$, $\omega \in \Omega$). Using the same argument as above to check measurability we define the operator $U^{(\varepsilon)}$ by

$$(3.11) \qquad (U^{(\varepsilon)}(\sigma, \tau)f)(z) = f(T_\sigma^\tau z) , \qquad\qquad f \in C(R_c^n) .$$

These propagators do not satisfy (P.i)–(P.iii) because the roles of σ and τ are interchanged. This is however an entirely superficial difference; it only changes the form of $\bar{V}$ since now σ and s are interchanged in (2.13). With this explanation in mind we record the new form of (P.i)–(P.iii):

(P.i)' The $U^{(\varepsilon)}$ of (3.11) are contraction operators on $L_0 \to L_0$ and they are strongly $\mathscr{F}_{\sigma/\varepsilon^2}^{\tau/\varepsilon^2}$ measurable. $U^{(\varepsilon)}(\tau, \sigma)$: $L_k \to L_k$, $k \geqq 1$.

(P.ii)' The following property holds:

$$(3.12) \qquad U^{(\varepsilon)}(\sigma, s) U^{(\varepsilon)}(s, \tau) = U^{(\varepsilon)}(\sigma, \tau) .$$

(P.iii)' Moreover,

$$(3.13) \qquad U^{(\varepsilon)}(\sigma, \tau)f = f + \frac{1}{\varepsilon} \int_\sigma^\tau V(s/\varepsilon^2) U^{(\varepsilon)}(s, \tau)f \, ds ,$$

$$(3.14) \qquad U^{(\varepsilon)}(\sigma, \tau)f = f + \frac{1}{\varepsilon} \int_\sigma^\tau U^{(\varepsilon)}(\sigma, s) V(s/\varepsilon^2)f \, ds , \qquad\qquad f \in L_1 .$$

We proceed next to the verification of hypotheses (1.i)–(1.vi) of Theorem 1. Hypothesis (1.i) follows from (2.ii). Hypothesis (1.ii) follows from (2.iii), the fact that the Cauchy problem (3.9) is well posed (cf. [17]), and the maximum principle. Note that (a_{ij}) is positive semidefinite as can be verified from its definition (3.6). The operator $\mathcal{V}$ is the second order (possibly degenerate) elliptic differential operator on the right side of (3.9), $\mathscr{D}_{\mathcal{V}}$ contains $C^2(R_c^n)$ and, since a_{ij}, b_j are smooth, $\mathscr{D}_{\mathcal{V}}^2$ contains $C^4(R_c^n)$. Thus $L_4 \cap \mathscr{D}_{\mathcal{V}}^2 \supset C^4(R_c^n)$ which is dense in $L_0 = C(R_c^n)$.

Hypothesis (1.iii) follows from (2.i). Hypothesis (1.vi) is included in (2.iii). Hypothesis (1.v) follows from the smoothness of the coefficients a_{ij}, b_j, and the results of Oleinik (cf. [17], Theorem 6, and [23], Appendix). It remains to show that (1.iv) holds. This means that we must show that

$$(3.15) \qquad \|f(T_\sigma^\tau z)\|_1 \leq C_1' \, e^{\alpha(\tau-\sigma)/\varepsilon} \, \|f\|_1 \qquad \text{a.s. ,}$$

and

$$(3.16) \qquad \|f(T_\sigma^\tau z)\|_2 \leq C_2'\left(1 + \frac{\tau - \sigma}{\varepsilon}\right) e^{\alpha(\tau-\sigma)/\varepsilon} \|f\|_2 \qquad \text{a.s. .}$$

We demonstrate (3.15) only, since (3.16) can be derived in the same manner. From the definition of the $\| \;\|_1$ norm it follows that it suffices to estimate

$$(3.17) \qquad \left\|\frac{\partial}{\partial z_j} f(T_\sigma^\tau z)\right\|_0 , \qquad\qquad j = 1, 2, \cdots, n .$$

Now for $f \in C^1[R_c^n]$ we have

$$(3.18) \qquad \left\|\frac{\partial}{\partial z_j} f(T_\sigma^\tau z)\right\| = \left\|\sum_{i=1}^n f_i(T_\sigma^\tau z) \frac{\partial(T_\sigma^\tau z)_i}{\partial z_j}\right\|$$
$$\leq \|f\|_1 \sup_{z \in R^n} \left\|\frac{\partial(T_\sigma^\tau z)}{\partial z}\right\| .$$

In (3.18), f_j, $j = 1, 2, \cdots, n$, denotes the partial derivative of f with respect to its j-th argument and $\| \; \|$ denotes some matrix norm. Also, $\partial(T_\sigma^\tau z)/\partial z$ denotes the matrix with entries $\partial(T_\sigma^\tau z)_i/\partial z_j$, $i, j = 1, 2, \cdots, n$. From (3.3) we have

$$(3.19) \qquad T_\sigma^\tau z = z + \frac{1}{\varepsilon} \int_\sigma^\tau F(T_\sigma^s z, s/\varepsilon^2, \omega) \, ds ,$$

and hence

$$(3.20) \qquad \frac{\partial(T_\sigma^r z)}{\partial z} = I + \frac{1}{\varepsilon} \int_\sigma^r \frac{\partial(T_\sigma^s z)}{\partial z} F^{(1)}(T_\sigma^s z, s/\varepsilon^2, \omega) \, ds \,.$$

Here I denotes the identity matrix and $F^{(1)}$ the matrix whose entries are the partial derivatives of F_i, $i = 1, \cdots, n$, with respect to its various arguments in the first (the z) position. From (3.20) and (2.i) it follows that

$$(3.21) \qquad \left\| \frac{\partial(T_\sigma^r z)}{\partial z} \right\| \leq 1 + \frac{c}{\varepsilon} \int_\sigma^r \left\| \frac{\partial(T_\sigma^s z)}{\partial z} \right\| \, ds \,,$$

so that

$$(3.22) \qquad \left\| \frac{\partial(T_\sigma^r z)}{\partial z} \right\| \leq e^{c(r-\sigma)/\varepsilon} \,.$$

The estimate (3.22) and (3.18) together imply (3.15). This completes the verification of the hypotheses and hence the proof of Theorem 2.

Let us observe the following.

1. If the operator $\bar{V}$ in (2.13) is unaltered by the interchange of s and σ in $E\{V(s)V(\sigma)f\}$, then $\bar{V}$ is symmetric in the sense that, for example, the coefficients $a_{ij}(z)$ in the concrete application of Theorem 2 are symmetric: $a_{ij} = a_{ji}$. In general, the fact that s and σ in (2.13) are interchangeable is called the microscopic reversibility (detailed balance) and the resulting symmetry of coefficients an Onsager relation. Some further comments on this can be found in [11].

2. If $F(z, t, \omega) = x(t, \omega) = x(t)$ independently of z, where $x(t)$ is an $\mathscr{F}_t^t$ measurable R^n-valued process, then

$$(3.23) \qquad z^{(\varepsilon)}(\tau) = z + \frac{1}{\varepsilon} \int_0^r x(s/\varepsilon^2, \omega) \, ds \,,$$

and

$$(3.24) \qquad V(t) = \sum_{j=1}^n x_j(t) \frac{\partial}{\partial z_j} \,.$$

Clearly then

$$(3.25) \qquad \bar{V} = \sum_{i,j=1}^n \left\{ \lim_{t \to \infty} \frac{1}{t} \int_{t_0}^{t_0+t} \int_{t_0}^s E\{x_i(\sigma)x_j(s)\} \, d\sigma \, ds \right\} \frac{\partial^2}{\partial z_i \, \partial z_j} \,,$$

and Theorem 2 implies that $z^{(\epsilon)}(\tau)$ converges weakly to a Gaussian random vector with covariance matrix

$$\lim_{t \to \infty} \frac{1}{t} \int_{t_0}^{t_0+t} \int_{t_0}^{t_0+t} E\{x_i(\sigma) x_j(s)\} \, d\sigma \, ds \, .$$

This is the classical central limit theorem for random vectors satisfying a strong mixing condition (cf. [16]). It is clearly a special case of Theorem 2. Since ϵ corresponds to $1/\sqrt{n}$ in the classical scaling (in discrete time), the error $O(\epsilon)$ corresponds to $O(1/\sqrt{n})$ which is best possible since $O(1/\sqrt{n})$ is achieved for sums of independent Bernoulli variables.

4. Application to Random Motion on a Lie Group

Let G denote a connected n-dimensional Lie group, I the identity element of G, and Λ the Lie algebra of G; that is, the tangent space of G at the identity. For concreteness (specifically so that we may write (4.6) directly) we assume that G is a matrix Lie group with elements g and the bracket operation in Λ is the matrix commutator, i.e., if $m_1, m_2 \in \Lambda$, then

$$(4.1) \qquad [m_1, m_2] = m_1 m_2 - m_2 m_1 \, .$$

Let $Y_1, Y_2, \cdots, Y_n$ denote a fixed basis in Λ and c_{ij}^k the structure constants

$$(4.2) \qquad [Y_i, Y_j] = \sum_{k=1}^{n} c_{ij}^k Y_k \, .$$

Let G_c denote the one-point compactification of G and $C(G_c)$ the separable Banach space of bounded continuous functions on G_c with the maximum norm $\| \ \|_0$. Let $C^1(G_c)$ denote the linear subspace of $C(G_c)$ for which the strong limits

$$(4.3) \qquad D_j f(g) \equiv \lim_{h \downarrow 0} \frac{f(e^{Y_j h} g) - f(g)}{h}, \qquad j = 1, 2, \cdots, n \, ,$$

exist. Here $e^{Y_j h}$ denotes the exponential mapping of Λ into G which is the ordinary exponential matrix since we are dealing with matrix groups. The operators $D_j, j = 1, 2, \cdots, n$, form a basis for the right invariant vector fields on G. The space $C^1(G_c)$ equipped with the norm

$$(4.4) \qquad \|f\|_1 = \|f\|_0 + \sum_{j=1}^{n} \|D_j f\|_0$$

becomes a Banach space which is dense in $C(G_e)$ (cf. [18]). Spaces $C^k(G_e)$, $k = 2, 3, \cdots$, can be defined along with norms $\| \ \|_k$ in a similar fashion. All spaces $C^k(G_e)$, $k = 1, 2, \cdots$, are dense in $C(G_e)$.

We wish to study a random motion on G. For this purpose let

$$(4.5) \qquad \tilde{m}(t, \omega) = \sum_{j=1}^{n} \tilde{m}_j(t, \omega) Y_j, \qquad\qquad \omega \in \Omega,$$

be a process with values in Λ. The R^n-valued process $(\tilde{m}_1(t), \tilde{m}_2(t), \cdots, \tilde{m}_n(t))$ has several properties which we state below. Let $A \in \Lambda$ be a fixed matrix and consider the matrix equation

$$(4.6) \qquad \frac{dg(t)}{dt} = Ag(t) + \varepsilon \tilde{m}(t)g(t), \qquad\qquad g(0) = g.$$

In order to study the asymptotic properties of the process $g(t)$ when $\varepsilon \to \infty$, $t \to \infty$, we let

$$(4.7) \qquad \tau = \varepsilon^2 t, \qquad g^{(\varepsilon)}(\tau) = e^{-A\tau/\varepsilon^2} g(\tau/\varepsilon^2).$$

Then $g^{(\varepsilon)}(\tau)$ satisfies the matrix equation

$$(4.8) \qquad \frac{dg^{(\varepsilon)}(\tau)}{d\tau} = \frac{1}{\varepsilon} m(\tau/\varepsilon^2)g^{(\varepsilon)}(\tau), \qquad\qquad g^{(\varepsilon)}(0) = g.$$

$$(4.9) \qquad m(t) = e^{-At} \tilde{m}(t) e^{At} = \sum_{j=1}^{n} m_j(t, \omega) Y_j.$$

We have the following result.

THEOREM 3. *Let* $g^{(\varepsilon)}(\tau)$ *be the process defined by (4.8) and let* $0 \leqq \tau \leqq \tau_0$. *Assume that the following conditions hold:*

(3.i) $|\tilde{m}_j(t)| \leqq C$ *a.s.,* $j = 1, \cdots, n$, *and* A *belongs to the compact part of* Λ *so that* e^{At} *remains in a compact subset of* G *containing the identity for all* $t \in (-\infty, \infty)$.

(3.ii) *For* $j = 1, 2, \cdots, n$,

$$(4.10) \qquad\qquad E\{\tilde{m}_j(t)\} = 0.$$

(3.iii) $\tilde{m}_j(t)$, $j = 1, \cdots, n$, *are* $\mathscr{F}_t^t$ *measurable and the σ-algebras* $\mathscr{F}_\bullet^t$ *satisfy all hypotheses above (2.1). In addition,*

$$\int_0^\infty \rho^{1/2}(s) \, ds < \infty.$$

(3.iv)   *There exist constants* a_{ij}, $i,j = 1, \cdots, n$, *such that*

$$(4.11) \qquad \sup_{t_0 \geq 0} \left| a_{ij} - \frac{1}{t} \int_{t_0}^{t_0+t} \int_{t_0}^{s} E\{m_i(\sigma)\, m_j(s)\}\, d\sigma\, ds \right| \leq \frac{C}{t},$$

$$t \to \infty, \; i,j = 1, \cdots, n.$$

Let $w^{(\varepsilon)}(\tau, g) = E\{f(g^{(\varepsilon)}(\tau))\}, f \in C^4(G_c)$. Then

$$(4.12) \qquad \sup_{0 \leq \tau \leq \tau_0} \sup_{g \in G_c} |w^{(\varepsilon)}(\tau, g) - w^{(0)}(\tau, g)| \leq C \|f\|_4 \cdot \varepsilon,$$

where C *is a constant depending on* τ_0, *and* $w^{(0)}$ *satisfies the Cauchy problem*

$$(4.13) \qquad \frac{\partial w^{(0)}}{\partial \tau} = \sum_{i,j=1}^{n} a_{ij}\, D_i\, D_j\, w^{(0)} \equiv \bar{V} w^{(0)}, \qquad w^{(0)}(0, g) = f(g).$$

Proof: As for Theorem 2 we transform the present problem so that Theorem 1 applies.

Let T_σ^τ denote the solution matrix of (4.8), $\tau \geq \sigma$, such that $T_\sigma^\sigma = I$. Define $U^{(\varepsilon)}$ and $V(t)$ by

$$(4.14) \qquad (U^{(\varepsilon)}(\sigma, \tau) f)(g) = f(T_\sigma^\tau\, g), \qquad\qquad f \in C(G_c),$$

$$(4.15) \qquad (V(t) f)(g) = \sum_{j=1}^{n} m_j(t) D_j f(g), \qquad\qquad f \in C^1(G_c).$$

It is easily verified that (P.i)′–(P.iii)′ of Section 3 hold again here. We may proceed therefore with the verification of the hypotheses and the identification of the objects with the ones of Theorem 1.

Clearly, $L_k = C^k(G_c)$, $k = 0, 1, 2, \cdots$. Hypothesis (1.i) follows from (3.ii). Hypothesis (1.ii) follows from (3.iv) and the results of K. Itô (cf. [19], [20]); a detailed treatment is given in [21]. In terms of semigroups, Itô's results are also given in [18]. $\mathcal{D}_{\bar{V}} \supset L_2$ and $L_4 \cap \mathcal{D}_{\bar{V}}^2 \supset C^4(G_c)$ which is dense in $C(G_c)$. Hypotheses (1.iii) and (1.vi) follow from (3.i) and (3.iii), respectively.

Next we verify hypothesis (1.iv). As in Theorem 2, it suffices to obtain (2.16) since (2.17) follows by similar considerations. We shall show that, for each $j = 1, 2, \cdots, n$,

$$(4.16) \qquad \|D_j\, U^{(\varepsilon)}(\sigma, \tau) f\|_0 \leq C e^{\alpha(\tau-\sigma)/\varepsilon} \|f\|_1,$$

which suffices for (2.16) to hold. We have

$$
\begin{aligned}
D_j\, U^{(\varepsilon)}(\sigma,\tau)f(g) &= \lim_{h\downarrow 0}\frac{U^{(\varepsilon)}(\sigma,\tau)f(e^{Y_j h}g) - U^{(\varepsilon)}(\sigma,\tau)f(g)}{h}\\[2mm]
&= \lim_{h\downarrow 0}\frac{f(T^\tau_\sigma\, e^{Y_j h}g) - f(T^\tau_\sigma g)}{h}\\[2mm]
&= \lim_{h\downarrow 0}\frac{f(T^\tau_\sigma\, e^{Y_j h}(T^\tau_\sigma)^{-1}\,T^\tau_\sigma g) - f(T^\tau_\sigma g)}{h}\,.
\end{aligned}
$$

(4.17)

For $g \in C$, let $Ad(g)$ be the linear transformation on Λ defined by

$$
(4.18)\qquad Ad(g)\,Y = \frac{d}{dh}\,(ge^{hY}\,g^{-1})\Big|_{h=0} = g\,Yg^{-1}\,,
$$

and let $(Ad(g))_{ij}$ denote the matrix elements of $Ad(g)$ relative to the basis $Y_1,\,Y_2,\,\cdots,\,Y_n$. Then we have

$$
(4.19)\qquad D_j\, U^{(\varepsilon)}(\sigma,\tau)f(g) = \sum_{l=1}^{n}\,(Ad(T^\tau_\sigma))_{jl}\,D_l f(T^\tau_\sigma g)\,,
$$

and hence

$$
(4.20)\qquad \|D_j\, U^{(\varepsilon)}(\sigma,\tau)f\|_0 \leqq \|f\|_1\,\|Ad(T^\tau_\sigma)\|\,.
$$

Here $\|\ \|$ denotes some matrix norm.

From (4.20) it follows that the desired estimate has been reduced to estimating $\|Ad(T^\tau_\sigma)\|$. For this purpose we proceed as follows. From (4.18) and (4.8) we obtain

$$
(4.21)\qquad \frac{\partial}{\partial\tau}\,Ad(T^\tau_\sigma) = \frac{1}{\varepsilon}\,[\,m(\tau/\varepsilon^2),\,Ad(T^\tau_\sigma)\,]\,,
$$

which in terms of the matrix elements of $Ad(T^\tau_\sigma)$ takes the form

$$
\frac{\partial}{\partial\tau}\,(Ad(T^\tau_\sigma))_{ij} = \frac{1}{\varepsilon}\sum_{k=1}^{n}\,(Ad(T^\tau_\sigma))_{ik}Q_{kj}(\tau/\varepsilon^2)\,,
$$

(4.22)

$$
Q_{kj}(t) = \sum_{l=1}^{n} m_l(t)c^j_{lk}\,,\qquad Ad(T^\sigma_\sigma) = I\,.
$$

178

In integral form, (4.22) becomes

$$(4.23) \qquad Ad(T_\sigma^\tau) = I + \frac{1}{\varepsilon} \int_\sigma^\tau Ad(T_\sigma^s) \cdot Q(s/\varepsilon^2) \, ds \, ,$$

so that

$$(4.24) \qquad \|Ad(T_\sigma^\tau)\| \leq 1 + \frac{C}{\varepsilon} \int_\sigma^\tau \|Ad(T_\sigma^s)\| \, ds \, .$$

Here we have used hypothesis (3.i) concerning $\tilde{m}$ and $A \in \Lambda$ and the boundedness of the structure constants. Clearly, (4.24) yields the desired estimate for $\|Ad(T_\sigma^\tau)\|$ and then (4.20) in turn yields hypothesis (1.iv).

We come now to the verification of hypothesis (1.v). From the results of Itô [20] and Hunt [18] it follows that there is a uniquely determined family of probability measures p_τ, $\tau \geq 0$, on the Borel sets of G_c such that

$$(4.25) \qquad e^{\tau V} f(g) = \int_{G_c} f(g'g) \, p_\tau(dg') \, , \qquad\qquad f \in C(G_c) \, .$$

Assuming that integrals exist we have now

$$(4.26) \qquad D_i \, e^{\tau V} f(g) = \int_{G_c} \sum_{j=1}^n (Ad(g'))_{ji} \, D_j \, f(g'g) \, p_\tau(dg') \, , \qquad g \in C^1(G_c) \, .$$

Thus, for any $j = 1, 2, \cdots, n$ and $f \in C^1(G_c)$,

$$(4.27) \qquad \|D_j \, e^{\tau V} f\|_0 \leq \|f\|_1 \int_{G_c} \|Ad(g')\| \, p_\tau(dg') \, .$$

Here $\| \cdot \|$ denotes some matrix norm. Since the function $\|Ad(g)\|$ does not belong to the class $C(G_c)$, we must show that the integral on the right side of (4.27) exists. Specifically we shall show that

$$(4.28) \qquad \sup_{0 \leq \tau \leq \tau_0} \int_{G_c} \|Ad(g)\| \, p_\tau(dg) \leq c(\tau_0) < \infty \, ,$$

where $c(\tau_0)$ is a constant depending on τ_0. Estimates for $\|D_i \, D_j \, e^{\tau V} f\|_0$ are obtained in the same manner. We proceed now with the proof of (4.28).

Let us observe that for any matrix norm there is a constant c such that if $A = (A_{ij})$, A_{ij} real, then

$$(4.29) \qquad \|A\| \leq c \left(\sum_{i,j=1}^n A_{ij}^2 \right)^{1/2} \equiv c \, \|A\|_E \, .$$

It suffices therefore to show (4.28) with $\| \ \|$ replaced by $\| \ \|_E$. Now, if $A \otimes A$ denotes the tensor (Kronecker) product of A with itself, then

$$(4.30) \qquad \|A\|_E = \{H(A \otimes A)\}^{1/2},$$

where $H(A \otimes A)$ is defined as $\sum_{i,j} A_{ij}^2$. Let us also note that the linear transformations $Ad(g)$, $g \in G$, constitute a finite-dimensional representation of G, the adjoint representation. The desired estimate (4.28), and hence the completion of the proof of Theorem 3, follow directly from the following result which is also of independent interest.

THEOREM 4. *Let* $\pi(g)$, $g \in G$, *be a finite-dimensional representation of* G. *Then*

$$(4.31) \qquad \int_{G_c} \pi(g) \, p_\tau(dg) = \exp \{\tau \bar{V} \pi(g)|_{g=I}\},$$

in the sense that the integral on the left exists and equals the expression on the right.

Proof: First note that $\pi(g)$ is analytic in g, being a finite-dimensional representation (cf. [22], p. 107). Thus the expression on the right side of (4.31) is well defined. We shall give the remainder of the proof after proving two lemmas.

LEMMA 4.1. *Let* $g(\tau)$ *be the diffusion (Brownian motion) corresponding to* $\bar{V}$ *starting at time* 0 *from the identity. Let* K_1 *be a compact neighborhood of the identity and let* σ_1 *be the first exit time from* K_1 *of the process* $g(\tau)$. *Then, with probability one,*

$$(4.32) \qquad \lim_{K_1 \uparrow G} \sigma_1 = \infty$$

Proof of Lemma 4.1: Define sets $K_2, K_3, \cdots$ as follows:

$$(4.33) \qquad \begin{aligned} K_2 &= \{g \in G \mid g = g_1 g_2, g_1, g_2 \in K_1\}, \\ K_3 &= \{g \in G \mid g = g_1 g_2, g_1, g_2 \in K_2\}, \cdots. \end{aligned}$$

Clearly, $K_1 \subset K_2 \subset K_3 \subset \cdots$ and $K_n \uparrow G$. Let $\sigma_1, \sigma_2, \cdots$ denote the exit times of $g(\tau)$ from $K_1, K_2, \cdots$ and let $g^{(j)} = g(\sigma_j)$, $j = 1, 2, \cdots$. Define $K_1^{(j)}$ by

$$(4.34) \qquad K_1^{(1)} = K_1, \qquad K_1^{(2)} = K_1 g^{(1)}, \qquad K_1^{(3)} = K_1 g^{(2)}, \qquad \cdots$$

The exit time of $g(\tau)$ from $K_1^{(j)}$, conditional on $g^{(j-1)}$ $(g^{(0)} = I), j = 1, 2, \cdots$, at $\tau = 0$, is denoted by $\sigma^{(1)}, \sigma^{(2)}, \cdots (\sigma^{(1)} = \sigma_1)$. In view of the right invariance of $g(\tau)$ and the strong Markov property, the positive random variables $\sigma^{(1)}, \sigma^{(2)}, \cdots$ are independent and identically distributed. Furthermore, since $K_1^{(2)} \subset K_2$ we have

$$(4.35) \qquad\qquad \sigma_2 \geqq \sigma^{(1)} + \sigma^{(2)} ,$$

and similarly

$$(4.36) \qquad\qquad \sigma_n \geqq \sigma^{(1)} + \sigma^{(2)} + \cdots + \sigma^{(n)} , \qquad\qquad n = 3, 4, \cdots .$$

From the strong law of large numbers it follows now that $\sigma^{(1)} + \sigma^{(2)} + \cdots + \sigma^{(n)} \uparrow \infty$ with probability one and hence, from (4.36), $\sigma_n \uparrow \infty$ with probability one. This then completes the proof of the lemma.

LEMMA 4.2. *Let* $\pi(g)$ *be any finite-dimensional representation of* G *and let* $\pi(g) \otimes \pi(g)$ *be the Kronecker product representation. Then*

$$(4.37) \quad H\{\exp\{\tau \overline{V}\pi(g) \otimes \pi(g)|_{g=I}\}\pi(g) \otimes \pi(g)\}$$

$$\geqq H\{\exp\{\tau \overline{V}\pi(g)|_{g=I}\}\pi(g) \otimes \exp\{\tau \overline{V}\pi(g)|_{g=I}\}\pi(g)\} \geqq 0 .$$

Proof of Lemma 4.2: First we note that there exists a family of probability measures $R_\varepsilon(dg)$ all having support in a compact neighborhood of I and such that

$$(4.38) \qquad (\overline{V}f)(I) = \lim_{\varepsilon \downarrow 0} \frac{1}{\varepsilon} \int [f(g) - f(I)]R_\varepsilon(dg) .$$

With $\varepsilon > 0$ fixed, define the operator $\overline{V}_\varepsilon$ by

$$(4.39) \qquad (\overline{V}_\varepsilon f)(g') = \frac{1}{\varepsilon} \int [f(gg') - f(g')]R_\varepsilon(dg) .$$

$\overline{V}_\varepsilon$ generates a Markov semigroup of Poisson type; for any function $f(g)$,

$$(e^{\overline{V}_\varepsilon \tau}f)(g') = \int f(gg')P_\varepsilon(\tau, dg) = \sum_{n=0}^{\infty} \frac{\tau^n}{n!}(\overline{V}_\varepsilon^n f)(g')$$

$$(4.40)$$

$$= \sum_{n=0}^{\infty} e^{-\tau/\varepsilon} \frac{(\tau/\varepsilon)^n}{n!} E\{f(g_n g_{n-1} \cdots g_1 g')\} .$$

Here $g_1, g_2, \cdots, g_n$ are independent random elements in G with common distribution $R_\varepsilon(dg)$.

If $\pi(g)$ is any finite-dimensional representation, then

$$(4.41) \qquad \sup_{g \in K} \|\pi(g)\|_E \leqq M < \infty \, ,$$

where K is the support of the measures $R_\varepsilon(dg)$. Clearly then, if f is replaced by π,

$$(4.42) \qquad \|(e^{V_\varepsilon \tau} \pi)(g')\|_E \leqq \|\pi(g')\|_E \sum_{n=0}^{\infty} e^{-\tau/\varepsilon} \left(\frac{\tau}{\varepsilon}\right)^n \frac{M^n}{n!} < \infty \, , \qquad g' \in K \, .$$

In view of (4.42) we find, by direct computation, that (4.31) holds with V_ε in place of V. But then (4.37), with V_ε in place of V, is just Schwartz's inequality:

$$(4.43) \qquad 0 \leqq \|E\{\pi(g_\varepsilon(\tau))\}\|_E^2 \leqq E\{\|\pi(g_\varepsilon(\tau))\|_E^2\} \, .$$

Here $g_\varepsilon(\tau)$ denotes the Markov process corresponding to V_ε and such that $g_\varepsilon(0) = I$. On letting $\varepsilon \to 0$ in (4.43) (which is (4.37) with V replaced by V_ε) we obtain the desired result.

Proof of Theorem 4:   Define

$$(4.44) \qquad \begin{aligned} U(\tau, g) &= \exp \{\tau V \pi(g)|_{g=I}\} \pi(g) \, , \\ W(\tau, g) &= \exp \{\tau V \pi(g) \otimes \pi(g)|_{g=I}\} \pi(g) \otimes \pi(g) \, , \\ w(\tau, g) &= H(W(\tau, g)) \, . \end{aligned}$$

From Lemma 4.2 we have

$$(4.45) \qquad w(\tau, g) \geqq \|U(\tau, g)\|_E^2 \geqq 0 \, .$$

Recalling that V is a right invariant differential operator and that $\pi(g)$ is a representation, we find by direct computation that

$$(4.46) \qquad \frac{\partial \theta(\tau, g)}{\partial \tau} = V \theta(\tau, g) \, , \qquad\qquad \tau > 0 \, ,$$

where θ is either w or any entry of U or W. Let K be any compact neighborhood of I and let σ_K be the first exit time for the process $g(\tau)$ corresponding to V with $g(0) = I$. Then for any $\tau \geqq 0$ we have, by Itô's formula,

$$(4.47) \quad \theta(\tau, I) = E\{\theta(0, g(\tau)), \tau \leqq \sigma_K\} + E\{\theta(\tau - \sigma_K, g(\sigma_K)), \tau > \sigma_K\} \, .$$

Taking $\theta = w$ in (4.47) and using (4.45) yields the inequality

$$(4.48) \qquad w(\tau, I) \geqq E\{w(0, g(\tau)), \tau \leqq \sigma_K\} .$$

We now let $K \uparrow G$, use Lemma 4.1, and Fatou's lemma, and obtain

$$(4.49) \qquad \infty > w(\tau, I) \geqq E\{W(0, g(\tau))\} = E\{\|\pi(g(\tau))\|_E^2\} .$$

Therefore, $E\{\|\pi(g(\tau))\|_E\} < \infty$.

For any entry $\pi_{ij}(g)$ of $\pi(g)$ we have

$$(4.50) \qquad \left|\int \pi_{ij}(g'g) P_\tau(dg')\right| \leqq \int |\pi_{ij}(g'g)| \, P_\tau(dg')$$

$$\leqq \left(\int \pi_{ij}^2(g'g) \, p_\tau(dg')\right)^{1/2} \leqq w^{1/2}(\tau, g) < \infty .$$

Thus,

$$(4.51) \qquad E\{\pi_{ij}(g(\tau))\} = \int \pi_{ij}(g) \, p_\tau(dg)$$

is finite.

We shall now show that (4.31) holds, i.e.,

$$(4.52) \qquad U(\tau, I) = E\{\pi(g(\tau))\} .$$

For this purpose we let $K \uparrow G$ in (4.47) with θ any of the entries of U. Since $P\{\tau \leqq \sigma_K\} \to 1$, $K \uparrow G$, and $\pi(g)$ is integrable with respect to $p_\tau(dg)$, the first term on the right side of (4.47) converges to

$$\int \pi(g) \, p_\tau(dg) .$$

It is therefore sufficient to show that if θ is any entry of U, then

$$(4.53) \qquad E\{\theta(\tau - \sigma_K, g(\sigma_K)), \tau > \sigma_K\} \to 0 \qquad \text{as} \qquad K \uparrow G .$$

But

$$(4.54) \qquad |E\{\theta(\tau - \sigma_K, g(\sigma_K)), \tau > \sigma_K\}|^2$$

$$\leqq E\{|\theta(\tau - \sigma_K, g(\sigma_K))|^2, \tau > \sigma_K\}P\{\tau > \sigma_K\}$$

$$\leqq E\{w(\tau - \sigma_K, g(\sigma_K)), \tau > \sigma_K\}P\{\tau > \sigma_K\}$$

$$\leqq w(\tau, I)P\{\tau > \sigma_K\} \to 0 , \qquad K \uparrow G .$$

The last inequality in (4.54) follows from (4.47), with $\theta = w$, in the same manner that (4.48) was obtained. This completes the proof of Theorem 4.

The integrability of $\pi(g)$ and hence (4.31) can also be shown by the following simple argument due to D. Stroock. Recall the definition of the sets

$$K_1, K_2, \cdots$$

in (4.33) and the first exit times $\sigma_1, \sigma_2, \sigma_3, \cdots$ from these sets. From (4.36) it follows that $P\{\sigma_k \leqq T\} \leqq (P\{\sigma_1 \leqq T\})^k$, $k = 1, 2, \cdots$. Let $M = \sup_{g \in K_1} \|\pi(g)\|$ and choose $T > 0$ such that $P\{\sigma_1 \leqq T\} \leqq 1/2M$. Then, with $\sigma_0 \equiv 0$, $g(0) = I$, and the representation property of $\pi(g)$,

$$E\{\|\pi(g(T))\|\} = \sum_{k=1}^{\infty} E\{\|\pi(g(T))\|, \sigma_{k-1} \leqq T < \sigma_k\}$$

$$\leqq \sum_{k=1}^{\infty} \frac{M^k}{(2M)^{k-1}} = 2M.$$

From the representation property again it follows that, for any $0 \leqq \tau < \infty$,

$$E\{\|\pi(g(\tau))\|\} \leqq (2M)^n, \qquad\qquad n = [\tau/T] + 1.$$

This proves the integrability of $\pi(g)$ from which (4.31) follows easily.

Let us comment briefly on the role of the term Ag in (4.6). Because of this term the process $m(t)$ in (4.9) is not stationary and one must allow for the generality we have here. Many applications, for example in [8], [9], present problems in the form (4.6). Frequently the presence of A, which is called the fast varying part of the process, simplifies the analysis of (4.13) considerably (see [8], [9]).

A central limit theorem on Lie groups due to Wehn is presented in [24]. A recent treatment of limit theorems on Lie groups is [25].

Bibliography

[1] Papanicolaou, G. C., and Hersh, R., *Some limit theorems for stochastic equations and applications*, Indiana Univ. Math. J., Vol. 21, 1972, pp. 815–840.

[2] Khasminskii, R. Z., *A limit theorem for the solutions of differential equations with random right-hand sides*, Theory Prob. Applications, Vol. 11, 1966, pp. 390–406.

[3] Stratonovich, R. L., *Conditional Markov Processes and their Application to the Theory of Optimal Control*, Elsevier, New York, 1968.

[4] Kubo, R., *Stochastic Liouville equation*, J. Math. Phys., Vol. 4, 1963, pp. 174–183.

[5] Lax, M., *Classical noise IV; Langevin methods*, Rev. Modern Phys., Vol. 38, 1966, 561–566.

[6] Papanicolaou, G., and Keller, J. B., *Stochastic differential equations with applications to random harmonic oscillators and wave propagation in random media*, SIAM J. Appl. Math., Vol. 20, 1971, pp. 287–305.

[7] Cogburn, R., and Hersh, R., *Two limit theorems for random equations*, Indiana Univ. Math. Jour., Vol. 22, 1973, pp. 1067–1089.

[8] Burridge, R., and Papanicolaou, G., *The geometry of coupled mode propagation in one dimensional random media*, Comm. Pure Appl. Math., Vol. 25, 1972, pp. 715–757.

[9] Papanicolaou, G., *A kinetic theory for power transfer in stochastic systems*, J. Math. Phys., Vol. 13, 1972, pp. 1912–1918.

[10] Griego, R., and Hersh, R., *Theory and application of random evolutions*, Trans. Amer. Math. Soc., Vol. 156, 1971, pp. 405–418.

[11] Hersh, R., and Papanicolaou, G., *Non-commuting random evolution and an operator valued Feynman-Kac formula*, Comm. Pure Appl. Math., Vol. 25, 1972, pp. 337–367.

[12] Pinsky, M., *Multiplicative operator functionals and their asymptotic properties*, in Advances in Probability, to appear.

[13] Kertz, R., *Northwestern University Thesis*, 1972.

[14] Hille, E., and Phillips, R., *Functional Analysis and Semigroups*, Am. Math. Soc. Coll. Publ. Vol. XXXI, Providence, R.I., 1957.

[15] Loève, M., *Probability Theory*, Van Nostrand, Princeton, 1963.

[16] Doob, J. L., *Stochastic Processes*, J. Wiley, New York, 1953.

[17] Oleinik, O., *Alcuni risultati sulle equazioni lineari e quasi lineari ellitticoparaboliche a derivate parziali del secondo ordine*, Rendiconti Accad. Naz., Vol. 40, 1966, pp. 775–784.

[18] Hunt, G. A., *Semigroups of measures on Lie groups*, Trans. Am. Math. Soc., Vol. 31, 1956, pp. 264–293.

[19] Itô, K., *Stochastic differential equations in a differentiable manifold*, Nagoya Math. J., Vol. 1, 1950, pp. 35–47.

[20] Itô, K., *Brownian Motion in a Lie Group*, Proc. Japan Acad., Vol. 26 (8), 1950, pp. 4–10.

[21] McKean, H. P. Jr., *Stochastic Integrals*, Academic Press, New York, 1969.

[22] Helgason, S., *Differential Geometry and Symmetric Spaces*, Academic Press, New York, 1962.

[23] Stroock, D. S., and Varadhan, S. R. S., *On degenerate elliptic-parabolic operators of second order and their associated diffusions*, Comm. Pure Appl. Math., Vol. 25, 1972, pp. 651–714.

[24] Grenander, U., *Probabilities on Algebraic Structures*, J. Wiley and Sons, New York, 1963.

[25] Stroock, D. S., and Varadhan, S. R. S., *Limit theorems on Lie groups*, to appear.

Received February, 1973.

Sankhya Ser. A 35 (1973) 277-294

LIMIT THEOREMS FOR RANDOM WALKS ON LIE GROUPS*

By DANIEL W. STROOCK

University of Colorado

and

S. R. S. VARADHAN

New York University, Courant Institute of Mathematical Sciences

SUMMARY. Limit Theorems for Random walks on Lie groups are prooved. Necessary and sufficient conditions analogous to Lindeberg's conditions are obtained.

1. INTRODUCTION

The aim of this paper is to prove the analogue of the classical central limit theorem under Lindeberg's conditions, when the random variables take values in a Lie group. Actually we consider the convergence of the associated random walks to the corresponding "Gaussian" process with independent increments. There have been various generalizations arrived at earlier. For instance the locally compact abelian group has been treated in Parthasarathy, Rao and Varadhan (1963). One can find some partial results in the noncommutative case in Parthasarathy (1964) and Wehn (1962). A survey of the results known until then appears in Sazanov and Tutubalin (1966). But there seems to be no complete theory.

In Stroock and Varadhan (1969b) we developed methods for examining the convergence of Markov chains to diffusion processes. These methods are applicable in the current context. One can exploit the special situation at hand and obtain complete results. We determine all "Gaussian" processes with independent increments on a Lie group G and obtain necessary and sufficient conditions for the convergence of random walks to any given "Gaussian" process with independent increments. Unlike the earlier works cited, the techniques do not require an intimate knowledge of the Lie group. In particular the representation theory is not used at all.

2. NOTATION AND SUMMARY

Let G be a connected Lie group of dimension d. A continuous process with independent left increments is a stochastic process $X(t) : 0 \leqslant t \leqslant T$, with values in G, satisfying the following.

A1. $X(0) = e$ with probability 1.

A2.   For $0 \leqslant t_1 \leqslant t_2 \leqslant ... \leqslant t_n \leqslant T$ and every n, the increments $X(t_1)$, $[X(t_1)]^{-1}X(t_2), ..., [X(t_j)]^{-1}X(t_{j+1}), ...$ are mutually independent.

---

* The research in this paper was supported by U.S. Air Force, Grant AF-AFOSR-72-2307.

A3. $X(t)$ is a continuous function of t with probability 1. Such a process will be said to be homogeneous in time if in addition to A1, A2 and A3 the process satisfies :

A4. The distribution of the increment $[X(t_1)]^{-1}X(t_2)$ for $0 \leqslant t_1 \leqslant t_2 \leqslant T$, as a function of t_1 and t_2 depends only on the difference $t_2 - t_1$.

Let Ω_c denote the space of continuous functions on $[0\ T]$ with values in G endowed with uniform convergence and the standard Borel structure. Let $x(s)$ denote the evaluation map at time s. Let M_t be the σ-field generated by $x(s)$ for $0 \leqslant s \leqslant t$. The total σ-field M is of course M_T. Let $\mathscr{M}$ denote the totality of all probability measures on $[\Omega_c, M]$. The convergence notion on $\mathscr{M}$ will be weak convergence. Processes satisfying assumptions A1, A2 and A3 will form a closed subset of $\mathscr{P}$ of $\mathscr{M}$. Those that satisfy in addition A4, will be a closed subset $\mathscr{P}_h$ of $\mathscr{P}$. One of the aims of this paper is to describe all the elements of $\mathscr{P}$ and parametrize them in a canonical fashion.

The theorem of Hunt (1956) describes $\mathscr{P}_h$ in a natural fashion. Let $\mathscr{G}$ be the Lie algebra of G. We shall view the elements ξ of $\mathscr{G}$ as left invariant vector fields on G. They act on the space $\mathscr{S}$ of C^∞ functions on $\mathscr{G}$ with compact support.

$$(\xi f)(x) = \frac{d}{dt} f(xe^{t\xi})\bigg|_{t=0} \qquad \ldots \ (2.1)$$

Let $\xi_1, \xi_2, \ldots, \xi_d$ be a basis for $\mathscr{G}$. Let $a = \{a_{ij}\}$ be any $d \times d$ symmetric nonnegative definite matrix and b_j any d vector. Let us denote by $L_{a,b}$ the operator

$$L_{a,b} = \frac{1}{2} \Sigma\, a_{ij}\xi_i\xi_j + \Sigma b_j\xi_j.$$

Then $L_{a,b}$ is the infinitesimal generator of a homogeneous Markov process, which is also a process with independent left increments. Fixing the starting point always at the identity e of the group G these processes exhaust all of $\mathscr{P}_h$ in a one to one manner as a and b are allowed to vary over their respective classes.

In some sense if one chose a and b to be time dependent, the processes corresponding to these should exhaust the bulk of $\mathscr{P}$. In the case of the additive group of real numbers the most general element of $\mathscr{P}$ is characterized by the variance function $\sigma^2(t)$ which is an arbitrary continuous nondecreasing function and the mean function $m(t)$ which is an arbitrary continuous function. The relation to a and b is

$$\sigma^2(t) = \int_0^t a(s)\,ds$$

$$m(t) = \int_0^t b(s)\,ds.$$

This shows that if $a(\,\cdot\,)$ and $b(\,\cdot\,)$ are chosen to be arbitrary integrable choices, then one gets most of $\mathscr{P}$ but not all of it. Hence by analogy, we define the following :

LIMIT THEOREMS FOR RANDOM WALKS ON LIE GROUPS

(i) Let V be the totality of all continuous mappings $A(t)$ from $[0, T]$ into nonnegative definite symmetric matrices such that $A(t) \geqslant A(s)$ for $t \geqslant s$ and $A(0) = 0$.

(ii) V_0 consists of absolutely continuous elements of V.

(iii) $\mathscr{F}$ consists of all continuous maps of $[0, T]$ into G.

(iv) $\mathscr{F}_0$ consists of absolutely continuous elements of $\mathscr{F}$.   ... (2.2)

The relevant topologies on V and $\mathscr{F}$ are those induced by uniform convergence of the elements. Let us take a function $A(t)$ in V_0 and a function $g(t) \epsilon \mathscr{F}_0$. We shall write

$$A(t) = \int_0^t a(s)\, ds$$

$$\dot{g}(t) = \Sigma b_j(t)\, \xi_j(g(t)). \qquad \qquad ...\ (2.3)$$

Corresponding to this we shall write down the family of operators L_s, $0 \leqslant s \leqslant T$ given by

$$L_s = \frac{1}{2}\, \Sigma a_{ij}(s)\xi_i\xi_j + \Sigma b_j(s)\xi_j.$$

We define a measure P on Ω_c by the relation

(i) $P\{x(0) = e\} = 1$

(ii) For each $f \epsilon \mathcal{S}$

$$f(x(t)) - \int_0^t (L_s f)(x(s))\, ds \qquad \qquad ...\ (2.4)$$

is a martingale relative to $[\Omega_c, M_t, P]$ for $0 \leqslant t \leqslant T$.

This determines a unique measure $P = P_{A(\cdot),\, g(\cdot)}$ for $A(\cdot) \epsilon V_0$ and $g(\cdot) \epsilon \mathscr{F}_0$. We have the following.

Theorem 2.1 : *The map* $(A(\cdot), g(\cdot)) \to P_{A(\cdot),\, g(\cdot)}$ *from* $V_0 \times \mathscr{F}_0 \to \mathcal{P}$ *extends by continuity into a map from all of* $V \times \mathscr{F} \to \mathcal{P}$, *which is one to one, onto and bicontinuous.*

This is the analog of the description in the simplest case of Gaussian processes with independent increments in terms of their mean functions and covariance functions. In such a case, a process $X(t)$ with variance $\sigma^2(t)$ and mean $m(t)$ can be written as

$$X(t) = Y(t) + m(t)$$

where $Y(t)$ is a process with mean 0 and variance $\sigma^2(t)$. The analog of this decomposition will yield a slightly better description of $\mathcal{P}$ than that provided by Theorem 2.1.

For each $A(\cdot) \epsilon V$, let us associate directly a process Q as follows :

(i) $Q[x(0) = e] = 1$

(ii) For each $f \epsilon \mathcal{S}$

$$f(x(t)) = \frac{1}{2}\, \Sigma \int_0^t (\xi_i \xi_j f)(x(s))\, dA_{ij}(s) \qquad \qquad ...\ (2.5)$$

is a martingale relative to $[\Omega_c, M_t, Q]$ for $0 \leqslant t \leqslant T$.

SANKHYĀ : THE INDIAN JOURNAL OF STATISTICS : Series A

We can show that $Q = Q_{A(\cdot)}$ exists and is unique. If $e(\cdot)$ is the element of $\mathscr{F}_0$ identically equal to e then one can easily identify $Q_{A(\cdot)}$, with $P_{A(\cdot),\,e(\cdot)}$ [if $A(\cdot)$ happens to be an element of $\mathscr{F}_0$]. If $X(t)$ is a process in $\mathscr{P}$ then so is $X(t)\,g(t)$ where $g(t)$ is an arbitrary element of $\mathscr{F}$. If $X(t)$ has distribution $Q_{A(\cdot)}$ we shall denote by $Q_{A(\cdot),\,g(\cdot)}$ the distribution of $X(t)\cdot g(t)$. We then have

Theorem 2.2 : *The map* $(A(\cdot),\,g(\cdot)) \to Q_{A(\cdot),\,g(\cdot)}$ *from* $V \times \mathscr{F} \to \mathscr{P}$ *is bicontinuous, one to one and onto.*

This of course leaves the question of reconciling the two parametrizations. Let $g \in G$ be arbitrary. Consider the induced inner automorphism

$$x \to gxg^{-1}$$

of G onto G. This in turn induces an automorphism Ad_g of the Lie algebra $\mathscr{G}$ on $\mathscr{G}$ by

$$e^{t(A_d{}^g)\xi} = ge^{t\xi}g^{-1}.$$

We can express this linear transformation by a matrix $\tau_{ij}(g)$ where

$$Ad_g\xi_j = \Sigma\tau_{ij}(g)\xi_j.$$

The following theorem links $P_{A(\cdot),\,g(\cdot)}$ to $Q_{B(\cdot),\,h(\cdot)}$.

Theorem 2.3 : $P_{A(\cdot),\,g(\cdot)} = Q_{B(\cdot),\,h(\cdot)}$ *if and only if*

(i) $g(t) \equiv h(t)$ *for all* $t \in [0, T]$.

and

(ii) $B_{rl}(t) = \Sigma \int_0^t \tau_{ir}(g(s))\tau_{jl}(g(s))\,dA_{ij}(s).$

We next turn to the convergence of random walks to processes with independent increments. For each n let k_n independent random variables $X_{n1},\,X_{n2},\,\ldots X_{nk_n}$ with values in G be given. Let $\alpha_{n1},\,\ldots,\,\alpha_{nk_n}$ be their respective distributions. For each n, divide the interval $[0, T]$ into k_n+1 parts with $0 = t_{n0} < t_{n1} < t_{n2} < \ldots < t_{nk_n} < t_{n,k_n+1} = T$ being the points of division. We assume that $\sup\limits_{0\,\leqslant\,j\,\leqslant\,k_n}|t_{nj}-t_{nj+1}|$ tends to zero with n.

Let us construct a process $X_n(t)$ by the following recipe :

$$
\begin{aligned}
X_n(t) &= e & &\text{for } 0 \leqslant t < t_{n1} \\
&= X_{n1}X_{n2}\ldots X_{nj} & &\text{for } t_{nj} \leqslant t < t_{n,j+1},\ \ j = 1, 2, \ldots, k_{n-1} \\
&= X_{n1}X_{n2}\ldots X_{nk_n} & &\text{for } t_{n,k_n} \leqslant t \leqslant T.
\end{aligned}
$$

Let Ω_d be the space of functions on $[0, T]$ with values in G which have discontinuities only of the first kind. The space is provided with the Skorohod topology. The process $X_n(t)$ induces a measure P_n on the space Ω_d. Since $\Omega_c \subset \Omega_d$ we can view any measure

LIMIT THEOREMS FOR RANDOM WALKS ON LIE GROUPS

$Pe\mathcal{P}$ as a measure on Ω_d concentrated on Ω_c. Given such a P we want to find the necessary and sufficient conditions for P_n to converge weakly to P. One necessary condition is obviously the Lindeberg condition

$$\lim_{n \to \infty} \sum_{j=1}^{k_n} \alpha_{nj}(U^c) = 0 \qquad \ldots \ (2.6)$$

for any neighbourhood U of e in G.

Let us take a coordinate system $\phi = (\phi_1, \ldots, \phi_d)$ given in a neighbourhood N of e. We shall assume that the coordinate system satisfies

and
(i) $\phi(e) = 0$

(ii) $(\xi_i \phi_j)(e) = 1$ for $i = j$ and 0 for $i \neq j$. $\qquad \ldots \ (2.7)$

Let us take a neighbourhood V with a compact closure contained in N and define the mean and covariance of α_{nj}. The mean g_{nj} is defined by

$$\phi(g_{nj}) = \int_V \phi(x)\, \alpha_{nj}(dx). \qquad \ldots \ (2.8)$$

It is well defined because (2.7) implies

$$\lim_{n \to \infty} \sup_{1 \leqslant j \leqslant k_n} \alpha_{nj}(U^c) = 0 \qquad \ldots \ (2.9)$$

for any neighbourhood U of e.

The covariance matrix a_{pq}^{nj} is defined by

$$a_{pq}^{nj} = \int_V [\phi_p(x) - \phi_p(g_{nj})][\phi_q(x) - \phi_q(g_{nj})]\alpha_{nj}(dx). \qquad \ldots \ (2.10)$$

We next define the cumulated means and variances by

and
$$h_{nj} = g_{n1} g_{n2} \ldots g_{nj} \qquad \ldots \ (2.11)$$

$$A^{nj} = a^{n1} + a^{n2} + \ldots + a^{nj}. \qquad \ldots \ (2.12)$$

We superimpose these on the time scale by defining

$$\tau_n(t) = 0 \qquad \text{for } 0 \leqslant t \leqslant t_{n1}$$
$$= j \qquad \text{for } t_{nj} \leqslant t < t_{n,j+1}, \ j = 1, \ldots, k_n - 1$$
$$= k_n \qquad \text{for } t_{nk_n} \leqslant t \leqslant T, \qquad \ldots \ (2.13)$$

$$h_n(t) = h_{n,\tau_n}(t), \qquad \ldots \ (2.14)$$

and

$$A^n(t) = A^{n,\tau_n(t)}. \qquad \ldots \ (2.15)$$

Given an element $Pe\mathcal{P}$, according to Theorem 2.1, we can write it in the form $P_{A(\cdot),\, g(\cdot)}$. In order to state the condition on $A^n(t)$ and $h_n(t)$ so that P_n may converge

SANKHYĀ : THE INDIAN JOURNAL OF STATISTICS : Series A

to P we need the following considerations. The means g_{nj} are sensitive to the coordinate system used. Equations (2.7) alone are not enough to determine g_{nj} uniquely. The representation $P_{A(.),g(.)}$ of P on the other hand is canonical. So if the means $h_n(t)$ in the ϕ coordinates converge to $h(t)$, there is a rule which connects $h(t)$ with $g(t)$. Let us define

$$\rho_i^{pq} = (\xi_p \xi_q \phi_i)(e) \qquad \ldots \text{(2.16)}$$

$$b_i(t) = -\frac{1}{2} \Sigma \, \rho_i^{pq} A_{pq}(t). \qquad \ldots \text{(2.17)}$$

Given $h(t)$, consider the solution $k(t)$ of the equation

$$dk(t) = \Sigma \, db_i(t) \tau_{il}(h(s)) \xi_l(k(t))$$
$$k(0) = e. \qquad \ldots \text{(2.18)}$$

Then

$$g(t) = k(t) \, h(t). \qquad \ldots \text{(2.19)}$$

To invert this, given $g(t)$, consider the solution $y(t)$ of the equation

$$dy(t) = -\Sigma \, db_i(t) \tau_{il}(g(t)) \, \xi_l(y(t))$$
$$y(0) = e. \qquad \ldots \text{(2.20)}$$

Then

$$h(t) = y(t) \, g(t). \qquad \ldots \text{(2.21)}$$

One can verify that $y(t) = k^{-1}(t)$. Indeed, one has only to remember the formula for left invariant differentials

$$dk^{-1}(t) = -\tau(k(t)) \, dk(t).$$

Therefore (2.20)-(2.21) and (2.18)-(2.19) express the same relation. It should be pointed out that if one were to use the canonical exponential coordinates, then $\rho_i^{pq} + \rho_i^{qp} = 0$ for all i, p and q. Therefore $h(t) \equiv g(t)$ and things are much simpler. We are now ready to state the theorem on the convergence of P_n.

Theorem 2.4 : *In order that P_n may converge to $P_{A(.),g(.)}$ it is necessary and sufficient that besides (2.6) the following hold.*

$$\lim_{\to \infty} A^n(t) = A(t) \ \text{uniformly on} \ [0, T] \qquad \ldots \text{(2.22)}$$

$$\lim_{n \to \infty} h_n(t) = h(t) \ \text{uniformly on} \ [0, T]. \qquad \ldots \text{(2.23)}$$

$h(t)$ and $g(t)$ *are related by* (2.18)-(2.19) *or equivalently by* (2.20)-(2.21).

3. Proofs

The first order of business is to describe how for any given set of integrable functions $\{a_{ij}(s)\}$ and $\{b_j(s)\}$ on $[0, T]$, one associates a process with independent increments. Recall that we define the operator

$$L_s = \frac{1}{2} \Sigma a_{ij}(s) \xi_i \xi_j + \Sigma b_j(s) \xi_j$$

LIMIT THEOREMS FOR RANDOM WALKS ON LIE GROUPS

and then look for a measure P on Ω_c such that

$$P[x(0) = e] = 1$$

and for each $f \epsilon \mathcal{S}$

$$f(x(t)) = \int_0^t (L_s f)(x(s)) \, ds \qquad \qquad \ldots (3.1)$$

is a martingale relative to $[\Omega_c, M_t, P]$ for $0 \leqslant t \leqslant T$.

To construct such a process globally on the manifold we construct the process locally and piece together. Let U be a neighbourhood of e in G, which admits a coordinate system and which has in addition a compact closure. Let V, W be neighbourhoods such that

$$V \subset \overline{V} \subset W \subset \overline{W} \subset U.$$

Let $y_1, y_2, \ldots, y_d$ be the coordinate system and let $\widetilde{V}$, $\widetilde{W}$ and $\widetilde{U}$ be the respective images of V, W and U in R^d. Choose ψ to be a smooth nonnegative cut-off function which is 1 on $\widetilde{V}$ and zero off $\widetilde{W}$. We can write L_s in local coordinates as

$$\tilde{L}_s = \frac{1}{2} \Sigma \tilde{a}_{ij}(s, y) \frac{\partial^2}{\partial y_i \partial y_j} + \Sigma \tilde{b}_j(s, y) \frac{\partial}{\partial y_j}$$

valid in the neighbourhood $\widetilde{U}$ of 0, which is assumed to be the image of e under the coordinate map. Since ξ_i are invariant operators, the $\tilde{a}_{ij}$ and $\tilde{b}_j$ are smooth functions of y for each fixed s. Thus $\psi \tilde{a}_{ij}$ and $\psi \tilde{b}_j$ are global C^∞ functions on R^d. Set

$$\hat{L} = \psi L_s = \frac{1}{2} \Sigma \hat{a}_{ij}(s, y) \frac{\partial^2}{\partial y_i \partial y_j} + \Sigma \hat{b}_j(s, y) \frac{\partial}{\partial y_j} .$$

We denote by $\varepsilon(s)$ the global bound in y for the $\hat{a}_{ij}$, their first and second order derivatives with respect to $y_1, \ldots, y_d$, $\hat{b}_j$ and their first order derivatives with respect to $y_1, \ldots, y_d$. We can estimate

$$\varepsilon(s) \leqslant C \Sigma[a_{ii}(s) + |b_i(s)|].$$

The constant C depends only on the neighbourhoods and coordinate system chosen. In the smaller neighbourhood $\widetilde{V}$, $\hat{L}_s$ is a representation of L_s in the given coordinates.

If we now take the positive semidefinite square root $\hat{\sigma}$ of $\hat{A}$, by a theorem of Phillips and Sarason (1967)

$$|\hat{\sigma}_{ij}(s, y) - \hat{\sigma}(s, y')| \leqslant C \sqrt{\varepsilon(s)} |y - y'|. \qquad \qquad \ldots (3.2)$$

In addition

$$|\hat{\sigma}_{ij}(s, y)| \leqslant C \sqrt{\varepsilon(s)}$$
$$|\hat{b}_j(s, y) - \hat{b}_j(s, y')| \leqslant C \varepsilon(s) |y - y'| \qquad \qquad \ldots (3.3)$$
$$|\hat{b}_j(s, y)| \leqslant C \varepsilon(s).$$

SANKHYĀ : THE INDIAN JOURNAL OF STATISTICS : Series A

In view of (3.2) and (3.3) the Itô equations

$$dy(t) = \hat{\sigma}(t, y(t))\, d\beta(t) + \hat{b}(t, y(t))dt, \ t \geqslant t_0$$
$$y(t_0) = y_0 \qquad \qquad \dots (3.4)$$

have a unique solution for any starting point y_0 and starting time t_0. By a theorem of Watanabe and Yamada [Proposition 1 on page 158, 1971] this implies the uniqueness of the solution to the martingale problem for $\hat{L}_s$. By the results in Stroock and Varadhan (1969a) (specifically Lemma 5.6 on page 386), this in turn implies that the solution to the martingale problem

$$\begin{cases} P\{x(t_0) = e\} = 1 \\[2mm] \text{for } f \in \mathcal{S} \\[2mm] f(x(t)) - \int\limits_{t_0}^{t} (L_s f)(x(s))\, ds, \ t_0 \leqslant t \leqslant T. \\[2mm] \text{is a martingale relative to } [\Omega_c, M_t^{t_0}, P] \end{cases} \qquad \dots (3.5)$$

exists and is unique until the first exit time $\tau_U^{t_0}$ from the set U after time t_0. If the starting point is any other point g of G, then the same observations hold by just translating by g on the left. The neighbourhood U_g of g will be gU. The same coordinates and the same cut-off function can be used. The resulting process is stochastically equivalent to $gx(t)$ where $x(t)$ is the process starting at the identity. We can now piece together the local solutions in the manner described in Stroock and Varadhan (1969a, Lemma 3.6 and Theorem 3.4 on pages 367-368), and we get a unique solution until the explosion time. That is, if K is a compact set and τ_K is the first exit time from K then the explosion time is

$$\zeta = \sup_K \tau_K.$$

We have to show that the explosion cannot occur prior to time T. In the compact case this is obvious. In general one has to make an argument. Let us define U_n to be the set of all elements which can be expressed as the product of n elements each of which is from U. Of course U_1 is the same as U. Let τ_n^s be the exit time from U_n for the process starting at the identity at time s. If the process does not exit before time T, τ_n^s is taken to be $+\infty$. Showing that the explosion time is infinite amounts to

$$\lim_{n \to \infty} E\left[e^{-\tau_n^s}\right] = 0 \quad \text{for } 0 \leqslant s \leqslant T. \qquad \dots (3.6)$$

We notice the obvious Markov inequality

$$E[e^{-\tau_n^s}] \leqslant \rho\, E[e^{-\tau_{n-1}^s}] \qquad \dots (3.7)$$

where

$$\rho = \sup_{s \leqslant t \leqslant T} E[e^{-\left(\tau_1^t - 1\right)}].$$

LIMIT THEOREMS FOR RANDOM WALKS ON LIE GROUPS

This is obvious because in order to get out of U_n the process has to get out of U_{n-1} at some point g and then get out of gU. Therefore, proving (3.6), amounts to proving that $\rho < 1$.

Let ϕ be a function in $\mathcal{S}$ which is 1 at the identity and zero off U. Then

$$\phi(x(t)) - \int_{t_0}^{t} (L_s\phi)(x(s))\,ds \qquad \ldots (3.8)$$

is a martingale. Given the function ϕ let

$$\theta(s) = \sup_{x \in G} |(L_s\phi)(x)|. \qquad \ldots (3.9)$$

One can verify that $\theta(s)$ is integrable on $[0, T]$.

It follows from (3.8) and (3.9) that

$$\phi(x(t)) + \int_{t_0}^{t} \theta(s)\,ds$$

is a submartingale. It is easy to deduce from this, using Doob's stopping theorem for submartingales, that

$$\text{Prob } [\tau_1^t \leqslant t+h] \leqslant \int_{t}^{t+h} \theta(s)\,ds. \qquad \ldots (3.10)$$

From (3.10) it follows that $\rho < 1$. We have therefore proved

Lemma 3.1 : *Given $\{a_{ij}(s)\}$ and $\{b_j(s)\}$ which are integrable on $[0, T]$ the solution to the martingale problem* (3.1) *exists and is unique.*

Lemma 3.2 : *Given a sequence $\{a_{ij}^{(n)}(s)\}$ and $\{b_j^{(s)}(s)\}$ of integrable functions such that the family*

$$E^n(t) = \int_{0}^{t} \Sigma[a_{ii}^{(n)}(s) + |b_i^{(s)}(s)|]\,ds$$

is equicontinuous, the corresponding family of solutions $P^{(n)}$ to the martingale problem (3.1) *is weakly compact on Ω_c.*

Proof : In order to prove the compactness of $P^{(n)}$ we have to check that for every neighbourhood V of e

$$\lim_{\delta \downarrow 0} \sup_{0 \leqslant t \leqslant T} \sup_n P^{(n)}[\tau_V^t - t < \delta] = 0 \qquad \ldots (3.11)$$

and

$$\sup_n \sup_{0 \leqslant t \leqslant T} E^{P^{(n)}} \left[e^{-\left(\tau_1^t - t\right)} \right] < 1. \qquad \ldots (3.12)$$

It is clear that we can estimate

$$\theta^{(n)}(s) \leqslant C\Sigma[a_{ii}^{(n)}(s) + |b_i^{(n)}(s)|]$$

and therefore (3.10) enables one to prove (3.11) and (3.12).

Lemma 3.3 : *Let P correspond to the operator*

$$L_s = \frac{1}{2} \Sigma a_{ij}(s)\xi_i\xi_j + \Sigma b_j(s)\xi_j$$

and Q correspond to

$$L_s^* = \frac{1}{2} \Sigma c_{ij}(s)\xi_i\xi_j$$

where

$$c_{ij}(t) = \Sigma \tau_{pi}(g(t))\tau_{qj}(g(t))a_{pq}(t) \qquad \ldots (3.13)$$

and g(t) is the solution of

$$g(t) = \Sigma b_j(t)\xi_j(g(t)) : g(0) = e. \qquad \ldots (3.14)$$

If X(t) is a process having P as its distribution then x(t) can be represented as

$$X(t) = Y(t) g(t)$$

where Y(t) is distributed according to Q.

Proof : $Y(t)$ has the property that for $f\epsilon s$

$$f(Y(t)) - \int_0^t (L_s^*f)(Y(s)) ds$$

is a martingale. It is easily seen that this implies

$$f(t, Y(t)) - \int_0^t \left(\frac{\partial}{\partial s} + L_s^*\right)(f)(s, Y(s)) ds \qquad \ldots (3.15)$$

is a martingale for functions $f(s, x)$ which are smooth. Let $h(x)$ be a function in the class $\mathcal{S}$ and let us take

$$f(t, y) = h(y\, g(t))$$

substituting in (3.15) we get

$$R(Y(t)\, g(t)) - \int_0^t \Phi(s, Y(s)) ds \qquad \ldots (3.16)$$

is a martingale. Here Φ is computed in terms of h as

$$\Phi(s, y) = \left(\frac{\partial}{\partial s} + L_s^*\right) h(y\, g(s)).$$

We can rewrite the martingales (3.16) as

$$h(x(t)) - \int_0^t \psi(s, x(s)) ds \qquad \ldots (3.17)$$

where

$$X(t) = Y(t)\, g(t).$$

LIMIT THEOREMS FOR RANDOM WALKS ON LIE GROUPS

A straightforward computation of ψ in terms of h is seen to lead to

$$\psi(s, x) = (L_s h)(x). \qquad \qquad \dots (3.28)$$

One has to keep in mind the definition of $\tau_{ij}(g)$ in terms of the inner automorphisms. The formula (3.18) identifies $X(t)$ as having distribution P.

Lemma 3.4 : *Let $A(t)$ be an element of V. Then there is a unique measure Q on Ω_c such that*

(i) $Q[x(0) = e] = 1$

(ii) *for $f \in \mathcal{S}$,*

$$f(x(t)) - \frac{1}{2} \Sigma \int\limits_0^t (\xi_i \xi_j f)(x(s)) \, dA_{ij}(s), \quad 0 \leqslant t \leqslant T$$

is a martingale relative to $[\Omega_c, M_t, Q]$.

Proof : By a transformation of the time scale one can always reduce the case of arbitrary $\{A_{ij}(t)\}$ to one where all the functios $\{A_{ij}(t)\}$ are absolutely continuous functions of t. Then Lemma 3.4 is exactly the same as Lemma 3.1.

Let us denote by $Q_{A(\cdot)}$ the measure corresponding to $A(\cdot)$ provided by Lemma 3.4. Recall that if $x(t)$ is distributed according to $Q_{A(\cdot)}$ we denote by $Q_{A(\cdot), g(\cdot)}$ the distribution of $X(t) g(t)$.

Lemma 3.5 : *The mapping $(A(\cdot), g(\cdot)) \to Q_{A(\cdot), g(\cdot)}$ is a continuous map of $V \times \mathcal{F} \to \mathcal{P}$.*

Proof : It is sufficient to prove that $Q_{A(\cdot)}$ is a continuous map of $V \to \mathcal{P}$. It follows easily from Lemma 3.2 that if $A_{n(\cdot)} \to A(\cdot)$ uniformly then $Q_{A^{n}(\cdot)}$ is compact.

It is then obvious that any limit point Q is characterized by Lemma 3.4 to be $Q_{A(\cdot)}$.

Lemma 3.6 : *The map $(A(\cdot), g(\cdot)) \to P_{A(\cdot), g(\cdot)}$ from $V_0 \times \mathcal{F}_0 \to \mathcal{P}$ defined by (2.3) and Lemma 3.1 extends by continuity to a map of $V \times \mathcal{F} \to \mathcal{P}$. Moreover*

$$P_{A(\cdot), g(\cdot)} = Q_{B(\cdot), g(\cdot)}$$

provided

$$B_{k, l}(t) = \Sigma \int\limits_0^t \tau_{ir}(g(s)) \, \tau_{jl}(g(s)) \, dA_{ij}(s).$$

Proof : This lemma follows immediately from Lemmas 3.3 and 3.5.

We now turn to the problem of proving the convergence of random walks to processes with independent increments. To simplify the notation we take $T = 1$. We shall also assume that for each n we are given $n-1$ random variables $X_{n1}, X_{n2}, \dots, X_{n, n-1}$ with values in G which are mutually independent. We construct a process $X_n(t)$ by defining

$$X_n(t) = X_{n1} X_{n2} \dots X_{n, j_n}(t)$$

where

$$j_n(t) = \begin{array}{ll} [nt] & \text{for } 0 \leqslant t < 1 \\ n-1 & \text{for } t = 1. \end{array}$$

SANKHYĀ : THE INDIAN JOURNAL OF STATISTICS : Series A

Although this is a special situation, the proof for the general case described in Section 2 is only different in terms of notation. The reason for not taking $j_n(1) = n$, is that convention dictates the functions in Ω_d be left continuous at the right end. We have the means g_{nj} defined by (2.8) and the covariances a_{pq}^{nj} defined by (2.9) . Let us define

$$X'_{nj} = X_{nj}g_{nj}^{-1} \qquad \ldots \text{(3.19)}$$

and

$$Y_{nj} = h_{n,j-1}X'_{nj}h_{n.j-1}^{-1}. \qquad \ldots \text{(3.20)}$$

Let α_{nj} be the distribution of X_{nj} and β_{nj} the distribution of Y_{nj}. Define the process $Y_n(t)$ by

$$Y_n(t) = Y_{n1}Y_{n2}\ldots Y_{nj_n}(t). \qquad \ldots \text{(3.21)}$$

Let us recall that

$$h_{nj} = g_{n1}\,g_{n2}\ldots g_{nj} \qquad \ldots \text{(3.22)}$$

and

$$h_n(t) = h_{nj_n}(t). \qquad \ldots \text{(3.23)}$$

One can easily verify that

$$X_n(t) = Y_n(t)h_n(t).. \qquad \ldots \text{(3.24)}$$

Let us also remember that

$$A^{nj} = a^{n1} + \ldots + a^{nj}, \qquad \ldots \text{(3.25)}$$

$$A^n(t) = A^{nj_n(t)} \qquad \ldots \text{(3.26)}$$

and Q_n, P_n are the distributions of $Y_n(\cdot)$, $X_n(\cdot)$ respectively We will now assume that

$$\lim_{n \to \infty} A^n(t) = A(t) \qquad \text{uniformly on } [0, T]$$

and

$$\lim_{n \to \infty} h_n(t) = h(t) \qquad \text{uniformly on } [0, T]. \qquad \ldots \text{(3.27)}$$

Lemma 3.7 : *Let (3.27) and (2.7) hold. Define the operators*

$$(\delta_{nj}f)(x) = \int [f(xy) - f(x)]\beta_{nj}(dy)$$

and

$$(L_{nj}f)(x) = \left(\frac{1}{2}\Sigma\, a_{pq}^{nj}\tau_{pr}(h_{n\,j-1})\tau_{ql}(h_{n\,j-1})\xi_r\xi_l\right)(f)(x) - \left(\frac{1}{2}\Sigma\, a_{pq}^{nj}\rho_i^{pq}\tau_{il}(h_{n\,j-1})\xi_l\right)(f)(x).$$

Then for any $f\epsilon S$ and for any $\epsilon > 0$ there is a constant C, a neighbourhood N of e and an integer n_0 such that for $n \geqslant n_0$ and for all j

$$\sup_{x \epsilon G} |(\delta_{nj}f)(x) - (L_{nj}f)(x)| \leqslant \epsilon \, \text{tr}\, a^{nj} + c\,\alpha_{nj}(N^c).$$

LIMIT THEOREMS FOR RANDOM WALKS ON LIE GROUPS

In particular since $\varepsilon > 0$ is arbitrary and $\overset{n-1}{\underset{j=1}{\Sigma}} \operatorname{Tr} a^{nj}$ is bounded,

$$\lim_{n \to \infty} \overset{n-1}{\underset{j=1}{\Sigma}} \sup_{x \in G} |(\delta_{nj}f)(x) - (L_{nj}f)(x)| = 0.$$

Proof: To avoid cumbersome notation in the proof of the lemma we shall denote h_{nj-1} by h and g_{nj} by g. We shall denote $(\delta_{nj}f)(e)$ by $\delta_{nj}(f)$ and $(L_{nj}f)(e)$ by $L_{nj}(f)$.

$$\delta_{nj}(f) = \int [f(hyg^{-1}h^{-1}) - f(e)]\alpha_{nj}(dy).$$

We shall expand $f(hyg^{-1}h^{-1})$ around $y = g$ using $\phi_1, ..., \phi_d$ as coordinates. We shall use the Taylor expansion up to the second order terms.

$$f(hyg^{-1}h^{-1}) - f(e) = \Sigma \theta_i [\phi_i(y) - \phi_i(g)] + \frac{1}{2} \Sigma \theta_{pq} [\phi_p(y) - \phi_p(g)][\phi_q(y) - \phi_q(y)] + R. \quad \dots \quad (3.28)$$

This is possible because for large n, g is uniformly close to the identity. In the above equation

$$\theta_i = \frac{\partial}{\partial \phi_i} f(hyg^{-1}h^{-1}) \Big|_{y=g}$$

$$\theta = \frac{\partial^2}{\partial \phi_p \partial \phi_q} f(hyg^{-1}h^{-1}) \Big|_{y=g}$$

Let us also define

$$\lambda_{pq} = \frac{\partial^2}{\partial \phi_p \partial \phi_q} f(hyh^{-1}) \Big|_{y=e}$$

We first note that because h ranges over a compact set for any $\varepsilon > 0$ there is a neighbourhood N of e such that for all n and j

$$|R| \leqslant \varepsilon \Sigma |\phi_p(y) - \phi_p(g)|^2.$$

Integrating (3.28) by splitting G into N and N^c yields

$$\left| \delta_{nj}(f) - \frac{1}{2} \Sigma \theta_{pq} a_{pq}^{nj} \right| \leqslant \varepsilon \operatorname{tr} a^{nj} + C\alpha_{nj}(N^c).$$

Since h ranges over a compact set and g is uniformly close to the identity,

$$|\theta_{pq} - \lambda_{pq}| < \varepsilon \text{ for } n \geqslant n_0 \text{ and for all } j.$$

We notice also that a computation of λ_{pq} in terms of $\xi_1, ..., \xi_d$, using (2.7) and (2.16), yields

$$\lambda_{pq} = (\Sigma \tau_{pr}(h) \tau_{ql}(h)\xi_r\xi_l - \Sigma \rho_i^{pq}\tau_{il}(h)\xi_l)(f)(e).$$

SANKHYĀ : THE INDIAN JOURNAL OF STATISTICS : Series A

Finally if $f \epsilon \mathcal{S}$, the family $f_x(y) = f(xy)$ for $x \epsilon G$ is uniformly bounded and uniformly smooth around e. Hence all the above considerations hold uniformly for them. We identify

$$(\delta_{nj}f)(x) = d_{nj}(f_x)$$

and the

$$(L_{nj}f)(x) = L_{nj}(f_x)$$

and this completes the proof of the lemma.

Lemma 3.8 : *Let* (3.27) *and* (2.7) *hold. Then* Q_n *converges weakly to a limit* Q. Q *is characterized by the fact that*

(i) $Q[x(0) \doteq e] = 1$

and

(ii)   *for* $f \epsilon \mathcal{S}$

$$f(x(t)) - \Sigma \frac{1}{2} [\tau_{pr}(h(s))\tau_{ql}(h(s))\xi_\gamma \xi_l - \rho_i^{pq}\tau_{tl}(h)\xi_l](f)(x(s)) \, dA_{pq}(s)$$

is a martingale.

Proof :   First we verify the compactness of Q_n. Let k be an integer $0 \leqslant k \leqslant n-1$. Consider the process $Y_n(t)$ and let Π_U^k be the number of steps needed for the Markov chain $\left[Y_n\left(\frac{k}{n}\right) \right]^{-1} Y_n\left(\frac{j}{n}\right) : j \geqslant k$, to exit from a neighbourhood U of e. We have to show that for any U

$$\lim_{\delta \downarrow 0} \ \limsup_{n \to \infty} \ \sup_{0 \leqslant t \leqslant 1} \ Q_n[\Pi_U^{[nt]} \leqslant n\delta] = 0. \qquad \ldots \ (3.29)$$

This is proved in the same way as Lemma 3.2. For any $f \epsilon s$ and k_0, denoting by $y(t)$ the function $[x(k_0/n)]^{-1}X(t)$

$$f\left(y\left(\frac{k}{n}\right)\right) - \sum_{k_0+1}^{k} \int \left[f\left(y\left(\frac{j-1}{n}\right)y\right) - f\left(y\left(\frac{j-1}{n}\right)\right) \right] d\beta_{nj}(dy) \qquad \ldots \ (3.30)$$

is a Q_n martingale. We can estimate

$$| \int [f(zy) - f(z)]\beta_{nj}(dy) | \leqslant \sup_{x \epsilon G} |(L_{nj}f)(x)| + \varepsilon_{nj}$$

where $\lim_{n=1} \sum_{j=1}^{n-1} \varepsilon_{nj} = 0$. Moreover

$$\sup_{.} |(L_{nj}f)(x)| \leqslant C \operatorname{Tr} a^{nj}.$$

Therefore there are numbers Δ_{nj} such that

$$f\left(y\left(\frac{k}{n}\right)\right) + \sum_{k_0+1}^{k} \Delta_{nj} \text{ is a } Q_n \text{ submartingale}$$

and

$$\lim_{\delta \downarrow 0} \ \limsup_{n \to \infty} \ \sup_{|t_1-t_2| \leqslant \delta} \ \sum_{[nt_1] \leqslant j \leqslant [nt_2]} \Delta_{nj} = 0.$$

LIMIT THEOREMS FOR RANDOM WALKS ON LIE GROUPS

Equation (3.29) is now proved in the same manner as Lemma 3.2 by taking f to be a function which is 1 at e and 0 off U. We then apply Doob's stopping theorem. Let Q be any limit point of the sequence $\{Q_n\}$. By taking $k_0 = 0$ in (3.30),

$$f\left(x\left(\frac{k}{n}\right)\right) - \sum_1^k \int \left[f\left(x\left(\frac{j-1}{n}\right)y\right) - f\left(x\left(\frac{j-1}{n}\right)\right)\right]\beta_{nj}(dy) \qquad \ldots \text{(3.31)}$$

is a martingale relative to $[\Omega_d, M_{k/q}, Q_n]$. From (2.7) it is clear that Q is concentrated on Ω_c. We look at the expression (3.31) as a functional $F_n(t, w)$ defined for t of the form k/n on Ω_d. From Lemma 3.7 it is clear that $F_n(t, w)$ is converging uniformly on a compact subset of Ω_d to the limit

$$F(t, w) = f(x(t)) - \Sigma \frac{1}{2} \int_0^t [\tau_{pr}(h(s))\tau_{ql}(h(s))\xi_y\xi_l - \rho_i^{pq}\tau_{il}(h(s))\xi_l] \cdot (f)(x(s))\, dA_{pq}(s).$$

Therefore for any $f \in \mathcal{S}$, $F(t, w)$ is a martingale relative to $[\Omega_c, M_t, Q]$. One can then verify that such a Q is unique. Hence the entire sequence Q_n converges to the limit Q.

Lemma 3.9 : P_n converges weakly to $P_{A(\cdot), g(\cdot)}$ where $g(\cdot)$ is given by (2.18) and (2.19).

Proof : From the convergence of Q_n to Q it follows that P_n converges to P where P is the distribution of $Y(t)h(t)$ and $Y(t)$ is distributed according to Q. We can use Lemmas 3.3 and 3.6 to compute what P is and verify that it is $P_{A(\cdot), g(\cdot)}$.

We now prove that the representations $P_{A(\cdot), g(\cdot)}$ and $Q_{A(\cdot), g(\cdot)}$ are unique.

Lemma 3.10 : If $P_{A(\cdot), g(\cdot)} = P_{B(\cdot), h(\cdot)}$ then $A(\cdot) = B(\cdot)$ and $g(\cdot) = h(\cdot)$. If $Q_{A(\cdot), g(\cdot)} = Q_{B(\cdot), h(\cdot)}$ then $A(\cdot) = B(\cdot)$ and $g(\cdot) = h(\cdot)$.

Proof : Since we can use Lemma 3.6 to go back and forth between "P" and the "Q" representations, it suffices to prove that if $Q_{A(\cdot), g(\cdot)} = Q_{B(\cdot), h(\cdot)}$ then $A(\cdot) = B(\cdot)$ and $g(\cdot) = h(\cdot)$. From the definition of $Q_{A(\cdot), g(\cdot)}$ find $Q_{B(\cdot), h(\cdot)}$ it follows that it is sufficient to prove that if $Q_{A(\cdot), g(\cdot)} = Q_{B(\cdot), e(\cdot)}$ then $g(\cdot) = e(\cdot)$ and $A(\cdot) = B(\cdot)$. Here $e(\cdot)$ is the function identically equal to e. For the moment let us suppose that $g(\cdot)$ is of bounded variation. Then we can write $Q_{A(\cdot), g(\cdot)}$ as $P_{C(\cdot), g(\cdot)}$ and $Q_{B(\cdot), e(\cdot)} = P_{B(\cdot), e(\cdot)} \cdot P_{C(\cdot), g(\cdot)}$ has the property that for $f \in \mathcal{S}$,

$$f(x(t)) - \int_0^t \frac{1}{2} \Sigma (\xi_p\xi_q f)(x(s))\, dC_{pq}(s) - \int_0^t \Sigma (\xi_i f)(x(s))\, db_1(s)$$

is a martingale. Here

$$dg(t) = \Sigma db_i(t)\, \xi_i(g(t)).$$

On the other hand with respect to $P_{B(\cdot), e(\cdot)}$

$$f(x(t)) - \int_0^t \frac{1}{2} \Sigma (\xi_p\xi_q f)(x(s))\, dB_{p_l}(s)$$

SANKHYĀ : THE INDIAN JOURNAL OF STATISTICS : Series A

is a martingale. If the two measures are the same, by the uniqueness of the Meyer decomposition for quasi-martingales $C(\cdot) = B(\cdot)$ and $db_i(\cdot) \equiv 0$. Therefore $A(\cdot) = B(\cdot)$. So all we have to do is prove that if $Q_{A(\cdot),\ g(\cdot)} = Q_{B(\cdot),\ e(\cdot)}$ then $g(\cdot)$ is of bounded variation. Let $X(t)$ be any stochastic process with P as its distribution. We can call P "regular" if for any $f \in \mathcal{S}$, $f(X(t))$ is a quasi-martingale, that is there is a process $F(t)$ such that $F(t)$ is of bounded variation and

$$f(X(t)) - F(t)$$

is a square integrable martingale. Suppose $X(t)$, $Y(t)$ are two "regular" processes which are possibly dependent, then Kunita-Watanabe's theory of square integrable martingales (Kunita and Watanabe, 1967) gurantees that the two processes are jointly regular, i.e., for any smooth function ϕ of two variables, $\phi(X(t), Y(t))$ is a quasi-martingale. Going back to our problem the process corresponding to $Q_{A(\cdot),\ g(\cdot)}$ is of the form $X(t)g(t)$ where $X(t)$ is regular. The process corresponding to $Q_{B(\cdot),\ e(\cdot)}$ is of course always regular. Since both processes are the same we have that $X(t)$ as well as $X(t)g(t)$ are regular processes. Therefore for smooth ϕ, $\phi(X(t), Y(t))$ is a quasi-martingale. In particular taking $\phi(x, y) = f(x^{-1}y)$ we get $f(g(t))$ to be a quasi-martingale. This forces $g(\cdot)$ to be of bounded variation.

Lemma 3.10 : *Let* $Q_{A^n(\cdot),\ g^n(\cdot)}$ *be compact on* Ω_c, *where* $A^n(\cdot)$, $g^n(\cdot)$ *are arbitrary sequences in* V *and* $\mathcal{F}$. *Then* $A^n(\cdot)$, $g^n(\cdot)$ *are equicontinuous at every point of* $[0, T]$.

Proof : Let us prove equicontinuity at 0 for $A^n(\cdot)$. Suppose the sequence is not equicontinuous. By taking a subsequence if need be we can assume that there is a sequence δ_n of positive numbers converging to 0 such that

$$\text{Tr } A^n(\delta_n) = \varepsilon > 0.$$

Let us make a monotone continuous transformation ϕ_n of $[0, \varepsilon]$ onto $[0, \delta_n]$ such that $A^n(\phi_n(t))$ is equicontinuous on $[0, \varepsilon]$. Denote $A^n(\phi_n(t))$ by $B^n(t)$, $g^n(\phi_n(t))$ by $h^n(t)$ and the measure corresponding to $x(\phi_n t)$ under $Q_{A^n(\cdot),\ g^n(\cdot)}$ on $[0, \varepsilon]$ by Q^n. Then Q^n on $[0, \varepsilon]$ is $Q_{B^n(\cdot),\ h^n(\cdot)}$. Since $B^n(\cdot)$ is equicontinuous we can assume that it converges to $B(\cdot)$ uniformly. Then $Q_{B^n(\cdot),\ e^n(\cdot)}$ will converge to $Q_{B(\cdot),\ e(\cdot)}$. Moreover since $\delta_n \to 0$ Q^n will converge to the process constantly equal to e. This forces $h^n(\cdot)$ to be equicontinuous and if $h^n(\cdot)$ were to converge to $h(\cdot)$ then $Q_{B(\cdot),\ h(\cdot)}$ is the trivial process. By the uniqueness of Lemma 3.9 this forces $B(\cdot)$ to be zero. But $\text{Tr } B(\varepsilon) = \lim \text{Tr } A^n(\phi_n(\varepsilon))$ $= \lim \text{Tr } A^n(\delta_n) = \varepsilon > 0$. Thus $A^n(\cdot)$ is equicontinuous at 0. By a similar proof it is equicontinuous at every point. Thus $Q_{A^n(\cdot),\ e^n(\cdot)}$ is compact. Since $Q_{A^n(\cdot),\ h^n(\cdot)}$ is also compact, $h^n(\cdot)$ is equicontinuous too.

Lemma 3.11 : *Let* P_n *be as in Lemma 3.9. If* P_n *is compact then* $A^n(\cdot)$ *and* $h_n(\cdot)$ *are equicontinuous on* $[0, T]$.

Proof : Again we shall prove equicontinuity only at $t = 0$. The same proof will work at any other point. Let U be a neighbourhood of e with a compact closure

LIMIT THEOREMS FOR RANDOM WALKS ON LIE GROUPS

and δ_n be the first time $h_n(t)$ exits from U. Let $\varepsilon > 0$ be an arbitrary positive number and let θ_n be the first time Tr $A^n(t)$ exceeds ε. We have to show that δ_n and θ_n are bounded below. Let us suppose that they are not. Then by a suitable choice of a subsequence we are reduced to one of the following cases.

Case 1 : $\theta_n \leqslant \delta_n$ and $\theta_n \to 0$.

Case 2 : $\delta_n \leqslant \theta_n$ and $\delta_n \to 0$.

Let us consider case 1. We can find monotone continuous maps $\phi_n(t)$ of $[0, \varepsilon]$ onto $[0, \theta_n]$ such that (a) $A_n(\phi_n(t))$ is equicontinuous on $[0, \varepsilon]$ and (b) the points t_{nj} where the jumps of the process $X(\phi_n(t))$ occur have the property $\lim_{n \to \infty} \sup_j |t_{nj} - t_{nj-1}| = 0$.

We can write $X(\phi_n(t))$ as $Y(\phi_n(t)) \cdot h_n(\phi(t))$ and we shall denote by $\tilde{P}_n$ and $\tilde{Q}_n$ the measures corresponding to $X(\phi_n(t))$ and $Y(\phi_n(t))$ on $[0, \varepsilon]$. Since $\delta_n \geqslant \theta_n$ we have that $h_n(\phi_n(t))$ is in U and therefore in a compact set during $[0, \varepsilon]$. Since $A^{(n)}(\phi_n(t))$ is equicontinuous and $h_n(\phi_n(t))$ ranges over a compact set $B^{(n)}(\phi_n(t))$ is also equicontinuous where

$$B_{rl}^{(n)}(t) = \sum_{j \leqslant [nt]} a_{pq}^{nj} \tau_{pr}(h_{n\ j-1}) \tau_{ql}(h_{n\ j-1}).$$

Moreover Tr $B^{(n)}(\phi_n(\varepsilon))$ is bounded from below because Tr $A^{(n)}(\phi_n(\varepsilon)) = \varepsilon$ and $h_n(\phi_n(t))$ ranges over a compact set. If we now choose a further subsequence and assume that $B^{(n)}(t) \to B(t)$ uniformly then $\tilde{Q}_n$ converges to a nontrivial limit $\tilde{Q}$. By that we man that the variance term is nonzero. Since $\theta_n \to 0$ we have $\tilde{P}_n$ converges to the trivial process. Since $\tilde{Q}_n$ and $\tilde{P}_n$ are compact so is $h_n(\phi_n(t))$. If we assume that $h_n(\phi_n(t))$ converges to a limit $h(t)$ then the trivial process $X(t) \equiv e$ is expressed as $Y(t)h(t)$ where $Y(t)$ has a nontrivial variance. This is a contradiction.

Case 2 : We proceed exactly as in case 1.

We conclude that the trivial process $X(t) \equiv e$ is expressed as $Y(t)h(t)$ where $h(\varepsilon) \epsilon U^c$ and therefore $h(\cdot)$ is a nontrivial continuous function. We are however unable to assert !that the $Y(\cdot)$ process has nontrivial variance. However if the $Y(t)$ process has trivial variance then $Y(\cdot) \equiv e$ with probability 1. (See Lemma 3.8 for the types of processes that come up.] So there is never any possibility of the $Y(\cdot)$ process exactly cancelling the function $h(t)$. This is again a contradiction.

We have now proved all the results claimed in the summary. Characterization of the compactness and uniqueness of the representation prove the continuity of the inverse maps in Theorems 2.1 and 2.2 and the necessity part of Theorem 2.4. Theorem 2.3 follows from the uniqueness of the representations and Lemma 3.6. The only remaining part is to prove that the maps in Theorems 2.1 and 2.2 are onto. This is proved below.

Lemma 3.12 : *If $P \in \mathcal{P}$ then $P = P_{A(\cdot),\ g(\cdot)}$ for some $A(\cdot)$ and $g(\cdot)$.*

SANKHYA : THE INDIAN JOURNAL OF STATISTICS : Series A

Proof : Let $X(t)$ be the process corresponding to P_0.

If we define random variables

$$X_{nj} = X\left(\frac{j-1}{n}\right)^{-1} X\left(\frac{j}{n}\right), \qquad j = 1, 2, \dots, n-1$$

then we can use Lemma 3.11 to conclude that $A^n(\cdot)$ and $h_n(\cdot)$ constructed as in Theorem 2.4 are equicontinuous because P_n in this context converges to P. By an appropriate choice of a subsequence we can assume that $A^n(\cdot) \to A(\cdot)$ and $h_n(\cdot) \to h(\cdot)$. Therefore $P = P_{A(\cdot),\, g(\cdot)}$ for some $A(\cdot)$ and $g(\cdot)$.

REFERENCES

HUNT, G. A. (1956) : Semi-groups of measures on Lie groups. *Trans. Amer. Math. Soc.*, 81(2), 264-293.

KUNITA, S. and WATANABE, S (1967) : On square integrable martingales. *Nagoya Math. J.*, 30, 209-245.

PATHASARATHY, K. R. (1964) : The central limit theorem for the rotation group. *Theory Prob. Applications*, 9, 248-257 (English translation).

PARTHASARATHY, K. R., RANGA RAO, R. and VARADHAN, S. R. (1963) : Probability distributions on locally compact abelian groups. *Illinois J. Math.*, 7, 337369.

PHILLIPS, R. S. and SARASON, L. (1967) : Elliptic-parabolic equations of the second order. *J. Math. Mech.* 17, 891-917.

SAZANOV, V. V. and TUTUBALIN, V. N. (1966) : Probability distributions on Topological groups. *Theory Prob. Applications*, 11, 1-45 (English translation).

STROOCK, D. W. and VARADHAN, S. R. S. (1969a) : Diffusion processes with continuous coefficients. I. *Comm. Pure Appl. Math.*, XII, 345-400.

——— (1969b) : Diffusion processes with continuous coefficients, II. *Comm. Pure Appl. Math.*, XII, 479-530.

YAMADA, T. and WATANABE, S. (1971) : On the uniqueness of solutions of stochastic differential equations. *J. Math. Kyoto Univ.*, 11, No. 1, 155-167.

WEHN, D. (1968) : Probabilities on Lie groups. *Proc. Nat. Zcad. Sci.*, U.S.A. 48, No. 5, 791-795.

Paper received : December, 1972.

Martingale Approach to Some Limit Theorems

G. C. Papanicolaou[†]

D. Stroock[††]

S. R. S. Varadhan[†]

[†]Courant Institute of Mathematical Sciences, New York University. Research supported by the National Science Foundation under Grant No. MCS75-09837.

[††]Department of Mathematics, University of Colorado.

This is an expanded version of a lecture given by one of us (G. P.) at the Conference on Turbulence at Duke University in April 1976.

Contents

1. Introduction.

The asymptotic behavior of stochastic processes under various circumstances is a central problem in probability theory and its applications. The theory of limiting distributions of sums of independent (or weakly dependent) random variables is among the best known examples.

Diffusion approximations for suitably normalized Markov chains have also been studied extensively starting with Khinchine [1] who employed a differential equations approach and continuing with Prohorov [2] and Skorohod [3] who employed compactness criteria that they developed. By formulating diffusion theory in terms of martingale problems, very general diffusion approximations were obtained in [4] under minimal regularity hypotheses. In [5] corresponding problems with boundary conditions were treated.

In a variety of applications one encounters asymptotic problems for stochastic processes that are somewhat different from the above. These problems are to some extent analogous to problems in differential equations that are suitable for application of the method of averaging [6] and can be described roughly as follows.

There are two processes given that vary on two different time scales. One, the rapidly varying one, is driving the other by entering as a coefficient in the equation defining the slowly varying process. The problem is to characterize

the asymptotic behavior of the driven process as the fluctuations of the driving process become more and more rapid. In
this work we analyze such problems, under a variety of different circumstances, by adapting the martingale approach of
[4,5] (cf. also [7]).

Among the many places where asymptotic problems of the
above form appear we mention the work of Stratonovich [8]
(on noise phenomena) and subsequent developments [9-12,14],
the diffusion limit of transport theory [13, 15 and references
cited there], random evolutions [16,17,18], waves in random
media [19], etc.

The following is organized in two parts. One deals with
the case of a driving process that has strong ergodic
properties. Sections 2.1 and 2.2 contain the formulation of a
theorem that is typical of what follows. Sections 2.4-2.8
contain results that may be considered as variations of the
main theme but are also of independent interest.

The second part deals with the case of a not so strongly
ergodic driving process and as a consequence the nature of
the limit process changes. Typical examples here are occupation time problems for Brownian motion [20,21,22]. Sections
3.1 and 3.4 contain the main theorems.

2.   Ergodic driving processes.

2.1   Formulation of the problem.

We shall first formulate a general problem and then give some examples which explain why we are interested in this problem.

Let $\varepsilon > 0$ be a parameter and let $(x^\varepsilon(t), y^\varepsilon(t))$ be the diffusion process in $R^n \times R^m$ defined by the following system of stochastic differential equations

$$(2.1.1) \quad dx_i^\varepsilon(t) = \left[\frac{1}{\varepsilon} F_i^{(1)}(x^\varepsilon(t), y^\varepsilon(t)) + G_i^{(1)}(x^\varepsilon(t), y^\varepsilon(t))\right] dt$$

$$+ \sum_{j=1}^{p} \sigma_{ij}^{(1)}(x^\varepsilon(t), y^\varepsilon(t)) d\beta_j^{(1)}(t) \ ,$$

$$x_i^\varepsilon(0) = x_i \ , \qquad i = 1, 2, \ldots, n \ ,$$

$$dy_k^\varepsilon(t) = \left[\frac{1}{\varepsilon^2} F_k^{(2)}(x^\varepsilon(t), y^\varepsilon(t)) + \frac{1}{\varepsilon} G_k^{(2)}(x^\varepsilon(t), y^\varepsilon(t))\right.$$

$$\left. + H_k^{(2)}(x^\varepsilon(t), y^\varepsilon(t))\right] dt$$

$$+ \sum_{\ell=1}^{q} \left[\frac{1}{\varepsilon} \sigma_{k\ell}^{(2)}(x^\varepsilon(t), y^\varepsilon(t))\right.$$

$$\left. + \sigma_{k\ell}^{(3)}(x^\varepsilon(t), y^\varepsilon(t))\right] d\beta_\ell^{(2)}(t) \ ,$$

$$y_k^\varepsilon(0) = y_k \ , \qquad k = 1, 2, \ldots, m \ .$$

Here $(\beta_1^{(1)}(t), \beta_2^{(1)}(t), \ldots, \beta_p^{(1)}(t))$ and $(\beta_1^{(2)}(t), \beta_2^{(2)}(t), \ldots, \beta_q^{(2)}(t))$ are p-dimensional and q-dimensional standard Brownian motions, respectively, and

(2.1.2) $\quad E\{\beta_j^{(1)}(t)\beta_\ell^{(2)}(s)\} = \min(t,s)b_{j\ell}$,

$$j = 1,\ldots,p , \qquad \ell = 1,\ldots,q .$$

The functions $F_i^{(1)}$, $G_i^{(1)}$, $\sigma_{ij}^{(1)}$, $i = 1,\ldots,n$, $j = 1,\ldots,p$ and $F_k^{(2)}$, $G_k^{(2)}$, $H_k^{(2)}$, $\sigma_{k\ell}^{(2)}$, $\sigma_{k\ell}^{(3)}$, $k = 1,\ldots,m$, $\ell = 1,\ldots,q$ are assumed to be bounded for $(x,y) \in R^n \times R^m$ and smooth i.e., C^∞ in both x and y. These hypotheses can be weakened considerably as we indicate briefly later. However, we assume them here in order to simplify the statement of results.

For $\varepsilon > 0$ and fixed the existence theory of Itô equations [23] provides a unique solution to (2.1.1) which is a continuous strong Markov process $(x^\varepsilon(t),y^\varepsilon(t))$. The infinitesimal generator $\mathscr{L}^\varepsilon$ of this process has the following form

(2.1.2) $\quad \mathscr{L}^\varepsilon = \dfrac{1}{\varepsilon^2}\,\mathscr{L}_1 + \dfrac{1}{\varepsilon}\,\mathscr{L}_2 + \mathscr{L}_3$

where

(2.1.3) $\quad \mathscr{L}_1 = \displaystyle\sum_{k=1}^{m} F_k^{(2)}(x,y)\,\dfrac{\partial}{\partial y_k} + \dfrac{1}{2}\sum_{k,\ell=1}^{m} a_{k\ell}^{(2)}(x,y)\,\dfrac{\partial^2}{\partial y_k \partial y_\ell}$,

(2.1.4) $\quad \mathscr{L}_2 = \displaystyle\sum_{i=1}^{n} F_i^{(1)}(x,y)\,\dfrac{\partial}{\partial x_i} + \sum_{k=1}^{m} G_k^{(2)}(x,y)\,\dfrac{\partial}{\partial y_k}$

$$+ \sum_{i=1}^{n}\sum_{k=1}^{m} a_{ik}^{(12)}(x,y)\,\dfrac{\partial^2}{\partial x_i \partial y_k}$$

$$+ \sum_{k,\ell=1}^{m} a_{k\ell}^{(23)}(x,y)\,\dfrac{\partial^2}{\partial y_k \partial y_\ell} ,$$

$$(2.1.5) \quad \mathcal{L}_3 = \sum_{i=1}^{n} G_i^{(1)}(x,y) \frac{\partial}{\partial x_i} + \sum_{k=1}^{m} H_k^{(2)}(x,y) \frac{\partial}{\partial y_k}$$

$$+ \frac{1}{2} \sum_{i,j=1}^{n} a_{ij}^{(1)}(x,y) \frac{\partial^2}{\partial x_i \partial x_j}$$

$$+ \sum_{i=1}^{n} \sum_{k=1}^{m} a_{ik}^{(13)}(x,y) \frac{\partial^2}{\partial x_i \partial y_k}$$

$$+ \frac{1}{2} \sum_{k,\ell=1}^{m} a_{k\ell}^{(3)}(x,y) \frac{\partial^2}{\partial y_k \partial y_\ell} \quad ,$$

and

$$a_{ij}^{(1)}(x,y) = \sum_{r=1}^{p} \sigma_{ir}^{(1)}(x,y)\sigma_{jr}^{(1)}(x,y) \quad , \quad i,j = 1,\ldots,n \quad ,$$

$$a_{k\ell}^{(2)}(x,y) = \sum_{r=1}^{q} \sigma_{kr}^{(2)}(x,y)\sigma_{\ell r}^{(2)}(x,y) \quad ,$$

$$a_{k\ell}^{(3)}(x,y) = \sum_{r=1}^{q} \sigma_{kr}^{(3)}(x,y)\sigma_{\ell r}^{(3)}(x,y) \quad , \quad k,\ell = 1,\ldots,m \quad ,$$

$$(2.1.6) \quad a_{ik}^{(12)}(x,y) = \sum_{r=1}^{p} \sum_{s=1}^{q} \sigma_{ir}^{(1)}(x,y)\sigma_{ks}^{(2)}(x,y)b_{rs} \quad ,$$

$$a_{ik}^{(13)}(x,y) = \sum_{r=1}^{p} \sum_{s=1}^{q} \sigma_{ir}^{(1)}(x,y)\sigma_{ks}^{(3)}(x,y)b_{rs} \quad ,$$

$$i = 1,\ldots,n \quad , \quad k = 1,\ldots,m \quad ,$$

$$a_{k\ell}^{(23)}(x,y) = \sum_{r=1}^{q} \sigma_{kr}^{(2)}(x,y)\sigma_{\ell r}^{(3)}(x,y) \quad ,$$

$$k,\ell = 1,2,\ldots,m \quad .$$

We now give some examples that motivate the above

formulation.

Example 1. $F^{(1)}$, $F^{(2)}$, $H^{(2)}$, $\sigma^{(2)}$ and $\sigma^{(3)}$ are identi-

cally zero (we omit subscripts for simplicity), $m = 1$ and $G^{(1)}$

and $\sigma^{(1)}$ are almost periodic functions of $y \in R^1$. Thus,

(2.1.1) reduces to

$$(2.1.7) \quad dx_i^\varepsilon(t) = G_i^{(1)}(x^\varepsilon(t), y + t/\varepsilon) dt$$

$$+ \sum_{j=1}^{p} \sigma_{ij}^{(1)}(x^\varepsilon(t), y + t/\varepsilon) d\beta_j^{(1)}(t) \ ,$$

$$x_i^\varepsilon(0) = x \ , \qquad i = 1, 2, \ldots, n \ .$$

The asymptotic analysis of the process $x^\varepsilon(t)$ defined by (2.1.7)

constitutes the <u>method of averaging</u> extended to Itô stochastic

differential equations [9,24]. This example is considered

further in Section 2.8.

Example 2. $F^{(1)} \equiv 0$ in (2.1.1). This example corresponds

to what may be called <u>generalized averaging</u> for Itô stochastic

differential equations. The deterministic equivalent was

analyzed by Volosov, Tichonov [25,26] and others, and the sto-

chastic problem by Khasminskii [12]. We return to this

example in Section 2.8.

Example 3. $n = m$, $F^{(1)} = F^{(2)}$, $G^{(1)} = G^{(2)}$, $\sigma^{(1)} = \sigma^{(2)}$,

$\beta^{(1)}(t) \equiv \beta^{(2)}(t)$ so that $b_{j\ell} = \delta_{j\ell}$ in (2.1.2), and all quan-

tities as functions of y_j are periodic functions of period

one, $j = 1, 2, \ldots, n$. This example corresponds to <u>homogenization</u>

[27] and is considered again in Section (2.7). If the initial

data are related by $y = x/\varepsilon$ then, in this example,

(2.1.8) $y^{\varepsilon}(t) \equiv x^{\varepsilon}(t)/\varepsilon$

and (2.1.1) becomes (cf. [37])

$$(2.1.9) \quad dx_i^{\varepsilon}(t) = \left[\frac{1}{\varepsilon} F_i^{(1)}\left(x^{\varepsilon}(t), \frac{x^{\varepsilon}(t)}{\varepsilon}\right) + G_i^{(1)}\left(x^{\varepsilon}(t), \frac{x^{\varepsilon}(t)}{\varepsilon}\right)\right] dt$$

$$+ \sum_{j=1}^{p} \sigma_{ij}^{(1)}\left(x^{\varepsilon}(t), \frac{x^{\varepsilon}(t)}{\varepsilon}\right) d\beta_j^{(1)}(t) ,$$

$$x_i^{\varepsilon}(0) = x , \quad i = 1,\ldots,n .$$

Example 4. $n = m$, $F^{(1)}(x,y) \equiv y$, $\sigma^{(1)} \equiv 0$ and $y^{\varepsilon}(t) = y(t/\varepsilon^2)$ where

$$(2.1.10) \quad dy_k(t) = -y_k(t)dt + d\beta_k(t) ,$$

$$y_k(0) = y_k , \quad k = 1,2,\ldots,n .$$

The asymptotic limit of (2.1.1) in this case is the Smoluchovski-Kramers limit of Brownian motion [28] and was analyzed by Ilin and Khasminskii [13] and Bromberg [29]. In the latter paper the process $x^{\varepsilon}(t)$ is confined to a bounded region in space with appropriate boundary conditions which give rise to boundary layers as $\varepsilon \to 0$.

From these examples it is clear that scaling by ε in (2.1.1) is intended to articulate the fact that in the pair $(x^{\varepsilon}(t), y^{\varepsilon}(t))$, $x^{\varepsilon}(t)$ is the slowly varying component of the process and $y^{\varepsilon}(t)$ the rapidly varying component. We also refer to $y^{\varepsilon}(t)$ as the driving process and to $x^{\varepsilon}(t)$ as the driven process. Note that, in general, the driven process, in turn, influences the driving process since the functions with superscript (2) and (3) depend on y. Thus, when $\varepsilon > 0$ is not small, the component processes $x^\varepsilon(t)$ and $y^\varepsilon(t)$ are on an equal footing; it is only when ε is small that the above terminology is meaningful. Problem (2.1.1) can also be considered as a random evolution problem [16,17]. In that context the driven component can be a very general object, not necessarily an R^n valued process. On the other hand the driving component is frequently taken to be a Markov chain. We return to this in Section (2.6) in connection with transport problems.

The process $(x^\varepsilon(t), y^\varepsilon(t))$, $\varepsilon > 0$ fixed, is also characterized by its infinitesimal generator $\mathscr{L}^\varepsilon$ acting on smooth functions. Under our hypotheses, almost any method can be used to construct the process from $\mathscr{L}^\varepsilon$ of (2.1.2)-(2.1.6). In particular the martingale method of [4]. In the martingale approach the measures $P^\varepsilon_{x,y}$ in the space of continuous trajectories in $R^n \times R^m$, with (x,y) initial point, corresponding to $(x^\varepsilon(t), y^\varepsilon(t))$, are characterized as follows:

(i) $P^\varepsilon_{x,y}\{x(0) = x, y(0) = y\} = 1$,

(ii) for any function $f(x,y)$ which is smooth and with compact support, the expression

(2.1.11) $\qquad M_f(t) = f(x(t), y(t)) - \int_0^t \mathscr{L}^\varepsilon f(x(s), y(s)) ds$

is a $P^\varepsilon_{x,y}$ martingale.

Here we do not use superscripts on $(x(t), y(t))$ since they appear as arguments of functionals while the measures are labeled by ε. If $\mathcal{F}_t$, $t \geq 0$, denotes the σ-algebras of events generated by the paths $(x(s), y(s))$ for $s \leq t$, then (ii) above says that $(P^\varepsilon$ a.s.$)$

$$(2.1.12) \qquad E^{P^\varepsilon} \{M_f(t) \mid \mathcal{F}_s\} = M_f(s) , \qquad 0 \leq s \leq t < \infty .$$

In the following we employ systematically the martingale approach. The reasons for this are that (i) it is well suited to asymptotics centering on weak convergence and provides a reasonably general and convenient scheme for a large class of problems, (ii) it is well suited to problems involving boundary conditions [5,30] (the characterization being formulated as a submartingale problem [5]) and (iii) it provides results with minimal regularity hypotheses on the coefficients of the processes under consideration [4].

2.2   The limit theorem.

In addition to the hypotheses introduced above regarding
(2.1.1), we shall require the following hypotheses for the
asymptotic analysis.

Let $Y(t;x)$ be the diffusion process in R^m generated by $\mathcal{L}_1$
of (2.1.3) with $x \in R^n$ regarded as a parameter. From (2.1.1)
it is clear that, roughly, one has $y^\varepsilon(t) \sim Y(t/\varepsilon^2;x)$ at least
for times t that are not too large. The asymptotic analysis
of (2.1.1) depends in an essential way on the ergodic properties
of $Y(t;x)$. Let $P(t,y,A;x)$ be the transition function of
$Y(t;x)$

(2.2.1) $P(t,y,A;x) = P\{Y(t;x) \in A | Y(0;x) = y\}$,

where $y \in R^m$ and A is a Borel subset of R^m. From (2.1.3) it
follows that $P(t,y,A;x)$ is a Feller transition function, depending smoothly upon $x \in R^n$, and such that the semigroup
associated with P maps functions that vanish at infinity to
functions that vanish at infinity.

We assume that $Y(t;x)$ is ergodic i.e., there exists a
unique invariant probability measure $\bar{P}(A;x)$, $A \subset R^m$,
$x \in R^n$, depending smoothly upon x, such that

$$\lim_{t \uparrow \infty} P(t,y,A;x) = \bar{P}(A;x)$$

uniformly in (x,y) on compact sets and $A \subset R^m$. Define the
recurrent potential kernel

$$(2.2.2) \quad \psi(y,A;x) = \int_0^\infty [P(t,y,A;x) - \bar{P}(A;x)]dt \ .$$

We assume that the kernel ψ exists, depends smoothly on x and transforms bounded smooth functions of y into themselves. These are the hypotheses regarding $Y(t;x)$.

We assume further that

$$(2.2.3) \quad \int_{R^m} F^{(1)}(x,y)\bar{P}(dy;x) \equiv 0 \ ,$$

and this is called the **centering** condition.

Let $f(x) \in C_o^\infty(R^n)$ and define $\psi_f^{(1)}$ by

$$(2.2.4) \quad \psi_f^{(1)}(x,y) = \int_{R^m} \psi(y,dz;x) \sum_{j=1}^n F_j^{(1)}(x,z) \frac{\partial f(x)}{\partial x_j} \ .$$

We define a diffusion operator $\mathscr{L}$ as follows

$$(2.2.5) \quad \mathscr{L}f(x) = \int_{R^m} \bar{P}(dy;x)A_f(x,y) \ ,$$

$$(2.2.6) \quad A_f(x,y) = \sum_{k=1}^m G_k^{(2)}(x,y) \frac{\partial \psi_f^{(1)}(x,y)}{\partial y_k}$$

$$+ \sum_{i=1}^n \sum_{k=1}^m a_{ik}^{(12)}(x,y) \frac{\partial^2 \psi_f^{(1)}(x,y)}{\partial x_i \partial y_k}$$

$$+ \sum_{k,\ell=1}^m a_{k\ell}^{(23)}(x,y) \frac{\partial^2 \psi_f^{(1)}(x,y)}{\partial y_k \partial y_\ell}$$

$$+ \sum_{i=1}^n F_i^{(1)}(x,y) \frac{\partial \psi_f^{(1)}(x,y)}{\partial x_i}$$

$$+ \sum_{i=1}^{n} G_i^{(1)}(x,y) \frac{\partial f(x)}{\partial x_i}$$

$$+ \frac{1}{2} \sum_{i,j=1}^{n} a_{ij}^{(1)}(x,y) \frac{\partial f(x)}{\partial x_i \partial x_j} .$$

As a consequence of our smoothness requirements, the coefficients of $\mathscr{L}$ are smooth and bounded functions of x. Thus, $\mathscr{L}$ generates a diffusion Markov process on R^n which is denoted by x(t), $t \geq 0$. In particular, the martingale problem associated with $\mathscr{L}$ has a unique solution P_x which for each x are probability measures in the space of continuous trajectories in R^n starting at x.

Theorem 2.1.

Under the above hypotheses the process $x^\varepsilon(t)$, $t \geq 0$ defined by (2.1.1) converges weakly in $C([0,T];R^n)$, $T < \infty$ but arbitrary, as $\varepsilon \to 0$ to the diffusion Markov process x(t), $t \geq 0$ generated by $\mathscr{L}$ of (2.2.5).

Remark 1.

We clarify now the role of smoothness in the above. Let $\mathscr{L}^\varepsilon$, $\varepsilon > 0$, be given by (2.1.2)-(2.1.6) and assume that the martingale problem for it has a unique solution. If $\mathscr{L}^\varepsilon$ is uniformly elliptic (in $R^n \times R^m$), then [4] it is enough for this that the diffusion coefficients be bounded and continuous and the drift coefficients bounded and measurable. If $\mathscr{L}^\varepsilon$ is not uniformly elliptic some smoothness is sufficient [31].

Suppose that $Y(t;x)$ is ergodic so that (2.2.2) is well defined and that (2.2.3) holds. Assume that for f smooth bounded, $\psi_f^{(1)}$ is bounded and has bounded derivatives of first order in x and second order in y. Assume further that $\mathscr{L}$ of (2.2.5) is uniformly elliptic with bounded continuous coefficients. If not uniformly elliptic smoothness is sufficient.

Then Theorem 2.1 is valid again.

Remark 2.

Concerning boundedness of the coefficients of $\mathscr{L}^\varepsilon$ and $\mathscr{L}$ we remark that one can have unbounded coefficients provided that $x(t)$, associated with $\mathscr{L}$ of (2.2.5), has infinite explosion time with probability one. In this generality, however, $x^\varepsilon(t)$ may exist only up to a finite explosion time $\tau^\varepsilon \to \infty$ as $\varepsilon \to 0$.

If convergence of moments of $x^\varepsilon(t)$ to moments of $x(t)$ is required, a convenient sufficient hypothesis is that the x-derivatives of all terms in (2.1.1) be uniformly bounded as functions of x and y.

Remark 3.

Concerning the strong ergodicity hypothesis about $Y(t;x)$ embodied in the existence of ψ in (2.2.2), we note that it can be weakened somewhat, as for example is done in Section (2.8) where the situation encountered is typical. The results of Part 3 show that distinctly different results arise without (2.2.2).

Specifically, Theorem 2.1 states that in the limit $\varepsilon \to 0$ the driven process $x^\varepsilon(t)$ decouples from the driving process $y^\varepsilon(t)$ and becomes a Markov process by itself. This is not true in the examples of Part 3.

2.3   Proof of Theorem 2.1.

The proof is divided into two steps. In Step 1 we prove that the processes $x^\varepsilon(\cdot)$ of (2.1.1) are relatively weakly compact in $D([0,T];R^n)$, $T < \infty$ but arbitrary. Since both $x^\varepsilon(\cdot)$ and the limit $x(\cdot)$ are supported in $C([0,T];R^n)$ this suffices for weak convergence in C. In Step 2 we identify the limit measure by showing that it solves the martingale problem for $\mathscr{L}$ of (2.2.5).

Both Step 1 and Step 2 are proved in a somewhat more general manner than required by the form (2.1.1). The advantage is that this method, which uses only the martingale formulation and not the Itô equations, applies equally well to processes with jumps (cf. Section 2.6) and/or boundary conditions   (cf. Section 2.5).

Let $f(x)$ be a smooth, bounded function of x and define $\psi_f^{(1)}(x,y)$ by (2.2.4). Let

(2.3.1) $f^\varepsilon(x,y) = f(x) + \varepsilon\psi_f^{(1)}(x,y)$.

By construction, $\psi_f^{(1)}$ satisfies

(2.3.2) $\mathscr{L}_1\psi_f^{(1)} + \mathscr{L}_2 f = 0$

since ψ of (2.2.2) is the kernel of the operator $-\mathscr{L}_1^{-1}$ acting on functions whose integrals relative to $\bar{P}$ are zero (Fredholm alternative). Thus, from (2.1.2)-(2.1.5) and the above we find that

(2.3.3) $\mathscr{L}^\varepsilon f^\varepsilon(x,y) = A_f(x,y) + \varepsilon \mathscr{L}_3 \psi_f^{(1)}(x,y)$

where $A_f(x,y)$ is given by (2.2.6).

From the martingale characterization of $(x^\varepsilon(t), y^\varepsilon(t))$
(cf. (2.1.11)) it follows that

(2.3.4) $M_{f^\varepsilon}(t) = f^\varepsilon(x^\varepsilon(t), y^\varepsilon(t)) - f^\varepsilon(x,y) - \int_0^t \mathscr{L}^\varepsilon f^\varepsilon(x^\varepsilon(s), y^\varepsilon(s)) ds$

$= f(x^\varepsilon(t)) + \varepsilon \psi_f^{(1)}(x^\varepsilon(t), y^\varepsilon(t)) - f(x) - \varepsilon \psi_f^{(1)}(x,y)$

$- \int_0^t \left[A_f(x^\varepsilon(s), y^\varepsilon(s)) + \varepsilon \mathscr{L}_3 \psi_f^{(1)}(x^\varepsilon(s), y^\varepsilon(s)) \right] ds$

is a bounded zero-mean martingale. Of course, we can express
$M_{f^\varepsilon}(t)$ explicitly as a Brownian stochastic integral but, as
remarked above, we prefer not to make use of this.

The increasing process associated with M_{f^ε} and denoted
by $\langle M_{f^\varepsilon}(t), M_{f^\varepsilon}(t) \rangle$ is given by

(2.3.5) $\langle M_{f^\varepsilon}(t), M_{f^\varepsilon}(t) \rangle = \int_0^t H_{f^\varepsilon}(x^\varepsilon(s), y^\varepsilon(s)) ds$

where

(2.3.6) $H_{f^\varepsilon}(x,y) = \left[\mathscr{L}^\varepsilon (f^\varepsilon)^2 - 2f^\varepsilon \mathscr{L}^\varepsilon f^\varepsilon \right](x,y)$

and, as can be verified by direct computation, $H_{f^\varepsilon}(x,y)$ is a
bounded function of x and y independently of ε (it is the
quadratic form associated with second derivatives in
(2.1.3)–(2.1.5)).

To prove compactness in $D([0,T];R^n)$ for $x^\varepsilon(\cdot)$, it suffices to show that [5]

$$(2.3.7) \quad \varlimsup_{\delta \downarrow 0} \varlimsup_{\varepsilon \downarrow 0} \sup_{|t-s| \leq \delta} \sup E\{|x^\varepsilon(t)-x^\varepsilon(s)|^2 \big| \mathcal{F}_s\} = 0 ,$$

where the sup next to the expectation is over the past up to time s i.e., over $x^\varepsilon(s) = x$, $y^\varepsilon(s) = y$. This follows immediately from $(2.3.4)-(2.3.6)$ upon taking $f(x) = x_i$, $i = 1,2,\ldots,n$, successively and squaring, etc. in $(2.3.4)$. Note that this choice of f is an unbounded function but $\psi_f^{(1)}$, A_f, $\mathcal{L}_3\psi_f^{(1)}$ and H_{f^ε} involve only drivatives of f and hence are bounded. Specifically, using vector notation, we have,

$$|x^\varepsilon(t)-x^\varepsilon(s)|^2 \leq 4\varepsilon\big[\psi_x^{(1)}(x^\varepsilon(t),y^\varepsilon(t)) - \psi_x^{(1)}(x^\varepsilon(s),y^\varepsilon(s))\big]^2$$

$$+ 4\left|\int_s^t \big[A_x(x^\varepsilon(\sigma),y^\varepsilon(\sigma)) + \varepsilon\mathcal{L}_3\psi_x^{(1)}(x^\varepsilon(\sigma),y^\varepsilon(\sigma))\big]d\sigma\right|^2$$

$$+ 4|M^\varepsilon(t)-M^\varepsilon(s)|^2 ,$$

where $M^\varepsilon(t)$ stands for $M_{f^\varepsilon}(t)$ with f^ε the vector function $x + \varepsilon\psi_x^{(1)}(x,y)$. Upon using

$$E\{|M^\varepsilon(t)-M^\varepsilon(s)|^2\big|\mathcal{F}_s\} = \int_s^t E\{|H^\varepsilon(x^\varepsilon(\sigma),y^\varepsilon(\sigma))|^2\big|\mathcal{F}_s\}d\sigma ,$$

with H^ε the vector function H_{f^ε} and f^ε as above, the result $(2.3.7)$ follows and Step 1 is complete.

For Step 2, the identification of the limit, we rewrite

(2.3.4), with f(x) a smooth bounded function, as follows

$$(2.3.8) \qquad f(x^\varepsilon(t)) - f(x^\varepsilon(s)) - \int_s^t \overline{\mathscr{L}} f(x^\varepsilon(\sigma)) d\sigma$$

$$= -\varepsilon \left[\psi_f^{(1)}(x^\varepsilon(t), y^\varepsilon(t)) - \psi_f^{(1)}(x^\varepsilon(s), y^\varepsilon(s)) \right]$$

$$+ \int_s^t \left[A_f(x^\varepsilon(\sigma), y^\varepsilon(\sigma)) - \overline{\mathscr{L}} f(x^\varepsilon(\sigma)) \right] d\sigma$$

$$+ \varepsilon \int_s^t \mathscr{L}_3 \psi_f^{(1)}(x^\varepsilon(\sigma), y^\varepsilon(\sigma)) d\sigma$$

$$+ M_{f^\varepsilon}(t) - M_{f^\varepsilon}(s) .$$

Multiplying (2.3.8) by any bounded continuous functional Φ_s

of $x(\cdot)$ which is $\mathscr{F}_s$ measurable, taking expectations and

passing to the limit $\varepsilon \to 0$ we obtain

$$(2.3.9) \qquad \lim_{\varepsilon \downarrow 0} E\left\{ \Phi_s \left[f(x^\varepsilon(t)) - f(x^\varepsilon(s)) - \int_s^t \overline{\mathscr{L}} f(x^\varepsilon(\sigma)) d\sigma \right] \right\}$$

$$= \lim_{\varepsilon \downarrow 0} E\left\{ \Phi_s \int_s^t \left[A_f(x^\varepsilon(\sigma), y^\varepsilon(\sigma)) - \mathscr{L} f(x^\varepsilon(\sigma)) \right] d\sigma \right\} .$$

The right hand side of (2.3.9) is zero by the definition

(2.2.5) of $\overline{\mathscr{L}} f(x)$ and the ergodicity of $Y(t;x)$ i.e., (2.2.2).

Since the coefficients are smooth here, this fact can be

seen immediately by defining first

(2.3.10) $\psi_f^{(2)}(x,y) = \int \psi(y,dz) A_f(x,z)$

so that

(2.3.11) $\mathscr{L}_1 \psi_f^{(2)} + A_f - \overline{\mathscr{L}} f = 0$.

Then we apply (2.3.4) with $f^\varepsilon = \varepsilon^2 \psi_f^{(2)}$ (using the smoothness
of the coefficients), multiply by Φ_s, take expectations and
pass to the limit $\varepsilon \to 0$ to obtain the result that the right
hand side of (2.3.9) is zero. If the coefficients are only
continuous, then the above is preceded by a preliminary
smoothing of the coefficients which is removed again at the
end.

In view of continuity of the functional on the left side
of (2.3.9) and the compactness it follows that for any limit
measure of the processes $x^\varepsilon(\cdot)$, the expression

$$f(x(t)) - f(x) - \int_0^t \overline{\mathscr{L}} f(x(s)) ds ,$$

is a martingale for all $f(x) \in C^\infty$ and bounded. But the solution to this martingale problem is unique by hypothesis hence
Step 2 and proof of the theorem is complete.

Remark.

Assuming smoothness as we do, the essential ingredient
of the above proof is the following formal computation.

Given $f(x)$ smooth and bounded let $\psi_f^{(1)}(x,y)$ and
$\psi_f^{(2)}(x,y)$ be two smooth functions to be chosen appropriately.
Let

(2.3.12) $f^\epsilon = f + \epsilon\psi_f^{(1)} + \epsilon^2\psi_f^{(2)}$.

Then

(2.3.13) $\mathscr{L}^\epsilon f^\epsilon = \left[\dfrac{1}{\epsilon^2}\mathscr{L}_1 + \dfrac{1}{\epsilon}\mathscr{L}_2 + \mathscr{L}_3\right]\left[f + \epsilon\psi_f^{(1)} + \epsilon^2\psi_f^{(2)}\right]$

$\qquad\qquad = \dfrac{1}{\epsilon}\left[\mathscr{L}_1\psi_f^{(1)} + \mathscr{L}_2 f\right] + \left[\mathscr{L}_1\psi_f^{(2)} + \mathscr{L}_2\psi_f^{(1)} + \mathscr{L}_3 f - \overline{\mathscr{L}}f\right]$

$\qquad\qquad\quad + \overline{\mathscr{L}}f + \epsilon\left[\mathscr{L}_2\psi_f^{(2)} + \mathscr{L}_3\psi_f^{(1)}\right] + \epsilon^2\mathscr{L}_3\psi_f^{(2)}$

$\qquad\qquad = \overline{\mathscr{L}}f + 0(\epsilon)$,

provided that

(2.3.14) $\mathscr{L}_1\psi_f^{(1)} + \mathscr{L}_2 f = 0$

and

(2.3.15) $\mathscr{L}_1\psi_f^{(2)} + \mathscr{L}_2\psi_f^{(1)} + \mathscr{L}_3 f - \overline{\mathscr{L}}f = 0$.

This is exactly how $\psi_f^{(1)}$ and $\psi_f^{(2)}$ were defined by (2.2.4)
(cf. (2.3.2)) and (2.3.11) (since $A_f \equiv \mathscr{L}_2\psi_f^{(1)} + \mathscr{L}_3 f$).

2.4   Stability and related questions.

In this section we discuss briefly how one can adapt, very simply in the present context, the usual Lyapounov methods for Markov processes (cf. for example [24]) to answer questions regarding the behavior of $x^\varepsilon(t)$ as $t \to \infty$, with ε fixed, $0 < \varepsilon \leq \varepsilon_o$ and ε_o sufficiently small. The objective here is to answer such questions by using only corresponding properties of the limit process $x(t)$ as $t \to \infty$. We assume, of course, that Theorem 2.1 holds.

Let $(x^\varepsilon(t), y^\varepsilon(t))$ be defined by (2.1.1) and assume that the hypotheses of Theorem 2.1 hold. Assume further that

$$(2.4.1) \quad F_i^{(1)}(0,y) = G_i^{(1)}(0,y) = 0 ,$$

$$\sigma_{ij}^{(1)}(0,y) = 0 , \quad i = 1,2,\ldots,n , \quad j = 1,2,\ldots,p ,$$

$$y \in R^m ,$$

which means that the origin is an equilibrium point of $x^\varepsilon(t)$. From (2.4.1) and (2.2.4)-(2.2.6) it follows that the origin is also an equilibrium point of $x(t)$ generated by $\overline{\mathscr{L}}$. A typical result is now the following.

Theorem 2.2.

Let V(x) be a smooth positive definite function on R^n i.e.,

$$(2.4.2) \quad V(x) \geq 0 , \quad V(x) = 0 \implies , \quad x = 0 .$$

Suppose that there is a constant $\gamma > 0$ and an open neighbor-hood K of the origin in R^n having compact closure, such that

(2.4.3) $\overline{\mathscr{L}}V(x) \leq -\gamma V(x)$, $x \in K$.

Then for $0 < \varepsilon \leq \varepsilon_o$, ε fixed and ε_o sufficiently small, $x^\varepsilon(t)$ is uniformly stochastically asymptotically stable as $t \to \infty$ i.e., for any $n_1 > 0$ and $n_2 > 0$, there is a $\delta > 0$ such that if $|x(0)| = |x| < \delta$ then

(2.4.4) $P_{x,y}\{|x^\varepsilon(t)| \leq n_2 e^{-\tilde{\gamma}t}$, $t \geq 0\} \geq 1 - n_1$

for all y and with $\tilde{\gamma} > 0$ a constant.

Proof.

Let $V(x)$ be as in the statement above and define $\psi_V^{(1)}(x,y)$ and $\psi_V^{(2)}(x,y)$ by (2.2.4) and (2.3.11), respectively, with $f = V$. Let

(2.4.5) $V^\varepsilon(x,y) = V(x) + \varepsilon\psi_V^{(1)}(x,y) + \varepsilon^2\psi_V^{(2)}(x,y)$.

From the smoothness and boundedness of the coefficients in (2.1.1) and (2.4.1), it follows that there is an ε_o suffi-ciently small and constants $\tilde{c}_1 > 0$ and $\tilde{c}_2 > 0$ such that for all $y \in R^m$ and $0 < \varepsilon \leq \varepsilon_o$

(2.4.6) $\tilde{c}_1 V(x) \leq V^\varepsilon(x,y) \leq \tilde{c}_2 V(x)$, $x \in K$.

Let $\hat{\gamma} > 0$ be a constant. From (2.3.13) with $f^\varepsilon = V^\varepsilon$ and (2.4.6) we have

(2.4.7) $(\mathscr{L}^\varepsilon + \hat{\gamma})v^\varepsilon = \hat{\gamma}v^\varepsilon + \overline{\mathscr{L}}v(x)$

$$+ \varepsilon\left[\mathscr{L}_2\psi_v^{(2)} + \mathscr{L}_3\psi_v^{(1)}\right] + \varepsilon^2\mathscr{L}_3\psi_v^{(2)}$$

$$\leq \left[\overline{\mathscr{L}} + \tilde{c}_2\hat{\gamma} + \varepsilon_o\tilde{c}_3\right]v , \quad x \in K ,$$

where $\tilde{c}_3$ is some constant. If we choose $\hat{\gamma}$ so that
$\tilde{c}_2\hat{\gamma} + \varepsilon_o\tilde{c}_3 \leq \gamma$ (ε_o being sufficiently small), we find that
(2.4.3) yields

(2.4.8) $(\mathscr{L}^\varepsilon + \hat{\gamma})v^\varepsilon(x,y) \leq 0 , \qquad x \in K , \qquad y \in R^m .$

Suppose that $x^\varepsilon(0) = x \in K$ and let τ be the first exit
time of $x^\varepsilon(t)$ from K. We use the identity (2.3.4) with
$f^\varepsilon = e^{t\hat{\gamma}}v^\varepsilon$ and the upper time limit equal to $\tau \wedge t = \min(t,\tau)$
to obtain

(2.4.9) $e^{\hat{\gamma}\tau \wedge t}v^\varepsilon(x^\varepsilon(t \wedge \tau),y^\varepsilon(t \wedge \tau)) = v^\varepsilon(x,y)$

$$+ \int_0^{t \wedge \tau} e^{s\hat{\gamma}}(\mathscr{L}^\varepsilon + \hat{\gamma})v^\varepsilon(x^\varepsilon(s),y^\varepsilon(s))ds$$

$$+ \tilde{M}_{v^\varepsilon}(t \wedge \tau) .$$

Here $\tilde{M}_{v^\varepsilon}(t \wedge \tau)$ is a zero mean bounded martingale with increasing process as in (2.3.5) with $f^\varepsilon = v^\varepsilon$, an extra factor
$e^{2\hat{\gamma}t}$ multiplying H_{v^ε} and $t \wedge \tau$ for upper limit of intergra-
tion in (2.3.5).

Using (2.4.6) and (2.4.8) in (2.4.9) we obtain

$(2.4.10) \quad 0 \le e^{\gamma \tau \wedge t} \tilde{c}_1 V(x^{\epsilon}(t \wedge \tau)) \le \tilde{c}_2 V(x) + \tilde{M}_{V^{\epsilon}}(t \wedge \tau)$.

Thus, $\tilde{c}_2 V + \tilde{M}_{V^{\epsilon}}(t \wedge \tau)$ is a nonnegative bounded martingale and by Kolmogorov's inequality we have that for each $\tilde{\eta}_2 > 0$

$(2.4.11) \quad P_{x,y} \left\{ \sup_{0 \le t \le T} e^{\hat{\gamma} t \wedge \tau} \tilde{c}_1 V(x^{\epsilon}(t \wedge \tau)) > \tilde{\eta}_2 \right\}$

$$\le P_{x,y} \left\{ \sup_{0 \le t \le T} (\tilde{c}_2 V(x) + \tilde{M}_{V^{\epsilon}}(t \wedge \tau)) > \tilde{\eta}_2 \right\}$$

$$\le \frac{\tilde{c}_2 V(x)}{\tilde{\eta}_2} .$$

Since the right hand side in (2.4.11) is independent of the set K and T, it follows that

$(2.4.12) \quad P_{x,y} \left\{ \sup_{t \ge 0} e^{\hat{\gamma} t} \tilde{c}_1 V(x^{\epsilon}(t)) > \tilde{\eta}_2 \right\} \le \frac{\tilde{c}_2 V(x)}{\tilde{\eta}_2}$.

The positive definiteness (2.4.2) of V(x) and estimate (2.4.12) yield, in the usual way, the result (2.4.4). The proof is complete.

Some more details regarding stability and related questions can be found in [32].

2.5 Boundary conditions.

Let $\phi(x)$, $x \in R^n$, be a smooth real-valued function such that

(2.5.1) $\mathscr{O} = \{x \in R^n \mid \phi(x) > 0\}$

is an open connected set in R^n with compact closure. We denote the boundary of $\mathscr{O}$ by $\partial\mathscr{O}$ and the unit inner normal at $x \in \partial\mathscr{O}$ by $\hat{n}(x)$. We assume that ϕ is such that $\nabla\phi(x) \equiv \hat{n}(x)$ for $x \in \partial\mathscr{O}$.

We shall consider the process $(x^\varepsilon(t), y^\varepsilon(t))$ with generator $\mathscr{L}^\varepsilon$ given by (2.1.2)-(2.1.5) in $\mathscr{O} \times R^m$ and we shall prescribe reflecting boundary conditions for $x^\varepsilon(t)$ on $\partial\mathscr{O}$. Throughout this section $\mathscr{L}^\varepsilon$ is assumed to be uniformly elliptic in $\bar{\mathscr{O}} \times R^m$.

Boundary conditions are given as follows. Let $B(x,y)$, $x \in \partial\mathscr{O}$, $y \in R^m$, be a smooth n-vector function pointing inward at the boundary. Specifically we assume that there is a $\delta > 0$ such that

(2.5.2) $\displaystyle\inf_{\substack{x \in \partial\mathscr{O} \\ y \in R^m}} \sum_{i=1}^{n} B_i(x,y) \frac{\partial\phi(x)}{\partial x_i} \geq \delta$.

Boundary conditions are prescribed so that $x^\varepsilon(t)$ is reflected obliquely at $\partial\mathscr{O}$ in the direction B. According to the results of [5] the reflected process in $\bar{\mathscr{O}} \times R^m$, again denoted by $(x^\varepsilon(t), y^\varepsilon(t))$, is under our hypotheses the unique solution of the following submartingale problem $(\mathscr{L}^\varepsilon, B)$. The measures $P^\epsilon_{x,y}$ in the space of continuous functions in $\overline{\mathscr{D}} \times R^m$ corresponding to $(x^\epsilon(t), y^\epsilon(t))$ satisfy

(i) $P^\epsilon_{x,y}\{x(0) = x, y(0) = y\} = 1$,

$$x \in \overline{\mathscr{D}} , \qquad y \in R^m ,$$

(2.5.3)   (ii)   for all bounded smooth functions on $\overline{\mathscr{D}} \times R^m$ such that

$$\sum_{i=1}^{n} B_i(x,y) \frac{\partial f(x,y)}{\partial x_i} \geq 0 ,$$

$$x \in \partial\mathscr{D} , \qquad y \in R^m ,$$

the expression

$$f(x(t), y(t)) - \int_0^t \mathscr{L}^\epsilon f(x(s), y(s)) ds$$

is a submartingale for $P^\epsilon_{x,y}$.

We shall assume, with no loss in generality in view of (2.5.2), that the vector field $B(x,y)$ is normalized so that

(2.5.4) $B_{\hat{n}}(x,y) \equiv \sum_{i=1}^{m} B_i(x,y) \frac{\partial \phi(x)}{\partial x_i} \equiv 1$.

The corresponding increasing process of $x^\epsilon(t)$ at the boundary will be denoted by $N^\epsilon(t)$. It is defined by the property that for all $f(x,y)$ smooth and bounded on $\overline{\mathscr{D}} \times R^m$,

(2.5.5) $f(x^\epsilon(t), y^\epsilon(t)) - \int_0^t \mathscr{L}^\epsilon f(x^\epsilon(s), y^\epsilon(s))$

$$- \int_0^t \sum_{i=1}^{n} B_i(x^\epsilon(s), y^\epsilon(s)) \frac{\partial f(x^\epsilon(s), y^\epsilon(s))}{\partial x_i} dN^\epsilon(s)$$

$$\equiv M_f(t)$$

231

is a martingale (when superscript ε appears on the paths we do not refer explicitly to the measure P^ε as in (2.5.3)). We refer to [5] for the details of the existence and uniqueness theory.

According to Theorem 2.1 the process $x^\varepsilon(t)$ converges, away from boundaries, to the diffusion process $x(t)$ generated by $\mathscr{L}$ of (2.2.5), (2.2.6). The question is: What happens to the boundary conditions as $\varepsilon \to 0$? Stated another way, we wish to characterize the diffusion process in $\mathscr{D}$ to which the reflected process $x^\varepsilon(t)$ converges weakly (under our hypotheses) and, of course, to prove weak convergence.

It is clear that the behavior of the reflected process $x^\varepsilon(t)$ must be analyzed in detail near the boundary $\partial\mathscr{D}$. For this purpose we introduce stretched boundary layer coordinates as follows.

Since $\partial\mathscr{D}$ is a compact (n-1)-dimensional submanifold of R^n, it can be covered by a finite system of neighborhoods in each of which we have a given coordinate patch. We fix attention to a single patch and let $x = \xi(\gamma)$, $\gamma \in R^{n-1}$ be the parametric equation of the surface $\partial\mathscr{D}$.

We introduce stretched coordinates $x \to (\gamma, \varepsilon\eta)$ near $\partial\mathscr{D}$ by setting

(2.5.6) $x = \xi(\gamma) + \varepsilon\eta\hat{n}(\gamma)$.

Here $\hat{n}(x)$, $x \in \partial\mathscr{D}$ is the unit inner normal and we think of it, alternatively, as a function $x \in \partial\mathscr{D}$ or of γ abbreviating $\hat{n}(\xi(\gamma)) = \hat{n}(\gamma)$. The scalar parameter $\eta \geq 0$ is confined to an interval depending on the size of the patch and goes to $+\infty$ as $\epsilon \to 0$. In fact we have

$$(2.5.7) \qquad \eta = \frac{(x-\xi(x)) \cdot \hat{n}(x)}{\epsilon} \; ,$$

where, in returning from $(\gamma, \epsilon \eta)$ to x, we write $\xi(x)$ instead of $\xi(\gamma(x))$, with $\gamma(x)$ being uniquely defined if the patch is small enough.

With $\xi \in \partial \mathcal{O}$ a fixed parameter we define the operator $\mathcal{L}_{BL}$, acting on functions of $y \in R^m$ and $\eta \geq 0$ as follows (recall that $\hat{n}(x) \equiv \nabla\phi(x)$ by hypothesis):

$$(2.5.8) \quad \mathcal{L}_{BL} = \sum_{k=1}^{m} F_k^{(1)}(\xi,y) \frac{\partial}{\partial y_k} + \frac{1}{2} \sum_{k,\ell=1}^{m} a_{k\ell}^{(2)}(\xi,y) \frac{\partial^2}{\partial y_k \partial y_\ell}$$

$$+ \sum_{i=1}^{n} F_i^{(1)}(\xi,y) \frac{\partial\phi(\xi)}{\partial x_i} \frac{\partial}{\partial \eta}$$

$$+ \sum_{k=1}^{m} \left[\sum_{i=1}^{n} a_{ik}^{(12)}(\xi,y) \frac{\partial\phi(\xi)}{\partial x_i} \right] \frac{\partial^2}{\partial \eta \partial y_k}$$

$$+ \frac{1}{2} \left[\sum_{i,j=1}^{n} a_{ij}^{(1)}(\xi,y) \frac{\partial\phi(\xi)}{\partial x_i} \frac{\partial\phi(\xi)}{\partial x_j} \right] \frac{\partial^2}{\partial \eta^2} \; .$$

This operator, the boundary layer operator, is characterized uniquely by the following property. With $\mathcal{L}^\epsilon$, $\mathcal{L}_1$, $\mathcal{L}_2$ and $\mathcal{L}_3$ defined by $(2.1.2)-(2.1.5)$ and with the change of variables $x \to (\gamma, \epsilon \eta)$ of $(2.5.6)$,

$$(2.5.9) \qquad \epsilon \left(\mathcal{L}^\epsilon - \frac{1}{\epsilon^2} \mathcal{L}_{BL} \right) f(\gamma,\eta)$$

is well defined near the boundary on smooth functions and is
bounded independently of ε.

We shall assume, as in Section 2.2, that $\mathscr{L}_{BL}$ has the
following ergodic properties. Their validity (and of the ones
in Section 2.2) is discussed in the Appendix.

There exists a unique probability measure $\bar{U}(A;\xi)$, A a
Borel subset of R^m, depending on $\xi \in \partial\mathscr{O}$ such that for any
bounded smooth function $\psi(y)$ on R^m with

$$(2.5.10) \quad \int_{R^m} \psi(y)\bar{U}(dy;\xi) = 0 ,$$

the boundary value problem

$$(2.5.11) \quad \begin{aligned} \mathscr{L}_{BL}v(\eta,y) &= 0 , \quad \eta > 0 , \quad y \in R^m , \\ \frac{\partial v(\eta,y)}{\partial \eta}\bigg|_{\eta=0} &= \psi(y) , \quad y \in R^m , \end{aligned}$$

has a smooth solution which can be made unique by imposing
the condition

$$(2.5.12) \quad \lim_{\eta \to +\infty} v(\eta,y) = 0 .$$

The limit zero in (2.5.12) is approached exponentially
fast, uniformly[†] in y, and so do derivatives of v.

Some additional information regarding $\mathscr{L}_{BL}$ is given in
the appendix. We comment briefly on the probabilistic significance of the above.

The operator $\mathscr{L}_{BL}$ with Neumann boundary condition at
$\eta = 0$ is the generator on $[0,\infty) \times R^m$ of a reflected process

---

[†]Uniformly on compact sets suffices in all considerations.

$(\eta(t), y(t))$. Let $N(t)$ be the increasing process of $y(t)$ at $\eta = 0$. Then the solution $v(\eta, y)$ of (2.5.11) has the representation

$$(2.5.13) \qquad v(\eta, y) = \lim_{t \uparrow \infty} E \left\{ \int_0^t \psi(y(s)) dN(s) \right\} ,$$

where $\eta(0) = \eta$ and $y(0) = y$ and provided the limit exists. Let

$$\tau_\theta = \sup\{s: s \geq 0, \ N(s) \leq \theta\} .$$

We show in the appendix that $\tau_\theta \to \infty$ as $\theta \to \infty$ a.s. and that we can write

$$(2.5.14) \qquad v(\eta, y) = \lim_{\theta \uparrow \infty} E \left\{ \int_0^\theta \psi(\tilde{y}(\zeta)) d\zeta \right\}$$

where $\tilde{y}(\zeta) = y(\tau_\zeta)$ is the boundary process [5]. This process is ergodic and the $\bar{U}$ introduced above is its stationary measure. This explains (2.5.10).

We return now to our main objective in this section which is the asymptotic analysis.

For each $x \in \partial \mathcal{D}$ we define the vector field

$$(2.5.15) \qquad \bar{B}(x) = \int_{R^m} B(x, y) \bar{U}(dy; x) , \qquad x \in \partial \mathcal{D} .$$

In view of (2.5.2), there is a $\delta > 0$ such that

$$(2.5.16) \qquad \inf_{x \in \partial \mathcal{D}} \sum_{i=1}^n \bar{B}_i(x) \frac{\partial \phi(x)}{\partial x_i} \geq \delta ,$$

and $\bar{B}(x)$ is smooth. Let $\mathcal{L}$ be defined by (2.2.5) and (2.2.6).

Theorem 2.3.

The reflected process $x^\varepsilon(t)$ under the above hypotheses converges weakly in $C([0,T];\overline{\mathcal{D}})$, $T < \infty$ but arbitrary, as $\varepsilon \to 0$ to the reflected diffusion process $x(t)$ in $\overline{\mathcal{D}}$ which is the unique solution of the submartingale problem ($\overline{\mathcal{L}}$, $\overline{B}$).

Proof.

As with Theorem 2.1, the proof is divided into two steps: (1) compactness in $D([0,T];\overline{\mathcal{D}})$ and (2) identification of the limit with ($\overline{\mathcal{L}}$, $\overline{B}$).

Let $f(x)$ be a smooth function on $\overline{\mathcal{D}}$ and define $\psi_f^{(1)}$ and ψ_f^{BL} as follows. The function $\psi_f^{(1)}(x,y)$ is defined by (2.2.4), as before, so that (2.3.2) holds. The function ψ_f^{BL} is a function of $y \in R^m$ and $\eta \geq 0$ with $\xi \in \partial\mathcal{D}$ a parameter (actually, a function of $\gamma \in R^{n-1}$ i.e. we think of ξ as $\xi(\gamma)$). It satisfies the problem

$$\mathcal{L}_{BL}\psi_f^{BL}(\eta,y;\xi) = 0 , \quad \eta > 0 , \quad y \in R^m , \quad \xi \in \partial\mathcal{D},$$

$$\frac{\partial}{\partial\eta} \psi_f^{BL}(\eta,y;\xi)\Big|_{\eta=0} = -\sum_{i=1}^{n} B_i(\xi,y)\frac{\partial f(\xi)}{\partial x_i}$$

(2.5.17)

$$+ \sum_{i=1}^{n} \overline{B}_i(\xi) \frac{\partial f(\xi)}{\partial x_i} ,$$

$$\lim_{\eta\to+\infty} \psi_f^{BL}(\eta,y;\xi) = 0 .$$

In view of definition (2.5.15) and (2.5.10)-(2.5.12), ψ_f^{BL} is well defined and, along with derivatives, decays to zero exponentially as $\eta \to \infty$.

Define

$$(2.5.18) \quad f^\varepsilon(x,y) = f(x) + \varepsilon \psi_f^{(1)}(x,y)$$

$$+ \varepsilon \zeta(x) \psi_f^{BL}\left[\frac{(x-\zeta(x)) \cdot \hat{n}(x)}{\varepsilon}, y; \xi(x)\right] ,$$

where $\zeta(x)$ is a C^∞ function identically equal to one in a fixed neighborhood θ_1 of $\partial \mathcal{D}$, covered by the coordinate patches, and identically equal to zero outside some fixed neighborhood $\theta_2 \supset \theta_1$. From the construction of $\psi_f^{(1)}$ and ψ_f^{BL} we have that

$$(2.5.19) \quad \mathcal{L}^\varepsilon f^\varepsilon(x,y) = A_f(x,y) + \varepsilon \mathcal{L}_3 \psi_f^{(1)}(x,y) + g_f^\varepsilon(x,y) ,$$

$$x \in \mathcal{D}, \quad y \in R^m ,$$

where A_f is defined by (2.2.6) and, because of (2.5.17),

$$(2.5.20) \quad g_f^\varepsilon = \varepsilon\left(\mathcal{L}^\varepsilon \zeta \psi_f^{BL} - \zeta \mathcal{L}^\varepsilon \psi_f^{BL}\right)$$

$$+ \varepsilon \zeta\left(\mathcal{L}^\varepsilon \psi_f^{BL} - \frac{1}{\varepsilon^2} \mathcal{L}_{BL} \psi_f^{BL}\right) .$$

The first term on the right side of (2.5.20) is identically equal to zero for $x \in \theta_1 \cap \overline{\mathcal{D}}$ and $x \notin \theta_2 \cap \overline{\mathcal{D}}$. In the rest of $\overline{\mathcal{D}}$ it decays to zero exponentially since ψ_f^{BL} and its derivatives decay to zero as $\eta \to \infty$. The second term on the right side of (2.5.20) is identically zero for $x \notin \theta_2 \cap \overline{\mathcal{D}}$ and uniformly bounded as $\varepsilon \downarrow 0$ in the rest of $\overline{\mathcal{D}}$; this follows from the characterization (2.5.9).

For $x \in \partial \mathcal{D}$ we have, by construction again and by (2.5.4),

$$(2.5.21) \quad \sum_{i=1}^{n} B_i(x,y) \left. \frac{\partial f^{\varepsilon}(x,y)}{\partial x_i} \right|_{x \in \partial \mathcal{D}}$$

$$= \sum_{i=1}^{n} \bar{B}_i(x) \left. \frac{\partial f(x)}{\partial x_i} \right|_{x \in \partial \mathcal{D}} + h_f^{\varepsilon}(x,y) \ .$$

Here $h_f^{\varepsilon} = 0(\varepsilon)$ because $\psi_f^{(1)}(x,y)$, $\psi_f^{BL}(\eta,y;\xi)$ and derivatives are bounded.

The above constructions and (2.5.5) with $f = f^{\varepsilon}$ lead to the following:

$$(2.5.22) \quad f(x^{\varepsilon}(t)) + \varepsilon \psi_f^{(1)}(x^{\varepsilon}(t),y^{\varepsilon}(t))$$

$$+ \ \varepsilon \zeta(x^{\varepsilon}(t)) \psi_f^{BL}\left[\frac{(x^{\varepsilon}(t)-\xi(x^{\varepsilon}(t))) \cdot \hat{n}(x^{\varepsilon}(t))}{\varepsilon}, y^{\varepsilon}(t); \xi(x^{\varepsilon}(t)) \right]$$

$$- \ f(x) \ - \ \varepsilon \psi_f^{(1)}(x,y) \ - \ \varepsilon \zeta(x) \psi_f^{BL}\left[\frac{(x-\xi(x)) \cdot \hat{n}(x)}{\varepsilon}, y; \xi(x) \right]$$

$$- \int_0^t \overline{\mathscr{L}} f(x^{\varepsilon}(s)) ds \ - \int_0^t (A_f - \overline{\mathscr{L}} f)(x^{\varepsilon}(s),y^{\varepsilon}(s)) ds$$

$$- \int_0^t \varepsilon \mathscr{L}_3 \psi_f^{(1)}(x^{\varepsilon}(s),y^{\varepsilon}(s)) ds \ - \int_0^t g_f(x^{\varepsilon}(s),y^{\varepsilon}(s)) ds$$

$$- \int_0^t \bar{B}(x^{\varepsilon}(s)) \cdot \frac{\partial f(x^{\varepsilon}(s))}{\partial x} \ dN^{\varepsilon}(s)$$

$$- \int_0^t h_f(x^{\varepsilon}(s),y^{\varepsilon}(s)) dN^{\varepsilon}(s)$$

$$\equiv M_{f^{\varepsilon}}(t) \ ,$$

where $M_{f^\varepsilon}(t)$ is a zero-mean martingale and $\bar{B} \cdot \frac{\partial f}{\partial x}$ stands for the inner product of $\bar{B}$ and the gradient of f (we shall see shortly that $N^\varepsilon(t)$ has finite second moments).

For Step 1, compactness, it suffices to show that (2.3.7) holds again. Suppose that $f(x) = x$, the vector of coordinate functions (letting f be a vector is all right, of course) in (2.5.22) and that the lower limit is not zero but s with $s \le t$. It is clear that (2.3.7) will hold provided we show that

$$(2.5.23) \qquad \overline{\lim_{\delta \downarrow 0}} \; \overline{\lim_{\varepsilon \downarrow 0}} \; \sup_{|t-s| \le \delta} \; \sup \; E\{ (N^\varepsilon(t) - N^\varepsilon(s))^2 | \mathcal{F}_s \} = 0 \; ,$$

where the sup next to the expectation is over all $x^\varepsilon(s) = x \in \overline{\mathcal{D}}$ and $y^\varepsilon(s) = y \in R^m$.

To prove (2.5.23) we note that the increasing process corresponding to $M_f(t)$ in (2.5.5) is given by

$$(2.5.24) \qquad <M_f(t), M_f(t)> = \int_0^t (\mathcal{L}^\varepsilon f^2 - 2f\mathcal{L}^\varepsilon f)(x^\varepsilon(s), y^\varepsilon(s))ds \; .$$

At this point, we do not know that $N^\varepsilon(t)$ has second moments; this will follow later. If we let τ_ν be the first time N(t) equals $\nu < \infty$, then τ_ν is a stopping time and the above is true with t replaced by $t \wedge \tau_\nu$. If the same is done with the calculations that follow, the desired estimate will be independent of ν so letting $\nu \to \infty$ in the end will produce the results. After this remark we continue with (2.5.23).

We apply (2.5.22) with $f(x) = \phi(x)$, the support function of the region $\mathcal{B}$. In view of (2.5.16) there is a $\delta_1 > 0$ so that for ε small enough

(2.5.25)
$$N^\varepsilon(t) - N^\varepsilon(s) \leq \frac{1}{\delta_1} \int_s^t \left[\bar{B} \cdot \frac{\partial \phi}{\partial x} + h_\phi^\varepsilon \right] (x^\varepsilon(\sigma), y^\varepsilon(\sigma)) dN^\varepsilon(\sigma)$$

$$= \frac{1}{\delta_1} \left[\phi(x^\varepsilon(t)) - \phi(x^\varepsilon(s)) - \left[M_{\phi^\varepsilon}(t) - M_{\phi^\varepsilon}(s) \right] \right]$$

$$+ 0(t-s) + 0(\varepsilon) .$$

Hence,

(2.5.26)
$$E\{ (N^\varepsilon(t) - N^\varepsilon(s))^2 | \mathcal{F}_s \}$$

$$\leq \frac{4}{\delta_1} E\{ (\phi(x^\varepsilon(t)) - \phi(x^\varepsilon(s)))^2 | \mathcal{F}_s \}$$

$$+ \frac{4}{\delta_1} E\{ (M_{\phi^\varepsilon}(t) - M_{\phi^\varepsilon}(s))^2 | \mathcal{F}_s \}$$

$$+ 0(t-s) + 0(\varepsilon) .$$

Using (2.5.24) with $f = \phi^\varepsilon$ (i.e., the f^ε that results in (2.5.18) when $f = \phi$) it follows that the second term on the right of (2.5.26) is $0(t-s)$. Hence it remains to consider the first term on the right.

Using (2.5.22) with $f = \phi^2$ it follows that

(2.5.27)
$$E\{ (\phi(x^\varepsilon(t)) - \phi(x^\varepsilon(s))^2 | \mathcal{F}_s \}$$

$$= E\{ \phi^2(x^\varepsilon(t)) - \phi^2(x^\varepsilon(s)) | \mathcal{F}_s \}$$

$$- 2\phi(x^\varepsilon(s)) E\{ \phi(x^\varepsilon(t)) - \phi(x^\varepsilon(s)) | \mathcal{F}_s \}$$

$$= 0(t-s) + 0(\varepsilon)$$

$$- 2\phi(x^\varepsilon(s))E\left\{\int_s^t \left[\bar{B}\cdot\frac{\partial\phi}{\partial x} + h_\phi^\varepsilon\right](x^\varepsilon(s),y^\varepsilon(s))dN^\varepsilon(s)\right\}.$$

But then the last term on the right is nonpositive for ε sufficiently small. Combining (2.5.27) and (2.5.26) yields (2.5.23).

Now we pass to Step 2 the idenfication. With $f(x)$ a smooth function on $\overline{\mathcal{D}}$ we rewrite (2.5.22) as follows

$$(2.5.28) \qquad f(x^\varepsilon(t)) - f(x) - \int_0^t \overline{\mathcal{L}}f(x^\varepsilon(s)ds$$

$$- \int_0^t \bar{B}(x^\varepsilon(s))\cdot\frac{\partial f(x^\varepsilon(s))}{\partial x} \, dN^\varepsilon(s)$$

$$= \int_0^t (A_f - \overline{\mathcal{L}}f)(x^\varepsilon(s),y^\varepsilon(s))ds$$

$$+ \int_0^t g_f^\varepsilon(x^\varepsilon(s),y^\varepsilon(s)ds + M_{f^\varepsilon}(t) + 0(\varepsilon).$$

As with Step 2 of Theorem 2.1, it suffices to show that the expectation of the right hand side of (2.5.28) tends to zero as $\varepsilon \to 0$. This implies that for any continuous $\mathcal{F}_s$ measurable functional Φ_s

$$(2.5.29) \qquad \lim_{\varepsilon\downarrow 0} E\left\{\Phi_s\left[f(x^\varepsilon(t)) - f(x^\varepsilon(s)) - \int_s^t \mathcal{L}f(x^\varepsilon(\sigma))d\sigma\right]\right\} \geq 0,$$

for all f smooth with $\bar{B}\cdot\frac{\partial f}{\partial x} \geq 0$. Hence the limit coincides with the solution $(\overline{\mathcal{L}},\bar{B})$ of the submartingale problem.

The first term on the right side of (2.5.28) has zero expectation as $\varepsilon \to 0$ by exactly the same argument following (2.3.9). Hence it remains to examine the second term.

Recall that g_f^ε is defined by (3.5.20) and is uniformly bounded in $\overline{\mathcal{O}}$ for all ε and goes to zero exponentially fast as $\varepsilon \to 0$ uniformly in any compact subset of $\mathcal{O}$. Let $\delta > 0$ be fixed and define

$$\mathcal{O}_\delta = \{x \in \overline{\mathcal{O}} \mid |x - \partial_a\mathcal{O}| < \delta\} \ .$$

Now from the remarks just made we have

$$(2.5.30) \quad \overline{\lim_{\delta \downarrow 0}} \ \overline{\lim_{\varepsilon \downarrow 0}} \ E\left\{ \int_0^t g_f^\varepsilon(x^\varepsilon(s), y^\varepsilon(s)) ds \right\}$$

$$\leq c \ \overline{\lim_{\delta \downarrow 0}} \ \overline{\lim_{\varepsilon \downarrow 0}} \ E\left\{ \int_0^t \chi_{\mathcal{O}_\delta}(x^\varepsilon(s)) ds \right\} \ ,$$

where c is a constant and $\chi_{\mathcal{O}_\delta}(x)$ is the characteristic function of the set $\mathcal{O}_\delta$.

To show that the right hand side of (2.5.30) is zero we proceed in a manner that is a bit more general than needed. Namely, we obtain the desired result using the uniform ellipticity of $\overline{\mathcal{L}}$ (the limit operator) only and not the uniform ellipticity of $\mathcal{L}^\varepsilon$ (which was used only to get existence and uniqueness of $(x^\varepsilon(t), y^\varepsilon(t))$).

Define $\eta_\delta(r)$ by

$$(2.5.31) \quad \eta_\delta(r) = \begin{cases} \dfrac{1}{\delta} (\delta-r)^3 \ , & 0 \leq r \leq \delta \\ 0 \ , & r \geq \delta \end{cases}$$

and

(2.5.32) $\hat{\phi}_\delta(x) = \eta_\delta(\phi(x)) = \eta_\delta \circ \phi(x)$,

where $\phi(x)$ is the support function of $\mathscr{D}$. Let $\psi^{(1)}_{\hat{\phi}_\delta}(x,y)$ be
the function defined by (2.2.4) with $f = \hat{\phi}$ and put

(2.5.33) $\hat{\phi}^\epsilon_\delta(x,y) = \hat{\phi}_\delta(x) + \epsilon \psi^{(1)}_{\hat{\phi}_\delta}(x,y)$.

Then, as in (2.5.19) with $\psi^{BL} \equiv 0$,

(2.5.34) $\mathscr{L}^\epsilon \hat{\phi}^\epsilon_\delta = \overline{\mathscr{L}} \hat{\phi}_\delta + A_{\hat{\phi}_\delta} - \overline{\mathscr{L}} \hat{\phi} + 0(\epsilon)$.

Also, let us write $\overline{\mathscr{L}}$ of (2.2.5) in the usual form

$$\overline{\mathscr{L}} = \frac{1}{2} \sum_{i,j=1}^{n} \bar{a}_{ij}(x) \frac{\partial^2}{\partial x_i \partial x_j} + \sum_{j=1}^{n} \bar{b}_j(x) \frac{\partial}{\partial x_j}$$

so that

(2.5.35) $\overline{\mathscr{L}} \hat{\phi}_\delta(x) = \eta_\delta''(\phi(x)) \frac{1}{2} \sum_{i,j=1}^{n} \bar{a}_{ij}(x) \frac{\partial \phi}{\partial x_i} \frac{\partial \phi}{\partial x_j}$

$$+ \eta_\delta'(\phi(x)) \overline{\mathscr{L}} \phi(x) .$$

We now use the above in the martingale (2.5.5) with
$f = \hat{\phi}^\epsilon_\delta$ to obtain the estimate

(2.5.36) $E\left\{ \int_0^t \left[\eta_\delta'' \circ \phi \sum_{i,j=1}^{n} \bar{a}_{ij} \frac{\partial \phi}{\partial x_i} \frac{\partial \phi}{\partial x_j} \right] (x^\epsilon(s)) ds \right\}$

$$= 0(\delta) + 0(\epsilon) + c_1 \left| E\left\{ \int_0^t (A_{\hat{\phi}_\delta} - \overline{\mathscr{L}}_{\hat{\phi}_\delta})(x^\epsilon(s), y^\epsilon(s)) ds \right\} \right|$$

$$+ c_2 \delta\, E\{N^\epsilon(t)\} .$$

From the remark following (2.5.28) and the estimate on moments of $N^\varepsilon(t)$, it follows that

$$(2.5.37) \quad \varliminf_{\delta\downarrow 0} \varliminf_{\varepsilon\downarrow 0} E\left\{\int_0^t \eta_\delta'' \circ \phi \sum_{i,j=1}^n \bar{a}_{ij} \frac{\partial\phi}{\partial x_i} \frac{\partial\phi}{\partial x_j} (x^\varepsilon(s))ds\right\} = 0 .$$

Thus, uniform ellipticity of $\mathcal{L}$, $|\nabla\phi| = 1$ and the construction of η_δ in (2.5.31) imply that for $t < \infty$

$$\varliminf_{\delta\downarrow 0} \varliminf_{\varepsilon\downarrow 0} E\left\{\int_0^t \chi_{\mathscr{B}_\delta} (x^\varepsilon(s))ds\right\} = 0 .$$

The proof of Theorem 2.3 is complete.

Remark 1.

The smoothness of the coefficients is a convenient assumption to avoid counting derivatives. The method used above and the results of [5] yield results under minimal regularity hypotheses.

Remark 2.

When smoothness is available, then one can construct full asymptotic expansions to solutions of the backward equation associated with the reflected $(x^\varepsilon(t), y^\varepsilon(t))$ process. In the context of transport problems (see next section) this is carried out in [30]. The methods used there extend directly to the problems of this section.

<u>Remark 3.</u>

More general boundary conditions, not just oblique reflection, can be treated by using the results of Anderson
[33].

2.6 Transport problems.

This is a class of processes $(x^\varepsilon(t), y^\varepsilon(t))$ scaled in the same manner as (2.1.1) but with $y^\varepsilon(t)$ being a jump process and not a diffusion. We shall consider a typical situation, not the most general, and we shall concentrate on the asymptotics.

Let $q(x,y) \geq 0$ be a bounded function smooth in x and let $\pi(x,y,A)$, A a Borel subset of R^m, be a probability measure for each $(x,y) \in R^n \times R^m$, smoothly depending on x. Define on bounded continuous functions on R^m the bounded operator Q by

$$(2.6.1) \qquad Qf(y) = q(x,y) \int_{R^m} \pi(x,y,dz) f(x) - q(x,y) f(y) .$$

Here $x \in R^n$ is merely a parameter. Let $F(x,y)$ and $H(x,y)$ be smooth in x and bounded n-vector functions on $R^n \times R^m$ and define a differential operator $\mathscr{L}^\varepsilon$ as follows

$$(2.6.2) \qquad \mathscr{L}^\varepsilon = \frac{1}{\varepsilon^2} Q + \frac{1}{\varepsilon} F \cdot \frac{\partial}{\partial x} + H \cdot \frac{\partial}{\partial x} ,$$

where

$$F \cdot \frac{\partial}{\partial x} = \sum_{i=1}^{n} F_i(x,y) \frac{\partial}{\partial x_i} .$$

Note that $\mathscr{L}^\varepsilon$ has the form (2.1.2) but $\mathscr{L}_1$, $\mathscr{L}_2$ and $\mathscr{L}_3$ are a bit different here.

It is elementary to show that $\mathscr{L}^\varepsilon$ of (2.6.2) is the infinitesimal generator of a Markov process $(x^\varepsilon(t), y^\varepsilon(t))$ with $\varepsilon > 0$ fixed. This process has very simple structure which

which motivates the name transport process: $x^\varepsilon(t)$ moves along the orbits of the vector field $\varepsilon^{-1}F + H$ in between jumps of $y^\varepsilon(t)$ which occur with frequency proportional to ε^{-2}. The scaling corresponds to the situation described after example 4 in Section 2.1 and is the usual diffusion limit of transport theory. The sample space of the process can be taken to be $C([0,\infty),R^n) \times D([0,\infty) \times R^m)$. It is a Feller process and hence a strong Markov process.

Let $Y(t;x)$ be the jump process in R^m generated by Q of (2.6.1) with $x \in R^n$ a parameter. We introduce here the same hypotheses regarding ergodicity of $Y(t;x)$ as in Section 2.2.

Assume that there are constants q_ℓ and q_u such that

(2.6.3) $0 < q_\ell \leq q(x,y) \leq q_u < \infty$,

and that $\pi(x,y,A)$ has a density $\tilde{\pi}(x,y,z)$ relative to a non-trivial probability measure on R^m such that

(2.6.4) $0 < \pi_\ell \leq \tilde{\pi}(x,y,z) \leq \pi_u < \infty$.

Let $P(t,y,A;x)$ be the transition probability function of $Y(t;x)$. It is easily seen that (2.6.3) and (2.6.4) imply that there exists a unique invariant probability measure $\bar{P}(A;x)$ and that

$$|P(t,y,A;x) - \bar{P}(A;x)| \leq e^{-\alpha t}$$

for some $\alpha > 0$, for t large enough and for all $y \in R^m$, $A \in R^n$ as well as $x \in R^n$. Consequently, the recurrent potential kernel

(2.6.5) $\psi(y,A;x) = \int_0^\infty [P(t,y,A;x) - \bar{P}(A;x)]dt$

is well defined and takes bounded continuous functions of y to themselves. Smoothness in x persists in the parametric dependence.

As in (2.2.3) we assume that $F(x,y)$ satisfies the centering condition

(2.6.6) $\int F(x,y)\bar{P}(dy;x) \equiv 0$,

and for $f \in C_0^\infty(R^n)$ we define

(2.6.7) $\psi_f^{(1)}(x,y) = \int_{R^m} \psi(y,dz;x) \sum_{j=1}^n F_j(x,y) \frac{\partial f(x)}{\partial x_j}$.

We define a diffusion operator $\overline{\mathcal{L}}$ (cf. (2.2.5) and (2.2.6)) as follows

(2.6.8) $\overline{\mathcal{L}}f(x) = \int_{R^m} \bar{P}(dy;x)A_f(x,y)$

(2.6.9) $A_f(x,y) = \sum_{i=1}^n F_i(x,y) \frac{\partial\psi_f^{(1)}(x,y)}{\partial x_i}$

$\qquad\qquad\qquad + \sum_{i=1}^n H_i(x,y) \frac{\partial f(x)}{\partial x_i}$.

From our smoothness requirements above, the coefficients of $\overline{\mathcal{L}}$ are smooth bounded functions of x and hence $\overline{\mathcal{L}}$ generates a diffusion process $x(t)$ in R^n.

Theorem 2.4.

Under the above hypotheses the process $x^\varepsilon(t)$, $t \geq 0$ of the pair $(x^\varepsilon(t), y^\varepsilon(t))$ generated by (2.6.2) converges weakly in $C([0,T];R^n)$, $T < \infty$ but arbitrary, as $\varepsilon \to 0$ to the diffusion Markov process $x(t)$, $t \geq 0$, generated by $\mathscr{L}$ of (2.6.8).

The proof of this theorem is identical to the one for Theorem 2.1. Regarding regularity assumptions, we refer to Remarks 1 and 2 following the statement of that theorem. Regarding the ergodicity of $Y(t;x)$, see Remark 3.

It is easily seen that one can consider more general pairs $(x^\varepsilon(t), y^\varepsilon(t))$ with generators $\mathscr{L}^\varepsilon$ having the same general structure as (2.6.2) and (2.1.2)-(2.1.5). The results do not change except for notation.

Let us now consider transport processes with boundary conditions. We limit discussion to one typical problem.

Let $\mathscr{O} \subset R^n$ be a bounded open set with smooth boundary $\partial\mathscr{O}$ and let $\phi(x)$ be its support function (cf. (2.5.1)) as described in the first two paragraphs of Section 2.5.

For each $x \in \partial\mathscr{O}$ let

$$(2.6.10) \quad S^\pm = \left\{ y \in R^m \mid F(x,y) \cdot \hat{n}(x) \begin{array}{c} < \\ > \end{array} 0 \right\},$$

where $\hat{n}(x) \equiv \nabla\phi(x)$ is the unit inward normal to $\partial\mathscr{O}$. In the usual moving particle model of a transport process where $x^\varepsilon(t)$ is position and $y^\varepsilon(t)$ is velocity, the spaces $S^\pm$ correspond to outward directed (+) and inward directed (-) velocities at the boundary. For each $x \in \partial\mathscr{O}$ and $y \in S^+$ let

$B(x,y,A)$, A a Borel subset of S^-, be a probability measure S^- depending smoothly on x. B is the reflection probability at the boundary. Let

(2.6.11) $G = (\mathscr{D} \times R^m) \cup (\partial\mathscr{D} \times S^-)$

(2.6.12) $\partial G = \partial\mathscr{D} \times S^+$

and consider the following operator on $\bar{G} = G \cup \partial G$ corresponding to instantaneously reflected[†] transport processes.

(2.6.13) $\mathscr{L}^\varepsilon = \frac{1}{\varepsilon^2} Q + \frac{1}{\varepsilon} F \cdot \frac{\partial}{\partial x}$, $(x,y) \in G$,

boundary operator $(B - I)$, $(x,y) \in \partial G$,

where B is the operator with kernel $B(x,y,A)$

(2.6.14) $Bf(x,y) = \int_{S^-} B(x,y,dz) f(x,z)$, $(x,y) \in \partial G$,

and I is the identity operator. It is easily seen that there exist processes on $\bar{G}$ with instantaneous reflection generated by $\mathscr{L}^\varepsilon$ of (2.1.13) with $\varepsilon > 0$ fixed. Moreover, in the interior G we have an ordinary transport process uniquely defined as above by (2.6.2) (with $H \equiv 0$ for simplicity here). To assure uniqueness for (2.6.13) we assume that given $\delta_1 > 0$ there is a $\delta_2 > 0$ such that

(2.6.15) $\displaystyle\inf_{x \in \partial\mathscr{D}} \inf_{y \in S^+} \int_{F(x,y) \cdot \hat{n}(x) < -\delta_1} B(x,y,dz) \geq \delta_2$.

[†]More precisely, the process characterized by (2.6.17) below.

The meaning of this condition is that upon reflection at the boundary the process has nontangential (grazing) velocity with high probability in a uniform sense.

Since the process is uniquely defined in the interior let for each $(x,y) \in G$, τ_1 be the first time ∂G is reached. It is instantaneously reflected there according to the law $B(x,y,A)$. Let τ_2 be the next time ∂G is reached, etc. Since τ_n is increasing we define

$$\tau_\infty = \lim_{n\uparrow\infty} \tau_n \; .$$

For uniqueness we must show $\tau_\infty = \infty$ with probability 1. It is easily shown using (2.1.15) that indeed, for $(x,y) \in G$,

$$\lim_{n\uparrow\infty} P_{x,y}\{\tau_n < t\} = 0 \; , \qquad 0 \le t < \infty \; .$$

Now we define the increasing process $N^\varepsilon(t)$ as follows (cf. (2.5.5)). With $\tau_0 = 0$, $\tau_n \uparrow \infty$,

$$(2.6.16) \quad N^\varepsilon(t) = n \; , \qquad \tau_n \le t < \tau_{n+1} \; , \qquad n = 0,1,2,\ldots \; .$$

With this definition and with $f(x,y)$ a smooth bounded function on $\bar{G}$,

$$(2.6.17) \quad M_f(t) = f(x^\varepsilon(t),y^\varepsilon(t)) - f(x,y) - \int_0^t \chi_G \mathcal{L}^\varepsilon f(x^\varepsilon(s),y^\varepsilon(s))\,ds$$

$$- \int_0^t \chi_{\partial G} \mathcal{L}^\varepsilon f(x^\varepsilon(s),y^\varepsilon(s))\,dN^\varepsilon(s)$$

is a right-continuous, zero mean martingale. Here χ_G is the characteristic function of the set G. Note that the existence and uniqueness theory for the reflected transport process $(x^\varepsilon(t), y^\varepsilon(t))$ on $\bar{G}$ is elementary by comparison to the theory needed in Section 2.5.

For the asymptotics we shall introduce one more hypothesis about the reflection operator B as follows. There is a nontrivial probability measure on S^- such that $B(x,y,A)$ has a density $\tilde{B}(x,y,z)$ relative to this measure and there are constants b_ℓ and b_u such that

$$(2.6.18) \quad 0 < b_\ell \leq \tilde{B}(x,y,z) \leq b_u < \infty ,$$

$$x \in \partial\mathcal{D}, \quad y \in S^+, \quad z \in S^- .$$

To characterize the limit of $x^\varepsilon(t)$ in $\overline{\mathcal{D}}$ we note that, clearly, in the interior the limit process should be governed by $\mathcal{L}$ of (2.6.8), (2.6.9) (with $H \equiv 0$). Thus, we must find the limiting form of the reflecting boundary conditions. As in Section 2.5, we must look closely at the appropriate boundary layer problem.

We introduce stretched coordinates $x \rightarrow (\gamma, \varepsilon\eta)$ as in (2.5.6) and carry over the various definitions and remarks made there. The operator $\mathcal{L}_{BL}$ and the boundary layer problem (2.5.11) take the following form here

$$(2.6.19) \quad \mathcal{L}_{BL} = \sum_{i=1}^{n} F_i(\xi,y) \frac{\partial\phi(\xi)}{\partial x_i} \frac{\partial}{\partial\eta} + Q\Big|_{x=\zeta}$$

where Q is defined by (2.6.1) and $x = \zeta \in \partial \mathcal{D}$ is a fixed
parameter, and (with $\zeta \in \partial \mathcal{D}$ fixed)

$$\mathcal{L}_{BL} v(\eta, y) = 0 , \quad \eta > 0 , \quad y \in R^m ,$$

$$\eta = 0 , \quad y \in s^- ,$$

(2.6.20)
$$\int_{s^-} B(\zeta, y, dz) v(0, z) - v(0, y) = \psi(y; x) , \quad y \in s^+ ,$$

$$\lim_{\eta \to +\infty} v(\eta, y) = 0 .$$

By results similar to the ones in the appendix (cf. [30]
for details), it follows that under the above hypotheses,
there is a unique probability measure $\bar{U}(A; x)$, $A \in s^+$,
$x \in \partial \mathcal{D}$, such that if

(2.6.21) $\displaystyle \int_{s^+} \bar{U}(dy; x) \psi(y; x) \equiv 0 , \quad x \in \partial \mathcal{D} ,$

then (2.6.20) has a unique solution that decays to zero as
$\eta \to \infty$ exponentially fast, uniformly in y.

Let $\psi(y, A; x)$ be defined by (2.6.5) and define the vector
field $\bar{B}(x)$, $x \in \partial \mathcal{D}$ by

(2.6.22) $\displaystyle \bar{B}(x) = \int_{s^+} \bar{U}(dy; x) \left[\int_{s^-} B(x, y, dz) \int_s \psi(z, dz'; x) F(x, z') \right.$

$$\left. - \int_s \psi(y, dz; x) F(x, z) \right] .$$

From our smoothness requirements in x it follows that $\bar{B}(x)$ is indeed a smooth vector field on $\partial \mathcal{D}$ and, also, it points inwards. We need to assume, however, the following stronger hypothesis. There is a $\delta > 0$ such that

$$(2.6.23) \qquad \sum_{i=1}^{n} \bar{B}_i(x) \; \frac{\partial \phi(x)}{\partial x_i} \geq \delta , \qquad x \in \partial \mathcal{D} .$$

This means that $\bar{B}$ is uniformly nontangential.

Theorem 2.5.

Under the above hypotheses the reflected process $x^\varepsilon(t)$ in $\overline{\mathcal{D}}$, of the pair $(x^\varepsilon(t), y^\varepsilon(t))$ in $\bar{G}$, converges weakly in $C([0,T]; \overline{\mathcal{D}})$, $T < \infty$ but arbitrary, as $\varepsilon \to 0$ to the reflected diffusion process $x(t)$ which is the unique solution of the submartingale problem $(\overline{\mathcal{L}}, \bar{B})$.

The proof of this theorem is nearly identical, step by step, to the one in Section 2.5. Only minor differences arise in showing that (2.5.23) holds here. We emply the theory of [5] to obtain the characterization of the limit problem $(\overline{\mathcal{L}}, \bar{B})$. The details are actually given in [30] where, in addition, absorbing boundary conditions and full expansions (not just weak convergence) are considered.

2.7 Homogenization.

An example of this was given in Section 2.1 (Example 3). Here we examine briefly the relation of homogenization problems to the ones of previous sections. We refer to Babuska's papers [27] for examples from continuum mechanics and many other nonprobabilistic stituations, to the work of De Giorgi-Spagnolo and Spagnolo [34, 35] for connections with the theory of G-convergence (other references can be found there) and to [36] for problems in variational inequalities and other areas, including probabilistic ones (see also [37]). A systematic treatment of these problems, boundary layers, etc., is given in [38].

Consider the following operator on $C_0^\infty(R^n)$

$$(2.7.1) \quad \mathscr{L}^\varepsilon = \frac{1}{2} \sum_{i,j=1}^n a_{ij}(x,\tfrac{x}{\varepsilon}) \frac{\partial^2}{\partial x_i \partial x_j}$$

$$+ \sum_{j=1}^n \left[\frac{1}{\varepsilon} b_j(x,\tfrac{x}{\varepsilon}) + c_j(x,\tfrac{x}{\varepsilon}) \right] \frac{\partial}{\partial x_j} ,$$

where we assume the following

(2.7.2) $(a_{ij}(x,y))$ is symmetric smooth and uniformly elliptic for $(x,y) \in R^n \times R^n$.

(2.7.3) $b_j(x,y)$ and $c_j(x,y)$ are smooth bounded functions on $R^n \times R^n$, $j = 1,2,\ldots,n$.

Clearly, these conditions imply immediately that $\mathscr{L}^\varepsilon$, $\varepsilon > 0$, generates a diffusion Markov process $x^\varepsilon(t)$, $t \geq 0$.

Let $\mathscr{L}_1$ be defined on $C_0^\infty(R^n)$ by

(2.7.4) $\mathscr{L}_1 = \frac{1}{2} \sum_{i,j=1}^{n} a_{ij}(x,y) \frac{\partial^2}{\partial y_i \partial y_j} + \sum_{j=1}^{n} b_j(x,y) \frac{\partial}{\partial y_j}$.

This operator is the generator of a process $Y(t;x) \in R^n$ where $x \in R^n$ is a parameter. As in Section 2.2 we assume that this process is ergodic with $\bar{P}(A;x)$ its invariant measure and $\psi(y,A;x)$, $y \in R^n$, $A \in R^n$, $x \in R^n$, as in (2.2.2) and well defined as described there.

Assume that

(2.7.5) $\int_{R^n} b_j(x,y) \bar{P}(dy;x) \equiv 0$, $x \in R^n$, $j = 1,2,\ldots,n$

and for $f(x) \in C_0^\infty(R_n)$ define $\psi_f^{(1)}(x,y)$ by

(2.7.6) $\psi_f^{(1)}(x,y) = \int_{R^n} \sum_{j=1}^{n} \psi(y,dz;x) b_j(x,y)$.

Define $\overline{\mathscr{L}}$ on $C_0^\infty(R^n)$ by

(2.7.7) $\overline{\mathscr{L}}f(x) = \int_{R^n} \bar{P}(dy;x) A_f(x,y)$,

where

(2.7.8) $A_f(x,y) = \sum_{i,j=1}^{n} a_{ij}(x,y) \frac{\partial^2 \psi_f^{(1)}(x,y)}{\partial x_i \partial y_i}$

$+ \sum_{j=1}^{n} b_j(x,y) \frac{\partial \psi_f^{(1)}(x,y)}{\partial x_j}$

$+ \sum_{j=1}^{n} c_j(x,y) \frac{\partial \psi_f^{(1)}(x,y)}{\partial y_j}$

$$+ \frac{1}{2} \sum_{i,j=1}^{n} a_{ij}(x,y) \frac{\partial^2 f(x)}{\partial x_i \partial x_j}$$

$$+ \sum_{j=1}^{n} c_j(x,y) \frac{\partial f(x)}{\partial x_j} .$$

Clearly the coefficients of $\mathcal{L}$ are smooth bounded functions of x.

Theorem 2.6.

Under the above hypotheses the process $x^{\varepsilon}(t)$ generated by $\mathcal{L}^{\varepsilon}$ in (2.7.1) converges weakly on $C([0,T];R^n)$, $T < \infty$ but arbitrary, as $\varepsilon \to 0$ to the diffusion process $x(t)$ generated by $\mathcal{L}$ of (2.7.7), (2.7.8).

Remark 1.

Concerning smoothness of the coefficients see Remark 1 and 2 following Theorem 2.1.

Remark 2.

If

$$(2.7.9) \quad b_j(x,y) = \frac{1}{2} \sum_{i=1}^{n} \frac{\partial a_{ij}(x,y)}{\partial y_i} ,$$

then $\mathcal{L}^{\varepsilon}$ in (2.7.1) can be put in divergence form

$$(2.7.10) \quad \mathcal{L}^{\varepsilon} = \frac{1}{2} \sum_{i,j=1}^{n} \frac{\partial}{\partial x_i} \left\{ a_{ij}(x,\frac{x}{\varepsilon}) \frac{\partial}{\partial x_j} \right\}$$

$$+ \sum_{j=1}^{n} \left[c_j(x,y) - \frac{1}{2} \sum_{i=1}^{n} \frac{\partial a_{ij}(x,\frac{x}{\varepsilon})}{\partial x_i} \right] \frac{\partial}{\partial x_j} .$$

In this case one can analyze the limit (in the appropriate sense) of the operator $\mathscr{L}^{\varepsilon}$ in a Hilbert space framework with $(a_{ij}(x,y))$ being only bounded measurable functions (and uniformly elliptic); see [34-36]. In particular, the case where $(a_{ij}(x,y))$ and $(c_j(x,y))$ are periodic functions of y has received most attention. For the case (2.7.10) with periodic y dependence, $\bar{P}(dy;x)$ is Lebesque measure, as can be seen easily, (2.7.5) is automatically satisfied and (2.7.7) simplifies a lot.

Proof of Theorem 2.6.

The proof is identical to the one of Theorem 2.1 after various objects are identified with the corresponding ones there.

The idea is to define

$$(2.7.11) \qquad y^{\varepsilon}(t) = \frac{x^{\varepsilon}(t)}{\varepsilon}$$

and then treat $(x^{\varepsilon}(t), y^{\varepsilon}(t))$ as a pair in the framework of Section 2.1. After all constructions are carried out we reestablish (2.7.11) which is only an analytical device. Implementing this we find that the generator of $(x^{\varepsilon}(t), y^{\varepsilon}(t))$ is of the form (2.1.2) with $\mathscr{L}_1$ as in (2.7.4) and

$$(2.7.12) \quad \mathscr{L}_2 = \sum_{i,j=1}^{n} a_{ij}(x,y) \frac{\partial^2}{\partial x_i \partial y_i} + \sum_{j=1}^{n} b_j(x,y) \frac{\partial}{\partial x_j}$$

$$+ \sum_{j=1}^{n} c_j(x,y) \frac{\partial}{\partial y_j},$$

$$(2.7.13) \qquad \mathcal{L}_3 = \frac{1}{2} \sum_{i,j=1}^{n} a_{ij}(x,y) \frac{\partial^2}{\partial x_i \partial x_j} + \sum_{j=1}^{n} c_j(x,y) \frac{\partial}{\partial x_j} \ .$$

From here on we proceed as in Theorem 2.1.

Remark 3.

Using the same device (2.7.11) one can treat processes $x^\varepsilon(t)$ with generator $\mathcal{L}^\varepsilon$ of (2.7.1) in $\mathcal{O}$, a bounded open set in R^n, and boundary conditions on $\partial \mathcal{O}$. The results are formally the same as the ones in Section 2.5. In the case of reflection, the only difference is that in the boundary layer problem (2.5.11) (or (2.5.17)) instead of having the normal derivative given, an appropriate oblique derivative is given at $\eta = 0$. The proof of the results analagous to the ones in the appendix is, however, different.

2.8 Averaging.

In the first two theorems that follow we deal with a variant of the process $(x^\varepsilon(t), y^\varepsilon(t))$ defined at the beginning of Section 2.6. Then we deal with a variant (2.1.1).

Let Q be defined by (2.6.1) and assume that (2.6.3) and (2.6.4) hold so that ψ in (2.6.5) is well defined as discussed there. Let $F(x,y,\tau)$ and $H(x,y,\tau)$ be n-vector functions defined on $R^n \times R^m \times R^1$, bounded, smooth in x and almost periodic in τ. Define for $t \geq 0$

$$(2.8.1) \quad \mathscr{L}^\varepsilon_t = \frac{1}{\varepsilon^2} Q + \frac{1}{\varepsilon} F(x,y,t/\varepsilon^2)\cdot\frac{\partial}{\partial x} + H(x,y,t/\varepsilon^2)\cdot\frac{\partial}{\partial x}.$$

This is a time inhomogeneous operator which, it is easily seen, generates a Markov process $(x^\varepsilon(t), y^\varepsilon(t))$, $t \geq 0$ in $R^n \times R^m$.

For the asymptotic analysis of $x^\varepsilon(t)$ we introduce the following hypotheses concerning $F(x,y,\tau)$ and $H(x,y,\tau)$.

$$(2.8.2) \quad \lim_{T\uparrow\infty} \frac{1}{T} \int_0^T d\tau \int \bar{P}(dy;x)F(x,y,\tau) \equiv 0,$$

and, more specifically, if

$$(2.8.3) \quad \bar{F}(x,\tau) \equiv \int \bar{P}(dy;x)F(x,y,\tau),$$

then

$$(2.8.4) \quad \int_0^\tau ds\, \bar{F}(x,s) \leq C, \qquad x \in R^n, \qquad \tau \geq 0,$$

where C is a constant. For $f \in C_0^\infty(R^n)$ let

(2.8.5) $\psi_f^{(1)}(x,y,\tau) = \int\limits_0^\infty ds \int\limits_{R^m} [P(s,y,dz;x)-\bar{P}(dz;x)]$

$$\cdot \sum_{j=1}^n \left[F_j(x,z,s+\tau)-\bar{F}_j(x,s+\tau)\right]\frac{\partial f(x)}{\partial x_j}$$

$$- \int\limits_0^\tau ds \sum_{j=1}^n \bar{F}_j(x,s)\frac{\partial f(x)}{\partial x_j} .$$

Let $A_f(x,y,\tau)$ be defined by

(2.8.6) $A_f(x,y,\tau) = \sum_{j=1}^n \left[F_j(x,y,\tau)\frac{\partial \psi_f^{(1)}(x,y,\tau)}{\partial x_j} + H_j(x,y,\tau)\frac{\partial f(x)}{\partial x_j}\right] .$

We assume that for $f \in C_o^\infty(R^n)$ fixed,

(2.8.7) $\mathscr{L}f(x) \equiv \lim_{T\uparrow\infty}\frac{1}{T}\int\limits_0^T ds\int \bar{P}(dy;x)A_f(x,y,\tau+s)$

exists uniformly in x and τ, it is independent of τ and the coefficients of the second order operator $\mathscr{L}$ (time homogeneous) are smooth and bounded functions of $x \in R^n$.

Theorem 2.7.

Under the above hypotheses $x^\varepsilon(t)$, $t \geq 0$ converges weakly on $C([0,T];R^n)$, $T < \infty$, as $\varepsilon \to 0$ to the time homogeneous diffusion process generated by $\mathscr{L}$ of (2.8.7).

Remark 1.

In the case F and H are independent of y we are dealing
with the case of averaging for ordinary differential equations [6], including second order effects (the effect of
$\varepsilon^{-1}F$). The orbits of the vector field $[\varepsilon^{-1}F(x,t/\varepsilon^2)$
$+ H(x,t/\varepsilon^2)] \cdot \frac{\partial}{\partial x}$ converge to the orbits of the vector field
$(F = \bar{F}$ here)

$$\mathcal{L} = - \lim_{T \uparrow \infty} \frac{1}{T} \int_0^T ds \int_0^s d\sigma \sum_{i,j=1}^n F_j(x,s) \frac{\partial F_i(x,\sigma)}{\partial x_j} \frac{\partial}{\partial x_i}$$

$$+ \lim_{T \uparrow \infty} \frac{1}{T} \int_0^T ds \sum_{i=1}^n H_i(x,s) \frac{\partial}{\partial x_i} .$$

Note that the coefficients of second order derivatives in
(2.8.7) are identically zero in this case. Clearly one cannot
get a diffusion process from deterministic vector fields even
when they oscillate rapidly (in an almost periodic fashion).

Remark 2.

In some applications [19] the interaction of averaging
with the diffusion approximation plays an important role.
This is one reason for presenting Theorem 2.7.

Proof of Theorem 2.7.

The proof is nearly the same as the one of Theorem 2.1.
It suffices to show how the scheme (2.3.12)-(2.3.15) must be
modified here.

In the spirit of the device (2.7.11) we introduce the "process"

(2.8.8) $\quad \tau^{\varepsilon}(t) = \tau + t/\varepsilon^2 , \quad \tau \in R^1 , \quad t \geq 0 .$

The augmented process $(x^{\varepsilon}(t), y^{\varepsilon}(t), \tau^{\varepsilon}(t))$ on $R^n \times R^m \times R^1$ is time homogeneous and its generator is given by

(2.8.9) $\quad \dfrac{1}{\varepsilon^2} (Q + \dfrac{\partial}{\partial \tau}) + \dfrac{1}{\varepsilon} F(x,y,\tau) \cdot \dfrac{\partial}{\partial x} + H(x,y,\tau) \cdot \dfrac{\partial}{\partial x}$

which is of the form (2.1.2).

Let $f(x) \in C_o(R^n)$, let $\psi_f^{(1)}(x,y,\tau)$ and $\psi_f^{(2,\lambda)}(x,y,\tau)$, $\lambda > 0$, be functions to be defined shortly and put

(2.8.10) $\quad f^{\varepsilon}(x,y,\tau) = f(x) + \varepsilon \psi_f^{(1)}(x,y,\tau) + \varepsilon^2 \psi_f^{(2,\lambda)}(x,y,\tau) .$

Then we have

(2.8.11) $\quad \left[\dfrac{1}{\varepsilon^2} (Q + \dfrac{\partial}{\partial \tau}) + \dfrac{1}{\varepsilon} F \cdot \dfrac{\partial}{\partial x} + H \cdot \dfrac{\partial}{\partial x} \right] \left(f + \varepsilon \psi_f^{(1)} + \varepsilon^2 \psi_f^{(2,\lambda)} \right)$

$$= \dfrac{1}{\varepsilon} \left[(Q + \dfrac{\partial}{\partial \tau}) \psi_f^{(1)} + F \cdot \dfrac{\partial f}{\partial x} \right]$$

$$+ \left[(Q + \dfrac{\partial}{\partial \tau}) \psi_f^{(2,\lambda)} + F \cdot \dfrac{\partial \psi_f^{(1)}}{\partial x} + H \cdot \dfrac{\partial f}{\partial x} - \mathscr{L} f - \lambda \tilde{\psi}_f^{(2,\lambda)} \right]$$

$$+ \lambda \tilde{\psi}_f^{(2,\lambda)} + \overline{\mathscr{L}} f + \varepsilon \left[F \cdot \dfrac{\partial \psi_f^{(2,\lambda)}}{\partial x} + H \cdot \dfrac{\partial \psi_f^{(1)}}{\partial x} \right]$$

$$+ \varepsilon^2 H \cdot \dfrac{\partial \psi_f^{(2,\lambda)}}{\partial x} .$$

We choose $\psi_f^{(1)}(x,y,\tau)$ so that

(2.8.12) $\quad (Q + \dfrac{\partial}{\partial \tau}) \psi_f^{(1)} + F \cdot \dfrac{\partial f}{\partial x} = 0 .$

With $\bar{F}$ defined by (2.8.3) we rewrite (2.8.12) as follows

(2.8.13) $(Q + \frac{\partial}{\partial\tau})\psi_f^{(1)}(x,y,\tau) + \left[F(x,y,\tau)-\bar{F}(x,\tau)\right]\cdot\frac{\partial f(x)}{\partial x}$

$+ \bar{F}(x,\tau)\cdot\frac{\partial f(x)}{\partial x} = 0$.

Because of the ergodic properties of Q (cf. (2.6.3)-(2.6.5)) it follows from (2.8.13) that $\psi_f^{(1)}(x,y,\tau)$ given by (2.8.5) satisfies (2.8.13) and it is bounded.

With $\psi_f^{(1)}$ so determined consider the second term on the right side of (2.8.11). We rewrite it as

(2.8.14) $(Q + \frac{\partial}{\partial\tau})\psi_f^{(2,\lambda)}(x,y,\tau) + A_f(x,y,\tau) - \bar{\bar{\mathscr{L}}}_\tau f(x)$

$+ \bar{\mathscr{L}}_\tau f(x) - \mathscr{L}f(x) - \lambda\tilde{\psi}_f^{(2,\lambda)} = 0$.

where

(2.8.15) $\bar{\mathscr{L}}_\tau f(x) = \int \bar{P}(dy;x)A_f(x,y,\tau)$,

with A_f given by (2.8.6). From (2.8.14) it is clear that

(2.8.16) $\psi_f^{(2,\lambda)}(x,y,\tau) = \tilde{\psi}_f^{(2,\lambda)}(x,y,\tau) + \int_0^\infty ds$

$\cdot \int [P(s,y,dz;x)-\bar{P}(dz;x)]$

$\cdot \left[A_f(x,z,s+\tau) - \bar{\mathscr{L}}_{s+\tau}f(x)\right]$

(2.8.17) $\tilde{\psi}_f^{(2,\lambda)}(x,\tau) = \int_0^\infty e^{-\lambda s}\left[\bar{\mathscr{L}}_{\tau+s}f(x) - \mathscr{L}f(x)\right]ds$,

$\lambda > 0$.

Collecting the above formulas we find that with $\mathscr{L}_t^\varepsilon$ given by (2.8.1) we have ($f \in C_o^\infty(R^n)$),

$$(2.8.18) \quad \left[\frac{\partial}{\partial t} + \mathscr{L}_t^\varepsilon\right]\left[f(x) + \varepsilon\psi_f^{(1)}(x,y,\frac{t}{\varepsilon^2}) + \varepsilon^2\psi_f^{(2,\lambda)}(x,y,\frac{t}{\varepsilon^2})\right]$$

$$= \overline{\mathscr{L}}f(x) + \lambda\tilde{\psi}_f^{(2,\lambda)}(x,\frac{t}{\varepsilon^2})$$

$$+ \varepsilon\left[F\cdot\frac{\partial\psi_f^{(2,\lambda)}}{\partial x} + H\cdot\frac{\partial\psi_f^{(1)}}{\partial x}\right](x,y,\frac{t}{\varepsilon^2})$$

$$+ \varepsilon^2 H\cdot\frac{\partial\psi_f^{(2,\lambda)}}{\partial x}(x,y,\frac{t}{\varepsilon^2}) \ .$$

Now because of (2.8.7) it follows that $\lambda\tilde{\psi}_f^{(2,\lambda)}(x,\tau)$ goes to zero as $\lambda \to 0$, uniformly in x and τ, while for $\lambda > 0$, $\psi_f^{(2,\lambda)}(x,y,t/\varepsilon^2)$ and x derivatives are bounded.

These are the analogs of the necessary facts (2.3.13)- (2.3.15) for the proof of Theorem 2.7. The rest follows as in Theorem 2.1 and the proof is complete.

A very similar situation to the above, which arises in a problem similar to the one in [19], is the following.

The time inhomogeneous process $(x^\varepsilon(t),y^\varepsilon(t))$ has generator

$$(2.8.19) \quad \mathscr{L}_t^\varepsilon = \frac{1}{\varepsilon^2} Q + \frac{1}{\varepsilon} F(x,y,t/\varepsilon)\cdot\frac{\partial}{\partial x} + H(x,y,t/\varepsilon)\cdot\frac{\partial}{\partial x} \ ,$$

the only difference being that F and H are functions of t/ε and not t/ε^2 as in (2.8.1). We now assume that

$$(2.8.20) \quad \int \bar{P}(dy;x)F(x,y,\tau) \equiv 0 \ , \qquad x \in R^n \ , \qquad \tau \in R^1 \ ,$$

and that $F(x,y,\tau)$ and $H(x,y,\tau)$ are also smooth functions of $\tau \in R^1$.

For $f \in C_o^\infty(R^n)$ let

(2.8.21) $\psi_f^{(1)}(x,y,\tau) = \int_{R^m} \psi(y,dz;x) \sum_{j=1}^n F_j(x,y,\tau) \cdot \dfrac{\partial f(x)}{\partial x}$

where ψ is defined by (2.6.5). Let

(2.8.22) $A_f(x,y,\tau) = \sum_{j=1}^n \left[F_j(x,y,\tau) \dfrac{\partial \psi_f^{(1)}(x,y,\tau)}{\partial x} + H_j(x,y,\tau) \dfrac{\partial f(x)}{\partial x_j} \right]$

and

(2.8.23) $\overline{\mathscr{L}}_\tau f(x) = \int_{R^m} \bar{P}(dy;x) A_f(x,y,\tau)$.

The limit operator $\overline{\mathscr{L}}$ is defined for $f \in C_o^\infty(R^n)$ by

(2.8.24) $\overline{\mathscr{L}} f(x) = \lim_{T \uparrow \infty} \dfrac{1}{T} \int_0^T ds \overline{\mathscr{L}}_{s+\tau} f(x)$,

where the limit is assumed to exist uniformly in (x,τ) and to be independent of τ as shown.

Theorem 2.8.

Under the above hypotheses the process $x^\varepsilon(t)$ associated with (2.8.19) converges weakly in $C([0,T];R^n)$, $T < \infty$ but arbitrary, as $\varepsilon \to 0$ to the diffusion process $x(t)$ generated by $\overline{\mathscr{L}}$ of (2.8.24).

Proof.

As in Theorem 2.7 we indicate the essential differences in the argument from Theorem 2.1.

With $f \in C_o^\infty(R^n)$ fixed define $\psi_f^{(1)}$ by (2.8.21) and

$$(2.8.25) \quad \tilde{\psi}_f^{(1,\lambda)}(x,\tau) = \int_0^\infty e^{-\lambda s}\left[\mathcal{L}_{\tau+s}f(x) - \overline{\mathcal{L}}f(x)\right]ds ,$$

$$\lambda > 0 ,$$

$$(2.8.26) \quad \psi_f^{(2,\lambda)}(x,y,\tau) = \int_{R^m} \psi(y,dz;x)\left[A_f(x,z,t) - \overline{\mathcal{L}}_\tau f(x)\right.$$

$$\left. + \frac{\partial\psi_f^{(1)}(x,z,\tau)}{\partial\tau} + F(x,z,\tau)\cdot\frac{\partial\tilde{\psi}^{(1,\lambda)}(x,\tau)}{\partial x}\right] .$$

From these definitions it follows that

$$(2.8.27) \quad (\partial_t + \mathcal{L}_t^\varepsilon)\left[f(x) + \varepsilon\psi_f^{(1)}(x,y,\tfrac{t}{\varepsilon}) + \varepsilon\tilde{\psi}_f^{(1,\lambda)}(x,\tfrac{t}{\varepsilon})\right.$$

$$\left. + \varepsilon^2\psi_f^{(2,\lambda)}(x,y,\tfrac{t}{\varepsilon})\right]$$

$$= \overline{\mathcal{L}}f(x) + \lambda\tilde{\psi}_f^{(1,\lambda)}(x,\tfrac{t}{\varepsilon}) + \varepsilon\left[\left[F\cdot\frac{\partial}{\partial x} + \frac{\partial}{\partial\tau}\right]\psi_f^{(2,\lambda)}(x,y,\tfrac{t}{\varepsilon})\right.$$

$$\left. + H\cdot\frac{\partial\psi_f^{(1)}}{\partial x}(x,y,\tfrac{t}{\varepsilon}) + H(x,y,\tfrac{t}{\varepsilon})\cdot\frac{\partial\tilde{\psi}_f^{(1,\lambda)}}{\partial x}(x,\tfrac{t}{\varepsilon})\right]$$

$$+ \varepsilon^2 H\cdot\frac{\partial\tilde{\psi}_f^{(2,\lambda)}}{\partial x}(x,y,\tfrac{t}{\varepsilon}) .$$

We note that coefficients of ε and ε^2 on the right side of (2.8.27) are bounded for $\lambda > 0$ and that, because of the uniform existence of (2.8.24), $\lambda\tilde{\psi}_f^{(1,\lambda)}(x,\tau) \to 0$ as $\lambda \to 0$, uniformly in x and τ. With these facts the proof goes exactly as the one of Theorem 2.1 and we omit further details.

As a final problem we consider the following system of stochastic equations, a special case of (2.1.1) (cf. example 2 in Section 2.1), also considered in [12].

$$(2.8.28) \quad dx_i^\varepsilon(t) = G_i^{(1)}(x^\varepsilon(t), y^\varepsilon(t))dt$$

$$+ \sum_{j=1}^{p} \sigma_{ij}^{(1)}(x^\varepsilon(t), y^\varepsilon(t))d\beta_j^{(1)}(t) \ ,$$

$$x_i^\varepsilon(0) = x_i \ , \qquad i = 1,2,\ldots,n \ ,$$

$$dy_k^\varepsilon(t) = \frac{1}{\varepsilon^2} F_k^{(2)}(x^\varepsilon(t), y^\varepsilon(t))$$

$$+ \frac{1}{\varepsilon} \sum_{\ell=1}^{q} \sigma_{k\ell}^{(2)}(x^\varepsilon(t), y^\varepsilon(t))d\beta_\ell^{(2)}(t)$$

$$y_k^\varepsilon(0) = y_k \ , \qquad k = 1,2,\ldots,m \ .$$

Here $(\beta_1^{(1)}(t), \beta_2^{(1)}(t),\ldots,\beta_p^{(1)}(t))$ and $(\beta_1^{(2)}(t), \beta_2^{(2)}(t),\ldots,\beta_q^{(2)}(t))$ are p-dimensional and q-dimensional standard Brownian motions, respectively, and (2.1.2) holds. The generator of $(x^\varepsilon(t), y^\varepsilon(t))$ has the form

$$(2.8.29) \quad \mathscr{L}^\varepsilon = \frac{1}{\varepsilon^2} \mathscr{L}_1 + \frac{1}{\varepsilon} \mathscr{L}_2 + \mathscr{L}_3$$

where $\mathscr{L}_1$ is given by (2.1.3) and

$$(2.8.30) \quad \mathscr{L}_2 = \sum_{i=1}^{n} \sum_{k=1}^{m} a_{ik}^{(12)}(x,y) \frac{\partial^2}{\partial x_i \partial y_k} \ ,$$

$$(2.8.31) \quad \mathscr{L}_3 = \sum_{i=1}^{n} G_i^{(1)}(x,y) \frac{\partial}{\partial x_i} + \frac{1}{2} \sum_{i,j=1}^{n} a_{ij}^{(1)}(x,y) \frac{\partial^2}{\partial x_i \partial x_j} \ .$$

The a's are defined by (2.1.6) and the smoothness and boundedness conditions of Section 2.1 hold here also.

Let $Y(t;x)$ be the process as R^m generated by $\mathscr{L}_1$ with $x \in R^n$ a fixed parameter as in Section 2.2. Let $P(t,y,A;x)$ be defined by (2.2.1). We shall not assume the strong ergodicity assumptions that lead to the existence of ψ in (2.2.2). Instead, we assume that for each $f(x) \in C_o^\infty(R^n)$,

$$(2.8.32) \qquad \overline{\mathscr{L}}f(x) = \lim_{T \uparrow \infty} \frac{1}{T} \int_0^T dt \int_{R^m} P(t,y,dz;x) \mathscr{L}_3 f(x,z) ,$$

exists uniformly in $x \in R^n$, $y \in R^m$, is independent of y as shown and the coefficients of $\overline{\mathscr{L}}$ are smooth and bounded functions of x.

Theorem 2.9.

Under the above hypotheses the process $x^\varepsilon(t)$ defined by (2.8.28) converges weakly on $C([0,T];R^n)$, $T < \infty$, but arbitrary, as $\varepsilon \to 0$ to the diffusion process $x(t)$ with generator $\overline{\mathscr{L}}$ given by (2.8.32).

Proof.

The proof is simpler than the one of Theorem 2.1 since compactness of $x^\varepsilon(t)$ is now immediate. For the identification, we define for $f(x) \in C_o^\infty(R^n)$, $\lambda > 0$,

(2.8.33) $\psi_f^{(1,\lambda)}(x,y) = \int\limits_0^\infty e^{-\lambda s} \int\limits_{R^m} P(s,y,dz;x)(\mathscr{L}_3 f(x,z) - \overline{\mathscr{L}} f(x))$.

From this definition and ($\mathscr{L}^\varepsilon$ given by (2.8.29))

$$\mathscr{L}^\varepsilon \left[f(x) + \varepsilon^2 \psi_f^{(1,\lambda)}(x,y) \right]$$

$$= \left[\mathscr{L}_1 \psi_f^{(1,\lambda)}(x,y) + \mathscr{L}_3 f(x,y) - \overline{\mathscr{L}} f(x) - \lambda \psi_f^{(1,\lambda)}(x,y) \right|$$

$$+ \overline{\mathscr{L}} f(x) + \lambda \psi_f^{(1,\lambda)}(x,y) ,$$

it follows that

$$\mathscr{L}^\varepsilon (f + \varepsilon^2 \psi_f^{(1,\lambda)}) = \overline{\mathscr{L}} f + \lambda \psi_f^{(1,\lambda)} .$$

In view of (2.8.32), $\lambda \psi_f^{(1,\lambda)}(x,y) \to 0$ as $\lambda \to 0$, uniformly in $(x,y) \in R^n \times R^m$ and this is all that is needed for the proof as usual.

Remark.

The special case $m = 1$, $F^{(2)} \equiv 1$, $\sigma^{(2)} \equiv 0$, is example 1 of Section 2.1 i.e., the method of averaging [6 ,25].

3. Non-ergodic driving process.

3.1 First order limit theorem.

We shall formulate first a special problem that displays the features that are characteristic in general and can be contrasted easily with those of previous problems, i.e., the case of ergodic driving processes.

Let $F_i(x,y)$, $x \in R^n$, $y \in R^1$, $i = 1,2,\ldots,n$ be smooth functions with compact support in y and let $a(y)$ be a smooth function on R^1 such that

(3.1.1) $0 < \alpha_\ell \leq a(y) \leq \alpha_u < \infty$, α_ℓ, α_u constants,

(3.1.2) $\lim_{y \to \pm\infty} a(y) = a_\pm > 0$.

Consider the system of stochastic differential equations

(3.1.3) $\dfrac{dx_i^\varepsilon(t)}{dt} = \dfrac{1}{\varepsilon} F_i\left(x^\varepsilon(t), \dfrac{y^\varepsilon(t)}{\varepsilon}\right)$,

$x_i^\varepsilon(0) = x_i$, $i = 1,2,\ldots,n$,

$dy^\varepsilon(t) = a^{1/2}\left(\dfrac{y^\varepsilon(t)}{\varepsilon}\right) d\beta(t)$,

$y^\varepsilon(0) = y$

Here $\beta(t)$ is the standard Brownian motion process on R^1.

The essential difference between system (3.1.3) and (2.1.1) is that the driving process $y^\varepsilon(t)$ is Brownian motion on R^1 with a time substitution and hence it is not ergodic.

It is null recurrent, however, and its local time is well defined. This is an important feature of (3.1.3). Note also that the scaling in (3.1.3) is different from (2.1.1).

Define the vector field $\bar{F}(x)$ by

$$(3.1.4) \qquad \bar{F}_i(x) = \int_{-\infty}^{\infty} \frac{F_i(x,y)}{a(y)} \, dy \, , \qquad i = 1,2,\ldots,n \, ,$$

and assume that it is not identically equal to zero (the case $\bar{F}(x) \equiv 0$ is considered in Section 3.4). Let $\bar{x}(t)$ be the solution of

$$(3.1.5) \qquad \frac{d\bar{x}(t)}{dt} = \bar{F}(\bar{x}(t)) \, , \qquad \bar{x}(0) = x \, .$$

Let

$$(3.1.6) \qquad \tilde{a}(y) = a_+ \chi_{y>0} + a_- \chi_{y \leq 0}$$

and let $y(t)$ be the process satisfying

$$(3.1.7) \qquad dy(t) = \tilde{a}^{1/2}(y(t)) d\tilde{\beta}(t) \, , \qquad y(0) = y$$

where $\tilde{\beta}(t)$ is another standard Brownian motion. Let $\ell_o(t)$ denote the local time at $y = 0$ of this process. The local time at zero is defined by

$$(3.1.8) \qquad \ell_o(t) = 2y_+(t) - 2y_+ - 2 \int_0^t H(y(s)) \, dy(s) \, ,$$

where

$$y_+ = \max(0,y) \; ,$$

(3.1.9)
$$H(y) = \begin{cases} 0 \; , & y < 0 \; , \\ 1 \; , & y \geq 0 \; . \end{cases}$$

It is clear that the left hand side of (3.1.8) is formally

$$\int_0^t \tilde{a}(y(s)) \, \delta(y(s)) \, ds$$

which explains the terminology. The right hand side of (3.1.8) is well defined as it stands and is referred to as Tanaka's representation of $\ell_0(t)$ [ ].

Define the process $x(t)$ in R^n by

(3.1.10) $x(t) = \bar{x}(\ell_0(t)) \; .$

Theorem 3.1.

Under the above hypotheses $(x^\varepsilon(t), y^\varepsilon(t))$ converges weakly in $C([0,T]; R^n \times R^1)$, $T < \infty$ but arbitrary, as $\varepsilon \to 0$ to the process $(x(t), y(t))$ defined by (3.1.7) and (3.1.5), (3.1.10).

Remark 1.

When $n = 1$ and $F(x,y) = \chi_A(y)$, A a bounded subset of R^1, then

$$\bar{F} = \int \frac{1}{a(y)} \, dy \; ,$$

273

is independent of x, and

$$x(t) = x + \bar{F}\ell_o(t) .$$

Thus,

$$\frac{1}{\varepsilon} \int_0^t \chi_A\left(\frac{y^\varepsilon(s)}{\varepsilon}\right) ds$$

converges weakly as $\varepsilon \to 0$ to $\bar{F}\ell_o(t)$ where $\ell_o(t)$ is the local time of $y(t)$ in (3.1.7). This is a well known result [20].

Remark 2.

Instead of (3.1.3), consider

$$dx_i^\varepsilon(t) = \frac{1}{\varepsilon} F_i\left(x^\varepsilon(t), \frac{y^\varepsilon(t)}{\varepsilon}\right) dt$$

$$+ \frac{1}{\sqrt{\varepsilon}} \sum_{j=1}^q \sigma_{ij}\left(x^\varepsilon(t), \frac{y^\varepsilon(t)}{\varepsilon}\right) d\beta_j^{(1)}(t)$$

(3.1.11)

$$x_i^\varepsilon(0) = x_i , \qquad i = 1,2,\ldots,n ,$$

$$dy^\varepsilon(t) = a^{1/2}\left(\frac{y^\varepsilon(t)}{\varepsilon}\right) d\beta(t) ,$$

$$y^\varepsilon(0) = y .$$

Here $(\sigma_{ij}(x,y))$ are bounded smooth functions on $R^n \times R^1$ and $(\beta_1^{(1)}(t), \beta_2^{(1)}(t), \ldots, \beta_q^{(1)}(t))$ is a q-dimensional Brownian motion, not necessarily independent of $\beta(t)$.

Define

$$a_{ij}^{(1)}(x,y) = \sum_{k=1}^{q} \sigma_{ik}(x,y)\sigma_{jk}(x,y) \ ,$$

and, assuming $a_{ij}^{(1)}(x,y)$ vanish for y outside a bounded set,

$$(3.1.12) \quad \bar{a}_{ij}(x) = \int \frac{a_{ij}^{(1)}(x,y)}{a(y)} \ dy \ , \qquad i,j = 1,2,\ldots,n \ .$$

Let $\mathscr{L}$ be defined on $C_o^\infty(R^n)$ by

$$(3.1.13) \quad \mathscr{L}f(x) = \frac{1}{2} \sum_{i,j=1}^{n} \bar{a}_{ij}(x) \frac{\partial^2 f(x)}{\partial x_i \partial x_j} + \sum_{j=1}^{n} \bar{F}_j(x) \frac{\partial f(x)}{\partial x_j} \ ,$$

where $\bar{F}(x)$ is defined by (3.1.4). From the smoothness of the coefficients of $\mathscr{L}$ it follows that it generates a diffusion Markov process $\bar{x}(t)$ in R^n. Let

$$(3.1.14) \quad x(t) = \bar{x}(\ell_o(t))$$

where $\ell_o(t)$ is the local time at $y = 0$ of $y(t)$ in (3.1.7).

The proof of Theorem 3.1 extends immediately to the above situation so that $(x^\varepsilon(t),y^\varepsilon(t))$ of (3.1.11) converges weakly as $\varepsilon \to 0$ to $(x(t),y(t))$ of (3.1.14) and (3.1.7). In fact, the proof of Section 3.2 deals with this example.

Remark 3.

The conclusion of Theorem 3.1 and Remark 2 should be contrasted with (2.1.1) and the theorems of Part 2. The driving process $y^\varepsilon(t)$ does not decouple in the limit and continues to drive the process x(t) through its local time at the origin.

Other examples and remarks are given in Section 3.3 after the proof.

3.2 Proof of Theorem 3.1.

As pointed out in Remark 2 we shall give the proof for (3.1.11) with $\overline{\mathscr{L}}$ defined by (3.1.13).

Before going into the actual proof, which is similar to the one of Theorem 2.1, we prove the following five lemmas that we shall need.

Lemma 3.1.

Let

$$(3.2.1) \qquad \ell_o^\alpha(t) = \frac{1}{\alpha} \int_0^t \tilde{a}(y(s)) \chi_{[0,\alpha]}(y(s)) ds \ ,$$

where $\alpha > 0$ is a parameter and $\chi_A(y)$ is the characteristic function of the set $A \subset R^1$. Let $\ell_o(t)$ be defined by (3.1.8). Then

$$(3.2.2) \qquad E\{(\ell_o^\alpha(t) - \ell_o(t))^2\} \leq c_5 \alpha^2 + c_6 \alpha \sqrt{t}$$

and

$$(3.2.3) \qquad \begin{aligned} E\{\ell_o^\alpha(t)\} &\leq c_1 \sqrt{t} \ , \qquad E\{(\ell_o^\alpha(t))^2\} \leq c_2 t \ , \\ E\{e^{\lambda \ell_o^\alpha(t)}\} &\leq c_3 e^{c_4 \lambda^2 t} \ , \qquad \lambda > 0 \ , \end{aligned}$$

where the c_i are constants independent of t, λ and α. Note that in view of (3.2.2), (3.2.3) holds for $\ell_o(t)$ also.

Proof of Lemma.

Define $g^\alpha(y)$ by

(3.2.4) $g^\alpha(y) = \dfrac{2}{\alpha} \displaystyle\int_{-\infty}^{y} d\eta \int_{-\infty}^{\eta} d\zeta \chi_{[0,\alpha]}(\zeta)$.

By Itô's formula and (3.2.1) we have

(3.2.5) $\ell_0^\alpha(t) = g^\alpha(y(t)) - g^\alpha(y) - \displaystyle\int_0^t \dfrac{dg^\alpha(y(s))}{dy} dy(s)$.

Thus

(3.2.6) $E\{\ell_0^\alpha(t)\} \leq 2E\{|y(t)|\} + E^{1/2}\left\{\left[\displaystyle\int_0^t \dfrac{dg^\alpha}{dy} dy(s)\right]^2\right\}$.

$\leq C_1\sqrt{t}$,

where C_1 depends only on $\tilde{a}(y)$ of (3.1.6). Similary we obtain
the bound for $E\{(\ell_0^\alpha(t))^2\}$.

From (3.2.5) it follows that

(3.2.7) $E\left\{e^{\lambda \ell_0^\alpha(t)}\right\} = e^{-\lambda g^\alpha(y)} E\left\{e^{\lambda g^\alpha(y(t))} e^{-\lambda \int_0^t \frac{dg^\alpha}{dy}(y(s))dy(s)}\right\}$

$= e^{-\lambda g^\alpha(y)} E\left\{e^{\lambda g^\alpha(y(t)) + \lambda^2 \int_0^t \tilde{a}(y(s))\left[\frac{dg^\alpha}{dy}(y(s))\right]^2 ds}\right.$

$\left. \cdot e^{-\lambda \int_0^t \frac{dg^\alpha}{dy}(y(s))dy(s) - \lambda^2 \int_0^t \tilde{a}(y(s))\left[\frac{dg^\alpha}{dy}(y(s))\right]^2 ds}\right\}$

$\leq E^{1/2}\left\{e^{4\lambda y_+(t) + 4\tilde{c}\lambda^2 t}\right\}$.

Here $\tilde{c}$ is the largest of a_+ and a_-. Clearly the desired exponential estimate in (3.2.3) follows from (3.2.7).

Subtracting (3.1.8) and (3.2.5) and squaring we obtain

$$\left[\ell_o^\alpha(t) - \ell_o(t)\right]^2 = 4\left[2y_+(t) - g^\alpha(y(t))\right] + 4\left[2y_+ - g^2(y)\right]^2$$
$$+ 4\left[\int_0^t \left(2H(y(s)) - \frac{dg^\alpha}{dy}(y(s))\right)dy(s)\right]^2 .$$

Hence

(3.2.8) $$E\left\{\left[\ell_o^\alpha(t) - \ell_o(t)\right]^2\right\} \leq 8\alpha^2 + 4E\left\{\int_0^t \tilde{c}\chi_{[0,\alpha]}(y(s))ds\right\}$$

$$= 8\alpha^2 + 4\tilde{c}\alpha E\left\{\ell_o^\alpha(t)\right\}$$

$$\leq 8\alpha^2 + 4\tilde{c}c_1\alpha\sqrt{t} ,$$

which is (3.2.2). The proof is complete.

Lemma 3.2.

Let $y(t)$ be defined by (3.1.7) and $\ell_o(t)$ by (3.1.8) and let $\bar{x}(t)$ be the diffusion in R^n generated by $\mathscr{L}$ of (3.1.13) (it is uniquely defined by smoothness of its coefficients). Let $x(t)$ be defined by (3.1.14). The measures $P_{x,y}$ on $C([0,\infty), R^n \times R^1)$ corresponding to $(x(t),y(t))$ are characterized by the following martingale problem.

(i) $P_{x,y}\{x(0) = x, y(0) = y\} = 1$·

(ii) <u>For each</u> $f(x,y) \in C_o^\infty(R^n \times R^1)$,

(3.2.9) $f(x(t), y(t)) - \dfrac{1}{2} \displaystyle\int_0^t \tilde{a}(y(s)) \dfrac{\partial^2 f(x(s), y(s))}{\partial y^2} \, ds$

$- \displaystyle\int_0^t \overline{\mathscr{L}} f(x(s), y(s)) \, d\ell_o(s)$

<u>is a</u> $P_{x,y}$ <u>martingale.</u>

In (ii) $\ell_o(t)$ is the local time at zero of the y component of the process as defined by (3.1.8).

<u>Proof.</u>

If $(x(t), y(t))$ is defined by (3.1.14) and (3.1.7) and if $f(x,y) \in C_o^\infty(R^n \times R^1)$, then we have

(3.2.10) $E_{x,y}\{f(x(t), y(t))\} = f(x,y) + E_{x,y}\left\{ \displaystyle\int_0^t \dfrac{1}{2} a(y(s) \dfrac{\partial^2 f(x(s), y(s))}{\partial y^2} ds \right\}$

$+ E_{x,y}\left\{ \displaystyle\int_0^t \overline{\mathscr{L}} f(x(s), y(s)) d\ell_o(s) \right\}$.

Here we use the fact that $\ell_o(t)$ is almost surely bounded, increasing and continuous $(t < \infty)$. The equation (3.2.10) implies (3.2.9).

Now suppose (3.2.9) holds. According to Theorem 2.1 in
[5] and Lemma 3.1 above, if (3.2.9) is a P martingale then
for any f(x,y) smooth, bounded and strictly positive,

$$(3.2.11) \quad f(x(t),y(t))e^{-\int_0^t \frac{Lf(x(s),y(s))}{f(x(s),y(s))} ds - \int_0^t \frac{\overline{\mathscr{L}}f(x(s),y(s))}{f(x(s),y(s))} d\ell_o(s)}$$

is also a P martingale and conversely. In (3.2.11) we have
defined

$$(3.2.12) \quad L = \frac{1}{2}\tilde{a}(y)\frac{\partial^2}{\partial y^2}.$$

Thus, it suffices to show that (3.2.11) being a martingale
for each f positive, bounded and smooth, implies that the
process $(x(\cdot),y(\cdot))$ is as defined by (3.1.14) and (3.1.7).

First, by taking f = f(y) in (3.2.11) and using the
uniqueness of the process associated with L, we conclude that
the restriction of the measure P to M_t, t ≥ 0, the σ-algebra
generated by events associated with y(s), s ≤ t, is the one
defined by (3.1.7).

Let $\mathcal{F}_t$ be the σ-algebra of events generated by (x(s),y(s)),
s ≤ t and let $M = \sigma(\bigcup_{t>0} M_t)$ be the σ-algebra of events
generated by y(s), s ≥ 0. Let $Q_{y(\cdot)}$ be a regular conditional

probability of P given M. We shall show that if for every $f(x) \in C_o^\infty(R^n)$ and strictly positive,

$$(3.2.13) \quad X_f(t) = f(x(t))e^{-\int_0^t \frac{\overline{\mathscr{L}}f(x(s))}{f(x(s))} \, d\ell_o(s)},$$

then $(X_f(t), \mathscr{F}_t, Q_{y(\cdot)})$ is a martingale for almost all (relative to P) paths $y(\cdot)$. Here, as above, $\ell_o(t)$ is the functional of $y(\cdot)$ defined by (3.1.8). The uniqueness of the process $\bar{x}(t)$ associated with $\mathscr{L}$ and (3.2.13), identify $x(t)$ as in (3.1.14) (or (3.1.10)).

To show that $(X_f(t), \mathscr{F}_t, Q_{y(\cdot)})$ is P a.s. a martingale we proceed as follows. Let $f(x)$ and $g(y)$ be strictly positive and smooth functions on R^n and R^1 respectively. Define $X_f(t)$ by (3.2.13) and

$$(3.2.14) \quad Y_g(t) = g(y(t))e^{-\int_0^t \frac{Lg(y(s))}{g(y(s))} \, ds}.$$

By (3.2.11) we have that $(X_f(t)Y_g(t), \mathscr{F}_t, P)$ is a martingale. For $s \geq 0$ fixed define a measure $P_{f,s}$ on $\mathscr{F}_t$, $t \geq s$ so that

$$(3.2.15) \quad \left. \frac{dP_{f,s}}{dP} \right|_{\mathscr{F}_t} = \frac{X_f(t)}{X_f(s)}, \quad t \geq s.$$

With this definition, $(Y_g(t), \mathscr{F}_t, P_{f,s}, t \geq s)$ is a martingale. Since however $Y_g(t)$ is M_t measurable it follows that $(Y_g(t), M_t, P_{f,s}, t \geq s)$ is also a martingale. Hence

$$E^{P_{f,s}}\{Y_g(t_2)|M_{t_1}\} = Y_g(t_1) \; ,$$

P a.s. for each g, f and s $\geq$ 0. Taking g $\equiv$ 1 it follows that

(3.2.16) $E^{P_{f,s}}\{1|M\} = 1$.

If $Q_{y(\cdot)}$ is a regular conditional probability of P given M, then (3.2.16) and (3.2.15) imply that

(3.2.17) $E^{Q_{y(\cdot)}}\{X_f(t)|\mathcal{F}_s\} = X_f(s)$, $t \geq s$,

P a.s.

We note that the null set involved in (3.2.17) depends on f, s and t. Since a countable dense set of f's suffices and $X_f(t)$ is P a.s. continuous it follows that (3.2.17) holds for a determining class of f's and hence (3.2.13) is a martingale as was to be shown.

Lemma 3.3.

The following are equivalent.

(i)   For every $f(x,y) \in C_o^\infty(R^n \times R^1)$,

(3.2.18) $f(x(t),y(t)) - \dfrac{1}{2} \displaystyle\int_0^t \tilde{a}(y(s)) \dfrac{\partial^2 f(x(s),y(s))}{\partial y^2} ds$

$$- \int_0^t \mathcal{L}f(x(s),y(s))d\ell_o(s)$$

is an $\mathcal{F}_t$, P martingale with $\ell_o(t)$ defined by (3.1.8).

282

(ii) <u>For every</u> $f(x,y) \in C_o^\infty(R^n \times R^1)$,

$$(3.2.19) \quad f(x(t),y(t)) - 2y_+(t)\overline{\mathscr{L}}f(x(t),y(t)) - \int_0^t \tilde{a}(y(s))$$

$$\cdot \left[\frac{1}{2} \frac{\partial^2 f(x(s),y(s))}{\partial y^2} - 2H(y(s))\overline{\mathscr{L}}\frac{\partial f(x(s),y(s))}{\partial y}\right.$$

$$\left. - y_+(s)\overline{\mathscr{L}}\frac{\partial^2 f(x(s),y(s))}{\partial y^2}\right] ds$$

<u>is an</u> $\mathcal{F}_t$, P <u>martingale where</u> y_+ <u>and</u> $H(y)$ <u>are as in</u>
(3.1.9).

Proof.

First we show that (i) implies (ii). Let $g^\alpha(y)$ be defined by (3.2.4). The operator $\overline{\mathscr{L}}$ of (3.1.13) is assumed to have smooth coefficients.   In what
follows we make use of this smoothness in a superficial way which can be avoided by a preliminary smoothing.

For $f(x,y)$ smooth and bounded let

$$(3.3.20) \quad f^\alpha(x,y) = f(x,y) - g^\alpha(y)\overline{\mathscr{L}}f(x,y) \ .$$

With f replaced by f^α in (3.2.18) we find that

$$(3.2.21) \quad f^\alpha(x(t),y(t)) - \int_0^t \frac{1}{2}\tilde{a}(y(s)) \frac{\partial^2 f(x(s),y(s))}{\partial y^2} \, ds$$

$$+ \int_0^t \frac{1}{2}\tilde{a}(y(s)) \left[\frac{d^2 g^\alpha(y(s))}{dy^2}\overline{\mathscr{L}}f(x(s),y(s))\right.$$

$$\left. + 2 \frac{dg^\alpha(y(s))}{dy}\overline{\mathscr{L}}\frac{\partial f(x(s),y(s))}{dy} + g^\alpha(y(s))\overline{\mathscr{L}}\frac{\partial^2 f(x(s),y(s))}{\partial y^2}\right]$$

$$- \int_0^t \overline{\mathscr{L}} f(x(s),y(s)) d\ell_0(s)$$

$$+ \int_0^t g^\alpha(y(s)) \overline{\mathscr{L}}^2 f(x(s),y(s)) d\ell_0(s)$$

is an $\mathscr{F}_t$, P martingale.

Using (3.2.4) and (3.2.1) we rewrite (3.2.21) as follows

(3.2.22) $f(x(t),y(t)) - 2y_+(t) \overline{\mathscr{L}} f(x(t),y(t))$

$$- \int_0^t \tilde{a}(y(s)) \left[\frac{1}{2} \frac{\partial^2 f(x(s),y(s))}{\partial y} - 2H(y(s)) \mathscr{L} \frac{\partial f(x(s),y(s))}{\partial y} \right.$$

$$\left. - y_+(s) \mathscr{L} \frac{\partial^2 f(x(s),y(s))}{\partial y^2} \right] ds$$

$$- (g^\alpha(y(t)) - 2y_+(t)) \overline{\mathscr{L}} f(x(t),y(t))$$

$$+ \int_0^t \overline{\mathscr{L}} f(x(s),y(s)) d(\ell_0^\alpha(s) - \ell_0(s))$$

$$+ \int_0^t \tilde{a}(y(s)) \left[\left[\frac{dg^\alpha(y(s))}{dy} - 2H(y(s)) \right] \mathscr{L} \frac{f(x(s),y(s))}{\partial y} \right.$$

$$\left. + \left[\frac{1}{2} g^\alpha(y(s)) - y_+(s) \right] \mathscr{L} \frac{\partial^2 f(x(s),y(s))}{\partial y^2} \right] ds$$

$$+ \int_0^t (g^\alpha(y(s)) - 2y_+(s)) \overline{\mathscr{L}}^2 f(x(s),y(s)) d\ell_0(s)$$

$$+ \int_0^t 2y_+(s) \overline{\mathscr{L}}^2 f(x(s),y(s)) d\ell_0(s) \ ,$$

is an $\mathcal{F}_t$, P martingale. Now we note first that the last term in (3.2.22) is identically zero since $\ell_o(t)$ increases only when $y(t) = 0$. The expectation of the absolute value of the other terms that involve α goes to zero as $\alpha \downarrow 0$. Thus from (3.2.22) we immediately get that (3.2.19) is a martingale by passing to the limit $\alpha \downarrow 0$. Since the $y(\cdot)$ process has finite moments, this limit is elementary. Only the term

$$(3.2.23) \qquad \int_0^t \overline{\mathcal{L}} f(x(s),y(s)) d(\ell_o^{\alpha}(s) - \ell_o(s))$$

requires comment since $\ell_o^{\alpha}(t) - \ell_o(t)$ does not converge in variation (over t) in probability to zero as $\alpha \downarrow 0$. However, we have (3.2.2) and (3.2.3) and this is easily seen to imply that the expectation of (3.2.23) tends to zero as $\alpha \downarrow 0$.

To prove that (ii) implies (i) we proceed as follows. Let $g^{\alpha}(y)$ be defined by (3.2.4). For $f(x,y)$ smooth and bounded let $f^{\alpha}(x,y)$ be defined by

$$(3.2.24) \qquad f^{\alpha}(x,y) - g^{\alpha}(y) \overline{\mathcal{L}} f^{\alpha}(x,y) = f(x,y) \ .$$

The function $f^{\alpha}(x,y)$ is well defined by (3.2.24) since $g^{\alpha}(y) \geq 0$ and $\overline{\mathcal{L}}$ is a generator; f^{α} is a smooth and bounded function of (x,y) for $\alpha > 0$ (y is a parameter in (3.2.24)).

We now use (3.2.19) with f replaced by f^{α} of (3.2.24) to find that

(3.2.25) $f^{\alpha}(x(t),y(t)) - 2y_+(t)\overline{\mathscr{L}}f^{\alpha}(x(t),y(t))$

$$-\int_0^t a(y(s))\left[\frac{1}{2}\frac{\partial^2 f(x(s),y(s))}{\partial y^2}\right.$$

$$- (2H(y(s)) - \frac{dg^{\alpha}}{dy}(y(s)))\overline{\mathscr{L}}\frac{f^{\alpha}(x(s),y(s))}{\partial y}$$

$$\div \frac{1}{2}(2y_+(s) - g^{\alpha}(y(s)))\overline{\mathscr{L}}\frac{\partial^2 f^{\alpha}(x(s),y(s))}{\partial y^2}$$

$$\left.+ \frac{1}{2}\frac{d^2 g^{\alpha}}{dy^2}(y(s))\overline{\mathscr{L}}f^{\alpha}(x(s),y(s))\right]ds$$

is an $\mathscr{F}_t$, P martingale. Note that $\overline{\mathscr{L}}$ acts on x dependence alone. Let $R_{g^{\alpha}}$ be defined by

(3.2.26) $R_{g^{\alpha}} = (I - g^{\alpha}\overline{\mathscr{L}})^{-1}$,

so that $||R_{g^{\alpha}}|| \leq 1$ and

(3.2.27) $f^{\alpha} = R_{g^{\alpha}}f$.

Using (3.2.24) and (3.2.5) we may rewrite (3.2.25) as follows

(3.2.28) $f(x(t),y(t)) - \int_0^t \frac{1}{2}\tilde{a}(y(s))\frac{\partial^2 f(x(s),y(s))}{\partial y^2} ds$

$$- \int_0^t \overline{\mathscr{L}}f(x(s),y(s))d\ell_0(s)$$

$$- (2y_+(t) - g^{\alpha}(y(t)))\overline{\mathscr{L}}R^{\alpha}f(x(t),y(t))$$

$$+ \int_0^t \tilde{a}(y(s)) \left[\left[2H(y(s)) - \frac{dg^\alpha}{dy}(y(s)) \right] \overline{\mathscr{L}} \frac{\partial f^\alpha(x(s),y(s))}{\partial y} \right.$$

$$\left. + \frac{1}{2}(2y_+(s) - g^\alpha(y(s))) \overline{\mathscr{L}} \frac{\partial^2 f^\alpha(x(s),y(s))}{\partial y^2} \right] ds$$

$$- \int_0^t \overline{\mathscr{L}} f(x(s),y(s)) d(\ell_o^\alpha(s) - \ell_o(s))$$

$$- \int_0^t g^\alpha(y(s)) \overline{\mathscr{L}}^2 f^\alpha(x(s),y(s)) d\ell_o^\alpha(s)$$

is an $\mathscr{F}_t$, P martingale. We must show that the expectation of the absolute value of the last four terms goes to zero as $\alpha \to 0$. This goes exactly as in the first part of the proof. Only the term that involves $\partial^2 f^\alpha / \partial y^2$ requires special consideration.

Differentiating (3.2.24) with respect to y we obtain

$$f_y^\alpha - g^\alpha \overline{\mathscr{L}} f_y^\alpha = f_y + \frac{dg^\alpha}{dy} \overline{\mathscr{L}} f^\alpha ,$$

hence

(3.2.29) $$f_y^\alpha = R_{g^\alpha} f_y + \frac{dg^\alpha}{dy} (R_{g^\alpha})^2 \overline{\mathscr{L}} f .$$

Differentiating once again we obtain

(3.2.30) $$f_{yy}^\alpha = R_{g^\alpha} f_{yy} + 2 \frac{dg^\alpha}{dy^2} R_{g^\alpha} \overline{\mathscr{L}} \left[R_{g^\alpha} f_y + \frac{dg^\alpha}{dy} (R_{g^\alpha})^2 \overline{\mathscr{L}} f \right]$$

$$+ \frac{d^2 g^\alpha}{dy^2} (R_{g^\alpha})^2 \overline{\mathscr{L}} f .$$

We note next that

(3.2.31) $(2y_+ - g^\alpha(y)) \frac{1}{2} \frac{d^2 g^\alpha(y)}{dy^2}$

is bounded by a linear function of y independently of $\alpha > 0$

and $\alpha \to 0$ it tends to zero pointwise.

Substituting (3.2.30) into the fifth term in (3.2.28)

and using what was just pointed out about (3.2.31) we

obtain the desired result (3.2.18) by taking the limit $\alpha \to 0$

in (3.2.28). The proof of the lemma is complete.

Lemma 3.4.

Let X be a complete separable metric space and let P_n

be sequence of probability measures on the Borel sets of X

that converge weakly to a probability measure P. Let

$f_n(x)$, $x \in X$ be a sequence of measurable real-valued functions

on X which are uniformly bounded. Let $\mathcal{D} \subset X$ be such that

(3.2.32) $f_n(x) \to f(x)$,   as $n \to \infty$, $x \in \mathcal{D}$,

and

(3.2.33) $P(\mathcal{D}) = 1$,

where f(x) is a bounded measurable function on X. Assume

in addition the following

(3.2.34)   If $x_n \in X$, $x_n \to x$ and $x \in \mathcal{D}$,

          then $f_n(x_n) \to f(x)$.

288

Then

(3.2.35) $\int_X f_n(x)\,dP_n(x) \to \int_X f(x)\,dP(x)$ as $n \to \infty$.

Proof.

According to a Lemma of Skorohod [3], there exists a probability space (Ω, Σ, μ) and random variables $\xi_n(\omega)$, $\xi(\omega)$ with values in X such that

(i) $\mu\{\xi_n \in A\} = P_n(A)$, $\mu\{\xi \in A\} = P(A)$,

for each Borel set $A \subset X$,

and

(ii) $\xi_n(\omega) \to \xi(\omega)$ for μ almost all ω.

It follows that if $E\{\cdot\}$ denotes integration relative to μ, (3.2.35) takes the form

(3.2.36) $E\{f_n(\xi_n)\} \to E\{f(\xi)\}$.

By (3.2.34), (3.2.33) and (ii), the set of points $\omega \in \Omega$ for which $f_n(\xi_n(\omega)) \to f(\xi(\omega))$ has μ measure equal to one. Since f_n and f are uniformly bounded, (3.2.36) follows by the bounded convergence theorem.

Lemma 3.5.

Let Y be a locally compact metric space and let $g_n(y)$ and $g(y)$ be uniformly bounded and measurable real-valued functions

on Y. Let G be an open subset of Y and suppose that the g_n
are continuous on G and

(3.2.37) $g_n(y) \to g(y)$ uniformly for $y \in K \subset G$,

for each compact subset K of G. Let $X = C([0,T];Y)$ and let

(3.2.38) $\mathcal{D} = \left\{ y(\cdot) \in X \mid m\{t \in [0,T] \mid y(t) \in G^c\} = 0 \right\}$,

where m is Lebesque measure on the line.

If $y_n(\cdot) \in X$ and $y_n(\cdot) \to y(\cdot) \in \mathcal{D}$ then

(3.2.39) $\displaystyle \int_0^T g_n(y_n(t))dt \to \int_0^T g(y(t))dt$,   as $n \to \infty$.

Proof.

Fix a path $y(\cdot) \in \mathcal{D}$ and let $t \in [0,T]$ be such that
$y(t) \in G$. The set of such points $t \in [0,T]$ has Lebesque
measure T, by definition of $\mathcal{D}$. Since $y_n(t) \to y(t)$ and
$y(t) \in G$, it follows that for n sufficiently large all $y_n(t)$
belong to a compact subset K of G. From the uniform convergence (3.2.37) it follows that $g_n(y_n(t)) \to g(y(t))$. Thus,
the bounded convergence theorem yields (3.2.39).

Proof of Theorem 3.1.

Let $\zeta^\varepsilon(t)$ be defined by

(3.2.40) $\zeta^\varepsilon(t) = \dfrac{y^\varepsilon(t)}{\varepsilon}$, and consider jointly the process $(x^\varepsilon(t), y^\varepsilon(t), \zeta^\varepsilon(t))$ whose generator $\mathscr{L}^\varepsilon$ has the form

$$(3.2.41) \quad \mathscr{L}^\varepsilon = \frac{1}{\varepsilon^2}\left[\frac{1}{2} a(\zeta) \frac{\partial^2}{\partial y^2}\right] + \frac{1}{\varepsilon}\left[a(\zeta) \frac{\partial^2}{\partial\zeta\partial y} + \sum_{k=1}^{n} F_k(x,\zeta) \frac{\partial}{\partial x_k}\right.$$

$$+ \frac{1}{2} \sum_{k,j=1}^{n} a_{kj}^{(1)}(x,\zeta) \frac{\partial^2}{\partial x_k \partial x_j}\right]$$

$$+ \frac{1}{2} a(\zeta) \frac{\partial^2}{\partial y^2} .$$

Here we have used the notation following (3.1.11) and we have assumed that $(\beta_1^{(1)}(t),\ldots,\beta_q^{(1)}(t))$ and $\beta(t)$ in (3.1.11) are independent. This entails no loss in generality and does not affect the limit. Note that $\mathscr{L}^\varepsilon$ has the form

$$(3.2.42) \quad \mathscr{L}^\varepsilon = \frac{1}{\varepsilon^2}\mathscr{L}_1 + \frac{1}{\varepsilon}\mathscr{L}_2 + \mathscr{L}_3$$

as in (2.1.2), with $\mathscr{L}_1, \mathscr{L}_2$ and $\mathscr{L}_3$ identified from (3.2.41).

Let $f(x,y)$ be a bounded smooth function and define $\psi_f^{(1)}(x,y,\zeta)$ and $\psi_f^{(2)}(x,y,\zeta)$ as follows

$$(3.2.43) \quad \psi_f^{(1)}(x,y,\zeta) = -2 \int_0^\zeta d\eta \int_{-\infty}^\eta d\xi \frac{1}{a(\xi)} A_f(x,y,\xi) ,$$

$$A_f(x,y,\zeta) = \sum_{k=1}^{n} F_k(x,\zeta) \frac{\partial f(x,y)}{\partial x_k}$$

$$+ \frac{1}{2} \sum_{j,k=1}^{n} a_{kj}^{(1)}(x,\zeta) \frac{\partial^2 f(x,y)}{\partial x_k \partial x_j} ,$$

$$(3.2.44) \quad \psi_f^{(2)}(x,y,\zeta) = -2 \int_0^\zeta d\eta \int_{-\infty}^\eta d\xi \frac{1}{a(\xi)} A_{\psi_f^{(1)}}(x,y,\xi) .$$

By construction we have

(3.2.45) $\mathscr{L}^{\varepsilon}(f + \varepsilon\psi_f^{(1)} + \varepsilon^2\psi_f^{(2)})(x,y,\zeta)$

$$= a(\zeta)\left[\frac{1}{2}\frac{\partial^2 f(x,y)}{\partial y^2} + \frac{\partial^2\psi_f^{(1)}(x,y,\zeta)}{\partial\zeta\partial y} + \frac{\varepsilon}{2}\frac{\partial^2\psi_f^{(1)}(x,y,\zeta)}{\partial y^2}\right]$$

$$+ \varepsilon\left[\sum_{k=1}^{n}F_k(x,\zeta)\frac{\partial\psi_f^{(2)}(x,y,\zeta)}{\partial x_k}\right.$$

$$+ \frac{1}{2}\sum_{k,j=1}^{n}a_{kj}^{(1)}(x,\zeta)\frac{\partial^2\psi_f^{(2)}(x,y,\zeta)}{\partial x_k\partial x_j}$$

$$\left.\cdot\; a(\zeta)\frac{\partial^2\psi_f^{(2)}(x,y,\zeta)}{\partial\zeta\partial y}\right]$$

$$+ \varepsilon^2\frac{1}{2}a(\zeta)\frac{\partial^2\psi_f^{(2)}(x,y,\zeta)}{\partial y^2}\;.$$

With $\overline{\mathscr{L}}$ defined by (3.1.13) we note that

(3.2.46) $\displaystyle\lim_{\varepsilon\downarrow 0}\varepsilon\psi_f^{(1)}(x,y,\frac{y}{\varepsilon}) = -2y_+\overline{\mathscr{L}}f(x,y)$,

uniformly on compact sets in $R^n \times R^1$. Also,

(3.2.47) $\displaystyle\lim_{\varepsilon\downarrow 0}\frac{\partial\psi_f^{(1)}(x,y,\frac{y}{\varepsilon})}{\partial\zeta} = -2H(y)\overline{\mathscr{L}}f(x,y)$,

uniformly on compact subsets of $R^n \times (R^1 - \{0\})$. Similarly,

(3.2.48) $\displaystyle\lim_{\varepsilon\downarrow 0}\varepsilon^2\psi_f^{(2)}(x,y,\frac{y}{\varepsilon}) = 0$, $\displaystyle\lim_{\varepsilon\downarrow 0}\varepsilon^2\frac{\partial^2\psi_f^{(2)}}{\partial y^2}(x,y,\frac{y}{\varepsilon}) = 0$,

$$\lim_{\varepsilon\downarrow 0}\varepsilon\frac{\partial^2\psi_f^{(2)}(x,y,\frac{y}{\varepsilon})}{\partial\zeta\partial y} = 0\;,$$

$$\lim_{\varepsilon \downarrow 0} \varepsilon \sum_{k=1}^{n} F_k(x, \tfrac{y}{\varepsilon}) \frac{\partial \psi_f^{(2)}(x, y, \tfrac{y}{\varepsilon})}{\partial x_k}$$

$$+ \frac{\varepsilon}{2} \sum_{k,j=1}^{n} a_{kj}^{(1)}(x, \tfrac{y}{\varepsilon}) \frac{\partial^2 \psi_f^{(2)}(x, y, \tfrac{y}{\varepsilon})}{\partial x_k \partial x_j} = 0 ,$$

uniformly on compact subsets of $R^n \times R^1$. Here we use the fact that $F_k(x,y)$ and $a_{kj}^{(1)}(x,y)$ have compact support in y and they are smooth.

On the basis of the above constructions and the fact that $y^\varepsilon(t)$ has bounded moments of all orders, for t in a finite interval, with bounds independent of ε, it follows as in Section 2.3 that

$$(3.2.49) \quad M_f^\varepsilon(t) = f(x^\varepsilon(t), y^\varepsilon(t)) + \varepsilon \psi_f^{(1)}(x^\varepsilon(t), y^\varepsilon(t), \tfrac{y^\varepsilon(t)}{\varepsilon})$$

$$+ \varepsilon^2 \psi_f^{(2)}(x^\varepsilon(t), y^\varepsilon(t), \tfrac{y^\varepsilon(t)}{\varepsilon}) - f(x,y)$$

$$- \varepsilon \psi_f^{(1)}(x, y, \tfrac{y}{\varepsilon}) - \varepsilon^2 \psi_f^{(2)}(x, y, \tfrac{y}{\varepsilon})$$

$$- \int_0^t a\left(\frac{y^\varepsilon(s)}{\varepsilon}\right) \left[\frac{1}{2} \frac{\partial^2 f}{\partial y^2} + \frac{\partial^2 \psi_f^{(1)}}{\partial \zeta \partial y} + \frac{\varepsilon}{2} \frac{\partial^2 \psi_f^{(1)}}{\partial y^2} \right]$$

$$\cdot (x^\varepsilon(s), y^\varepsilon(s), \tfrac{y^\varepsilon(s)}{\varepsilon}) \, ds$$

$$- \varepsilon \int_0^t a\left(\frac{y^\varepsilon(s)}{\varepsilon}\right) \left[\sum_{k=1}^{n} F_k \frac{\partial \psi_f^{(2)}}{\partial x_k} + \frac{1}{2} \sum_{k,j=1}^{n} a_{kj}^{(1)} \frac{\partial^2 \psi_f^{(2)}}{\partial x_k \partial x_j} \right.$$

$$\left. + \frac{\varepsilon}{2} \frac{\partial^2 \psi_f^{(2)}}{\partial y^2} \right] (x^\varepsilon(s), y^\varepsilon(s), \tfrac{y^\varepsilon(s)}{\varepsilon}) \, ds$$

is an integrable zero-mean martingale. The increasing process corresponding to $M_f^\varepsilon(t)$ has the form

$$(3.2.50) \quad <M_f^\varepsilon(t),M_f^\varepsilon(t)> = \int_0^t H_f^\varepsilon(x^\varepsilon(s),y^\varepsilon(s),\frac{y^\varepsilon(s)}{\varepsilon})ds ,$$

where, one can verify it easily by direct computation,
$H_f^\varepsilon(x,y,\frac{y}{\varepsilon})$ is bounded by a constant times one plus a power of
$|y|$ independently of ε.

As in Section 2.3 we conclude from the above that the
process $(x^\varepsilon(\cdot),y^\varepsilon(\cdot))$ is relatively weakly compact in
$D([0,T]; R^n \times R^1)$. It remains now to identify the limit
process.

We rewrite (3.2.49) as follows.

$$(3.2.51) \quad f(x^\varepsilon(t),y^\varepsilon(t)) - 2y_+^\varepsilon(t)\overline{\mathscr{L}}f(x^\varepsilon(t),y^\varepsilon(t))$$

$$- f(x^\varepsilon(\sigma),y^\varepsilon(\sigma)) + 2y_+^\varepsilon(\sigma)\overline{\mathscr{L}}f(x^\varepsilon(\sigma),y^\varepsilon(\sigma))$$

$$- \int_\sigma^t \tilde{a}(y^\varepsilon(s))\left[\frac{1}{2}\frac{\partial^2 f}{\partial y^2} - 2H(y)\overline{\mathscr{L}}\frac{\partial f}{\partial y} - y_+\overline{\mathscr{L}}\frac{\partial^2 f}{\partial y^2}\right](x^\varepsilon(s),y^\varepsilon(s))ds$$

$$= M_f^\varepsilon(t) - M_f^\varepsilon(\sigma) - \left[2y_+^\varepsilon(t)\overline{\mathscr{L}}f(x^\varepsilon(t),y^\varepsilon(t)) - \varepsilon\psi_f^{(1)}(x^\varepsilon(t),y^\varepsilon(t),\frac{y^\varepsilon(t)}{\varepsilon})\right]$$

$$+ \left[2y_+^\varepsilon(\sigma)\overline{\mathscr{L}}f(x^\varepsilon(\sigma),y^\varepsilon(\sigma)) - \varepsilon\psi_f^{(1)}(x^\varepsilon(\sigma),y^\varepsilon(\sigma),\frac{y^\varepsilon(\sigma)}{\varepsilon})\right]$$

$$- \int_\sigma^t \left[a(\frac{y^\varepsilon(s)}{\varepsilon}) - \tilde{a}(y^\varepsilon(s))\right]\left[\frac{1}{2}\frac{\partial^2 f}{\partial y^2} - 2H(y)\overline{\mathscr{L}}\frac{\partial f}{\partial y} - y_+\overline{\mathscr{L}}\frac{\partial^2 f}{\partial y^2}\right]$$

$$\cdot (x^\varepsilon(s),y^\varepsilon(s))ds$$

$$
- \int_{\sigma}^{t} a\left(\frac{y^{\varepsilon}(s)}{\varepsilon}\right) \left[\frac{\partial^2 \psi_f^{(1)}}{\partial \zeta \partial y} + 2H(y)\overline{\mathscr{L}}\frac{\partial f}{\partial y} + \frac{\varepsilon}{2}\frac{\partial^2 \psi_f^{(1)}}{\partial y^2} + y_+ \overline{\mathscr{L}}\frac{\partial^2 f}{\partial y^2}\right]
$$

$$
\cdot \; (x^{\varepsilon}(s), y^{\varepsilon}(s), \frac{y^{\varepsilon}(s)}{\varepsilon}) ds
$$

$$
- \varepsilon \int_{\sigma}^{t} a\left(\frac{y^{\varepsilon}(s)}{\varepsilon}\right) \left[\sum_{k=1}^{n} F_k \frac{\partial \psi^{(2)}}{\partial x_k} + \frac{1}{2}\sum_{k,j=1}^{n} a_{kj}^{(1)}\frac{\partial^2 \psi_f^{(2)}}{\partial x_k \partial x_j} + \frac{\varepsilon}{2}\frac{\partial^2 \psi_f^{(2)}}{\partial y^2}\right]
$$

$$
\cdot \; (x^{\varepsilon}(s), y^{\varepsilon}(s), \frac{y^{\varepsilon}(s)}{\varepsilon}) ds \; .
$$

We intend to show that the absolute value of the terms after M_f^{ε} on the right hand side have expectation that tends to zero as $\varepsilon \to 0$ along a subsequence for which P^{ε}, the measure corresponding to $(x^{\varepsilon}(\cdot), y^{\varepsilon}(\cdot))$, converges weakly to a measure P (by compactness).

Let Φ_{σ} be a bounded continuous and $\mathscr{F}_{\sigma}$ measurable functional of $(x(\cdot), y(\cdot)) \in C([0,T]; R^n \times R)$. We multiply both sides of (3.2.51) by Φ_{σ}, take expectations relative to P^{ε} and let $\varepsilon \to 0$ along the chosen subsequence. We have

$$
(3.2.52) \quad \lim_{\varepsilon \downarrow 0} E^{P^{\varepsilon}}\left\{\Phi_{\sigma}\left[\overline{f(x(t), y(t))} - 2y_+(t)\overline{\mathscr{L}}f(x(t), y(t))\right.\right.
$$

$$
- \, f(x(\sigma), y(\sigma)) + 2y_+(\sigma)\overline{\mathscr{L}}f(x(\sigma), y(\sigma))
$$

$$
\left.\left.- \int_{\sigma}^{t}\tilde{a}(y(s))\left[\frac{1}{2}\frac{\partial^2 f}{\partial y^2} - 2H(y)\overline{\mathscr{L}}\frac{\partial f}{\partial y} - y_+\overline{\mathscr{L}}\frac{\partial^2 f}{\partial y^2}\right](x(s), y(s))ds\right]\right\}
$$

$$
= \lim_{\varepsilon \downarrow 0} E^{P^{\varepsilon}}\left\{-\Phi_{\sigma}\left[\int_{\sigma}^{t}\left(a\left(\frac{y(s)}{\varepsilon}\right) - \tilde{a}(y(s))\right)\left[\frac{1}{2}\frac{\partial^2 f}{\partial y^2} - 2H(y)\overline{\mathscr{L}}\frac{\partial f}{\partial y} - y_+\overline{\mathscr{L}}\frac{\partial^2 f}{\partial y^2}\right]\right.\right.
$$

$$
\left.\left.\cdot \; (x(s), y(s))ds\right]\right\}
$$

$$+ \lim_{\varepsilon \downarrow 0} E^{P^{\varepsilon}} \left\{ -\Phi_{\sigma} \left[\int_{\sigma}^{t} a(\frac{y(s)}{\varepsilon}) \left(\frac{\partial^2 \psi_f^{(1)}}{\partial \zeta \partial y} + 2H(y) \mathscr{L} \frac{\partial f}{\partial y} \right. \right. \right.$$

$$\left. \left. \left. + \frac{\varepsilon}{2} \frac{\partial^2 \psi_f^{(1)}}{\partial y} + y_+ \mathscr{L} \frac{\partial^2 f}{\partial y^2} \right) (x(s), y(s), \frac{y(s)}{\varepsilon}) ds \right] \right\} .$$

Here we have used the compactness of the measures P^{ε} and
(3.2.46), (3.2.48) to drop some terms on the right side of
(3.2.52) by the dominated convergence theorem.

In the notation of Lemma 3.5 we let $X = C([0,T],Y)$,
$Y = R^n \times R$, $G = R - \{0\}$ and

$$\mathscr{D} = \{ (x(\cdot), y(\cdot)) \in X \mid m\{t \in [0,T] \mid y(t) = 0\} = 0 \} .$$

From the existence on the local time for $y^{\varepsilon}(t)$ it follows
that for any $\varepsilon > 0$

(3.2.53) $P^{\varepsilon}(\mathscr{D}) = 0$,

and hence $P(\mathscr{D})$ for any weak limit of the measures P^{ε}.

The functional inside the expectation on the left hand
side of (3.2.52) is P a.s. continuous for any limit measure
by Lemma 3.5 (with $g_n \equiv g$). By Lemma 3.4 we can equate the
left side to the expectation of the functional with respect
to the weak limit P. On the right side we use the compactness
of the measures P^{ε}, (3.2.53), (3.2.46), (3.2.47) and Lemmas
3.5 and 3.4 to conclude it is zero.

From the fact that the right side of (3.2.52) is zero,
Lemmas 3.3 and 3.2 and the uniqueness of the limit process

described in Theorem 3.1 (and Remark 2) the identification step is complete. The proof of Theorem 3.1 is complete.

Remark.

As in the case of ergodic driving, the above proof extends immediately to a variety of similar situations of merely changing (3.2.43), (3.2.44) and (3.2.45) appropriately. Examples are given in the following sections.

3.3 Some simple consequences.

Instead of (3.1.3) (or (3.1.11)) we may consider the asymptotic behavior as $\varepsilon \to 0$ of the following system

$$(3.3.1) \qquad dx_i^\varepsilon(t) = \frac{1}{\varepsilon} \sum_{p=1}^{N} F_i^{(p)}\left(x^\varepsilon(t), \frac{y^\varepsilon(t) - y_p}{\varepsilon}\right),$$

$$x_i^\varepsilon(0) = x_i, \qquad i = 1, 2, \ldots, n$$

$$dy^\varepsilon(t) = a^{1/2}\left(\frac{y^\varepsilon(t)}{\varepsilon}\right) d\beta(t), \qquad y^\varepsilon(0) = y.$$

Here $F^{(1)}, F^{(2)}, \ldots, F^{(N)}$ are smooth bounded and with compact support in y n-vector functions of (x,y) and $y_1, y_2, \ldots, y_N$ are fixed points on the line. The function $a(y)$ satisfies (3.1.1) and (3.1.2) and is smooth.

It is easily seen that the methods of Section 3.2 apply immediately to (3.3.1). Let

$$(3.3.2) \qquad \bar{F}^{(p)}(x) = \int_{-\infty}^{\infty} \frac{F^{(p)}(x,y)}{a(y)}\, dy,$$

and

$$\overline{\mathscr{L}}_p = \sum_{i=1}^{n} \bar{F}_i^{(p)}(x)\, \frac{\partial}{\partial x_i}, \qquad p = 1, 2, \ldots, N.$$

Under our hypotheses, there is a unique family of probability measures $P_{(x,y)}$ on $C([0,\infty), R^n \times R^1)$ for $(x,y) \in R^n \times R^1$ such that

298

(i) $P_{(x,y)} \{x(0) = x, y(0) = y\}$

(ii) For each $f(x,y) \in C_o^\infty (R^n \times R^1)$

(3.3.3) $f(x(t),y(t)) - \int_0^t \frac{1}{2} \tilde{a}(y(s)) \frac{\partial^2 f(x(s),y(s))}{\partial y^2} ds$

$$- \sum_{p=1}^N \int_0^t \overline{\mathscr{L}}_p f(x(s),y(s)) d\ell (s,y_p)$$

is a P martingale. Here $\tilde{a}(y)$ is defined by (3.1.6) and $\ell(t,\bar{y})$ is the local time of $y(t)$ of (3.1.7) at $\bar{y} \in R^1$. The local time is defined by

(3.3.4) $\ell(t,\bar{y}) = 2(y(t)-\bar{y})_+ - 2(y-\bar{y})_+$

$$- 2 \int_0^t H(y(s)-\bar{y}) dy(s) .$$

This is the natural generalization of (3.1.8).

Theorem 3.2.

Under the above hypotheses $(x^\varepsilon (t), y^\varepsilon (t))$ of (3.3.1) converges weakly in $C([0,T];R^n \times R^1)$, $T < \infty$, to the process $(x(t),y(t))$ characterized by the martingale problem (3.3.3).

As   an   application consider the following problem which is similar to one considered by Isaacson [39].

Let $y(t)$ be Brownian motion on R^1 and $F(y)$ be a smooth nonnegative function of compact support such that

299

$$(3.3.5) \qquad \int_{-\infty}^{\infty} F(y) \, dy = 1 \; .$$

Consider the Cauchy problem (f(y) smooth bounded)

$$(3.3.6) \qquad \frac{\partial u^\varepsilon(t,y)}{\partial t} = \frac{1}{2} \frac{\partial^2 u^\varepsilon(t,y)}{\partial y^2} - \frac{1}{2\varepsilon}\left[F(\tfrac{y+1}{\varepsilon}) + F(\tfrac{y-1}{\varepsilon})\right] u^\varepsilon(t,y) \; ,$$

$$u^\varepsilon(0,y) = f(y) \; .$$

It follows from Theorem 3.2 that

$$(3.3.7) \qquad \lim_{\varepsilon \downarrow 0} u^\varepsilon(t,y) = E_y\left\{ e^{-\frac{1}{2}[\ell(t,+1)+\ell(t,-1)]} f(y(t)) \right\} \; .$$

If u(t,y) stands for the right hand side of (3.3.7), formally, it satisfies the Cauchy problem

$$(3.3.8) \qquad \frac{\partial u(t,y)}{\partial t} = \frac{1}{2} \frac{\partial^2 u(t,y)}{\partial y^2} - \frac{1}{2}\Big(\delta(y+1) + \delta(y-1) \Big) u(t,y) \; ,$$

$$u(0,y) = f(y) \; .$$

To see in more detail how (3.3.7) follows introduce $x^\varepsilon(t)$ by

$$x^\varepsilon(t) = x + \frac{1}{2\varepsilon} \int_0^t \left[F\left(\frac{y(s)+1}{\varepsilon}\right) + F\left(\frac{y(s)-1}{\varepsilon}\right) \right] ds$$

so that

$$(3.3.9) \qquad \frac{dx^\varepsilon(t)}{dt} = \frac{1}{2\varepsilon}\left[F\left(\frac{y(s)+1}{\varepsilon}\right) + F\left(\frac{y(s)-1}{\varepsilon}\right) \right]$$

$$x^\varepsilon(0) = x \; .$$

By the Feynman-Kac formula,

$$u^\varepsilon(t,y) = E_y\left\{e^{-(x^\varepsilon(t)-x)}f(y(t))\right\}$$

$$= e^x E_{x,y}\left\{e^{-x^\varepsilon(t)}f(y(t))\right\} .$$

Thus, the quantity of interest $u^\varepsilon(t,y)$ is a bounded continuous functional of $(x^\varepsilon(t),y(t))$ to which Theorem 3.2 applies directly.

Theorem 3.1 (or Theorem 3.2) can be extended to the case where $y^\varepsilon(t)$, the driving process, is not, essentially, Brownian motion as we have assumed (cf. (3.1.3)).

One possibility is to allow $a(y)$ to violate (3.1.1) so that $\int_0^y \frac{1}{a(\zeta)}\, d\zeta$ tends to ∞ as $y \to \pm\infty$ at a rate other than $|y|$ (the Brownian case). Another is to allow $y^\varepsilon(t)$ to be a symmetric stable process and to adjust the scaling appropriately. When $F(x,y)$ in (3.1.3) is a function of y only, such problems are considered in [20,21]. In connection with results analogous to the one of the next section see [22].

We note finally that if $F(x,y)$ is a periodic function of y (or an appropriate almost periodic function) in (3.1.3), then, even though the driving process is not ergodic, we are effectively in the ergodic case; in fact in a situation similar to the one of Section 2.7.

3.4 Second order limit theorem.

We shall now consider a rescaled form of (3.1.3) (or (3.1.11)) where $\bar{F}(x) \equiv 0$ in (3.1.4). We call the asymptotic limit a second order limit because the first order result, Theorem 3.1, is trivial in case $\bar{F}(x) \equiv 0$ and we have to go to higher order (rescale) to get the first nontrivial result.

Let $F_i^{(1)}(x,y)$, $F_i^{(2)}(x,y)$ and $\sigma_{ij}(x,y)$, $i = 1,2,\ldots,n$, $j = 1,2,\ldots,q$, $x \in R^n$, $y \in R^1$ be smooth bounded functions with compact support in y and let $a(y)$ satisfy (3.1.1) and (3.1.2). Consider the system of stochastic differential equations

$$(3.4.1) \quad dx_i^\varepsilon(t) = \left[\frac{1}{\varepsilon^3} F_i^{(1)}\left(x^\varepsilon(t),\frac{y^\varepsilon(t)}{\varepsilon^2}\right) + \frac{1}{\varepsilon^2} F^{(2)}\left(x^\varepsilon(t),\frac{y^\varepsilon(t)}{\varepsilon^2}\right)\right]dt$$

$$+ \frac{1}{\varepsilon}\sum_{j=1}^q \sigma_{ij}\left(x^\varepsilon(t),\frac{y^\varepsilon(t)}{\varepsilon^2}\right)d\beta_j^{(1)}(t)$$

$$x_i^\varepsilon(0) = x_i, \qquad i = 1,2,\ldots,n,$$

$$dy^\varepsilon(t) = a^{1/2}\left[\frac{y^\varepsilon(t)}{\varepsilon^2}\right]d\beta(t), \qquad y^\varepsilon(0) = y.$$

Here $(\beta_1^{(1)}(t),\beta_2^{(1)}(t),\ldots,\beta_q^{(1)}(t))$ is a q-dimensional Brownian motion, not necessarily independent of the Brownian motion $\beta(t)$.

The distinguishing hypothesis about (3.4.1) is that

$$(3.4.2) \quad \int_{-\infty}^{\infty} \frac{F^{(1)}(x,y)}{a(y)} dy \equiv 0 .$$

The scaling of (3.4.1) should be compared with (3.1.3) (or (3.1.11)); replacing ε by $\sqrt{\varepsilon}$ in (3.4.1) identifies the $F^{(1)}$ drift as a more singular term than usual which explains the necessity of (3.4.2).

Define

$$a_{ij}^{(1)}(x,y) = \sum_{k=1}^{q} \sigma_{ik}(x,y)\sigma_{jk}(x,y)$$

and

(3.4.3) $$\bar{a}_{ij}(x) = \int_{-\infty}^{\infty} \frac{a_{ij}^{(1)}(x,y)}{a(y)}\, dy \ ,$$

$$\bar{F}_i^{(2)}(x) = \int_{-\infty}^{\infty} \frac{\bar{F}_i^{(2)}(x,y)}{a(y)}\, dy \ , \qquad i,j = 1,2,\ldots,n \ .$$

For $f(x) \in C_o^{\infty}(R^n)$, define $\tilde{\psi}_f^{(1)}(x,\zeta)$, $\zeta \in R^1$, by

(3.4.4) $$\tilde{\psi}_f^{(1)}(x,\zeta) = -2 \int_0^{\zeta} d\eta \int_{-\infty}^{\eta} d\zeta \sum_{j=1}^{n} \frac{F_j^{(1)}(x,\xi)}{a(\xi)} \frac{\partial f(x)}{\partial x_j} \ .$$

Similary define

(3.4.5) $$\tilde{\psi}_f(x) = \int_{-\infty}^{\infty} d\zeta \sum_{i=1}^{n} \frac{F_i^{(1)}(x,\zeta)}{a(\zeta)} \frac{\partial \psi_f^{(1)}(x,\zeta)}{\partial x_i}$$

$$= 2 \int_{-\infty}^{\infty} d\zeta \int_{-\infty}^{\zeta} d\eta \sum_{i=1}^{n} \frac{F_i^{(1)}(x,\eta)}{a(\eta)} \frac{\partial}{\partial x_i}$$

$$\cdot \left[\int_{-\infty}^{\zeta} d\xi \sum_{j=1}^{n} \frac{F_j^{(1)}(x,\xi)}{a(\xi)} \frac{\partial f(x)}{\partial x_j} \right] \ .$$

Note that because of (3.4.2), (3.4.5) is well defined and smooth for $f(x) \in C_o^{\infty}(R^n)$.

Let $\overline{\mathscr{L}}$ be the second order elliptic operator

$$(3.4.6) \quad \overline{\mathscr{L}}f(x) = \tilde{\psi}_f(x) + \frac{1}{2}\sum_{i,j=1}^{n} \bar{a}_{ij}(x) \frac{\partial^2 f(x)}{\partial x_i \partial x_j}$$

$$+ \sum_{j=1}^{n} \bar{F}_j^{(1)}(x) \frac{\partial f(x)}{\partial x_j} .$$

By hypothesis it has smooth coefficients and so it generates a diffusion process $\bar{x}(t)$ on R^n.

Let $y(t)$ be the process defined by (3.1.7) and let $\ell_o(t)$ be the local time of $y(t)$ at zero, defined by (3.1.8). Let $x(t)$ be the diffusion process $\bar{x}(\cdot)$ run at the local time of $y(\cdot)$ i.e., let (3.1.10) hold.

Theorem 3.3.

Under the above hypotheses, the process $(x^\varepsilon(t), y^\varepsilon(t))$ defined by (3.4.1) converges weakly on $C([0,T];R^n \times R^1)$, $T < \infty$, to the process $(x(t), y(t))$ described above.

Proof.

The proof of this theorem is nearly identical to the one of Theorem 3.1. It suffices to show how the constructions centering around (3.2.45) change here.

To simplify writing we shall assume that $F^{(2)} \equiv 0$, $\sigma \equiv 0$ and $F^{(1)} = F$ with $n = 1$.

As in (3.2.40) put

(3.4.7) $\zeta^{\varepsilon}(t) = \dfrac{y^{\varepsilon}(t)}{\varepsilon^2}$,

and consider $(x^{\varepsilon}(t), y^{\varepsilon}(t), \zeta^{\varepsilon}(t))$ jointly so that its generator $\mathscr{L}^{\varepsilon}$ has the form

(3.4.8) $\mathscr{L}^{\varepsilon} = \dfrac{1}{\varepsilon^4} \left[\dfrac{1}{2} a(\zeta) \dfrac{\partial^2}{\partial \zeta^2} \right] + \dfrac{1}{\varepsilon^3} F(x, \zeta) \dfrac{\partial}{\partial x}$

$\qquad + \dfrac{1}{\varepsilon^2} a(\zeta) \dfrac{\partial^2}{\partial \zeta \partial y} + \dfrac{1}{2} a(\zeta) \dfrac{\partial^2}{\partial y^2}$.

Define $\psi_f^{(1)}(x, y, \zeta)$, $\psi_f^{(2)}(x, y, \zeta)$ and $\psi_f^{(3)}(x, y, \zeta)$ for each $f \in C_o^{\infty}(R^n \times R^1)$ as follows

(3.4.9) $\psi_f^{(1)}(x, y, \zeta) = -2 \displaystyle\int_0^{\zeta} d\eta \int_{-\infty}^{\eta} d\xi \, \dfrac{F(x, \xi)}{a(\xi)} \dfrac{\partial f(x, y)}{\partial x}$,

(3.4.10) $\psi_f^{(2)}(x, y, \zeta) = -2 \displaystyle\int_0^{\zeta} d\eta \int_{-\infty}^{\eta} d\xi \, \dfrac{F(x, \xi)}{a(\xi)} \dfrac{\partial \psi_f^{(1)}(x, y, \xi)}{\partial x}$,

(3.4.11) $\psi_f^{(3)}(x, y, \zeta) = -2 \displaystyle\int_0^{\zeta} d\eta \int_{-\infty}^{\eta} d\xi \left[\dfrac{F(x, \xi)}{a(\xi)} \dfrac{\partial \psi_f^{(2)}(x, y, \xi)}{\partial x} \right.$

$\qquad\qquad\qquad \left. + \dfrac{\partial^2 \psi_f^{(1)}(x, y, \xi)}{\partial \zeta \partial y} \right]$.

We have the following

$$(3.4.12) \quad \mathscr{L}^\varepsilon \left[f + \varepsilon \psi_f^{(1)} + \varepsilon^2 \psi_f^{(2)} + \varepsilon^3 \psi_f^{(3)} \right] (x,y,\zeta)$$

$$= a(\zeta) \left[\frac{1}{2} \frac{\partial^2 f(x,y)}{\partial y^2} + \frac{\partial^2 \psi_f^{(2)}(x,y,\zeta)}{\partial \zeta \partial y} + \frac{\varepsilon^2}{2} \frac{\partial^2 \psi_f^{(2)}(x,y,\zeta)}{\partial y^2} \right]$$

$$+ F(x,\zeta) \frac{\partial \psi_f^{(3)}(x,y,\zeta)}{\partial x} + \varepsilon a(\zeta) \frac{\partial^2 \psi_f^{(3)}(x,y,\zeta)}{\partial \zeta \partial y}$$

$$+ \frac{\varepsilon}{2} a(\zeta) \frac{\partial^2 \psi_f^{(1)}(x,y,\zeta)}{\partial y^2} + \frac{\varepsilon^3}{2} a(\zeta) \frac{\partial^2 \psi_f^{(3)}(x,y,\zeta)}{\partial y^2} .$$

From (3.4.2) it follows that $\psi_f^{(1)}(x,y,\zeta)$ is bounded and it is easily seen that the following limits hold.

$$(3.4.13) \quad \lim_{\varepsilon \downarrow 0} \varepsilon^2 \psi_f^{(2)}(x,y,\frac{y}{\varepsilon^2}) = -2y_+ \overline{\mathscr{L}} f(x,y)$$

uniformly on compact subsets of $R \times R$,

$$(3.4.14) \quad \lim_{\varepsilon \downarrow 0} \frac{\partial \psi_f^{(2)}(x,y,\frac{y}{\varepsilon^2})}{\partial \zeta} = -H(y) \overline{\mathscr{L}} f(x,y)$$

uniformly on compact subsets of $R \times (R-[0])$,

$$(3.4.15) \quad \lim_{\varepsilon \downarrow 0} F(x,\frac{y}{\varepsilon^2}) \frac{\partial \psi_f^{(3)}(x,y,\frac{y}{\varepsilon^2})}{\partial x} = 0 ,$$

$$\lim_{\varepsilon \downarrow 0} \varepsilon \frac{\partial \psi_f^{(3)}(x,y,\frac{y}{\varepsilon^2})}{\partial \zeta} = 0 ,$$

$$\lim_{\varepsilon \downarrow 0} \varepsilon^3 \psi_f^{(3)}(x,y,\frac{y}{\varepsilon^2}) = 0 ,$$

uniformly on compact subsets of $R \times R$.

The above parts correspond to $(3.2.46)-(3.2.48)$. The proof now proceeds exactly as in Theorem 2.1.

3.5 A jump driving process.

We shall restrict attention to the situation that is comparable to Theorem 3.1. It is easily extended to the one of Theorem 3.3.

Let $Y_0 = 0$ and $Y_1, Y_2, \ldots$, be independent identically distributed random variables with distribution $\phi(dz)$ such that

$$(3.5.1) \qquad \int_{-\infty}^{\infty} z\phi(dz) = 0 , \qquad \int_{-\infty}^{\infty} z^2\phi(dz) = 1$$

and for some $\alpha > 0$

$$(3.5.2) \qquad \int_{-\infty}^{\infty} e^{\alpha z}\phi(dz) < \infty .$$

Let $N(t)$ be an independent Poisson process with parameter 1 and let $N^\varepsilon(t)$ denote the same process rescaled so that its parameter is $1/\varepsilon^2$. Define

$$(3.5.3) \qquad y^\varepsilon(t) = y + \varepsilon \int_0^t Y_{N^\varepsilon(s)} \, dN^\varepsilon(s) ,$$

which is simply the sum of a Poisson number of random variables. The scaling is such that $y^\varepsilon(t)$ converges weakly to Brownian motion as $\varepsilon \to 0$. This can be verified very easily.

Let $F(x,y)$ be a smooth and bounded n-vector function on $R^n \times R$ with compact support in y. Consider the process $x^\varepsilon(t)$ on R^n defined by

(3.5.4) $\quad \frac{dx^{\varepsilon}(t)}{dt} = \frac{1}{\varepsilon} F(x^{\varepsilon}(t), \frac{y^{\varepsilon}(t)}{\varepsilon})$, $\quad x^{\varepsilon}(0) = x$.

We shall study the asymptotic behavior of $(x^{\varepsilon}(t), y^{\varepsilon}(t))$ as

$\varepsilon \to 0$.

Let

(3.5.5) $\quad \overline{F}(x) = \int\limits_{-\infty}^{\infty} F(x,y)$

and define $\overline{x}(t)$ by

(3.5.6) $\quad \frac{d\overline{x}(t)}{dt} = \overline{F}(\overline{x}(t))$, $\quad \overline{x}(0) = x$.

Let $y(t)$ be Brownian motion with $y(0) = y$ and let $\ell_0(t)$ be

its local time at $y = 0$. Define $x(t)$ by

(3.5.7) $\quad x(t) = \overline{x}(\ell_0(t))$.

Theorem 3.4.

Under the above hypotheses, $(x^{\varepsilon}(t), y^{\varepsilon}(t))$ converges

weakly in $C([0,T]; R^n \times R)$, $T < \infty$, to the process $(x(t), y(t))$

defined above.

Proof.

Again the proof is very similar to the one of Theorem

3.1. We outline the constructions centering around (3.2.45).

We take $n = 1$ for simplicity.

Define

(3.5.8) $\qquad\qquad\qquad \zeta^{\varepsilon}(t) = \frac{y^{\varepsilon}(t)}{\varepsilon}$

so that $(x^\epsilon(t), y^\epsilon(t), \zeta^\epsilon(t))$ are jointly Markov with generator

$$(3.5.9) \quad \mathscr{L}^\epsilon f(x,y,\zeta) = \frac{1}{\epsilon^2}\left[\int_{-\infty}^{\infty} f(x,y+\zeta z,\epsilon+z)\phi(dz) - f(x,y,\zeta)\right]$$

$$+ \frac{1}{\epsilon} F(x,\zeta) \frac{\partial}{\partial x} .$$

We rewrite this as follows:

$$(3.5.10) \quad \mathscr{L}^\epsilon f(x,y,\zeta) = \frac{1}{\epsilon^2}\left[\int_{-\infty}^{\infty} f(x,y,\zeta+z)\phi(dz) - f(x,y,\zeta)\right]$$

$$+ \frac{1}{\epsilon} F(x,\zeta) \frac{\partial f(x,y,\zeta)}{\partial x} + \frac{1}{\epsilon}\int \frac{\partial f(x,y,\zeta+z)}{\partial y} z\phi(dz)$$

$$+ \frac{1}{2}\int \frac{\partial^2 f(x,y,\zeta+z)}{\partial y^2} z^2\phi(dz) + S^\epsilon f(x,y,z) .$$

Here

$$(3.5.11) \quad S^\epsilon f(x,y,\zeta) = \frac{1}{\epsilon^2}\int_{-\infty}^{\infty}\left[f(x,y+\epsilon z,\zeta+z) - f(x,y,\zeta+z)\right.$$

$$\left. - \epsilon \frac{\partial f(x,y,\zeta+z)}{\partial y} z - \frac{\epsilon^2}{2}\frac{\partial^2 f(x,y,\zeta+z)}{\partial y^2} z^2\right]\phi(dz)$$

and it easily seen that for f smooth $S^\epsilon f \to 0$ as $\epsilon \to 0$ uniformly on compact sets.

Now we construct $\psi_f^{(1)}(x,y,\zeta)$ for each $f(x,y) \in C_o^\infty$ so that

$$(3.5.12) \quad \int \psi_f^{(1)}(x,y,\zeta+z)\phi(dz) - \psi_1(x,y,\zeta) + F(x,\zeta)\frac{\partial f(x,y)}{\partial x} = 0 .$$

If $\hat{\phi}(k)$ denotes the characteristic function of $\phi(dz)$ and $^\wedge$, in general, Fourier transform, we have

(3.5.13) $\hat{\psi}_f^{(1)}(x,y,k) = \dfrac{\hat{F}(x,k)\frac{\partial f(x,y)}{\partial x}}{1 - \hat{\phi}(k)}$

which is well defined by (3.5.1), (3.5.2) and the smoothness and compact support of $F(x,y)$. Furthermore, we have

(3.5.14) $\displaystyle\lim_{\varepsilon \downarrow 0} \varepsilon\psi_f^{(1)}(x,y,\tfrac{y}{\varepsilon}) = -2y_+ \, \bar{F}(x) \, \dfrac{\partial f(x,y)}{\partial x}$

uniformly on compact sets,

and

(3.5.15) $\displaystyle\lim_{\varepsilon \downarrow 0} \dfrac{\partial\psi_f^{(1)}(x,y,\tfrac{y}{\varepsilon})}{\partial y} = -2H(y)\bar{F}(x) \, \dfrac{\partial f(x,y)}{\partial x}$

uniformly on compact subsets of $R \times (R-[0])$.

We define, similarly, $\psi_f^{(2)}(x,y,\zeta)$ as solutions of

(3.5.16) $\displaystyle\int_{-\infty}^{\infty} \psi_f^{(2)}(x,y,\zeta+z)\phi(dz) - \psi_f^{(2)}(x,y,\zeta)$

$+ F(x,\zeta) \, \dfrac{\partial\psi_f^{(1)}(x,y,\zeta)}{\partial x} = 0$

and let $\mathcal{L}^\varepsilon$ of (3.5.10) act on $f(x,y) + \varepsilon\psi_f^{(1)}(x,y,\zeta)$ $+ \varepsilon^2\psi_f^{(2)}(x,y,\zeta)$ as usual. Upon collecting terms, using (3.5.12), (3.5.16), the smallness of S^ε and (3.5.14), (3.5.15) we arrive, after some additional simple manipulations, to the usual set-up of the proof of Theorem 3.1.

3.6   Interaction with ergodic driving.

Clearly one can consider combined forms of Theorem 2.1 and Theorem 3.1. The situation is somewhat analogous to the one of Section 2.8. We shall confine attention to one example.

Let $y(t)$ be Brownian motion on R^1 with $y(0) = y$ and let $z(t)$ be an ergodic Markov process on R^m with exponentially convergent recurrent potential kernel (cf. Section 2.1 where $z(t)$ corresponds to $Y(t)$). Let $F(x,y,z)$ be a smooth and bounded n-vector function on $R^n \times R \times R^m$ with compact support in y. Let

$$(3.6.1) \qquad z^\varepsilon(t) = z(t/\varepsilon^2)$$

and define $x^\varepsilon(t)$ by

$$(3.6.2) \qquad \frac{dx^\varepsilon(t)}{dt} = \frac{1}{\varepsilon} F(x^\varepsilon(t), \frac{y(t)}{\varepsilon}, z^\varepsilon(t)) , \qquad x^\varepsilon(0) = x .$$

In view of the Brownian scaling $y(t/\varepsilon^2) \sim y(t)/\varepsilon$. If $y^\varepsilon(t) = y(t/\varepsilon^2)$ we may write (3.6.2) as

$$(3.6.3) \qquad \frac{dx^\varepsilon(t)}{dt} = \frac{1}{\varepsilon} F(x^\varepsilon(t), y^\varepsilon(t), z^\varepsilon(t)) , \qquad x^\varepsilon(0) = x .$$

Let

$$(3.6.4) \qquad \bar{F}(x) = \int_{-\infty}^{\infty} dy \int_{R^m} \bar{P}(z) F(x,y,z) ,$$

where $\bar{P}(A)$, $A \subset R^m$ is the invariant probability measure of $z(t)$. Let $\bar{x}(t)$ be defined by

(3.6.5) $\dfrac{d\bar{x}(t)}{dt} = \bar{F}(\bar{x}(t))$, $\bar{x}(0) = x$,

and let

(3.6.6) $x(t) = \bar{x}(\ell_o(t))$

where $\ell_o(t)$ is the local time of $y(t)$ at $y = 0$.

Theorem 3.5.

The process $x^\epsilon(t)$ converges weakly on $C([0,T];R^n)$, $T < \infty$, to the process defined by (3.6.5) and (3.6.6) (relative to a Brownian motion $y(t)$).

The proof of this result is again entirely analogous to the one of Theorem 3.1. We omit further details.

Appendix. Ergodic Properties of $\mathscr{L}_{BL}$.

We shall discuss briefly here the hypotheses concerning
ergodicity of the process generated by $\mathscr{L}_1$ in Section (2.2)
($\mathscr{L}_1$ is defined by (2.1.3)) and the process generated by $\mathscr{L}_{BL}$
in Section (2.5) (defined by (2.5.8)). These hypotheses were
sufficient for the asymptotic analysis.

It is convenient to restrict attention to the case where
the diffusion process Y(t) generated[†] by $\mathscr{L}_1$ is confined to a
compact subset S of R^m. This can be done by imposing reflecting boundary conditions on ∂S, assumed smooth. In the sequel
Y(t) will be this process with $\mathscr{L}_1$ its generate in the interior
(the boundary conditions will not be introduced explicitly)
and with transition function P(t,y,A), $y \in S$, $A \subset S$.

A simple sufficient condition for ergodicity and existence of the recurrent potential (2.2.2) is the following
well-known Doeblin's condition: there is a reference probability measure μ on S a $t_o > 0$ and a c > 0 such that for all
Borel subsets A of S,

(A.1) $P(t_o,y,A) \geq c\mu(A)$, $y \in S$.

One easily deduces from this that there exists a unique invariant measure $\bar{P}(A)$ on S and that it is approached exponentially fast by P(t,y,A), as $t \to \infty$, uniformly in y and A.

[†]Parametric dependence on other, frozen variables will
not be shown in the Appendix.

A slightly different formulation of this result, useful in the context of differential equations and boundary value problems for $P(t,y,A)$ as here, is the following: $P(t,y,A)$ is uniformly Feller continuous i.e., for each $t > 0$

$$(A.2) \qquad \sup_{A \subset S} |P(t,y,A) - P(t,y',A)| \to 0 \quad \text{as} \quad |y-y'| \to 0 \ ,$$

and for any x and y in S and some $t_o > 0$ the measures $P(t_o,x,\cdot)$ and $P(t_o,y,\cdot)$ are not orthogonal.

To see how this implies ergodicity with exponential approach let x and y be points in S and let $H_{x,y}$ be a positive Hahn decomposition of S for $P(t_o,x,\cdot) - P(t_o,y,\cdot)$ then,

$$\sup_{A \subset S} |P(t_o,x,A) - P(t_o,y,A)|$$

$$= P(t_o,x,H_{x,y}) - P(t_o,y,H_{x,y}) \equiv \rho(x,y)$$

and by the orthogonality hypothesis $\rho(x,y) < 1$ for each $x,y \in S$. By the uniform Feller property $\rho(x,y)$ is continuous on $S \times S$ which is compact. Hence

$$\sup_{(x,y) \in S \times S} \sup_{A \subset S} |P(t_o,x,A) - P(t_o,y,A)| = \bar{\rho} < 1 \ ,$$

from which exponential ergodicity follows.

In the context of diffusions on S that are uniformly elliptic, the uniform Feller property follows from interior in t estimates of the y derivatives of $P(t,y,A)$ and the orthogonality property from the strong maximum principle.

Let us next consider the operator $\mathscr{L}_{BL}$ of (2.5.8) which we rewrite as follows.

(A.3) $\quad \mathscr{L}_{BL} = \mathscr{L}_1 + F(y)\, \dfrac{\partial}{\partial \eta} + \displaystyle\sum_{k=1}^{n} \bar{b}_k(y)\, \dfrac{\partial^2}{\partial \eta \partial y_k} + \dfrac{1}{2}\, \bar{a}(y)\, \dfrac{\partial^2}{\partial \eta^2} .$

Here $\mathscr{L}_1$ is defined by (2.1.3) (with x-dependence omitted) and acts on functions on S; the reflecting boundary conditions are again not written explicitly. The coefficients $F(y)$, $\bar{b}_k(y)$ and $\bar{a}(y)$ are identified from (2.5.8) and their form is not important here except that

(A.4) $\quad \displaystyle\int_S F(y)\bar{P}(dy) = 0 \qquad$ (inherited from (2.2.3))

and that they are smooth functions of y, independent of η, with $\bar{a}(y)$ strictly positive on S.

The generator $\mathscr{L}_{BL}$ above defines uniquely a process $(Y(t), H(t))$ on $S \times [-\infty, \infty]$, with $Y(t)$ in S as before. Suppose that $H(0) = \eta > 0$ and let τ be the first time $H(0) = 0$. We want to show that τ is a finite random variable with probability one so that for $A \subset S$

(A.5) $\quad Q(\eta, y, A) = P\{Y(\tau) \in A \,|\, Y(0) = y,\ H(0) = \eta\}$,

$$\eta \geq 0\ ,\qquad y \in S\ ,$$

is a well defined transition probability function. Moreover, Q satisfies[†]

[†]The reflecting boundary conditions on ∂S are again not shown.

(A.6) $\mathscr{L}_{BL}Q = 0$, $\eta > 0$, $y \in S$,

$$Q(0,y,A) = \chi_A(y) .$$

Let $N > 0$ be an integer and let τ_N be the first time H(t) equals N, starting from the left of N. The function[†]

(A.7) $Q_N(\eta,y,A) = P_{\eta,y}\{Y(\tau) \in A, \tau \prec \tau_N\}$

satisfies the boundary value problem

(A.8)
$$\mathscr{L}_{BL}Q_N = 0 , \qquad 0 < \eta < N , \qquad y \in S$$

$$Q_N(0,y,A) = \chi_A(y) , \qquad Q_N(N,y,A) = 0$$

as can be seen by Itô's formula applied to the smooth solution of (A.8) (the latter follows from results in differential equations). Now

$$Q(\eta,y,A) = Q_N(\eta,y,A)$$

$$+ E_{\eta,y}\{\chi_A(Y(\tau))|\tau \geq \tau_N\}P_{\eta,y}\{\tau \geq \tau_N\}$$

so that if we show that $P_{\hat{\eta},y}(\tau \geq \tau_N) \to 0$ as $N \to \infty$ then, we wil have (A.5) and (A.6) and $Q(\eta,y,A)$ will be, in fact a transition probability function on S for $\eta \geq 0$.

Since (A.4) holds (cf. (2.2.2))

(A.9) $\psi_F(y) = \int \psi(y,dz)F(z)$

[†]Subscripts denote the starting points as usual.

is well defined and

(A.10) $\mathcal{L}_1 \psi_F + F = 0$.

It follows that $\eta + \psi_F(y)$ is a harmonic function for $\mathcal{L}_{BL}$

(A.11) $\mathcal{L}_{BL}(\eta + \psi_F(y)) = 0$.

Thus,

$$E_{\eta,y}\{H(\tau \wedge \tau_N) + \psi_F(Y(\tau \wedge \tau_N))\} = \eta + \psi_F(y)$$

and also

$$NP_{\eta,y}\{\tau_N \geq \tau\} + E_{\eta,y}\{\psi_F(Y(\tau \wedge \tau_N))\} = \eta + \psi_F(y)$$

from which we conclude that $P_{\eta,y}\ \{\tau_N \geq \tau\} \to 0$ as $N \to \infty$.

Since the operator $\mathcal{L}_{BL}$ is uniformly elliptic, its co-efficients are smooth and S is compact the previous considerations imply that $Q(\eta,y,A)$ is an ergodic transition function with a unique invariant measure $\bar{Q}(A)$ approached exponentially fast as $\eta \to \infty$ uniformly in $y \in S$ and $A \in S$. $\bar{Q}(A)$ is the limiting distribution of $Y(\tau)$ as $\eta \to \infty$.

The exponential ergodicity implies the validity of the Fredholm alternative for the Dirichlet problem

(A.11) $$\begin{cases} \mathcal{L}_{BL}u(\eta,y) = 0 , & \eta > 0 , \quad y \in S , \\ u(0,y) = g(y) , & y \in S \\ \lim_{\eta \to \infty} u(\eta,y) = 0 . \end{cases}$$

Namely, (A.11) has a unique solution u provided

(A.12) $\int_S g(y)\bar{Q}(dy) = 0$

and u decays to zero exponentially fast as $\eta \to \infty$, uniformly in $y \in S$.

The Fredholm alternative for the Neumann problem

$$(A.13) \quad \begin{cases} \mathscr{L}_{BL}v(\eta,y) = 0 , \quad \eta > 0 , \quad y \in S \\[2mm] \frac{\partial v}{\partial \eta}(0,y) = g(y) , \\[2mm] \lim_{\eta\to\infty} v(\eta,y) \to 0 , \end{cases}$$

can be obtained simply[†] from (A.11). Indeed, if $u(\eta,y)$ is the solution of (A.11) then

$$(A.14) \quad v(\eta,y) = -\int_\eta^\infty u(\zeta,y)\,d\zeta = \int_0^\infty u(\eta+\zeta,y)\,d\zeta ,$$

satisfies (A.13) so that (A.12) is its solvability condition.

The reason why (A.12) serves as solvability condition for both (A.11) and (A.13) can be understood better perhaps by noting that if $T_Q(\eta)$ denotes the semigroup generated by $Q(\eta,y,A)$ on continuous functions on S then, the infinitesimal generator of $T_Q(\eta)$ is the generator of the boundary process associated with $(Y(t),H^R(t))$, the reflected process at $\eta = 0$

[†]In the present special case of nonoblique derivative condition at $\eta = 0$.

corresponding to $(Y(t),H(t))$, on $S \times [0,\infty)$ (cf. [5]). This observation and the remarks concerning (2.5.13) and (2.5.14) explain the relation between (A.11) and (A.13).

References

[1]   A. Khinchine, Asymptotische Gesetze der Wahrscheinlich-
       keitsrechnung, Springer, Berlin, 1933.

[2]   Yu. V. Prohorov, Convergence of random processes and
       limit theorems, Theor. Probability Appl., 9 (1956),
       pp. 157-214.

[3]   A. V. Skorohod, Studies in the theory of random processes,
       Addison Wesley, Reading, Mass., 1964.

[4]   D. W. Stroock and S. R. S. Varadhan, Diffusion processes
       with continuous coefficients I, II, Comm. Pure Appl.
       Math., 22 (1969), pp. 345-400 and 479-530.

[5]   D. W. Stroock and S. R. S. Varadhan, Diffusion processes
       with boundary conditions, Comm. Pure Appl. Math., 24
       (1971), pp. 147-225.

[6]   N. N. Bogoliubov and Ju. A. Mitropolski, Asymptotic
       methods in the theory of nonlinear oscillations, Gordon
       and Breach, New York, 1961.

[7]   T. G. Kurtz, Semigroups of conditioned shifts and approxi-
       mation of Markov processes, Annals of Probab., 3 (1975),
       pp. 618-642.

[8]   R. L. Stratonovich, Topics in the theory of random noise,
       Vols. I, II, Gordon and Breach, New York, 1963.

[9]   R. Z. Khasminskii, The averaging principle for parabolic
       and elliptic differential equations and Markov processes
       with small diffusion. Theor. Prob. Appl., 8 (1963),
       pp. 1-21.

[10] R. Z. Khasminskii, On stochastic processes defined by differential equations with a small parameter, Theor. Prob. Appl., 11 (1966), pp. 211-228.

[11] R. Z. Khasminskii, A limit theorem for solutions of differential equations with a random right hand side, Theor. Prob. Appl., 11 (1969), pp. 211-228.

[12] R. Z. Khasminskii, On the principle of averaging for Itô's stochastic differential equations, Kybernetica (Prague), 4 (1968), pp. 260-279.

[13] A. M. Ilin and R. Z. Khasminskii, On the equations of Brownian motion, Theor. Prob. Appl., 9 (1964), pp. 421-444.

[14] G. C. Papanicolaou and W. Kohler, Asymptotic theory of mixing stochastic ordinary differential equations, Comm. Pure and Appl. Math., 27 (1974), pp. 641-668.

[15] G. C. Papanicolaou, Asymptotic analysis of transport processes, Bull. A.M.S., 81 (1975), pp. 330-392.

[16] R. Hersh, Random evolutions: a survey of results and problems, Rocky Mountain Math. Journal, 14 (1974), pp. 443-496.

[17] M. Pinsky, Multiplicative operator functionals and their asymptotic properties, in Advances in Probability III, Marcel Dekker, New York, 1974.

[18] T. G. Kurtz, A limit theorem for perturbed operator semigroups with applications to random evolution, J. Funct. Analysis, 12 (1973), pp. 55-67.

[19] W. Kohler and G. C. Papanicolaou, Power statistics for
      waves in one dimension I, II, J. Math. Phys., 14
      (1973), pp. 1733-1745 and 15 (1974), pp. 2186-2197.

[20] D. A. Darling and M. Kac, On occupation times for Markov
      processes, Trans. Amer. Math. Soc., 84 (1957),
      pp. 444-458.

[21] C. Stone, Limit theorems for random walks, birth and
      death processes and diffusion processes, Ill. J. Math.,
      7 (1963), pp. 638-660.

[22] Y. Kasahara, A limit theorem for occupation times for
      Markov processes, to appear.

[23] H. P. McKean, Stochastic Integrals, Academic Press,
      New York, 1968.

[24] I. I. Gihman and A. V. Skorohod, Stochastic Differential
      Equations, Springer, Berlin-Heidelberg-New York, 1972.

[25] V. M. Volosov, Averaging in systems or ordinary dif-
      ferential equations, Uspehi Mat. Nauk, 17 (1962), #6
      (108), pp. 3-126 (Russian Math. Surveys, 17 (1962),
      pp. 1-126).

[26] A. N. Tichonov, Systems of differential equations con-
      taining a small parameter multiplying the highest
      derivatives, Mat. Sb., 73 (1952), pp. 575-585.

[27] I. Babuška, Several reports from the Department of Math.,
      University of Maryland, College Park, Maryland.

[28] H. Kramers, Brownian motion in a field of force and the diffusion model of chemical reactors, Physica, 7 (1940), pp. 284-304.

[29] N. Bromberg, N.Y.U. dissertation, 1974.

[30] A. Bensoussan, J. L. Lions and G. C. Papanicolaou, Boundary layers and homogenization of transport processes, to appear.

[31] D. W. Stroock and S. R. S. Varadhan, On degenerate elliptic-parabolic operators of second order and their associated diffusions, Comm. Pure Appl. Math., 25 (1972), pp. 651-713.

[32] G. Blankenship and G. C. Papanicolaou, Stability and control of stochastic systems with wide-band noise disturbances I, II, SIAM J. Appl. Math., to appear.

[33] R. F. Anderson, Diffusions with second order boundary I, II, Indiana Univ. Math. J., 25, pp. 367-395 and 403-441.

[34] E. De Giorgi and S. Spagnolo, Sulla convergenza degli integrali dell' energia per operatsii ellittici del secondo ordine, Boll. U.M.I., 8 (1973), pp. 391-411.

[35] S. Spagnolo, Convergence in energy for elliptic operators, in Numerical solution of partial differential equations III, Academic Press, New York, 1976.

[36] A. Bensoussan, J. L. Lions, G. C. Papanicolaou, C. R. Acad. Sc. Paris, 281 (A) (1975), pp. 89-94 and pp. 317-322; also, 282 (A) (1976), pp. 143-147 and pp. 1277-1282.

[37] A. Bensoussan, J. L. Lions, G. C. Papanicolaou, Homogenization and ergodic theory, to appear.

[38] A. Bensoussan, J. L. Lions, G. C. Papanicolaou, book in preparation.

[39] D. Isaacson, N.Y.U. dissertation, 1976.

Commun. Math. Phys. 104, 1–19 (1986)

Communications in
**Mathematical
Physics**
© Springer-Verlag 1986

Central Limit Theorem for Additive Functionals of Reversible Markov Processes and Applications to Simple Exclusions

C. Kipnis[1] and S. R. S. Varadhan[2]*

1 Centre de Mathematiques Appliqueés, Ecole Polytechnique, F-91128 Palaiseau, France
2 Courant Institute, New York University, New York 10012, USA

Abstract. We prove a functional central limit theorem for additive functionals of stationary reversible ergodic Markov chains under virtually no assumptions other than the necessary ones. We use these results to study the asymptotic behavior of a tagged particle in an infinite particle system performing simple excluded random walk.

Introduction

In a recent work Lebowitz and Spohn [4] proved that diffusion of color for "mechanically" identical particles with two colors as well as convergence to local equilibrium in the hydrodynamical limit were related to the asymptotics, after proper rescaling, of the movements of one or more tagged particles of the system.

Since for the moment purely mechanical systems seem to be out of reach, several models which are to lesser or greater extent stochastic have been proposed. In [3] Kipnis et al. considered the case of a one dimensional system of hard rods with stochastic collisions. In this paper we consider the case of the so-called simple exclusion process.

The intuitive description of the process is the following: Infinitely many particles move on $\mathbb{Z}^d$ according to a simple random walk with exponential (mean one) holding time at each site and jump law $p(x)$ which is symmetric, i.e. satisfies $p(x) = p(-x)$ for all x's. However if a particle attempts a transition to a site already occupied, the jump is suppressed.

The key remark is that, due to invariance of the mechanics under translations, the evolution of the rest of the medium *seen from an observer sitting on a tagged particle* follows a Markovian evolution (which possesses many reversible probabilities) and that the movement of the tagged particle (the observer) in the absolute frame is in a certain sense "driven" by this Markov process.

* Supported by NSF Grant MCS-8301364, ONR Contract N00014-81-K-0012 and a Fellowship from John S. Guggenheim Memorial Foundation

In Sect. 1 we prove an abstract theorem on weak convergence to Brownian motion of a process driven by a reversible Markov process.

In Sect. 2 we prove that our tagged particle satisfies the conditions described in Sect. 1 to conclude that if x_t denotes the position of the tagged particle then $\varepsilon x_{t\varepsilon^{-2}}$ converges to a Brownian motion with finite diffusion constant D. (Notice that before taking the limit the random process x_t is not Markovian due to the interactions!)

However it has been shown by Arratia [1] that when $d = 1$ and $p(1) = p(-1) = 1/2$, the diffusion constant D is equal to zero and the correct scaling needed to obtain a nontrivial limit is $\varepsilon x_{t\varepsilon^{-4}}$. We therefore prove that if $d \geq 2$ or $d = 1$ and $p(x)$ is concentrated on more than two points then D is strictly positive.

In Sect. 3. we prove that two different tagged particles become independent in the limit contrary to what happens in [3] where they remain correlated.

In Sect. 4. we apply our theorem to a jump process in random environment obtained by first choosing (randomly) rates for each bond of nearest-neighbor points of the lattice $\mathbb{Z}^d$ and letting the particle jump across a bond with the aforementioned rate.

1. Convergence to Brownian Motion of an Arbitrary Reversible Velocity Process

The question of proving a central limit theorem for the partial sums

$$X_n = \sum_{j=1}^{n} Z_j$$

of a sequence $Z_1, Z_2, \ldots, Z_n, \ldots$ of random variables has been studied exhaustively in probability theory for at least fifty years. We will limit our discussion here to the situation where $\{Z_j\}$ forms a stationary sequence in the strict sense. If the Z_j's are independent with mean zero and variance σ^2 then the classical central limit theorem asserts that the distribution of $X_n/\sqrt{n}$ converges to a normal distribution with mean 0 and variance σ^2. The invariance principle asserts that the distribution of the stochastic process $X_n(t) = 1/\sqrt{n} \, X_{[nt]}$ converges weakly in the Skorohod space to the distribution of a Brownian motion with variance σ^2.

These results have been generalized to the situation where the independence of the random variables $\{Z_j\}$ is replaced by asymptotic independence of one type or other. The assumptions are often difficult to check in specific circumstances.

One special situation where analogs of the classical results are valid without additional assumptions is the following:

Theorem 1.1. *Let* $\{Z_j: -\infty < j < \infty\}$ *be a stationary ergodic process such that* $E[Z_{n+1} | F_n] = 0$ *a.e., where* F_n *is the σ-field generated by* Z_j *for* $j \leq n$. *For such a martingale difference sequence the distribution of* $X_n(t) = 1/\sqrt{n} \, [Z_1 + \cdots + Z_{[nt]}]$ *converges weakly to the distribution of Brownian motion with variance* σ^2 *provided* $EZ_n^2 = \sigma^2 < \infty$.

Proof. The above theorem is deduced easily from Theorem 3.2 of [2] where condition (b) can be checked easily in our situation.

Remark 1.2. If $\{Z_j\}$ is a vector valued sequence one has an identical theorem where one obtains in the limit a multidimensional Brownian motion with a covariance matrix identical to that of any Z_j.

Suppose $\{y_j\}$, $-\infty < j < \infty$ is a stationary Markov chain on a state space X and $Z_j = V(y_j)$. Let us suppose that $\pi(dx)$ is the common invariant distribution and $q(x, dy)$ is the transition probability of the chain. Let us assume that the chain is ergodic. We denote by the symbol q the operator $(qf)(x) = \int f(y)\, q(x, dy)$ as well. Suppose $V(x) = f(x) - (qf)(x)$ for some bounded function $f(x)$. Then $V(x)$ has necessarily mean 0 with respect to π and it is natural to ask if the central limit theorem is valid for

$$X_n = \sum_{j=1}^{n} V(y_j).$$

If we define

$$Z'_j = f(y_j) - (qf)(y_{j-1}),$$

then an elementary computation shows that

$$E[Z'_{n+1}|F_n] = E[f(y_{n+1})|F_n] - (qf)(y_n) = (qf)(y_n) - (qf)(y_n) = 0.$$

Moreover if

$$X'_n = \sum_{j=1}^{n} Z'_j,$$

then

$$X_n = \sum_{j=1}^{n} f(y_j) - \sum_{j=1}^{n} (qf)(y_j) = \sum_{j=1}^{n} Z'_j + (qf)(y_0) - (qf)(y_n).$$

We write $X_n = X'_n + X''_n$, where $X''_n = (qf)(y_0) - (qf)(y_n)$. Since $X''_n/\sqrt{n}$ is negligible as $n \to \infty$, the central limit theorem for X_n follows from the central limit theorem for X'_n which is an application of Theorem 1.1.

Although the approach is quite elementary, when the underlying Markov chain $\{y_j\}$ is reversible the method is powerful enough to establish the central limit theorem and the invariance principle for the most general V that one can expect the theory to hold.

Let $q(x, dy)$ be a translation probability on a state space X with $\pi(dx)$ as a reversible stationary probability. We will assume that the stationary Markov process P with π as marginal distribution and $q(x, dy)$ with transition probability is ergodic. Let $V(y)$ be a function on X with $\int V^2(x)\, \pi(dx) < \infty$, and we will assume that $\int V(x)\, \pi(dx) = 0$ and

$$\lim_{n \to \infty} \frac{1}{n} E^P[(V(y_1) + \cdots + V(y_n))^2] = \sigma^2 < \infty.$$

If we denote by $e_V(d\lambda)$, the spectral measure of V corresponding to the selfadjoint operator q on $L_2(X, \pi)$, we obtain

$$\sigma^2 = \int \frac{1 + \lambda}{1 - \lambda} e_V(d\lambda) < \infty. \tag{1.1}$$

The condition that $\sigma^2 < \infty$ is therefore equivalent to the condition $\int 1/1 - \lambda \, e_V(d\lambda) < \infty$ or $V \in \text{Range}(I-q)^{1/2}$. This can be verified by checking the inequality

$$\left| \int V(y)\phi(y)\pi(dy) \right| \leqq C(\int ((I-q)\phi)(y)\phi(y)\pi(dy))^{1/2} \tag{1.2}$$

for all test functions ϕ in $L_2(X, \pi)$. If we introduce the inner products

$$\langle f_1, f_2 \rangle = \int f_1(y) f_2(y)\pi(dy),$$
$$\langle f_1, f_2 \rangle_1 = \int ((I-q)f_1)(y) f_2(y)\pi(dy),$$

then our assumption can be restated as $\langle V, V \rangle_{-1} < \infty$, where $\langle \quad \rangle_{-1}$ is dual to $\langle \quad \rangle_1$ under the pairing $\langle \quad \rangle$.

We denote by Ω the space of all X-valued sequences $\{y_j\}$, $-\infty < j < \infty$ and by F_n the σ-field generated by y_j for $j \leqq n$. We have the measure P on (Ω, F), where

$$F = \bigvee_n F_n.$$

Theorem 1.3. *Let V in $L_2(x, \pi)$ satisfy condition (1.1). Then the sequence*

$$X_n = \sum_{j=1}^{n} V(y_j)$$

can be written as $X_n = M_n + \xi_n$, where M_n is a martingale relative to (Ω, F_n, P) and

$$\lim_{n \to \infty} \frac{1}{\sqrt{n}} \sup_{1 \leqq j \leqq n} |\xi_j| = 0 \text{ in probability } (P). \tag{1.3}$$

Moreover

$$\lim_{n \to \infty} \frac{1}{n} E^P |\xi_n|^2 = 0. \tag{1.4}$$

Proof. We denote by u_ε the solution of the equation

$$(1 + \varepsilon)u_\varepsilon - qu_\varepsilon = V \tag{1.5}$$

for $\varepsilon > 0$. We will investigate the behavior of u_ε as $\varepsilon \to 0$. We can rewrite (1.5) as

$$u_\varepsilon - qu_\varepsilon = V - \varepsilon u_\varepsilon. \tag{1.6}$$

Therefore for every $n \geqq 0$,

$$E^P\{u_\varepsilon(y_{n+1}) - u_\varepsilon(y_n) - \varepsilon u_\varepsilon(y_n) + V(y_n) | F_n\} = 0.$$

If we define

$$M_n^\varepsilon = \sum_{j=0}^{n-1} [u_\varepsilon(y_{j+1}) - u_\varepsilon(y_j) + V_\varepsilon(y_j)],$$

where $V_\varepsilon = V - \varepsilon u_\varepsilon$, then for each $\varepsilon > 0$, M_n^ε if a martingale relative to (Ω, F_n, P). We can now express for each $\varepsilon > 0$, $X_n = M_n^\varepsilon + \xi_n^\varepsilon + \eta_n^\varepsilon$, where

$$\xi_n^\varepsilon = -\sum_{j=0}^{n-1} [u_\varepsilon(y_{j+1}) - u_\varepsilon(y_j)],$$

329

and

$$\eta_n^\varepsilon = \sum_{j=0}^{n-1} (\varepsilon u_\varepsilon)(y_j).$$

We shall now establish that for each $n \geq 1$,

$$\lim_{\varepsilon \to 0} M_n^\varepsilon = M_n \text{ exists in } L_2(\Omega, P), \tag{1.7}$$

$$\lim_{\varepsilon \to 0} \zeta_n^\varepsilon = \zeta_n \text{ exists in } L_2(\Omega, P), \tag{1.8}$$

$$\lim_{\varepsilon \to 0} \eta_n^\varepsilon = 0 \text{ in } L_2(\Omega, P). \tag{1.9}$$

It follows then that $X_n = M_n + \zeta_n$. We will then establish (1.3) and (1.4) for the sequence $\{\zeta_n\}$. Since M_n^ε is a martingale with stationary increments in order to show that M_n^ε has a limit in $L_2(\Omega, P)$, it is sufficient to check that

$$\lim_{\varepsilon \to 0} \{u_\varepsilon(y_2) - u_\varepsilon(y_1) + V_\varepsilon(y_1)\} \text{ exists in } L_2(\Omega, P). \tag{1.10}$$

From relation (1.6) the above limit is of the form

$$\lim_{\varepsilon \to 0} \{u_\varepsilon(y_2) - (qu_\varepsilon)(y_1)\}.$$

An easy computation yields

$$E^P[f(y_2) - (qf)(y_1)]^2 = \langle f, (I - q^2)f \rangle.$$

Therefore we need to check only that

$$\lim_{\substack{\varepsilon_1 \to 0 \\ \varepsilon_2 \to 0}} \langle u_{\varepsilon_1} - u_{\varepsilon_2}, (I - q^2)(u_{\varepsilon_1} - u_{\varepsilon_2}) \rangle = 0. \tag{1.11}$$

Equation (1.11) can be calculated in terms of spectral measure $e_V(d\lambda)$ and the formula reads

$$\langle u_{\varepsilon_1} - u_{\varepsilon_2}, (I - q^2)(u_{\varepsilon_1} - u_{\varepsilon_2}) \rangle = \int_{-1}^{1} (1 - \lambda^2) \left(\frac{1}{1 + \varepsilon_1 - \lambda} - \frac{1}{1 + \varepsilon_2 - \lambda} \right)^2 e_V(d\lambda)$$

$$= \int_{-1}^{1} \frac{(\varepsilon_2 - \varepsilon_1)^2 (1 - \lambda^2)}{(1 + \varepsilon_1 - \lambda)^2 (1 + \varepsilon_2 - \lambda)^2} e_V(d\lambda).$$

If we assume that $\varepsilon_2 \geq \varepsilon_1 > 0$ the integrand is dominated by $(\varepsilon_2^2(1 - \lambda^2)/(1 - \lambda)^2 \varepsilon_2^2) = (1 + \lambda)/(1 - \lambda)$. Since our basic hypothesis is that $\int (1 + \lambda)/(1 - \lambda) \, e_V(d\lambda) < \infty$, we have (1.7) by the dominated convergence theorem. By a similar argument we also obtain

$$\lim_{\substack{\varepsilon_1 \to 0 \\ \varepsilon_2 \to 0}} \langle u_{\varepsilon_1} - u_{\varepsilon_2}, (I - q)(u_{\varepsilon_1} - u_{\varepsilon_2}) \rangle = 0. \tag{1.11a}$$

A similar calculation also yields

$$\lim_{\varepsilon \to 0} \langle u_\varepsilon, (I - q^2) u_\varepsilon \rangle = \int \frac{1+\lambda}{1-\lambda} e_V(d\lambda) = \sigma^2.$$

Equation (1.9) involves proving

$$\lim_{\varepsilon \to 0} \varepsilon \| u_\varepsilon \| = 0.$$

We will actually prove the stronger version,

$$\lim_{\varepsilon \to 0} \varepsilon \langle u_\varepsilon, u_\varepsilon \rangle = 0. \tag{1.12}$$

Again by spectral calculation

$$\varepsilon \langle u_\varepsilon, u_\varepsilon \rangle = \int_{-1}^{1} \frac{\varepsilon}{(1+\varepsilon-\lambda)^2} e_V(d\lambda).$$

The integrand goes to zero and is dominated by $1/(1-\lambda)$, which is integrable with respect to $e_V(d\lambda)$. Since $X_n = M_n^\varepsilon + \xi_n^\varepsilon + \eta_n^\varepsilon$ and is independent of $\varepsilon > 0$, (1.7) and (1.9) clearly imply (1.8).

We now concentrate on proving (1.3) and (1.4) for the sequence $\{\xi_n\}$. We first prove (1.4). Clearly for every $\varepsilon > 0$,

$$\xi_n = M_n^\varepsilon - M_n + \xi_n^\varepsilon + \eta_n^\varepsilon.$$

Since $M_n^\varepsilon - M_n$ is a martingale with stationary increments,

$$\frac{1}{n} E^P |\xi_n|^2 \leq \frac{3}{n} E^P |M_n^\varepsilon - M_n|^2 + \frac{3}{n} E^P |\xi_n^\varepsilon|^2 + \frac{3}{n} E^P |\eta_n^\varepsilon|^2$$

$$= 3 E^P |M_1^\varepsilon - M_1|^2 + \frac{3}{n} E^P |\xi_n^\varepsilon|^2 + \frac{3}{n} E^P |\eta_n^\varepsilon|^2.$$

We have

$$\lim_{\varepsilon \to 0} E^P |M_1^\varepsilon - M_1|^2 = 0.$$

Therefore it is enough to choose $\varepsilon = 1/n$ and then show that

$$\lim_{n \to \infty} \frac{1}{n} E^P |\xi_n^{1/n}|^2 = 0 \quad \text{and} \quad \lim_{n \to \infty} \frac{1}{n} E^P |\eta_n^{1/n}|^2 = 0.$$

Clearly

$$E^P |\xi_n^{1/n}|^2 = E^P |u_{1/n}(y_n) - u_{1/n}(y_0)|^2 \leq 4 E^P |u_{1/n}(y_0)|^2$$

$$= 4 \langle u_{1/n}, u_{1/n} \rangle = o(n) \quad \text{by (1.12)}.$$

Moreover

$$E^P |\eta_n^{1/n}|^2 = E^P \left[\frac{1}{n} \sum_{j=0}^{n-1} u_{1/n}(y_j) \right]^2 \leq E^P |u_{1/n}(y_0)|^2$$

$$= \langle u_{1/n}, u_{1/n} \rangle = o(n) \quad \text{again by (1.12)}.$$

Now we turn to (1.3). We need Lemma 1.4 which we will use but prove after completing the argument;

$$\sup_{1 \leq j \leq n} |\xi_j| \leq \sup_{1 \leq j \leq n} |M_j^{1/n} - M_j| + \sup_{1 \leq j \leq n} |\xi_j^{1/n}| + \sup_{1 \leq j \leq n} |\eta_j^{1/n}|.$$

Now by Doob's inequality

$$P[\sup_{1 \leq j \leq n} |M_j^{1/n} - M_j| \geq \delta \sqrt{n}] \leq \frac{1}{n\delta^2} E^P[M_n^{1/n} - M_n]^2$$

$$= \frac{1}{\delta^2} E^P[M_1^{1/n} - M_1]^2 = o(1) \quad \text{as} \quad n \to \infty.$$

$$P[\sup_{1 \leq j \leq n} |\eta_j^{1/n}| \geq \delta \sqrt{n}] \leq P\left[\frac{1}{n}\sum_0^{n-1} |u_{1/n}(y_j)| \geq \delta \sqrt{n}\right]$$

$$\leq \frac{1}{n\delta^2} E^P\left[\frac{1}{n}\sum_0^{n-1} |u_{1/n}(y_j)|\right]^2 \leq \frac{1}{n\delta^2} \langle u_{1/n}, u_{1/n} \rangle$$

$$= o(1) \quad \text{as} \quad n \to \infty.$$

Finally

$$\sup_{1 \leq j \leq n} |\xi_j^{1/n}| = \sup_{1 \leq j \leq n} |u_{1/n}(y_j) - u_{1/n}(y_0)| \leq \sup_{1 \leq j \leq n} |u_{1/n}(y_j)| + |u_{1/n}(y_0)|,$$

$$P[|u_{1/n}(y_0)| \geq \delta \sqrt{n}] \leq \frac{1}{n\delta^2} E^P[u_{1/n}(y_0)]^2 = \frac{1}{n\delta^2} \langle u_{1/n}, u_{1/n} \rangle$$

$$= o(1) \quad \text{as} \quad n \to \infty.$$

As for $\sup_{1 \leq j \leq n} |u_{1/n}(y_j)|$, we write

$$\sup_{1 \leq j \leq n} |u_{1/n}(y_j)| \leq \sup_{1 \leq j \leq n} |u_\varepsilon(y_j) - u_{1/n}(y_j)| + \sup_{1 \leq j \leq n} |u_\varepsilon(y_j)|$$

for some $\varepsilon > 0$.

$$P[\sup_{1 \leq j \leq n} |u_\varepsilon(y_j)| \geq \delta \sqrt{n}] \leq nP[|u_\varepsilon(y_0)| \geq \delta \sqrt{n}] = o(1),$$

because $E^P|u_\varepsilon(y_0)|^2 < \infty$.

We use Lemma 1.4 to estimate

$$P[\sup|u_\varepsilon(y_j) - u_{1/n}(y_j)| \geq \delta \sqrt{n}]$$

$$\leq \frac{3}{\delta \sqrt{n}}(n \langle u_\varepsilon - u_{1/n}, (I - q)(u_\varepsilon - u_{1/n}) \rangle + \langle u_\varepsilon - u_{1/n}, u_\varepsilon - u_{1/n} \rangle)^{1/2},$$

letting $n \to \infty$,

$$\limsup_{n \to \infty} P[\sup_{1 \leq j \leq n} |u_\varepsilon(y_j) - u_{1/n}(y_j)| \geq \delta \sqrt{n}]$$

$$\leq \frac{3}{\delta} \lim_{n \to \infty} \langle u_\varepsilon - u_{1/n}, (I - q)(u_\varepsilon - u_{1/n}) \rangle^{1/2}.$$

If we let $\varepsilon \to 0$, we obtain

$$\lim_{\varepsilon \to 0} \limsup_{n \to \infty} P[\sup_{1 \leq j \leq n} |u_\varepsilon(y_j) - u_{1/n}(y_j)| \geq \delta \sqrt{n}]$$

$$\leq \frac{3}{\delta} \lim_{\varepsilon \to 0} \lim_{n \to \infty} \langle u_\varepsilon - u_{1/n}, (I - q)(u_\varepsilon - u_{1/n}) \rangle$$

$$= 0 \text{ by (1.1a)},$$

and we are done.

Lemma 1.4. *Let $f(x)$ be an arbitrary function in $L_2(X, \pi)$. Let*

$$\langle f, f \rangle = a, \quad \langle f, (I - q)f \rangle = b.$$

Then

$$P\{\sup_{0 \leq j \leq n} |f(y_j)| \geq l\} \leq \frac{3}{l} \sqrt{a + nb}. \tag{1.13}$$

Proof. Let $G = \{x : |f(x)| \geq l\}$ and let τ be the hitting time of G, i.e. $\tau = \{\inf j : y_j \in G\}$. Then

$$P\{\sup_{0 \leq j \leq n} |f(y_j)| \geq l\} = P\{\tau \leq n\}.$$

Let us consider the function $\phi_\rho(x) = E^{P_x}\{\rho^\tau\}$, where P_x is the Markov measure on the space of sequences $\{y_j\}$ for $j \geq 0$ initialized to start from x at time 0. Then

$$P\{\tau \leq y\} \leq \rho^{-n} \int \phi_\rho(x)\pi(dx) \leq \rho^{-n}[\int \phi_\rho^2(x)\pi(dx)]^{1/2}.$$

To estimate the last term we note that

$$\phi_\rho = \rho q \phi_\rho \text{ on } G^c, \quad \phi_\rho = 1 \text{ on } G.$$

By the Dirichlet principle

$$\int \phi_\rho^2(x)d\pi \leq \langle \phi_\rho, \phi_\rho \rangle + \frac{\rho}{1 - \rho} \langle \phi_\rho, (I - q)\phi_\phi \rangle \leq \langle \omega, \omega \rangle + \frac{\rho}{1 - \rho} \langle \omega, (I - q)\omega \rangle$$

for any ω which is 1 on G. The function $(|f(x)| \wedge l/l) = \bar\omega$ is an admissible choice;

$$\langle \bar\omega, \bar\omega \rangle \leq \frac{1}{l^2} \langle f, f \rangle = \frac{a}{l^2}, \quad \langle \bar\omega, (I - q)\bar\omega \rangle \leq \frac{1}{l^2} \langle f, (I - q)f \rangle = \frac{b}{l^2}.$$

Therefore

$$P[\tau \leq n] \leq \rho^{-n}\left(\frac{a}{l^2} + \frac{\rho}{1 - \rho}\frac{b}{l^2}\right)^{1/2}.$$

Taking $\rho = e^{-1/n}$ we obtain (1.13).

Corollary 1.5. *If V satisfies (1.1), then the distribution of $X_n(t) = 1/\sqrt{n} \, X_{[nt]} = (1/\sqrt{n}) \sum_{j=1}^{[nt]} V(y_j)$ under P converges to Brownian motion with variance σ^2 weakly in the Skorohod space on any finite interval. Moreover σ^2 is given by (1.1).*

Remark 1.6. If V is vector valued then essentially the same result is valid. The vector valued process $X_n(t)$ converges weakly to a multidimensional Brownian motion with the corresponding covariance matrix.

Remark 1.7. If we replace P by P_x the Markov measure starting from x at time 0, then the distribution of $X_n(t)$ under P_x will again converge to the correct Brownian motion but the convergence is in measure as functions of x relative to the measure π. The question of almost sure convergence is open.

In the case of continuous time Markov processes we have the transition probabilities $p(t, x, dy)$ and an invariant probability measure $\pi(dx)$ with respect to which the process is assumed to be reversible and ergodic. We also have the infinitesimal generator L of the process. L is self adjoint on $L_2(\pi)$ with a nonpositive spectrum and a simple eigenvalue of 0, constants being the eigenspace corresponding to that eigenvalue.

If one is interested in proving a central limit theorem for an additive functional of the form

$$X(t) = \int_0^t V(y(s)) ds,$$

where $\int V \, d\pi = 0$, then

$$\lim_{t \to \infty} \frac{1}{t} E^P[X^2(t)] = 2 \int_0^\infty E^P[V(y(s))V(y(0))] ds = 2 \int_0^\infty \langle T_s V, V \rangle ds$$

$$= 2 \langle -L^{-1}V, V \rangle,$$

where T_s is the semigroup induced by the transition probabilities.

The natural assumption on V then is that $V \in D((-L)^{-1/2})$ or equivalently an estimate of the form

$$|\langle V, \phi \rangle| \leq c \langle -L\phi, \phi \rangle^{1/2} \tag{1.14}$$

for all ϕ in the domain of L. We will assume that V in $L_2(\pi)$ satisfies (1.14). When we say domain of L we mean the domain as the generator of a semigroup in $L_2(\pi)$.

We now state the analog of Theorem 1.3.

Theorem 1.8. *Let $y(t)$ be a Markov process, reversible with respect to a probability measure π, and let us suppose that the reversible stationary process P with π as invariant measure is ergodic. Let V be a function on the state space in $L_2(\pi)$ satisfying $\int V \, d\pi = 0$ and condition (1.14). Let F_t be the σ-field generated by the process up to time t. There exists a square integrable Martingale M_t relative to (Ω, F_t, P) for $f \geq 0$ such that M_t has stationary increments and*

$$\lim_{t \to \infty} \frac{1}{\sqrt{t}} \sup_{0 \leq s \leq t} |x(s) - M_s| = 0$$

in probability with respect to P, where $X(0) = M_0 = 0$ and

$$X(t) = \int_0^t V(y(s)) ds.$$

Moreover

$$\lim_{t \to \infty} \frac{1}{t} E^P |X(t) - M_t|^2 = 0.$$

Corollary 1.9. *It follows now from results is Helland* [2] *that* $1/\sqrt{\lambda} \, X(\lambda t)$ *satisfies a functional central limit theorem relative to P and the limiting variance* σ^2 *is given by*

$$\sigma^2 = \lim_{t \to \infty} \frac{1}{t} EX^2(t) = 2 \int_0^\infty \langle T_t V, V \rangle \, dt = 2 \langle -L^{-1} V, V \rangle.$$

Remark 1.10. If we denote by P_x the Markov process starting from a point x in the state space and by $Q_{\lambda,x}$ the measure on the space of continuous functions induced by $1/\sqrt{\lambda} \, X(\lambda t)$ from P_x and by Q the Brownian motion measure with variance σ^2 on the same function space, then we have besides $\int Q_{\lambda,x} \pi(dx) \Rightarrow Q$ actually $Q_{\lambda,x} \Rightarrow Q$ in measure with respect to π. The question of almost sure convergence is open.

Remark 1.11. If V is a vector of functions one has analogous results regarding convergence to the corresponding vector Brownian motion with appropriate covariance matrix. The basic approximation in Theorem 1.8 can be done for each component and in [2] the multidimensional version of the central limit theorem for vector martingales can be found.

Comments on Proofs. The proof follows the discrete time situation very closely. The only tricky point is the analog of Lemma 1.4. Since the $X(t)$ process is almost surely continuous and M_t can be taken to be almost surely right continuous, one replaces the supremum in

$$\sup_{0 \le s \le t} |X(s) - M_s|$$

by a supremum over diadic points in $[0, t]$. Denoting by D_N the set of diadic points of the form $j/2^N$, it is clearly sufficient to prove

Lemma 1.12. *Let* f *satisfy* $\langle f, f \rangle = a$, $\langle f, -Lf \rangle = b$. *Then for every* t, N *and* l

$$P[\sup_{0 \le s \le t, s \in D_N} |f(y(s))| \ge l] \le \frac{3}{l} \sqrt{a + tb}.$$

Proof. If we denote by $q_N(x, dy)$ the transition probability $p(1/2^N, x, dy)$ and by q_N the operator $T_{1/2^N}$, then by the spectral theorem,

$$\langle f, (I - q_N)f \rangle = \int (1 - e^{-\lambda/2^N}) E_f(dy) \le \frac{1}{2^N} \langle f, -Lf \rangle = \frac{b}{2^N}.$$

We now apply Lemma 1.4 for the discretely sampled process with $n = [t2^N]$, and we obtain the validity of the lemma.

2. Asymptotics for a Single Test Particle in Simple Exclusion

In this section we study the behavior of one test particle in the symmetric simple exclusion process. We will denote by $\xi \in \{0, 1\}^{\mathbb{Z}^d}$ the state of such a process

$$\xi(u) = \begin{cases} 1 & \text{if the site } u \text{ has a particle} \\ 0 & \text{if the site } u \text{ is free.} \end{cases}$$

For any state ξ and any two sites u, v we denote by $\xi^{u,v}$ the new state obtained by

$$\xi^{u,v}(z) = \begin{cases} \xi(z) & \text{if } z \neq u \text{ or } v \\ \xi(u) & \text{if } z = v \\ \xi(v) & \text{if } z = u. \end{cases}$$

The evolution of the process $\xi(t, \cdot)$ is governed by its infinitesimal generator L

$$(LF) = \sum_{u \neq v} [F(\xi^{u,v}) - F(\xi)]p(u - v). \tag{2.1}$$

The summation here is over unordered pairs (u, v) with $u \neq v$. For the process $\xi(t, \cdot)$ the Bernoulli measures P_θ, where θ is the probability that any site is occupied (independently of all the other sites) are reversible ergodic measures provided the random walk $p(x - y)$ is irreducible.

If we have a single test particle in the system, it is more convenient to describe the system in terms of the location x of the test particle and the configuration η of the entire system seen from the test particle:

$$\eta(u) = \xi(x + u) \quad \text{for} \quad u \neq 0, \tag{2.2}$$

and the state space for the system is $x \in \mathbb{Z}^d$ and $\eta \in \{0, 1\}^{\mathbb{Z}^d - \{0\}}$. The infinitesimal generator for the $(x, \eta)_t$ process is given by

$$(\tilde{L}F)(x, \eta) = \sum_{z \neq 0} p(z)(1 - \eta(z))[F(x + z, \tau_{-z}\eta) - F(x, \eta)]$$

$$+ \sum_{u \neq v \neq 0} [F(x, \eta^{u,v}) - F(x, \eta)]p(u - v), \tag{2.3}$$

where

$$(\tau_{-z}\eta)(u) = \eta(u + z) \quad \text{for} \quad u \neq -z$$
$$= 0 \quad \text{for} \quad u = -z. \tag{2.4}$$

From the form of (2.3) we see that η by itself is a Markov process with generator

$$(L_0 F)(\eta) = \sum_z p(z)(1 - \eta(z))[F(\tau_{-z}\eta) - F(\eta)] + \sum_{u \neq v \neq 0} [F(\eta^{u,v}) - F(\eta)]p(u - v)$$

$$= (L_1 F)(\eta) + (L_2 F)(\eta). \tag{2.5}$$

The invariant measures for the η process will turn out to be Bernoulli product measures P_θ with common probability θ, and these will be ergodic and reversible for L_0. [Although for the η process there is no site at 0 we will continue to denote by P_θ the Bernoulli measure over all the other sites.]

Lemma 2.1. The η_t process is reversible and ergodic with respect to P_θ.

Proof. We need only check that

$$\int f L_0 g \, dP_\theta = \int g L_0 f \, dP_\theta,$$

where f and g are of the form

$$f(\eta) = \prod_{u \in A} \eta(u), \quad g(\eta) = \prod_{u \in B} \eta(u),$$

where $A, B \subset \mathbb{Z}^d - \{0\}$.

From the form of the generator (2.5) we need only check that

$$\int f(\eta^{u,v}) g(\eta) dP_\theta = \int g(\eta^{u,v}) f(\eta) dP_\theta, \tag{2.6}$$

$$\int f(\eta) g(\tau_{-z}\eta)(1 - \eta(z)) dP_\theta = \int g(\eta) f(\tau_z\eta)(1 - \eta(-z)) dP_\theta. \tag{2.7}$$

Equation (2.6) follows from the invariance of P_θ under the map $\eta \to \eta^{u,v}$ and (2.7) follows from the fact that $\eta \to \tau_z\eta$ maps $(1 - \eta(z))dP_\theta$ to $(1 - \eta(-z))dP_\theta$. It is easy to see that P_θ is reversible and ergodic for L_2. Since any harmonic function ϕ such that $(L_1 + L_2)\phi = 0$ has the property $\int \phi L_1 \phi dP_\theta = -\int \phi L_2 \phi \, dP_\theta$ and $\int \phi L_i \phi \, dP_\theta \leq 0$ for $i = 1, 2$ we have $\int \phi L_1 \phi dP_\theta = \int \phi L_2 \phi dP_\theta = 0$. But L_2 is reversible and ergodic. Therefore ϕ is a constant almost surely with respect to P_θ. This completes the proof of the lemma.

The x_t process is driven by the η_t process. If we pick a vector l and consider $(x \cdot l)$, then

$$L_0(x \cdot l) = \sum (z \cdot l) p(z)(1 - \eta(z)) = \phi_l(\eta),$$

so that

$$(x_t \cdot l) - \int_0^t \phi_l(\eta_s) ds = N_t^l \tag{2.7}$$

is a Martingale relative to the natural σ-fields. We now apply Theorem 1.3 to $\phi_l(\eta)$ and write

$$\int_0^t \phi_l(\eta_s) ds = M_t^l + E_t^l,$$

where M_t^l is another martingale and E_t^l is negligible. Taking l to be the various basis vectors, we can approximate each component of x_t by a sum of two martingales. The central limit theorem now follows from what we said earlier. We need to show that the conditions of Theorem 1.3 apply and that the limiting covariance of the sum $M_t + N_t$ is nondegenerate. We assume that

$$\sum |z|^2 p(z) < \infty. \tag{2.8}$$

Lemma 2.2. *For any l the function*

$$\phi_l(\eta) = \sum (z \cdot l)(1 - \eta(z)) p(z)$$

satisfies the condition

$$\left| \int \phi_l(\eta) F(\eta) dP_\theta \right| \leq C(-\int (L_0 F)(\eta) F(\eta) dP_\theta)^{1/2}.$$

Proof. We can take F to be a function depending on a finite number of coordinates and then

$$\int \phi_l(\eta) F(\eta) dP_\theta = \int (\sum_z (z, l)(1 - \eta(z)) p(z)) F(\eta) dP_\theta$$

$$= \tfrac{1}{2} \int (\sum_z (z \cdot l)[(1 - \eta(z)) - (1 - \eta(-z))] F(\eta) p(z)) dP_\theta$$

$$= \tfrac{1}{2} \int \sum_z (z \cdot l)(1 - \eta(z))[F(\tau_{-z}\eta) - F(\eta)] p(z) dP_\theta$$

$$\leq \tfrac{1}{2} (\sum p(z)(z \cdot l)^2)^{1/2} (\int \sum (1 - \eta(z))[F(\tau_{-z}\eta) - F(\eta)]^2 p(z) dP_\theta)^{1/2}$$

$$\leq C(\int (-L_1 F)(\eta) F(\eta) dP_\theta)^{1/2}$$

$$\leq C(-\int (L_0 F)(\eta) F(\eta) dP_\theta)^{1/2}. \qquad (2.9)$$

For the next lemma which is required to prove the nondegeneracy of the limiting covariance we will assume that under the random walk $p(u - v)$ one can get from any nonzero site to any other nonzero site without going through the origin. Notice that this rules out only the nearest neighbor random walk in one dimension.

Lemma 2.3. *Under the above assumption we have an estimate of the form*

$$|\int \phi_l(\eta) F(\eta) dP_\theta| \leq C(-\int (L_2 F)(\eta) F(\eta) dP_\theta)^{1/2}.$$

Proof. It is clearly sufficient to prove (see (2.9))

$$|\int [(1 - \eta(u)) - (1 - \eta(v))] F(\eta) dP_\theta| \leq C|u - v|(-\int (L_2 F)(\eta) F(\eta) dP_\theta)^{1/2}.$$

In view of the triangle inequality is to enough to establish

$$\sup_{|u-v| \leq k} |\int [(1 - \eta(u)) - (1 - \eta(v))] F(\eta) dP_\theta| \leq C(-\int (L_2 F)(\eta) F(\eta) dP_\theta)^{1/2} \quad (2.10)$$

The estimate is going to be derived through a random walk path from u to v that avoids zero. Since for a fixed path of finite length, translations will produce only at most a finite number that will go through zero the supremum in (2.10) comes for free provided we establish (2.10) for each fixed u and v (with "translation invariant" C, see below). Let

$$u = z_0, \quad z_1, \ldots z_n = v$$

be a path that avoids 0, and let $p(z_i - z_{i-1}) > 0$ for $i = 1, 2, \ldots, n$. Again because of the triangle inequality we can concentrate on

$$|\int [(1 - \eta(z_i)) - (1 - \eta(z_{i-1}))] F(\eta) dP_\theta| = |\int (1 - \eta(z_{i-1}))(F(\eta^{z_{i-1}, z_i}) - F(\eta)) dP_\theta|$$

$$\leq (\int [F(\eta^{z_{i-1}, z_i}) - F(\eta)]^2 dP_\theta)^{1/2}$$

$$\leq \left(\frac{1}{p(z_i - z_{i-1})}\right)^{1/2} (-\int (L_2 F)(\eta) F(\eta) dP_\theta)^{1/2}. \qquad \square$$

We are finally ready to state and prove our main theorem: Let us consider a test particle in simple exclusion where the underlying random walk probabilities $p(u - v)$ are assumed to be symmetric and satisfying $\sum |z|^2 p(z) < \infty$. Moreover the random walk is assumed to be irreducible and *not* the one dimensional nearest neighbor random walk. The test particle is assumed to start at 0. The rest of the configuration is distributed according to the invariant Bernoulli product measure P_θ. Then

Theorem 2.4. *The distribution of* $(1/\sqrt{\lambda})x_{\lambda t}$ *converges to a nondegenerate diffusion with covariance* σ *in the Skorohod space. Moreover the covariance* σ *is also the limit of the covariances of* $(1/\sqrt{\lambda})x_{\lambda}$ *as* $\lambda \to \infty$.

Proof. In view of the earlier comments we need only prove nondegeneracy. There are many martingales associated with the $(x, \eta)_t$ process;

$$v_t^{u,v} = \sum_{s \leq t} \chi_{(\eta_{s-} = \eta_s^{u,v})} |\eta_s(u) - \eta_s(v)| - \int_0^t |\eta_s(u) - \eta_s(v)| - \int_0^t |\eta_s(u) - \eta_s(v)| p(u-v)ds,$$

$$v_t^z = \sum_{s \leq t} \chi_{(\eta_s = \tau_{-z}\eta_{s-})} - \int_0^t p(z)(1 - \eta_s(z))ds$$

are basic martingales, and any other martingale is expressible in terms of these. For instance N_t given by (2.7) can be expressed as

$$N_t = \sum_{z \neq 0} z v_t^z,$$

For any function F the martingale

$$M_t^F = F(\eta(t) - F(\eta(0)) - \int_0^t (L_0 F)(\eta(s))ds$$

can be expressed as

$$M_t^F = \int_{u,v \neq 0} \int_0^t [F(\eta^{u,v}(s)) - F(\eta(s))]dv_s^{u,v} + \sum_z \int_0^t [F(\tau_{-z}\eta(s)) - F(\eta(s))]dv_s^z.$$

Moreover the martingales $\{v_t^z v_t^{u,v}\}$ are mutually orthogonal in the sense that the product of any two distinct ones is again a martingale. For any l the martingale $\langle N_t, l \rangle$ is of the form

$$\langle N_t, l \rangle = \sum_{z \neq 0} (z, l) v_t^z,$$

and $\langle M_t, l \rangle$ of the decomposition

$$\int_0^t \phi_l(\eta_s)ds = M_t^l + E_t^l$$

comes from $M_t^{F_\lambda}$ as $\lambda \to 0$, where $F_\lambda = (\lambda I - L_0)^{-1}\phi_l$. From the orthogonality of these martingales it is clear that the nondegeneracy of $M_t + N_t$ is assured if we establish that M_t should involve at least some $v_t^{u,v}$ component so that it cannot cancel N_t. In other words we need to prove

$$\lim_{\lambda \to 0} E^{P_\theta}\left\{ \sum_{u \neq v \neq 0} p(u-v)[F_\lambda(\eta^{u,v}(s)) - F_\lambda(\eta(s)]^2 \right\} \neq 0,$$

i.e.

$$\lim_{\lambda \to 0} \langle -L_2(\lambda I - L_0)^{-1}\phi_l, \ (\lambda I - L_0)^{-1}\phi_l > \neq 0.$$

Lemma 2.3 assures us that

$$|\langle \phi_l, \psi \rangle| \leq C \langle -L_2 \psi, \psi \rangle^{1/2}$$

or

$$\langle -L_2^{-1} \phi_l, \phi_l \rangle \leq C.$$

We can estimate

$$\langle \phi_l, (\lambda I - L_0)^{-1} \phi_l \rangle = \langle (-L_2^{-1/2} \phi_l, (-L_2)^{1/2} (\lambda I - L_0)^{-1} \phi_l \rangle$$
$$\leq \langle -L_2^{-1} \phi_l, \phi_l \rangle^{1/2} \langle -L_2 (\lambda I - L_0)^{-1} \phi_l, (\lambda I - L_0)^{-1} \phi_l \rangle$$
$$\leq C \langle -L_2 (\lambda I - L_0)^{-1} \phi_l, (\lambda I - L_0)^{-1} \phi_l \rangle.$$

Since $\langle \phi_l, (\lambda I - L_0)^{-1} \phi_l \rangle$ can go to zero only if $\phi_l = 0$, we are done.

Corollary 2.5. *Suppose the initial configuration in the sites $u \neq 0$ is different from P_θ but is some measure $Q \ll P_\theta$. Then the basic central limit theorem is still valid and the limiting distribution is the same as before. If dQ/dP_θ is bounded, then the covariance of $(1/\sqrt{\lambda})x_\lambda$ converges to the covariance of the limiting Brownian motion.* The proof follows easily from Remark 1.10. The boundedness of dQ/dP_θ guarantees that the error is still negligible in the sense of mean square.

3. Asymptotic Independence of Two Test Particles

In this section we study the behavior of two test particles located initially at points x and y.

Theorem 3.1. *The scaled positions of two test particles converge jointly to two independent Brownian motions.*

Proof. First of all we note that we are working with an initial distribution conditioned to having sites x and y occupied. Since the conditioning set has nonzero measure by Corollary 2.5, the two individual positions converge to Brownian motions separately and only asymptotic independence has to be proved. In view of the results in Helland [2] in order to prove that two Martingales satisfying the central limit theorem separately converge jointly, after rescaling, to independent Brownian motions we need only show that

$$\lim_{t \to \infty} \frac{1}{t} E M_{i,t}^{(1)} M_{j,t}^{(2)} = 0,$$

where $M_{i,t}^{(j)}$ is the ith component of the jth martingale. By standard formula one can compute

$$E M_{1t}^{(1)} M_{jt}^{(2)} = E \int_0^t \psi_{ij}(s) ds$$

for some suitable function $\psi_{ij}(s)$.

We will carry out these computations in the next several lemmas and establish that $E\psi_{ij}(s) \to 0$ as $s \to \infty$. This will prove the result.

First some notation. The probability that a typical site is occupied is of course θ and is fixed. We will not specify it hereafter. We have the initial measures P, P_x, P_y,

P_{xy} corresponding to no conditions, conditioned for site x, site y and both sites x and y to be occupied respectively at time 0. We also have the measures for the entire process ξ_t of the infinite particle system corresponding to these initial conditions which we denote by Q, Q_x, Q_y and Q_{xy} respectively.

Lemma 3.2. *The positions x_t and y_t of two tagged particles starting from x and y at time $t = 0$ can be represented as*

$$x_t = x + M_t^{(1)} + E_t^{(1)}, \quad y_t = y + M_t^{(2)} + E_t^{(2)},$$

where $M_t^{(1)}$ and $M_t^{(2)}$ are martingales relative to Q_x and Q_y respectively. They are both martingales relative to $Q_{x,y}$. Further $E_t^{(1)}$, $E_t^{(2)}$ are negligible relative to $Q_{x,y}$ in the mean square sense.

Proof. The representation of x_t relative to Q_x and y_t relative to Q_y follow from Sect. 2. Since all martingale properties are valid relative to almost all initial starting points and $P_{x,y} \ll P_x$ as well as $P_{x,y} \ll P_y$, the martingale property holds true for $Q_{x,y}$ as well. Moreover $dQ_{x,y}/dQ_x$ and $dQ_{x,y}/dQ_y$ are bounded. So all the negligibility results are valid in the mean square sense as well.

Let us look at the complete class of martingales for the ξ_t process for any $a, b \in \mathbb{Z}^d$ and $a \neq b$:

$$n(t, a, b) = N_t(a, b) - \int_0^t p(b - a)\xi_s(a)(1 - \xi_s(b)) \, ds$$

is a martingale where $N_t(a, b)$ is the number of transitions in time $[0, t]$ from site a to site b. Since any martingale is expressible in terms of these basic ones

$$M_{i,t}^{(1)} = \sum_{a \neq b} \int_0^t \phi_{i,a,b}^{(1)}(s)dn(s, a, b) \tag{3.1}$$

and

$$M_{i,t}^{(2)} = \sum_{a \neq b} \int_0^t \phi_{i,a,b}^{(2)}(s)dn(s, a, b). \tag{3.2}$$

Lemma 3.3. *For each $t > 0$*

$$E^{Q_{x,y}}M_{i,t}^{(1)}M_{j,t}^{(2)} = E^{Q_{x,y}}\int_0^t \psi_{ij}(s)ds,$$

where

$$\psi_{ij}(s) = \sum_{a \neq b} \phi_{i,a,b}^{(1)}(s)\phi_{j,a,b}^{(2)}(s)\eta_s(a)(1 - \eta_s(b))p(b - a).$$

Proof. Follows from the fact that unless $a = a'$ and $b = b'$ the martingales $n(t, a, b)$ and $n(t, a', b')$ are orthogonal and

$$En^2(t, a, b) - \int_0^t \eta_s(a)(1 - \eta_s(b))p(b - a)ds$$

is a martingale.

Lemma 3.4. *For any $L < \infty$,*

$$\lim_{t \to \infty} Q_{x,y}[|x(t) - y(t)| \leq L] = 0.$$

Proof. Let us consider the system consisting of the first tagged particle being the origin of reference and the rest of the world including the second tagged particle being viewed from the first tagged particle. For this system there is a σ-finite invariant measure with mass one corresponding to each location of the second tagged particle. Given the location of the second tagged particle the rest of the universe (excluding also the origin) is provided a Bernoulli product measure. One checks that except for the one dimensional nearest neighbor model the system is ergodic. Moreover the system is reversible so that if T_t is the evolution semigroup for the system $\langle f, T_t g \rangle \to 0$ by the spectral theorem provided f and g are square integrable with respect to the invariant measure. Picking $g = \chi_{\{|u| \leq L\}}(\cdot)$ and $f = \chi_y$, i.e. the indicators of the sets where the second tagged particle is within a distance L of the origin and at y respectively we obtain our result.

Lemma 3.5.

$$\lim_{t \to \infty} E^{Q_{x,y}} \psi_{ij}(t) = 0.$$

Proof. We use the simple inequality

$$|\psi_{ij}(t)| \leq \chi_{E_1(t) \cap E_2(t) \neq \phi} \cdot [Q_i^{(1)}(t)]^{1/2}[Q_j^{(2)}(t)]^{1/2}$$
$$+ [Q_i^{(1)}(t)]^{1/2}[Q_{j,E_2(t)}^{(2)}(t)]^{1/2} + [Q_{i,E_1(t)}^{(1)}(t)]^{1/2}[Q_j^{(2)}(t)]^{1/2}, \qquad (3.3)$$

where $L < \infty$ is fixed, and

$$E_1(t) = \{(u,v) \in \mathbb{Z}^d \times \mathbb{Z}^d, \; u \neq v \text{ and } \max[|u - x_t|, |v - x_t|] \leq L\},$$
$$E_2(t) = \{(u,v) \in \mathbb{Z}^d \times \mathbb{Z}^d, \; u \neq v \text{ and } \max[|u - y_t|, |v - y_t|] \leq L\},$$

$$Q_i^{(r)}(t) = \sum_{u,v} |\phi_{i,u,v}^{(r)}(t)|^2 \eta_t(u)(1 - \eta_t(v))p(v - u),$$

$$Q_{i,A}^{(r)}(t) \sum_{(u,v) \in A} |\phi_{i,u,v}^{(r)}(t)|^2 \eta_t(u)(1 - \eta_t(v))p(v - u).$$

We write for each $L < \infty$

$$E^{Q_{x,y}} |\psi_{ij}(t)| \leq a_L(t) + b_L(t) + c_L(t),$$

corresponding to the estimate (3.3), which we think of as

$$\psi_{ij}(t) \leq T_{1,L}(t) + T_{2,L}(t) + T_{3,L}(t).$$

Step 1.

$$a_L(t) = E^{Q_{x,y}} T_{1,L}(t) \leq \int_{|x_t - y_t| \leq 2L} [Q_i^{(1)}(t)Q_j^{(2)}(t)]^{1/2} dQ_{x,y}$$
$$\leq \left(\int_{|x_t - y_t| \leq 2L} Q_i^{(1)}(t) dQ_{x,y} \right)^{1/2} \left(\int_{|x_t - y_t| \leq 2L} Q_j^{(2)}(t) dQ_{x,y} \right)^{1/2}.$$

To show that $a_L(t) \to 0$ as $t \to 0$, it suffices to know that $Q_i^{(r)}(t)$ are uniformly integrable with respect to $Q_{x,y}$. But $dQ_{x,y}/dQ_x$ and $dQ_{x,y}/dQ_y$ are bounded and $Q_i^{(1)}(t)$ and $Q_i^{(2)}(t)$ are stationary processes relative to Q_x and Q_y respectively.

Step 2.

$$b_L(t) = E^{Q_{x,y}} T_{2,L}(t) \leq (E^{Q_{x,y}} [Q_i^{(1)}(t)])^{1/2} (E^{Q_{x,y}} [Q_{j,E_2(t)}^{(2)}(t)])^{1/2}$$
$$\leq C(E^{Q_x} [Q_i^{(1)}(t)])^{1/2} (E^{Q_y} [Q_{j,E_2(t)}^{(2)}(t)]).$$

$E^{Q_x} [Q_i^{(1)}(t)]$ is independent of t and so is $E^{Q_y} [Q_{j,E_2(t)}^{(2)}(t)]$. Moreover the last quantity tends to zero as $L \to \infty$, and we are done.

Step 3. The estimate for $c_L(t)$ is identical to step 2 with the roles of the two tagged particles reversed.

4. Application to a Bond Diffusion Process

In this section we want to study the movement of a particle in a random environment. Consider for each bond b of Z^d a positive random variable a_b and suppose that these random variables are i.i.d.. Once this random environment is chosen, let a particle evolve according to the jump process with generator $Lf(x) = \sum_{y \sim x} a_{yx} [f(y) - f(x)]$, where $y \sim x$ means that y and x are nearest-neighbors and a_{yx} is the rate of the corresponding bond. We also assume that the a_b's have a uniform upper and lower bound. Call x_t the position of the particle at time t, and $\varepsilon x_{t/\varepsilon^2}$ the rescaled position. Then

Theorem 4.1. *After rescaling the position of the particle converges to Brownian motion.*

Proof. The only quantities that matter in this problem are the values of the rates given to each bond (which we call an environment).

It is also clear that the process ξ_t of the environment seen from the position of the particle at time t is a Markov process whose generator is

$$\tilde{L}f(\xi) = \sum_e a_e(\xi) [f(\tau_e \xi) - f(\xi)].$$

Therefore to prove our central limit theorem we only have to check that, for any $\vec{\alpha} \in R^d$, the function

$$\phi(\xi) = \sum_e a_e(\xi)(\vec{e} \cdot \vec{\alpha})$$

satisfies the condition of Theorem 1.3, because we easily check that the original product measure (say $m(d\xi)$) is reversible for this process;

$$\int a_e(\xi)(e \cdot \alpha) g(\xi) m(d\xi) = \int a_{-e}(\tau_e \xi)(\vec{e} \cdot \vec{\alpha}) g(\xi) m(d\xi)$$
$$= -\int a_{-e}(\eta)(-\vec{e} \cdot \vec{\alpha}) g(\tau_{-e}\eta) m(d\eta).$$

Therefore

$$\sum_e \int g(\xi) a_e(\xi)(\vec{e} \cdot \vec{\alpha}) m(d\xi) = \tfrac{1}{2} \int \sum_e a_e(\xi)(\vec{e} \cdot \vec{\alpha})(f(\tau_e \xi) - f(\xi)) m(d\xi)$$

$$\leq \tfrac{1}{2}[\sum_e \int a_e(\xi)(\vec{e}\cdot\vec{\alpha})^2 m(d\xi)][\sum_e \int a_e(\xi)(f(\tau_e\xi)-f(\xi))^2 m(d\xi)],$$

which is the bound that we were looking for.

Remark 4.2. With some additional work one can prove the nondegeneracy of the limiting Brownian motion.

Acknowledgements. C. K. acknowledges the warm hospitality of the CIMS where this work was done. The authors wish to thank S. Goldstein and H. Spohn for valuable discussions as well as detailed comments on an earlier version of the paper which are responsible for this improved version.

References

1. Arratia, R.: The motion of a tagged particle in the simple symmetric exclusion system in Z. Ann. Probab. **11** 362–373 (1983)
2. Helland, I.: Central limit theorems for martingales with discrete or continuous time. Scand. J. Stat., **9**, 79–94 (1982)
3. Kipnis, C., Lebowitz, J. L., Presutti, E., Spohn, H.: J. Stat. Phys. **30**, 107–121 (1983)
4. Lebowitz, J. L., Spohn, H.: J. Stat. Phys. **28**, 539–556 (1982)

Communicated by J. L. Lebowitz

Received August 10, 1984; in revised form July 2, 1985

Adv. Appl. Prob. **21**, 842–860 (1989)
Printed in N. Ireland
© *Applied Probability Trust* 1989

BOUNDING FUNCTIONS OF MARKOV PROCESSES AND THE SHORTEST QUEUE PROBLEM

JOHN A. GUBNER,* *University of Maryland*
B. GOPINATH,** *Bell Communications Research*
S. R. S. VARADHAN,*** *Courant Institute of Mathematical Sciences*

Abstract

We prove a theorem which can be used to show that the expectation of a non-negative function of the state of a time-homogeneous Markov process is uniformly bounded in time. This is reminiscent of the classical theory of non-negative supermartingales, except that our analog of the supermartingale inequality need not hold almost surely. Consequently, the theorem is suitable for establishing the stability of systems that evolve in a stabilizing mode in most states, though from certain states they may jump to a less stable state. We use this theorem to show that 'joining the shortest queue' can bound the expected sum of the squares of the differences between all pairs among N queues, *even under arbitrarily heavy traffic*.

STABILITY; LYAPUNOV FUNCTION; SUPERMARTINGALE; LOAD BALANCING; PARALLEL QUEUES

1. Introduction

Controlled Markov processes often arise when modeling the behavior of asynchronous algorithms, queuing systems, and computer and communication networks. In these models it is often the case that the control policy considers only the immediate future of the process in an attempt to stabilize it. While in general a myopic policy will not guarantee the long-term stability of the process, in this paper we shall give sufficient conditions on the one-step behavior of a process to guarantee its long-term stability.

Suppose we are given a time-homogeneous Markov process, $\{X_n\}$. Consider an analysis in which we choose a non-negative function u such that if the expected value of $u(X_n)$ is suitably bounded, then the process is stable in some sense. For example, in Section 3 our process will be the state of a system of N parallel queues. Our control policy will direct the traffic arriving at time $n + 1$ to the queue whose

Received 10 December 1987; revision received 21 October 1988.

* Present address: Department of Electrical and Computer Engineering, University of Wisconsin, Madison, WI 53706, USA.

** Present address: Department of Electrical Engineering, Rutgers University, New Brunswick, NJ 08903, USA.

*** Postal address: Courant Institute of Mathematical Sciences, New York University, 251 Mercer Street, New York, NY 10012, USA.

This work was supported by Bell Communications Research. The research of the first author was also supported in part by the University of Maryland Systems Research Center under NSF Grant OIR-85-00108.

length was shortest at time n. We shall show that this policy can stabilize the system in the sense that the expected sum of the squares of the differences between all pairs among the N queues is uniformly bounded for all time.

To motivate our results, consider the following example of a single-server queue. Let $\{Y_n, n = 0, 1, 2, \cdots\}$ be a discrete-time Markov chain whose transition probabilities are as follows. For $i = 0$,

$$\mathbb{P}(Y_{n+1} = j \mid Y_n = i) = \begin{cases} \lambda, & \text{if } j = 1, \\ 1 - \lambda, & \text{if } j = 0, \end{cases}$$

and for $i \geqq 1$,

$$\mathbb{P}(Y_{n+1} = j \mid Y_n = i) = \begin{cases} \lambda, & \text{if } j = i+1, \\ 1 - (\lambda + \mu), & \text{if } j = i, \\ \mu, & \text{if } j = i-1, \end{cases}$$

where $0 \leqq \lambda < \mu \leqq 1$ and $\lambda + \mu \leqq 1$. Our concern here is the long-term behavior of $\mathbb{E}[Y_n \mid Y_0 = i]$. In particular, we should like to know if

$$\limsup_{n \to \infty} \mathbb{E}[Y_n \mid Y_0 = i] < \infty.$$

In this simple example, since $\lambda < \mu$, it is very easy to compute the stationary distribution (see Billingsley (1979), Theorem 8.6, p. 106),

$$\lim_{n \to \infty} \mathbb{P}(Y_n = j \mid Y_0 = i) = \left(\frac{\lambda}{\mu}\right)^j \left(1 - \frac{\lambda}{\mu}\right).$$

Recalling that for $|z| < 1$, $\sum_{j=1}^{\infty} j z^{j-1} = (1 - z)^{-2}$, it is not hard to compute that indeed,

$$\lim_{n \to \infty} \mathbb{E}[Y_n \mid Y_0 = i] = \frac{\lambda}{\mu - \lambda} < \infty,$$

assuming that we can take the limit inside the expectation.

How could we study the behavior of $\mathbb{E}[Y_n \mid Y_0 = i]$ if the stationary distribution were not available to us? Recall that in the field of differential equations, a 'Lyapunov function' can sometimes be used to study the behavior of solutions to a differential equation without actually solving it. In this spirit, our paper generalizes the preceding example and gives conditions on the one-step behavior of a process that will yield useful information about its long-run (and short-run) behavior.

We now make a precise statement of our model. Let $\{X_n, n = 0, 1, 2, \cdots\}$ be a time-homogeneous Markov process defined on a probability space $(\Omega, \mathcal{F}, \mathbb{P})$ and taking values in a measurable space $(\mathbb{X}, \mathcal{X})$. Let $u : \mathbb{X} \to [0, \infty)$ be a given measurable function. (Here $[0, \infty)$ is equipped with the usual Borel σ-field.) For $x \in \mathbb{X}$, let

(1.1) $\qquad \hat{u}_n(x) \triangleq \mathbb{E}[u(X_n) \mid X_0 = x], \qquad n = 0, 1, 2, \cdots.$

In this paper we give sufficient conditions to answer the following question: 'Does there exist a positive, finite constant D such that

$$\limsup_{n \to \infty} \hat{u}_n(x) \leqq D$$

for all $x \in \mathbb{X}$?' To answer this question, we construct a sequence of functions $\{u_n\}$ such that for all $n \geqq 0$,

(1.2) $\hat{u}_n(x) \leqq u_n(x), \quad \forall x \in \mathbb{X},$

and such that $\{u_n\}$ is easy to analyze. To suggest how one might establish (1.2), consider the following result.

Proposition 1. *Set* $u_0(x) \triangleq u(x)$, *and let* $\{u_n\}_{n \geqq 1}$ *be given. If one can show that for all* $n \geqq 0$, *and all* $x \in \mathbb{X}$,

(1.3) $\mathbb{E}[u_n(X_1) \mid X_0 = x] \leqq u_{n+1}(x),$

then $\hat{u}_n(x) \leqq u_n(x)$ *for all* $n \geqq 0$, *and all* $x \in \mathbb{X}$.

Proof. We proceed by induction. Clearly,

$$\hat{u}_0(x) \triangleq \mathbb{E}[u(X_0) \mid X_0 = x] = u(x) = u_0(x),$$

and so (1.2) holds for $n = 0$. Suppose (1.2) holds for some $n \geqq 0$. Then

$$\begin{aligned}
\hat{u}_{n+1}(x) &\triangleq \mathbb{E}[u(X_{n+1}) \mid X_0 = x] \\
&= \mathbb{E}[\mathbb{E}[u(X_{n+1}) \mid X_1, X_0] \mid X_0 = x] \\
&= \mathbb{E}[\mathbb{E}[u(X_{n+1}) \mid X_1] \mid X_0 = x], \quad \text{by the Markov property,} \\
&= \mathbb{E}[\hat{u}_n(X_1) \mid X_0 = x], \quad \text{by time homogeneity,} \\
&\leqq \mathbb{E}[u_n(X_1) \mid X_0 = x], \quad \text{by the induction hypothesis,} \\
&\leqq u_{n+1}(x), \quad \text{by (1.3).}
\end{aligned}$$

The difficulty is to establish conditions under which (1.3) holds. If we let $p(x, B)$ denote a regular version of $\mathbb{P}(X_1 \in B \mid X_0 = x)$, then we can rewrite (1.3) as

(1.4) $\displaystyle\int_{\mathbb{X}} u_n(y) p(x, dy) \leqq u_{n+1}(x).$

Remark. The careful reader will see later that since $u_n(x)$ will have the form $u_n(x) = f_n(u(x))$, in order to derive the results in this paper, we do not need a regular version of $\mathbb{P}(X_1 \in B \mid X_0 = x)$, but only a regular version of $\mathbb{P}(u(X_1) \in C \mid X_0 = x)$. Since u is real valued, such a regular version always exists; see Billingsley (1979), Theorem 33.3, p. 390.

In Section 2 we give sufficient conditions under which (1.4) will hold. These conditions are summarized in Theorem 2. In Section 3 we describe the problem of 'joining the shortest queue'. We then apply Theorem 2 to show that the servers are equally loaded, even in heavy traffic. In Section 4 we prove Theorem 2, in Section 5 we present an extension of Theorem 2, and in Section 6 we compare Theorem 2 to a result of Hajek (1982).

2. Statement of main result

As pointed out in Section 1, the problem of finding a sequence $\{u_n\}$ satisfying (1.2) can be reduced to the problem of finding a sequence satisfying (1.4). We now present our main result, Theorem 2, which gives conditions under which this can be done. The proof is given in Section 4.

Theorem 2. Fix $b \geqq 0$ and $0 < \varepsilon < 1$. Suppose that there exist positive constants C and v satisfying

$$(2.1) \qquad \frac{(b+1)(b+2)}{2} \frac{(1+\varepsilon)^b}{(1-\varepsilon)^{b+3}} C \leqq 2v,$$

and such that for some L,

$$(2.2) \qquad |u(y) - u(x)| \leqq \varepsilon u(x), \; p(x, \cdot)\text{-a.s.}, \quad if \quad u(x) > L,$$

$$(2.3) \qquad \mathbb{E}[u(X_{n+1}) - u(x) \mid X_n = x] \leqq -v, \quad if \quad u(x) > L,$$

and

$$(2.4) \qquad \mathbb{E}[|u(X_{n+1}) - u(x)|^2 \mid X_n = x] \leqq Cu(x), \quad if \quad u(x) > L.$$

If one can find a finite constant A, depending on b and L, such that

$$(2.5) \qquad \mathbb{E}[u(X_{n+1})^{b+2} \mid X_n = x]^{1/(b+2)} \leqq A, \quad if \quad 0 \leqq u(x) \leqq L,$$

then the sequence $\{u_n\}$ given by

$$(2.6) \qquad u_n(x) \triangleq \frac{u(x)^{b+2}}{\left[u(x) + \dfrac{v}{b+1} n\right]^{b+1}} + A + \frac{A^{b+2}}{\left(\dfrac{v}{b+1}\right)^{b+1}} \sum_{k=1}^{n-1} \frac{1}{k^{b+1}},$$

satisfies (1.4), and so by Proposition 1,

$$(2.7) \qquad \mathbb{E}[u(X_n) \mid X_0 = x] \leqq u_n(x), \quad n \geqq 0, x \in \mathbb{X}.$$

Of course, when $n = 0$, Equation (2.6) is to be interpreted as $u_0(x) \triangleq u(x)$, and when $n = 1$, it is to be interpreted as

$$u_1(x) \triangleq \frac{u(x)^{b+2}}{\left[u(x) + \dfrac{v}{b+1}\right]^{b+1}} + A.$$

Remark. When $b = 1$, (2.6) becomes

(2.8)
$$u_n(x) = \frac{u(x)^3}{\left[u(x) + \frac{v}{2}n\right]^2} + A + 4\frac{A^3}{v^2}\sum_{k=1}^{n-1}\frac{1}{k^2}.$$

In fact, since $\sum_{k=1}^{\infty}k^{-2} = \pi^2/6$ (see Papadimitriou (1973) for an elementary proof of this fact; see Stark (1974) for more general series), it is easy to see that

(2.9)
$$\lim_{n\to\infty} u_n(x) = A\left(1 + \frac{2}{3}\left(\frac{\pi A}{v}\right)^2\right),$$

and that

(2.10)
$$u_n(x) \leq u(x) + A\left(1 + \frac{2}{3}\left(\frac{\pi A}{v}\right)^2\right).$$

Note that the limit in (2.9) is independent of x, whereas the bound in (2.10) is independent of n.

Remark. The inequality (2.3) is very similar to the definition of a Markovian supermartingale. To see this, recall that if $\mathscr{F}_n \triangleq \sigma(X_0, \cdots, X_n)$, then $\{u(X_n)\}$ is an $\{\mathscr{F}_n\}$-supermartingale if and only if for all $n \geq 0$,

(2.11)
$$\mathbb{E}[u(X_{n+1}) \mid \mathscr{F}_n] \leq u(X_n).$$

Since $\{X_n\}$ is Markovian, this can be rewritten as

$$\mathbb{E}[u(X_{n+1}) \mid X_n] \leq u(X_n),$$

or

(2.12)
$$\mathbb{E}[u(X_{n+1}) - u(x) \mid X_n = x] \leq 0, \qquad \forall x \in \mathbb{X}.$$

To compare (2.3) with (2.12), observe that (2.3) is a stronger inequality, but holds only when $u(x) > L$. On the other hand, (2.12) is a weaker inequality, but holds for *every* $x \in \mathbb{X}$. Now, observe that if (2.11) were to hold, then we could immediately write

$$\mathbb{E}[u(X_{n+1}) \mid X_0] \leq \mathbb{E}[u(X_n) \mid X_0],$$

from which it would follow by induction that for all $n \geq 0$,

$$\mathbb{E}[u(X_n) \mid X_0] \leq u(X_0),$$

which we can rewrite as

$$\mathbb{E}[u(X_n) \mid X_0 = x] \leq u(x).$$

Compare this with the conclusion of Theorem 2 when $b = 1$; substituting (2.10) into (2.7) yields

$$\mathbb{E}[u(X_n) \mid X_0 = x] \leq u(x) + A\left(1 + \frac{2}{3}\left(\frac{\pi A}{v}\right)^2\right).$$

Example. Let us apply Theorem 2 to the single-server queue described in Section 1. Since Y_n takes values in $[0, \infty)$, we take $u(Y_n) = Y_n$. Now, set $b = 1$ and fix $0 < \varepsilon < 1$. Make the following observations. First, it is easy to calculate

$$\mathbb{E}[Y_{n+1} \mid Y_n = i] = \begin{cases} \lambda, & \text{if } i = 0, \\ i - (\mu - \lambda), & \text{if } i > 0; \end{cases}$$

rewrite this as

(2.13) $$\mathbb{E}[Y_{n+1} - i \mid Y_n = i] = \begin{cases} \lambda, & \text{if } i = 0, \\ -(\mu - \lambda), & \text{if } i > 0. \end{cases}$$

Clearly, we should take $v = \mu - \lambda > 0$. Set

$$C = \tfrac{2}{3} v \frac{(1 - \varepsilon)^4}{(1 + \varepsilon)}.$$

Now note that whenever $Y_n = i$,

(2.14) $$|Y_{n+1} - i| \leq 1,$$

and so

(2.15) $$\mathbb{E}[|Y_{n+1} - i|^2 \mid Y_n = i] \leq 1.$$

To put (2.14) and (2.15) into the form of (2.2) and (2.4), we can require that

$$1 \leq \varepsilon i \quad \text{and} \quad 1 \leq Ci.$$

This will happen whenever

(2.16) $$i \geq \max \left\{ \frac{1}{\varepsilon}, \frac{3(1 + \varepsilon)}{2v(1 - \varepsilon)^4} \right\}.$$

Let L denote the right-hand side of (2.16). Finally, note that since (2.14) implies

$$Y_{n+1} \leq i + 1,$$

we have

$$\mathbb{E}[Y_{n+1}^3 \mid Y_n = i]^{\frac{1}{3}} \leq i + 1$$

$$\leq L + 1, \quad \text{if } i \leq L.$$

Hence, we may take $A = L + 1$. It should now be clear that Theorem 2 applies to the single-server queue if $\mu > \lambda$. In this case,

$$\limsup_{n \to \infty} \mathbb{E}[Y_n \mid Y_0 = i] \leq (L + 1) + \frac{2}{3} \left(\frac{\pi}{\mu - \lambda} \right)^2 (L + 1)^3 < \infty.$$

We remark here that the preceding bound depends on L, which by (2.16), depends on ε. To reduce L, we should try to choose $0 < \varepsilon < 1$ so that

$$\frac{1}{\varepsilon} = \frac{3(1 + \varepsilon)}{2v(1 - \varepsilon)^4},$$

or equivalently,

(2.17) $$2v(1 - \varepsilon)^4 - 3\varepsilon(1 + \varepsilon) = 0.$$

In other words, to reduce L, we must find the zeros of a fourth-degree polynomial in ε. Observe that there is at least one solution of (2.17) in $(0, 1)$.

3. Joining the shortest queue

In this section we investigate the balancing of N queues to N identical servers under a range of traffic conditions. Consider a system of N identical servers, each with its own infinite waiting room, indexed by $i = 1, \cdots, N$. Suppose also that we are given $N + 1$ arrival processes indexed by $i = 1, \cdots, N, N + 1$. Customers from arrival processes $i = 1, \cdots, N$ are assigned to the corresponding waiting room for server i. Customers from the $(N + 1)$th arrival process may in general be assigned to any waiting room; however, we consider only what happens when we employ the policy of assigning these customers to the waiting room with the fewest waiting customers. In case of a tie, the waiting room of smallest index is selected.

Let $Q_n^{(i)}$ denote the size of the queue for server i at time $n = 0, 1, 2, \cdots$. This includes customers in the waiting room as well as the customer being served. Let

$$Q_n \triangleq (Q_n^{(1)}, \cdots, Q_n^{(N)}).$$

We want to show that the policy of joining the shortest queue will ensure, even under arbitrarily heavy traffic conditions, that for all $i \neq j$,

(3.1) $$\mathbb{E}[|Q_n^{(j)} - Q_n^{(i)}| \mid Q_0 = q]$$

is uniformly bounded for all time. To apply the theory of Section 2, we need to find a suitable function u. Below we show that the function defined in (3.2) will suffice.

Before proceeding, we point out that the continuous-time analog of this model has been studied by a number of authors, though, to our knowledge, none has investigated the behavior of the continuous-time analog of (3.1). For example, Haight (1958), Kingman (1961), Flatto and McKean (1977), and Conolly (1984) characterized the stationary distribution of the state of the system. Foschini and Salz (1978) studied a diffusion approximation. Blanc (1987) outlined a numerical method for calculating the state probabilities and moments of the queue-length distribution. Brumelle (1971) bounded the expected waiting time. Wolff (1977) bounded moments of the delay distribution (see also Wolff (1987) for corrections and comments). Halfin (1985) bounded the probability distribution of the number of customers in the system, and its expected value in equilibrium. Winston (1977) and Weber (1978) proved that the policy of 'joining the shortest queue' is optimal in the sense of maximizing the customer throughput. Ephremides et al. (1980) proved that this policy is also optimal in the sense of minimizing the expected total time for completing service on all customers which arrive before a fixed time.

We now make a precise statement of our model. Using the N-fold cartesian product of the non-negative integers as the state space for $\{Q_n\}$, we take $\{Q_n\}$ to be a Markov process with transition probabilities defined as follows. Suppose that

$$Q_n = q = (q^{(1)}, \cdots, q^{(N)}),$$

where the $q^{(\cdot)}$ are non-negative integers. Let $\mu, \lambda, \lambda_1, \cdots, \lambda_N$ be a sequence of elements from the interval $[0, 1]$ such that $\lambda + \sum_{j=1}^{N} \lambda_j + N\mu \leq 1$. Suppose that the N-vector Δq_j is defined by $(0, \cdots, 0, 1, 0, \cdots, 0)$, where the 1 occupies the jth coordinate. Let i denote the smallest index such that $q^{(i)} \leq q^{(j)}$ holds for $j = 1, \cdots, N$. Let $\delta_i(j) = 1$ if $j = i$, and 0 otherwise. Let $\mu(t) = \mu$ if $t \geq 1$ and $\mu(t) = 0$ if $t = 0$. We assume that

$$\mathbb{P}(Q_{n+1} = q' \mid Q_n = q) = \begin{cases} \lambda_j + \lambda \delta_i(j), & \text{if } q' = q + \Delta q_j, \\ \mu(q^{(j)}), & \text{if } q' = q - \Delta q_j, \\ 1 - \lambda - \sum_{j=1}^{N} [\lambda_j + \mu(q^{(j)})], & \text{if } q' = q. \end{cases}$$

Let $u(q)$ denote the sum of the squares of the differences between all pairs of components of $q = (q^{(1)}, \cdots, q^{(N)})$. More precisely,

$$(3.2) \qquad u(q) \triangleq \sum_{\alpha=1}^{N-1} \sum_{\beta=\alpha+1}^{N} |q^{(\beta)} - q^{(\alpha)}|^2.$$

Theorem 3. Without loss of generality, assume that $\lambda_1 \leq \cdots \leq \lambda_N$. If

$$(3.3) \qquad \lambda > \sum_{k=1}^{N-1} (\lambda_N - \lambda_k),$$

then there exist finite positive constants v and A such that

$$\mathbb{E}[u(Q_n) \mid Q_0 = q] \leq \frac{u(q)^3}{\left[u(q) + \dfrac{v}{2} n\right]^2} + A\left(1 + 4\left(\frac{A}{v}\right)^2 \sum_{k=1}^{n-1} \frac{1}{k^2}\right),$$

from which it follows that

$$(3.4) \qquad \mathbb{E}[u(Q_n) \mid Q_0 = q] \leq u(q) + A\left(1 + \frac{2}{3}\left(\frac{\pi A}{v}\right)^2\right),$$

and

$$\limsup_{n \to \infty} \mathbb{E}[u(Q_n) \mid Q_0 = q] \leq A\left(1 + \frac{2}{3}\left(\frac{\pi A}{v}\right)^2\right) < \infty.$$

Remarks. (i) If $\lambda > (N-1)(\lambda_N - \lambda_1)$ then (3.3) holds. (ii) Since the inequality in (3.3) does not depend on μ, the interpretation of (3.4) is that the policy of 'joining the shortest queue' will ensure that even under arbitrarily heavy traffic, $\mathbb{E}[u(Q_n) \mid Q_0 = q]$ does not 'blow up'. By Jensen's inequality, (3.4) also implies that for every $i \neq j$,

$$\mathbb{E}[|Q_n^{(j)} - Q_n^{(i)}| \mid Q_0 = q] \leq \mathbb{E}[|Q_n^{(j)} - Q_n^{(i)}|^2 \mid Q_0 = q]^{\frac{1}{2}}$$
$$\leq \mathbb{E}[u(Q_n) \mid Q_0 = q]^{\frac{1}{2}}$$

is bounded for all time. In this sense, each queue has 'approximately' the same number of customers; that is, the queues are 'equally loaded' or 'balanced'. The reader may find it interesting to contemplate various special cases such as $\mu = 0$, $\lambda_N = \lambda_1 > 0$, or $\lambda_N = \lambda_1 = 0$.

Proof. We show below that (3.3) implies

$$(3.5) \qquad |u(q \pm \Delta q_j) - u(q)| \leq (N-1)(1 + 2\sqrt{u(q)}),$$

$$(3.6) \qquad \mathbb{E}[u(Q_{n+1}) - u(q) \mid Q_n = q] \leq (N-1) - 2\left(\lambda - \sum_{k=1}^{N-1}(\lambda_N - \lambda_k)\right)\sqrt{\frac{u(q)}{2N-1}},$$

and

$$(3.7) \qquad \mathbb{E}[|u(Q_{n+1}) - u(q)|^2 \mid Q_n = q] \leq (N-1)^2(1 + 4\sqrt{u(q)} + 4u(q)).$$

First, however, we use (3.3) and (3.5)–(3.7) to show that the hypotheses of Theorem 2 are satisfied. Set $b = 1$, and fix any $0 < \varepsilon < 1$. Now make the following observations. Since

$$\lim_{v \to \infty} \frac{(N-1)(1 + 2\sqrt{v})}{v} = 0,$$

it is clear that for all sufficiently large v,

$$(N-1)(1 + 2\sqrt{v}) \leq \varepsilon v.$$

Now, fix any $C > 4(N-1)^2$. Since

$$\lim_{v \to \infty} \frac{(N-1)^2(1 + 4\sqrt{v} + 4v)}{v} = 4(N-1)^2,$$

for all sufficiently large v,

$$(N-1)^2(1 + 4\sqrt{v} + 4v) \leq Cv.$$

Next, choose (recall (2.1))

$$(3.8) \qquad v \geq \tfrac{3}{2}C\frac{(1+\varepsilon)}{(1-\varepsilon)^4};$$

since (3.3) holds, we have

$$\lim_{v \to \infty}\left\{(N-1) - 2\left(\lambda - \sum_{k=1}^{N-1}(\lambda_N - \lambda_k)\right)\sqrt{\frac{v}{2N-1}}\right\} = -\infty.$$

It follows that for all sufficiently large v, we also have

$$(N-1) - 2\left(\lambda - \sum_{k=1}^{N-1} (\lambda_N - \lambda_k)\right) \sqrt{\frac{v}{2N-1}} \leq -v.$$

It should now be clear that if $b = 1$, $0 < \varepsilon < 1$, $C > 4(N-1)^2$, and v satisfies (3.8), then there is some L such that (2.2)–(2.4) hold. To find a finite constant A satisfying (2.5), observe that if $u(q) \leq L$, then (3.5) implies

$$|u(q \pm \Delta q_j) - u(q)| \leq (N-1)(1 + 2\sqrt{L}),$$

and so

$$u(q \pm \Delta q_j) \leq (N-1)(1 + 2\sqrt{L}) + L.$$

Hence, we may take

$$A = (N-1)(1 + 2\sqrt{L}) + L.$$

We conclude that if (3.3) holds, (3.5)–(3.7) are sufficient to apply Theorem 2.

We now establish (3.5)–(3.7). Clearly, (3.5) implies (3.7). To establish (3.5) and (3.6), we proceed as follows. We denote by i the smallest integer in $\{1, \cdots, N\}$ such that $q^{(i)} \leq q^{(j)}$ holds for $j = 1, \cdots, N$. Setting $l_j \triangleq q^{(j)} - q^{(i)}$, we can write

$$q = (q^{(i)} + l_1, \cdots, q^{(i)} + l_N).$$

Note that each $l_j \geq 0$, and $l_i = 0$. Fixing any $j \in \{1, \cdots, N\}$, we have

(3.9)
$$u(q) = \sum_{\beta \neq j} |q^{(\beta)} - q^{(j)}|^2 + \sum_{\substack{\alpha=1 \\ \alpha \neq j}}^{N-1} \sum_{\substack{\beta=\alpha+1 \\ \beta \neq j}}^{N} |q^{(\beta)} - q^{(\alpha)}|^2$$

$$= \sum_{\beta \neq j} |l_\beta - l_j|^2 + \sum_{\substack{\alpha=1 \\ \alpha \neq j}}^{N-1} \sum_{\substack{\beta=\alpha+1 \\ \beta \neq j}}^{N} |l_\beta - l_\alpha|^2,$$

and

$$u(q \pm \Delta q_j) = \sum_{\beta \neq j} |q^{(\beta)} - (q^{(j)} \pm 1)|^2 + \sum_{\substack{\alpha=1 \\ \alpha \neq j}}^{N-1} \sum_{\substack{\beta=\alpha+1 \\ \beta \neq j}}^{N} |q^{(\beta)} - q^{(\alpha)}|^2$$

$$= \sum_{\beta \neq j} |(l_\beta - l_j) \mp 1|^2 + \sum_{\substack{\alpha=1 \\ \alpha \neq j}}^{N-1} \sum_{\substack{\beta=\alpha+1 \\ \beta \neq j}}^{N} |l_\beta - l_\alpha|^2,$$

so that

(3.10)
$$u(q \pm \Delta q_j) - u(q) = \sum_{\beta \neq j} (1 \mp 2(l_\beta - l_j)).$$

It follows from (3.9) that for any $\beta \neq j$,

$$|l_\beta - l_j| \leq \sqrt{u(q)}.$$

This with (3.10) implies (3.5).

To establish (3.6), use (3.10) to show that

$$\mathbb{E}[u(Q_{n+1}) - u(q) \mid Q_n = q] = \sum_{j=1}^{N} \left\{ \left(\sum_{\beta \neq j} (1 - 2(l_\beta - l_j)) \right) (\lambda_j + \lambda \delta_i(j)) \right.$$

$$+ \left(\sum_{\beta \neq j} (1 + 2(l_\beta - l_j)) \right) \mu(q^{(i)} + l_j) \right\}$$

$$= \sum_{j=1}^{N} \left\{ (N-1)[\mu(q^{(i)} + l_j) + \lambda_j + \lambda \delta_i(j)] \right.$$

$$+ 2 \sum_{\beta \neq j} (l_\beta - l_j)[\mu(q^{(i)} + l_j) - \lambda_j - \lambda \delta_i(j)] \right\}$$

$$\leq v_0 + 2 \sum_{j=1}^{N} \sum_{\beta \neq j} (l_\beta - l_j)[\mu(q^{(i)} + l_j) - \lambda_j - \lambda \delta_i(j)],$$

where $v_0 \triangleq (N-1)[N\mu + \sum_{j=1}^{N} \lambda_j + \lambda] \leq (N-1)$. For the second term, write

$$v_1 \triangleq \sum_{j=1}^{N} \sum_{\beta \neq j} (l_\beta - l_j)[\mu(q^{(i)} + l_j) - \lambda_j - \lambda \delta_i(j)].$$

Break this into two terms. In the first, change the order of summation so that

$$v_1 = \sum_{\beta=1}^{N} l_\beta \sum_{j \neq \beta} [\mu(q^{(i)} + l_j) - \lambda_j - \lambda \delta_i(j)] - \sum_{j=1}^{N} l_j \sum_{\beta \neq j} [\mu(q^{(i)} + l_j) - \lambda_j - \lambda \delta_i(j)].$$

Using the fact that $l_i = 0$, that $\mu(\cdot) \leq \mu$, and that the minimum value of $\sum_{j \neq \beta} \lambda_j$ occurs when $\beta = N$,

$$v_1 \leq \sum_{\beta \neq i} l_\beta \sum_{j \neq \beta} [\mu - \lambda_j - \lambda \delta_i(j)] - \sum_{j \neq i} l_j [\mu(q^{(i)} + l_j) - \lambda_j - \lambda \delta_i(j)](N-1)$$

$$= \sum_{\beta \neq i} l_\beta \left[(N-1)\mu - \lambda - \sum_{j \neq \beta} \lambda_j \right] - \sum_{j \neq i} l_j [\mu(q^{(i)} + l_j) - \lambda_j](N-1)$$

$$\leq \sum_{\beta \neq i} l_\beta \left[(N-1)\mu - \lambda - \sum_{k=1}^{N-1} \lambda_k \right] - \sum_{j \neq i} l_j [\mu(q^{(i)} + l_j) - \lambda_j](N-1)$$

$$= \sum_{j \neq i} l_j \left[(N-1)\mu - \lambda - \sum_{k=1}^{N-1} \lambda_k \right] - \sum_{j \neq i} l_j [\mu - \lambda_j](N-1),$$

where the last step follows by observing that $l_j \mu(q^{(i)} + l_j) = l_j \mu$ for all j. Continuing, we see that

$$v_1 \leq \left[(N-1)\lambda_N - \sum_{k=1}^{N-1} \lambda_k - \lambda \right] \sum_{j \neq i} l_j$$

(3.11)

$$= - \left[\lambda - \sum_{k=1}^{N-1} (\lambda_N - \lambda_k) \right] \sum_{j \neq i} l_j.$$

Now, observe that according to (3.9), if we set $j = i$ and use the fact that $l_i = 0$, then

$$u(q) = \sum_{\substack{\beta \neq i}} l_\beta^2 + \sum_{\substack{\alpha=1 \\ \alpha \neq i}}^{N-1} \sum_{\substack{\beta=\alpha+1 \\ \beta \neq i}}^{N} |l_\beta - l_\alpha|^2$$

$$\leq \sum_{\substack{\beta \neq i}} l_\beta^2 + \sum_{\substack{\alpha=1 \\ \alpha \neq i}}^{N-1} \sum_{\substack{\beta=\alpha+1 \\ \beta \neq i}}^{N} (l_\beta^2 + l_\alpha^2)$$

$$\leq (2N - 1) \sum_{\substack{\beta \neq i}} l_\beta^2,$$

and so

(3.12)
$$\sqrt{\frac{u(q)}{2N - 1}} \leq \sum_{\beta \neq i} l_\beta.$$

If (3.3) holds, substituting (3.12) into (3.11) now yields

(3.13)
$$v_1 \leq - \left(\lambda - \sum_{k=1}^{N-1} (\lambda_N - \lambda_k)\right) \sqrt{\frac{u(q)}{2N - 1}}.$$

Finally, we have

$$\mathbb{E}[u(Q_{n+1}) - u(q) \mid Q_n = q] \leq v_0 + 2v_1$$

$$\leq (N - 1) - 2\left(\lambda - \sum_{k=1}^{N-1} (\lambda_N - \lambda_k)\right) \sqrt{\frac{u(q)}{2N - 1}},$$

which is exactly (3.6). This completes the proof of Theorem 3.

We now briefly discuss the continuous-time analog of this model. Suppose that $\{Q_n\}$ is the jump chain corresponding to a continuous-time queuing process $\{\hat{Q}_t\}$ with instantaneous arrival rates $\hat{\lambda}, \hat{\lambda}_1, \cdots, \hat{\lambda}_N$, and instantaneous departure rate $\hat{\mu}$ for each of the N servers. If we let $d(q) \triangleq \hat{\lambda} + \sum_{j=1}^{N} [\hat{\lambda}_j + \hat{\mu}(q^{(j)})]$, then the jump chain parameters are

(3.14)
$$\lambda = \hat{\lambda}/d(q),$$
$$\lambda_j = \hat{\lambda}_j/d(q),$$
$$\mu(q^{(j)}) = \hat{\mu}(q^{(j)})/d(q).$$

Clearly, the transition probabilities given that $Q_n = q$ now depend on q in a slightly different way. However, assuming that $\hat{\lambda}_1 \leq \cdots \leq \hat{\lambda}_N$, a review of the proof of Theorem 3 through (3.11) when (3.14) is in force shows that the crucial question is whether or not

$$\frac{\hat{\lambda}}{d(q)} > \frac{1}{d(q)} \sum_{k=1}^{N-1} (\hat{\lambda}_N - \hat{\lambda}_k).$$

This condition obviously holds if and only if

(3.15)
$$\hat{\lambda} > \sum_{k=1}^{N-1} (\hat{\lambda}_N - \hat{\lambda}_k).$$

If (3.15) holds, we can express (3.13) in terms of the infinitesimal rates:

$$v_1 \leqq - \frac{\hat{\lambda} - \sum\limits_{k=1}^{N-1} (\hat{\lambda}_N - \hat{\lambda}_k)}{d(q)} \sqrt{\frac{u(q)}{2N-1}}.$$

Since $\max_q d(q) = \hat{\lambda} + \sum_{j=1}^{N} \hat{\lambda}_j + N\hat{\mu}$,

$$v_1 \leqq - \frac{\hat{\lambda} - \sum\limits_{k=1}^{N-1} (\hat{\lambda}_N - \hat{\lambda}_k)}{\hat{\lambda} + \sum\limits_{j=1}^{N} \hat{\lambda}_j + N\hat{\mu}} \sqrt{\frac{u(q)}{2N-1}}.$$

Thus, for jump chains, (3.6) should be replaced by

$$\mathbb{E}[u(Q_{n+1}) - u(q) \mid Q_n = q] \leqq (N-1) - 2 \frac{\hat{\lambda} - \sum\limits_{k=1}^{N-1} (\hat{\lambda}_N - \hat{\lambda}_k)}{\hat{\lambda} + \sum\limits_{j=1}^{N} \hat{\lambda}_j + N\hat{\mu}} \sqrt{\frac{u(q)}{2N-1}},$$

and (3.3) should be replaced by (3.15).

4. Proof of Theorem 2

Before proving Theorem 2, we introduce the following notation. Let

(4.1)
$$s \triangleq \frac{v}{b+1},$$

and set

$$B \triangleq \frac{A^{b+2}}{s^{b+1}}$$

so that we can use (2.5) to write

(4.2)
$$\frac{1}{s^{b+1}} \int_{\mathsf{X}} u(y)^{b+2} p(x, dy) \leqq B, \quad \text{if } 0 \leqq u(x) \leqq L.$$

Now, for $v \geqq 0$, let $\varphi_0(v) \equiv v$, and for $n \geqq 1$ set

(4.3)
$$\varphi_n(v) \triangleq \frac{v^{b+2}}{[v + sn]^{b+1}}.$$

With this notation we can write (recall (2.6))

(4.4)
$$u_0(x) = \varphi_0(u(x)),$$
$$u_1(x) = \varphi_1(u(x)) + A,$$
$$u_n(x) = \varphi_n(u(x)) + A + B \sum_{k=1}^{n-1} \frac{1}{k^{b+1}}, \quad n \geq 2.$$

For later reference, note that differentiating φ_n with respect to v yields

(4.5)
$$\varphi_n'(v) = \frac{v^{b+2} + sn(b+2)v^{b+1}}{[v+sn]^{b+2}},$$

and

(4.6)
$$\varphi_n''(v) = \frac{(b+1)(b+2)s^2n^2v^b}{[v+sn]^{b+3}}.$$

Proof of Theorem 2. Fix $x \in \mathbb{X}$. We first show that (1.4) holds if $0 \leq u(x) \leq L$. Then we show that (1.4) holds if $u(x) > L$. Suppose $0 \leq u(x) \leq L$ and $n = 0$. Use Jensen's inequality and (2.5) to get

$$\int_{\mathbb{X}} u_0(y)p(x, dy) - u_1(x) = \int_{\mathbb{X}} u(y)p(x, dy) - \varphi_1(x) - A$$
$$\leq \left(\int_{\mathbb{X}} u(y)^{b+2}p(x, dy)\right)^{1/(b+2)} - \varphi_1(x) - A$$
$$\leq A - \varphi_1(x) - A \leq 0.$$

Now for $n \geq 1$, keep in mind (4.2) while using (4.4) to write

$$\int_{\mathbb{X}} u_n(y)p(x, dy) - u_{n+1}(x) = \int_{\mathbb{X}} \varphi_n(u(y))p(x, dy) - \varphi_{n+1}(u(x)) - \frac{B}{n^{b+1}}$$
$$= \int_{\mathbb{X}} \frac{u(y)^{b+2}}{[u(y)+sn]^{b+1}}p(x, dy) - \varphi_{n+1}(u(x)) - \frac{B}{n^{b+1}}$$
$$\leq \frac{1}{(sn)^{b+1}} \int_{\mathbb{X}} u(y)^{b+2}p(x, dy) - \varphi_{n+1}(u(x)) - \frac{B}{n^{b+1}}$$
$$\leq \frac{B}{n^{b+1}} - \varphi_{n+1}(u(x)) - \frac{B}{n^{b+1}}$$
$$\leq 0.$$

We next establish (1.4) when $u(x) > L$ and $n \geq 1$. The case $n = 0$ follows the same pattern (without Taylor's theorem, since $\varphi_0(v) \equiv v$) and is left to the reader. Using Taylor's theorem, we have

$$\int_{\mathbb{X}} u_n(y)p(x, dy) = u_n(x) + \int_{\mathbb{X}} [\varphi_n(u(y)) + \varphi_n(u(x))]p(x, dy)$$

(4.7)
$$\leq u_n(x) + \int_{\mathbb{X}} [\varphi_n'(u(x))(u(y) - u(x)) + K(u(y) - u(x))^2]p(x, dy),$$

where K is the upper bound on $\frac{1}{2}\varphi_n''$ given by

$$K \triangleq \frac{(b+1)(b+2)}{2}\frac{(1+\varepsilon)^b}{(1-\varepsilon)^{b+3}}\frac{u(x)^b s^2 n^2}{[u(x)+sn]^{b+3}}.$$

To verify that K is an upper bound, observe that (2.2) implies that $p(x, \cdot)$-a.s.,

$$u(x)(1-\varepsilon) \leq u(y) \leq u(x)(1+\varepsilon).$$

Now, using (4.6),

$$\sup_{u(x)(1-\varepsilon) \leq v \leq u(x)(1+\varepsilon)} \frac{1}{2}\varphi_n''(v) \leq \frac{(b+1)(b+2)s^2 n^2 u(x)^b (1+\varepsilon)^b}{2[u(x)(1-\varepsilon)+sn]^{b+3}}$$

$$\leq K,$$

since $[u(x)(1-\varepsilon)+sn] \geq (1-\varepsilon)[u(x)+sn]$. Returning to (4.7), we use (2.3) and (2.4) to see that

$$\int_X u_n(y)p(x, dy) \leq u_n(x) + \varphi_n'(u(x))(-v) + KCu(x).$$

Using the definition of K and (2.1) together with (4.5),

$$\int_X u_n(y)p(x, dy) \leq u_n(x) - v\frac{u(x)^{b+2} + sn(b+2)u(x)^{b+1}}{[u(x)+sn]^{b+2}} + \frac{2vu(x)^{b+1}s^2 n^2}{[u(x)+sn]^{b+3}}$$

$$= u_n(x) - \frac{vu(x)^{b+2}}{[u(x)+sn]^{b+2}} - \frac{vsnu(x)^{b+1}}{[u(x)+sn]^{b+3}}[(b+2)u(x)+bsn]$$

(4.8)

$$\leq u_n(x) - \frac{vu(x)^{b+2}}{[u(x)+sn]^{b+2}}$$

$$= u_{n+1}(x) + u_n(x) - u_{n+1}(x) - \frac{vu(x)^{b+2}}{[u(x)+sn]^{b+2}}.$$

Observe that

$$u_n(x) - u_{n+1}(x) = \varphi_n(u(x)) - \varphi_{n+1}(u(x)) - \frac{B}{n^{b+1}}$$

$$\leq \varphi_n(u(x)) - \varphi_{n+1}(u(x))$$

(4.9)

$$= u(x)^{b+2}\left(-\int_n^{n+1} \frac{\partial}{\partial\theta}\left(\frac{1}{[u(x)+s\theta]^{b+1}}\right)d\theta\right)$$

$$= u(x)^{b+2}(b+1)s\int_n^{n+1} \frac{d\theta}{[u(x)+s\theta]^{b+2}}$$

$$\leq \frac{vu(x)^{b+2}}{[u(x)+sn]^{b+2}},$$

where the last step follows by setting $\theta = n$ in the integrand and then using (4.1). Combining (4.8) and (4.9) yields (1.4) when $u(x) > L$.

5. Extension of Theorem 2

It is possible to weaken the hypotheses of Theorem 2 by replacing (2.2)–(2.4) with (5.1)–(5.4) below.

Suppose that for each $x \in \mathbb{X}$, there is a subset, B_x, with

(5.1) $$B_x \subset \{y \in \mathbb{X} : u(y) \leq u(x)\},$$

and such that

(5.2) $\quad I_{B_x^c}(y) |u(y) - u(x)| \leq \varepsilon u(x), \qquad p(x, \cdot)\text{-a.s.,} \quad \text{if} \quad u(x) > L,$

(5.3) $\quad \mathbb{E}[I_{B_x^c}(X_{n+1})(u(X_{n+1}) - u(x)) \mid X_n = x] \leq -v, \quad \text{if} \quad u(x) > L,$

and

(5.4) $\quad \mathbb{E}[I_{B_x^c}(X_{n+1}) |u(X_{n+1}) - u(x)|^2 \mid X_n = x] \leq Cu(x), \quad \text{if} \quad u(x) > L.$

Here, B_x^c denotes the complement of B_x and $I_{B_x^c}$ denotes the indicator function of the set B_x^c.

To prove that we can weaken the hypotheses as claimed, we proceed as follows. Observe that (4.5) implies $\varphi_n' \geq 0$, and so φ_n is a non-decreasing function on $[0, \infty)$. Hence, (5.1) implies

$$\int_{B_x} [\varphi_n(u(y)) - \varphi_n(u(x))] p(x, dy) \leq 0.$$

From this and Taylor's theorem,

$$\int_{\mathbb{X}} u_n(y) p(x, dy) = u_n(x) + \int_{\mathbb{X}} [\varphi_n(u(y)) - \varphi_n(u(x))] p(x, dy)$$

(5.5) $$\leq u_n(x) + \int_{B_x^c} [\varphi_n(u(y)) - \varphi_n(u(x))] p(x, dy)$$

$$\leq u_n(x) + \int_{B_x^c} [\varphi_n'(u(x))(u(y) - u(x)) + K(u(y) - u(x))^2] p(x, dy).$$

If we replace (4.7) with (5.5), then the derivation given in Section 4 will prove the modification of Theorem 2 stated above.

6. Relation to other work

In this section, we consider systems to which we can apply both Theorem 2 and the results of Hajek (1982). We point out that Theorem 2 applies only to non-negative functions of time-homogeneous Markov processes while Hajek's results apply to real-valued functions of non-Markov processes. The purpose of this section is to show that when considering a non-negative function of a time-homogeneous Markov process such that (2.2) holds, whenever Hajek's results apply, so does Theorem 2. Further, if the parameter ρ below is greater than or

equal to 1, then Hajek's results do not yield useful asymptotic information, while
Theorem 2 still does.

To apply Hajek's results to $\{u(X_n)\}$ requires that the following hold.
There exist positive constants v, η, ρ, and δ such that for some $L < \infty$,

$$(6.1) \qquad E[u(X_{n+1}) - u(x) \mid X_n = x] \leq -v, \quad \text{if} \quad u(x) > L,$$

$$(6.2) \qquad E[\exp(\eta(u(X_{n+1}) - u(x))) \mid X_n = x] \leq \rho, \quad \text{if} \quad u(x) > L,$$

and

$$(6.3) \qquad E[\exp(\eta(u(X_{n+1}) - L)) \mid x_n = x] \leq \delta, \quad \text{if} \quad u(x) \leq L.$$

Remarks. (i) If $\rho < 1$, applying Jensen's inequality with the natural logarithm
function to (6.2) yields (6.1) with $-v = (1/\eta) \ln \rho$. (ii) Observe that if (6.1) and
(6.2) hold for some L, they also hold for any $L' \geq L$. As for (6.3), rewrite (6.2) and
(6.3) as

$$E[\exp(\eta u(X_{n+1})) \mid X_n = x] \leq \rho e^{\eta u(x)}, \quad \text{if} \quad u(x) > L,$$

and

$$E[\exp(\eta u(X_{n+1})) \mid X_n = x] \leq \delta e^{\eta L}, \quad \text{if} \quad u(x) \leq L.$$

For any $L' \geq L$,

$$E[\exp(\eta u(X_{n+1})) \mid X_n = x] \leq \max\{\delta e^{\eta L}, \rho e^{L'}\}, \quad \text{if} \quad u(x) \leq L',$$
$$\leq \max\{\delta, \rho\} e^{\eta L'}, \quad \text{if} \quad u(x) \leq L'.$$

It now follows that if (6.1)–(6.3) hold for some L, they hold with L replaced by any
$L' \geq L$ and δ replaced by $\max\{\delta, \rho\}$.

Hajek proves by a simple induction argument that (6.2) and (6.3) imply

$$(6.4) \qquad E[\exp(\eta u(X_n)) \mid X_0 = x] \leq \rho^n e^{\eta u(x)} + \frac{1 - \rho^n}{1 - \rho} \delta e^{\eta L}.$$

To relate this to $E[u(X_n) \mid X_0 = x]$, apply Jensen's inequality to (6.4) to get

$$(6.5) \qquad E[u(X_n) \mid X_0 = x] \leq \frac{1}{\eta} \ln\left[\rho^n e^{\eta u(x)} + \frac{1 - \rho^n}{1 - \rho} \delta e^{\eta L}\right],$$

which yields both

$$(6.6) \qquad E[u(X_n) \mid X_0 = x] \leq \frac{1}{\eta} \ln\left[e^{\eta u(x)} + \frac{1}{1 - \rho} \delta e^{\eta L}\right], \quad \text{if} \quad \rho < 1,$$

and

$$(6.7) \qquad \limsup_{n \to \infty} E[u(X_n) \mid X_0 = x] \leq \frac{1}{\eta} \ln\left(\frac{\delta}{1 - \rho}\right) + L, \quad \text{if} \quad \rho < 1.$$

Under the hypotheses of Theorem 2 (with $b = 1$ and $0 < \varepsilon < 1$),

(6.8) $$\mathbb{E}[u(X_n) \mid X_0 = x] \leq \frac{u(x)^3}{[u(x) + (v/2)n]^2} + A + 4\frac{A^3}{v^2} \sum_{k=1}^{n-1} \frac{1}{k^2},$$

(6.9) $$\mathbb{E}[u(X_n) \mid X_0 = x] \leq u(x) + A\left(1 + \frac{2}{3}\left(\frac{\pi A}{v}\right)^2\right),$$

and

(6.10) $$\limsup_{n \to \infty} \mathbb{E}[u(X_n) \mid X_0 = x] \leq A\left(1 + \frac{2}{3}\left(\frac{\pi A}{v}\right)^2\right) < \infty.$$

Clearly, both approaches yield very similar results. We now compare the respective assumptions which each approach requires. First, (6.1) is identical to (2.3). Second, we show below that (6.3) implies the existence of a *finite* constant A satisfying (2.5). Third, we show below that if (2.2) holds, then (6.2) implies (2.4) with a constant C satisfying (2.1). It follows that if there exist positive constants v, η, ρ, δ, and $0 < \varepsilon < 1$ such that for some $L < \infty$, (6.1), (6.2), and (6.3) hold, and if we have $p(x, \cdot)$-a.s.,

$$|u(y) - u(x)| \leq \varepsilon u(x), \quad \text{if} \quad u(x) > L,$$

then both (6.5) and (6.8) hold. However, if $\rho \geq 1$, the uniform bounds (6.6) and (6.7) are not available, while the uniform bounds (6.9) and (6.10) still hold.

The remainder of this section is devoted to establishing the preceding claims. Suppose that (6.3) holds. Rewrite it as

$$\int_X e^{\eta u(y)} p(x, dy) \leq \delta e^{\eta L}, \quad \text{if} \quad u(x) \leq L.$$

Now, for any *integer* $b \geq 0$, since $\eta u(y) \geq 0$ implies

$$\frac{[\eta u(y)]^{b+2}}{(b+2)!} \leq e^{\eta u(y)},$$

we have

$$\int_X u(y)^{b+2} p(x, dy) \leq \frac{(b+2)!}{\eta^{b+2}} \delta e^{\eta L}, \quad \text{if} \quad u(x) \leq L,$$

providing a finite constant A satisfying (2.5).

Suppose that (6.2) and (2.2) hold. Rewrite (2.2) as

$$(1 - \varepsilon)u(x) \leq u(X_{n+1}) \leq (1 + \varepsilon)u(x).$$

Taylor's theorem implies

$$\exp\left(\eta(u(X_{n+1}) - u(x))\right) = 1 + \eta[u(X_{n+1}) - u(x)] + \tfrac{1}{2}\eta^2 e^{\eta v^*}(u(X_{n+1}) - u(x))^2,$$

where $(1 - \varepsilon)u(x) \leqq v^* \leqq (1 + \varepsilon)u(x)$. Thus,

$$\exp\left(\eta(u(X_{n+1}) - u(x))\right) \geqq 1 - \eta\varepsilon u(x) + \tfrac{1}{2}\eta^2 \exp\left(\eta(1 - \varepsilon)u(x)\right)(u(X_{n+1}) - u(x))^2,$$

and then (6.2) implies

$$\mathbb{E}[|u(X_{n+1}) - u(x)|^2 \mid X_n = x] \leqq [\eta\varepsilon u(x) - 1 + \rho]\frac{2}{\eta^2}\exp\left(-\eta(1 - \varepsilon)u(x)\right).$$

Now, setting

$$a(v) \triangleq [\eta\varepsilon v - (1 - \rho)]\frac{2}{\eta^2}\exp\left(-\eta(1 - \varepsilon)v\right),$$

we have $\lim_{v\to\infty} a(v)/v = 0$. If we choose any positive C satisfying (2.1) we can then take L large enough so that whenever $v > L$, $a(v) \leqq Cv$.

Acknowledgement

We thank Professor A. M. Makowski for his careful reading of an earlier version of this paper and his suggestions for improving its presentation. We thank R. B. Sowers for his suggestions for improving the revised version of this paper.

References

BILLINGSLEY, P. (1979) *Probability and Measure*. Wiley, New York.

BLANC, J. P. C. (1987) A note on waiting times in systems with queues in parallel. *J. Appl. Prob.* **24**, 540–546.

BRUMELLE, S. L. (1971) Some inequalities for parallel-server queues. *Operat. Res.* **19**, 402–413.

CONOLLY, B. W. (1984) The autostrada queueing problem. *J. Appl. Prob.* **21**, 394–403.

EPHREMIDES, A., VARAIYA, P., AND WALRAND, J. (1980) A simple dynamic routing problem. *IEEE Trans. Autom. Contr.* **25**, 690–693.

FLATTO, L. AND MCKEAN, H. P. (1977) Two queues in parallel. *Comm. Pure Appl. Math.* **30**, 255–263.

FOSCHINI, G. J. AND SALZ, J. (1978) A basic dynamic routing problem and diffusion. *IEEE Trans. Comm.* **26**, 320–327.

HAIGHT, F. A. (1958) Two queues in parallel. *Biometrika* **45**, 401–410.

HAJEK, B. (1982) Hitting-time and occupation-time bounds implied by drift analysis with applications. *Adv. Appl. Prob.* **14**, 502–525.

HALFIN, S. (1985) The shortest queue problem. *J. Appl. Prob.* **22**, 865–878.

KINGMAN, J. F. C. (1961) Two similar queues in parallel. *Ann. Math. Statist.* **32**, 1314–1323.

PAPADIMITRIOU, I. (1973) A simple proof of the formula $\sum_{k=1}^{\infty} k^{-2} = \pi^2/6$. *Amer. Math. Monthly* **80**, 424–425.

STARK, E. L. (1974) The series $\sum_{k=1}^{\infty} k^{-s}$, $s = 2, 3, 4, \cdots$, once more. *Mathematics Mag.* **47**, 197–202.

WEBER, R. R. (1978) On the optimal assignment of customers to parallel servers. *J. Appl. Prob.* **15**, 406–413.

WINSTON, W. (1977) Optimality of the shortest line discipline. *J. Appl. Prob.* **14**, 181–189.

WOLFF, R. W. (1977) An upper bound for multi-channel queues. *J. Appl. Prob.* **14**, 884–888.

WOLFF, R. W. (1987) Upper bounds on work in system for multichannel queues. *J. Appl. Prob.* **24**, 547–551.

Rev. Math. Phys. 6 (1994) 621-648

FINITE APPROXIMATIONS TO QUANTUM SYSTEMS

TROND DIGERNES*

Department of Mathematics
University of Trondheim
7034 Trondheim-NTH
Norway

V. S. VARADARAJAN

Department of Mathematics
University of California at Los Angeles
Los Angeles, CA 90024
USA

S. R. S. VARADHAN**

Courant Institute of Mathematical Sciences
New York University
New York, N. Y. 10012
USA

Received 8 November 1993

1. Introduction

1.1. In his studies on the foundations of quantum mechanics in the 1960's Julian Schwinger had considered quantum systems in finite dimensional spaces as approximations to quantum systems in infinite dimensional spaces. For example, to study the motion of a single particle in the one dimensional space **R**, Schwinger's idea [1], which is already present in Weyl [2] (pp. 277–280), consists in replacing the configuration space **R** by a large finite grid $X(N) \subset \mathbf{R}$:

$$X(N) = \{ r\varepsilon_N \mid r = 0, \pm 1, \pm 2, \ldots, \pm N^0 \} \qquad \varepsilon_N = \left(\frac{2\pi}{N} \right)^{1/2}, \qquad N = 2N^0 + 1 .$$

The Hilbert space of the system is $L^2(X(N))$; the position operator q_N is multiplication by the coordinate; the momentum operator p_N is the Fourier transform of q_N, the Fourier transform being the one defined for the finite group $\mathbf{Z}(N) = \mathbf{Z}\varepsilon_N / N \cdot \mathbf{Z}\varepsilon_N$ with which $X(N)$ can be identified. Schwinger's idea is to regard the system with the Hamiltonian $H = \frac{1}{2}(p^2 + V(q))$ on $L^2(\mathbf{R})$ as approximated for large N by the system on $L^2(X(N))$ with Hamiltonian $H_N^{(s)} = \frac{1}{2}(p_N^2 + V_N)$ $(V_N = V(q_N))$.

*Supported by the Research Council of Norway. The first two authors would like to express their gratitude to the departments of Mathematics of the University of Trondheim, Norway, and the University of California at Los Angeles, Ca, USA, for their hospitality.
**Supported by NSF grant DMS–920122 and ARO grant DAAL 03–92–G–0317.

Schwinger did not analyse this approximation in detail. For the harmonic oscillator where $H = \frac{1}{2}(p^2 + q^2)$ and the spectrum of H simple, consisting of

$$\frac{1}{2}, \frac{3}{2}, \frac{5}{2}, \cdots$$

a computation with *Mathematica* showed that even for relatively small values of N, the eigenvalues and eigenfunctions of $H_N^{(s)}$ were remarkably close to those of H: for $N = 41$, the first 10 eigenvalues of $H_N^{(s)}$ agree with those of H up to 12 decimal places, and the graphs of the eigenfunctions could hardly be differentiated. This suggested that one should try to prove that for a large class of Hamiltonians, the spectral data of $H_N^{(s)}$ converge in a very strong fashion to the spectral data of H.

In this paper we shall treat this problem in arbitrary dimension d for Hamiltonians $H = -\Delta + V$ where Δ is the Laplacian in $\mathbf{R}^d$ and V is a potential which is continuous, ≥ 0, and $V(x) \longrightarrow \infty$ as $|x| \longrightarrow \infty (x \in \mathbf{R}^d)$. The resolvent operators are then compact so that H has a pure discrete spectrum $0 \leq h_1 < h_2 < \ldots, h_n, \ldots, h_n \longrightarrow \infty$, with finite multiplicities. We then prove, using analytic arguments in Sec. 2, that the eigenvalues and eigenfunctions of H_N converge to those of H; here $H_N = -\Delta_N + V_N$ is defined as above using finite grids in the lattice $(\mathbf{Z}\varepsilon_N)^d$, and Δ_N is either a discrete Laplacian (i.e., a second difference operator) with suitable modifications at the boundary of the grid, or the Schwinger Laplacian associated to the grid by finite Fourier transform. In Secs. 3 through 6 we use path integral methods based on the Feynman–Kac formula to prove that if the potential V blows up at infinity fast enough so that the operators $e^{-tH}(t > 0)$ are of trace class, then, for each $t > 0$ the operator e^{-tH_N} converges to the operator e^{-tH} in the very strong trace norm.

Both the analytic and probabilistic techniques used in this paper are well-known to specialists so that the paper as a semi-expository character. The analytic problem is one of studying the stability and convergence of difference schemes for solving eigenvalue problems. Kreiss [3] has treated this question for ODE's ($d = 1$) on a finite interval. The present case, when the potential becomes infinite at infinite distances, does not go very much beyond what is done in [3] since the large size of the potential at infinity introduces a natural cutoff. The Schwinger approximation is however not a true difference scheme; but it has good comparison properties with periodic difference schemes, and so can be handled. Nevertheless it is in some sense the best approximation to the infinite system. It is desirable to explore this aspect more.

In the probabilistic approach $\frac{1}{2}\Delta$ is the infinitesimal generator of Brownian motion in $\mathbf{R}^d$ while $\frac{1}{2}\Delta_N$ and its variants are infinitesimal generators of random walks on the lattice $(\mathbf{Z}\varepsilon_N)^d$ and its finite grids. In the continuum limit when the inter-lattice distance ε_N goes to 0, the random walks will converge to Brownian motion in the sense of measures in function space. The trace convergence of the dynamical operators will then follow from this on applying the Feynman–Kac formula. As before the Schwinger Hamiltonians are handled by comparison with the periodic ones.

1.2. Notation. For any real nonnegative continuous function V on $\mathbf{R}^d$ the operator $H = -\Delta + V$ is essentially self-adjoint on $C_c^\infty(\mathbf{R}^d)$, hence on $C_c^\infty(\mathbf{R}) \otimes \cdots \otimes C_c^\infty(\mathbf{R})$ ([4], Theorem X. 28, pp. 183–184). Suppose now that $V(x) \longrightarrow \infty$ as $|x| \longrightarrow \infty$. The formula

$$(Hf, f) = \sum_{1 \leq i \leq d} \|\partial f / \partial x_i\|^2 + \|V^{1/2} f\|^2 \qquad (f \in C_c^\infty(\mathbf{R}^d))$$

shows that $H \geq 0$ and moreover yields the compactness of the resolvent $(I+H)^{-1}$, as is classically known. Indeed, if $(I+H)f = g$ where $f \in C_c^\infty(\mathbf{R}^d)$ and $\|g\| \leq 1$, then $\|f\| \leq 1$ and so $((I+H)f, f) \leq 1$, giving $\|\partial f / \partial x_i\| \leq 1 (1 \leq i \leq d), \|V^{1/2}f\| \leq 1$. The first estimate implies, in view of the Rellich imbedding [5], that f varies in a compact set in each $L^2(K)$, K compact in $\mathbf{R}^d$, while the second estimate shows that

$$\int_{|x|>a} |f|^2 dx \leq \left(\inf_{|x|>a} V(x) \right)^{-1} \longrightarrow 0 \qquad (a \longrightarrow \infty)$$

thus proving that f varies in a compact set in $L^2(\mathbf{R}^d)$. The analytical proof of the main approximation theorem given in Sec. 2 is based on the fact that the above argument remains uniform when ε goes to 0.

For any $\varepsilon > 0$, let $G(\varepsilon)$ denote the lattice $(\mathbf{Z}\varepsilon)^d$. For $h \in G(\varepsilon)$ the translation operator T_h acts on functions on $G(\varepsilon)$ by $(T_h f)(x) = f(x + h)$. If $e_i (1 \leq i \leq d)$ are the standard basis vectors in $\mathbf{R}^d$, the partial difference operators $D_i(\varepsilon)^{\pm}$ are defined by

$$D_i(\varepsilon)^+ = \varepsilon^{-1}(T_{\varepsilon e_i} - I), \qquad D_i(\varepsilon)^- = \varepsilon^{-1}(I - T_{-\varepsilon e_i}) \quad (1 \leq i \leq d)$$

and the discrete Laplacian $\Delta(\varepsilon)$ by

$$\Delta(\varepsilon) = \sum_{1 \leq i \leq d} D_i(\varepsilon)^- D_i(\varepsilon)^+.$$

On $L^2(G(\varepsilon))$ the T_h are unitary, $D_i(\varepsilon)^- = -(D_i(\varepsilon)^+)^t$ and $\Delta(\varepsilon)$ are bounded, and

$$(-\Delta(\varepsilon)f, f) = \sum_{1 \leq i \leq d} \|D_i(\varepsilon)^+ f\|^2 \qquad (f \in L^2(G(\varepsilon))) .$$

The Hamiltonian $H(\varepsilon)$ on $L^2(G(\varepsilon))$ is given by $H(\varepsilon) = -\Delta(\varepsilon) + V_\varepsilon$, the operator V_ε acting as multiplication by the restriction of V to $G(\varepsilon)$. It is clear that $H(\varepsilon) \geq 0$ and $(I + H(\varepsilon))^{-1}$ is compact.

Let $N^0 = N^0(\varepsilon)$ be an integer ≥ 1 depending on ε; $\varepsilon N^0 \longrightarrow \infty$ as $\varepsilon \longrightarrow 0$; $N = N(\varepsilon) = 2N^0 + 1$. $G(\varepsilon)^0$ is the grid $X(\varepsilon)^d$ in $G(\varepsilon)$ where $X(\varepsilon)$ is the set $\{r\varepsilon \mid r = 0, \pm 1, \dots, \pm N^0\}$. There are several variants of $\Delta(\varepsilon)$ depending on the treatment of boundary terms.

The periodic case: $H(\varepsilon)^{(p)} = -\Delta(\varepsilon)^{(p)} + V_\varepsilon$. We identify $G(\varepsilon)^0$ with the finite group $G(\varepsilon)/N(\varepsilon) \cdot G(\varepsilon)$ and hence $L^2(G(\varepsilon)^0)$ with the space of functions on $G(\varepsilon)$

invariant under $T_h, h \in N(\varepsilon) \cdot G(\varepsilon)$; $D_i(\varepsilon)^{\pm(p)}$ are the restrictions of $D_i(\varepsilon)^{\pm}$ to this space; and

$$\Delta(\varepsilon)^{(p)} = \sum_{1 \leq i \leq d} D_i(\varepsilon)^{-(p)} D_i(\varepsilon)^{+(p)}, \quad (-\Delta(\varepsilon)^{(p)} f, f) = \sum_{1 \leq i \leq d} \|D_i(\varepsilon)^{+(p)} f\|^2.$$

The Hamiltonians $H(\varepsilon)^{(0)} = -\Delta(\varepsilon)^{(0)} + V_\varepsilon$. We denote by $\Delta(\varepsilon)^{(0)}$ any member of a class of operators obtained from $\Delta(\varepsilon)$ by boundary modifications. The simplest of these is defined as $\Delta(\varepsilon)^0 = P_\varepsilon \Delta(\varepsilon) P_\varepsilon$ where P_ε is multiplication by the characteristic function of $G(\varepsilon)^0$ and we identify $L^2(G(\varepsilon)^0)$ as a subspace of $L^2(G(\varepsilon))$ in the obvious way. More generally, $\Delta(\varepsilon)^{(0)}$ is defined so that it satisfies the following conditions:

(a) $\Delta(\varepsilon)^{(0)}$ is symmetric, real, and ≥ 0.

(b) For any $b > 0$ let $C(b)$ be the cube of all $x = (x_1, \ldots, x_d) \in \mathbf{R}^d$ with $|x_i| < b$ for all i; if ε is small enough, $C(b) \cap G(\varepsilon) \subset G(\varepsilon)^0$. We then have, for all $f \in L^2(G(\varepsilon)^0)$ with support contained in $C(b) \cap G(\varepsilon)$,

$$\Delta(\varepsilon)^{(0)} f = \Delta(\varepsilon) f .$$

This will be true if for a suitable $N' = N'(\varepsilon) < N^0, N'/N^0 \longrightarrow 1$ as $\varepsilon \longrightarrow 0$, the matrix elements of $\Delta(\varepsilon)^{(0)}$ and $\Delta(\varepsilon)$ (in the delta function basis) satisfy

$$\Delta(\varepsilon)^{(0)}(r, s) = \Delta(\varepsilon)(r, s) \qquad (|r|, |s| \leq N^0, \min(|r|, |s|) \leq N') .$$

(c) For any $b > 0$ and all sufficiently small ε, with the suffix indicating that the norm is computed over $C(b) \cap G(\varepsilon)$,

$$(-\Delta(\varepsilon)^{(0)} f, f) \geq \sum_{1 \leq i \leq d} \|D_i(\varepsilon)^+ f\|_b^2 .$$

It is clear that $\Delta(\varepsilon)^0 = P_\varepsilon \Delta(\varepsilon) P_\varepsilon$ satisfies (a)–(c). To get another example let $d = 1$ and define D_ε^+ by

$$(D_\varepsilon^+ f)(k\varepsilon) = \varepsilon^{-1}[f((k+1)\varepsilon) - f(k\varepsilon)] \, (-N^0 \leq k < N^0), \, (D_\varepsilon^+ f)(N^0\varepsilon) = 0 .$$

We then take $-\Delta(1 : \varepsilon) = (D_\varepsilon^+)^\dagger D_\varepsilon^+$. For $d \geq 1, \Delta(d : \varepsilon) := \sum_{1 \leq i \leq d} \Delta_i(1 : \varepsilon)$ where $\Delta_i(1 : \varepsilon)$ is just $\Delta(1 : \varepsilon)$ acting only on the ith variable. It is clear that $\Delta(d : \varepsilon)$ verifies (a)–(c).

The Schwinger Hamiltonians $H(\varepsilon)^{(s)} = -\Delta(\varepsilon)^{(s)} + V_\varepsilon$. In this case we relate ε to N by the equation $\varepsilon = (2\pi/N)^{1/2}$ and write $\mathbf{F}_\varepsilon$ for the finite Fourier transform on $L^2(G(\varepsilon)/N \cdot G(\varepsilon))$ transferred to $L^2(G(\varepsilon)^0)$. It is a unitary operator with matrix elements $\mathbf{F}_\varepsilon(rs)(r, s \in G(\varepsilon)^0)$ in the basis $(u_\varepsilon^r)_{r \in G(\varepsilon)^0}$, u_ε^r being the delta function at r, given by

$$\mathbf{F}_\varepsilon(rs) = N^{-d/2} \, e^{ir \cdot s} .$$

The operator $q_{i,\epsilon}$ is the multiplication by the ith coordinate and $p_{i,\epsilon}$ is its Fourier transform, while $-\Delta(\epsilon)^{(s)}$ is the sum of squares of the $p_{i,\epsilon}$:

$$(q_{i,\epsilon}f)(x) = x_i f(x) \quad (x \in G(\epsilon)), \quad p_{i,\epsilon} = \mathbf{F}_\epsilon q_{i,\epsilon} \mathbf{F}_\epsilon^{-1}, \quad -\Delta(\epsilon)^{(s)} = \sum_{1 \leq k \leq d} p_{i,\epsilon}^2 .$$

From our remarks it is clear that for $* = p, 0, s$,

$$(-\Delta(\epsilon)^{(*)}f, f) \geq \sum_{1 \leq i \leq d} \|D_i(\epsilon)^+ f\|_b^2 \quad (f \in L^2(G(\epsilon)^0)) .$$

The validity of this for $* = s$ is due to the fact that *the Schwinger Laplacian (with the negative sign) is bigger than the periodic version*:

$$(-\Delta(\epsilon)^{(s)}f, f) \geq (-\Delta(\epsilon)^{(p)}f, f) \quad (f \in L^2(G(\epsilon)^0)) .$$

Indeed, it is enough to verify this when $d = 1$ and in that case it follows from the inequality $1 - \cos x \leq \frac{1}{2}x^2 (x \in \mathbf{R})$ and the formula

$$\mathbf{F}_\epsilon^{-1}(-(1/2)\Delta(\epsilon)^{(p)})\mathbf{F}_\epsilon = \epsilon^{-2}(1 - \cos q_\epsilon \epsilon) .$$

Since it is necessary to work in a single Hilbert space to formulate the approximation theorems neatly we shall introduce certain natural imbeddings of the $L^2(G(\epsilon))$ (hence also of the $L^2(G(\epsilon)^0)$)) into $L^2(\mathbf{R}^d)$. For any $x \in \mathbf{R}^d$ let

$$R(x) = \{y = (y_1, \ldots, y_d) \mid x_i - \epsilon/2 \leq y_i < x_i + \epsilon/2, 1 \leq i \leq d\} .$$

Then the map $f \mapsto f^\#$ is an isometric imbedding of $L^2(G(\epsilon))$ into $L^2(\mathbf{R}^d)$ where

$$f^\# = \epsilon^{-d/2} \sum_{x \in G(\epsilon)} f(x)\chi_{R(x)} \quad (\chi_E \text{ is the characteristic function of } E) .$$

Thus we may view operators on $L^2(G(\epsilon))$ as operators on $L^2(\mathbf{R}^d)$ by the identification $A \to Q_\epsilon A Q_\epsilon$ where Q_ϵ is the orthogonal projection on the image of $L^2(G(\epsilon))$. If $\theta_\epsilon = \varphi_\epsilon \otimes \cdots \otimes \varphi_\epsilon, \varphi_\epsilon = \epsilon^{-1}\chi_{[-\epsilon/2, \epsilon/2]}$, then we have,

$$Q_\epsilon f =: f_\epsilon = \sum_{x \in G(\epsilon)} (\theta_\epsilon * f)(x)\chi_{R(x)} \quad (f \in L^2(\mathbf{R}^d)) .$$

If Q_ϵ^0 is the orthogonal projection on the image of $L^2(G(\epsilon)^0)$, then $Q_\epsilon^0 f \longrightarrow f$ for all $f \in L^2(\mathbf{R}^d)$; this is trivial for $f \in C_c^1(\mathbf{R}^d)$ and follows for all f by approximation.

2. Analytic Method: Convergence of Eigenvalues and Eigenfunctions When $V(x) \longrightarrow \infty$ For $|x| \longrightarrow \infty$

2.1. Let $\Delta(\epsilon)^{(*)}$ be any of $\Delta(\epsilon), \Delta(\epsilon)^{(p)}, \Delta(\epsilon)^{(0)}, \Delta(\epsilon)^{(s)}$, and $H(\epsilon)^{(*)} = -\Delta(\epsilon)^{(*)} + V_\epsilon$. Assuming that $V(x)$ is continuous, ≥ 0, and $\longrightarrow \infty$ as $|x| \longrightarrow \infty$ we shall give in this section an analytic proof that the eigenvalues and eigenfunctions of $H(\varepsilon)^{(*)}$ converge to those of $H = -\Delta + V$, as $\varepsilon \longrightarrow 0$. There are two steps to the proof. The first proves that the resolvents $(I + H(\varepsilon)^{(*)})^{-1}$ converge in the strong operator topology to the resolvent $(I + H)^{-1}$. The second step is to show that the operators $(I + H(\varepsilon)^{(*)})^{-1}$ are uniformly compact in a suitable sense.

2.2. The strong convergence of the resolvents and exponentials will follow if we show that $H(\varepsilon)^{(*)}f \longrightarrow Hf$ for all f in a common core for all the operators $H, H(\varepsilon)^{(*)}$ ([6], Theorem VIII. 20 and VIII. 25). It follows from [4] (Theorem X. 28) that $C_c^\infty(\mathbf{R}^d)$ is a common core, hence also $C_c^\infty(\mathbf{R}) \otimes \cdots \otimes C_c^\infty(\mathbf{R})$. This is true even if V has singularities of a certain type and is in $L^2(\mathbf{R}^d)^{\mathrm{loc}}$, for instance if V is an atomic potential ([4], Example 2, p. 167 and Theorem X. 26). For the strong convergence of the resolvents and exponentials it is enough to suppose only the following.

(a) $-\Delta + V$ is essentially self-adjoint on $C_c^\infty(\mathbf{R}^d)$.
(b) $V = \sum_j V_j$, where each V_j is continuous away from a linear subspace L_j of codimension ≥ 3 and V_j is bounded by constant times (distance from L_j)$^{-1}$ on compact sets of $\mathbf{R}^d$.

Then $C_c^\infty(\mathbf{R}) \otimes \cdots \otimes C_c^\infty(\mathbf{R})$ is a common core for all $H, H(\varepsilon)^{(*)}$.

Proposition 1. *Let V be as above. Then the resolvents of $H(\varepsilon)^{(*)}$ converge strongly to those of H. In particular, if V is ≥ 0 and continuous, $(I + H(\varepsilon)^{(*)})^{-1}$ converges strongly to $(I+H)^{-1}$, and we also have the strong convergence of $e^{-tH(\varepsilon)^{(*)}}$ to e^{-tH} for all $t > 0$.*

Proof. Since we deal only with f with compact support, boundary modifications do not matter. Hence there are only two cases to be treated: $H(\varepsilon)$ and $H(\varepsilon)^{(s)}$. For the convergence of the potential part we may suppose that $V = V_j$. If V_ε is the function which is $V(x)$ if the cube $R(x)(x \in G(\varepsilon))$ does not meet L_j and is 0 otherwise, we have the estimate $|V_\varepsilon(x)| \leq$ const. (distance of x to L_j)$^{-1}$ for all x, and so, as $V_\varepsilon f_\varepsilon \longrightarrow Vf$ pointwise away from L_j, we have, by the dominated convergence theorem the same limit relation in $L^2(\mathbf{R}^d)$. $\qquad\square$

To prove that $\Delta(\varepsilon)f_\varepsilon \longrightarrow \Delta f$ for f in $L^2(\mathbf{R}^d)$ it is clear that we need only work in dimension 1. Then $\Delta(\varepsilon)f_\varepsilon$ is the function $\sum_{x \in G(\varepsilon)} \theta_\varepsilon^{[3]} * f''(x)\chi_{R(x)}$, where $\theta_\varepsilon^{[3]} = \theta_\varepsilon * \theta_\varepsilon * \theta_\varepsilon$. It is then easy to check that $\Delta(\varepsilon)f_\varepsilon \longrightarrow f''$.

To show that $\Delta(\varepsilon)^{(s)}f_\varepsilon \longrightarrow \Delta f$ requires a little more effort. As before it is enough to treat the case $d = 1$. Let $\mathbf{F}_\varepsilon$ denote the operator which is identified with the finite Fourier transform on the image of $L^2(G(\varepsilon)^0)$ and the identity on the orthogonal complement. Let $\mathbf{F}$ be the *unitary* Fourier transform on $\mathbf{R}$ so that

$$\mathbf{F}f = \frac{1}{\sqrt{2\pi}}\hat{f}, \qquad \hat{f}(x) = \int e^{ixy} f(y)dy .$$

It is easily seen to be a question of showing that, as $\varepsilon \longrightarrow 0$, for all $f \in C_c^\infty(\mathbf{R})$,

$$q_\varepsilon^r(\mathbf{F}_\varepsilon^{-1}f_\varepsilon - (\mathbf{F}^{-1}f)_\varepsilon) \longrightarrow 0 \qquad (r \text{ an integer} \geq 0)$$

(we need this only for $r = 0, 2$). Recall that $\varepsilon = (2\pi/N)^{1/2}, N \longrightarrow \infty, N^0 = (N-1)/2$. An elementary calculation gives

$$\|q_\varepsilon^r \mathbf{F}_\varepsilon^{-1} f_\varepsilon - q_\varepsilon^r (\mathbf{F}_\varepsilon^{-1} f)_\varepsilon\|^2 = (2\pi)^{-1} \sum_{|k| \leq (N-1)/2} (k\varepsilon)^{2r} |\alpha_{k\varepsilon} - A_{k\varepsilon}|^2$$

where, writing $h = \theta_\varepsilon * f$,

$$\alpha_{k\varepsilon} = \varepsilon^{3/2} \sum_{|m| \leq (N-1)/2} e^{-ik\varepsilon m\varepsilon} h(m\varepsilon) = \varepsilon^{3/2} \sum_{m \in \mathbf{Z}} e^{-ik\varepsilon m\varepsilon} h(m\varepsilon)$$

$$A_{k\varepsilon} = \varepsilon^{-1/2} \int_{(k-\frac{1}{2})\varepsilon}^{(k+\frac{1}{2})\varepsilon} \hat{f}(-x)dx = \varepsilon^{1/2} \int e^{-ik\varepsilon y} f(y)a(\varepsilon y)dy$$

with $a(u) = \sin(u/2)/(u/2)$. Applying the *Poisson Summation Formula* to the function $y \mapsto h(y)e^{-ik\varepsilon y}$ we get

$$\alpha_{k\varepsilon} = \varepsilon^{1/2} a(-k\varepsilon^2)\hat{f}(-k\varepsilon) + \varepsilon^{1/2} \sum_{|m| \geq 1} \hat{h}(Nm\varepsilon - k\varepsilon) \ .$$

We next rewrite $A_{k\varepsilon}$ as

$$A_{k\varepsilon} = \varepsilon^{1/2} \hat{f}(-k\varepsilon) + \varepsilon^{1/2} \int e^{-ik\varepsilon y}(a(\varepsilon y) - 1)f(y)dy \ .$$

Then $A_{k\varepsilon} - \alpha_{k\varepsilon} = \sum_{p=0}^{p=2} E_{k\varepsilon}^{(p)}$ where

$$E_{k\varepsilon}^{(0)} = \varepsilon^{1/2} \hat{f}(-k\varepsilon)(1 - a(-k\varepsilon^2)), \qquad E_{k\varepsilon}^{(1)} = \varepsilon^{1/2} \int e^{-ik\varepsilon y}(a(\varepsilon y) - 1)f(y)dy$$

$$E_{k\varepsilon}^{(2)} = -\varepsilon^{1/2} \sum_{|m| \geq 1} \hat{h}(Nm\varepsilon - k\varepsilon) \ .$$

It is to be shown that as $\varepsilon \longrightarrow 0$,

$$J^{(p)} := \sum_{|k| \leq (N-1)/2} (k\varepsilon)^{2r} |E_{k\varepsilon}^{(p)}|^2 \longrightarrow 0 \qquad (p = 0, 1, 2) \ .$$

$p = 0$. $J^{(0)}$ is seen to go to 0 in view of the estimate (uniform in g)

$$\varepsilon \sum_{|k| \leq (N-1)/2} g(k\varepsilon) - \int_{|x| \leq N\varepsilon/2} gdx = O(N^{-1/2} \|g''\|_\infty)$$

for $g(x) = |\hat{f}(-x)|^2 x^{2r} |(1 - a(-\varepsilon x))|^2$, and the remark that the error term is $O(N^{-1/2})$ while the main term goes to 0 by the dominated convergence theorem. Here we must note that a and all its derivatives are bounded on $\mathbf{R}$.

$p = 1$. We argue as before with $g = u_\epsilon$ where

$$u_\epsilon(x) = x^{2r} \int e^{-ixy}(a(\epsilon y) - 1)f(y)dy .$$

It is clear that $u_\epsilon \longrightarrow 0$ pointwise. Since

$$(u_\epsilon(x)x^m)^{(p)} = (i)^{2r+m} \int e^{-ixy}(a(\epsilon y) - 1)f(y))^{(2r+m)}(-iy)^p dy$$

it is clear that $u_\epsilon^{(p)}(x)x^m = O(1)$ as $|x| \longrightarrow \infty$, uniformly in ϵ, for each p, m.

$p = 2$. We have $|\hat{h}| = |a\hat{f}| \leq$ const. $|\hat{f}|$, so that

$$(k\epsilon)^{2r}|E_{k\epsilon}^{(2)}|^2 \leq \text{const.}\epsilon(k\epsilon)^{2r}\left(\sum_{|m|\geq 1} |\hat{f}(Nm\epsilon - k\epsilon)|\right)^2 .$$

Since $|Nm\epsilon - k\epsilon| \geq N|m|\epsilon/2(|k| \leq (N-1)/2)$, the expression within the parentheses is $O(N^{-t})$ for every $t > 0$ uniformly in k and so $J^{(2)} \longrightarrow 0$.

Corollary 2. *For any $t > 0$ and any operator K in $L^2(\mathbf{R}^d)$ of finite rank, as $\epsilon \longrightarrow 0$*

$$\text{Tr}(e^{-tH(\epsilon)^{(\bullet)}}K) \longrightarrow \text{Tr}(e^{-tH}K) .$$

2.3. We now suppose that V is continuous, ≥ 0 and goes to infinity at infinite distances and prove the uniform compactness of $(I + H(\epsilon)^{(*)})^{-1}$ as $\epsilon \longrightarrow 0$. This is based on the uniformity of the lattice versions of the Rellich imbeddings as $\epsilon \longrightarrow 0$, which is formulated as follows.

Lemma 1. *Suppose $f_\epsilon \in L^2(G(\epsilon))$ satisfy the following conditions: for some $a > 0$ there is a constant $A > 0$ such that for all sufficiently small ϵ,*

$$\|f_\epsilon\|_a \leq A, \qquad \|D_i(\epsilon)^+ f_\epsilon\|_a \leq A \quad (1 \leq i \leq d)$$

where the suffix means that the norm is computed over $C(a) \cap G(\epsilon)$, $C(a)$ being the open cube where all coordinates are $< a$ in absolute value. If (ϵ_n) is a sequence tending to 0, then for any $b > 0$ with $0 < b < a$, the sequence $(f_{\epsilon_n}^\#)$ is relatively compact in $L^2(C(b))$.

Proof. We first reduce the proof to the case where all the f_ϵ vanish outside a cube $C(c)$ with $b < c < a$. Select $u \in C^1(\mathbf{R}^d)$ with $u = 1$ on $C(b), u = 0$ outside $C(c)$, and $0 \leq u \leq 1$. It is then immediate that $\|(uf_\epsilon)\| \leq A', \|D_i(\epsilon)^+(uf_\epsilon)\| \leq A'$ for all sufficiently small ϵ, A' being a suitable constant. So we assume that $f_\epsilon = 0$ outside $C(c)$. For small ϵ we see that the $f_\epsilon^\#$ vanish outside $C(a)$ and hence their Fourier transforms $F_\epsilon(\xi)$ are entire functions of ξ; moreover for any $t > 0$ and all ξ with $|\text{Im } \xi| \leq t$ and all sufficiently small ϵ we have the estimate $|F_\epsilon(\xi)| \leq B_t$ where $B_t > 0$ is a constant depending only on A, a, t. So passing to a subsequence we may suppose that $F_\varepsilon \longrightarrow g$ uniformly on compact subsets of $\mathbf{C}^d$ for some entire function g. The Fourier transform of $D_k(\varepsilon)^+ f_\varepsilon^\#$ is

$$-i\xi_k e^{-i\varepsilon\xi_k|^2} \Phi(\varepsilon\xi_k) F_\varepsilon(\xi) \,, \qquad \Phi(z) = \frac{\sin(z/2)}{(z/2)}$$

Hence, for all sufficiently small $\varepsilon > 0$,

$$\int_{\mathbf{R}^d} |\xi|^2 |\Phi(\varepsilon\xi_1) \ldots \Phi(\varepsilon\xi_d) F_\varepsilon(\xi)|^2 d\xi \leq A^2 d \,.$$

If $g_\varepsilon(\xi) = \Phi(\varepsilon\xi_1) \ldots \Phi(\varepsilon\xi_d) F_\varepsilon(\xi)$ it now follows that (g_{ε_n}) is relatively compact in $L^2(\mathbf{R}^d)$. As g_ε is the Fourier transform of $\theta_\varepsilon * f_\varepsilon^\#$ we see that $\theta_{\varepsilon_n} * f_{\varepsilon_n}^\#$ is relatively compact in $L^2(\mathbf{R}^d)$. But, it is easy to see that there is a fixed finite set $F \subset \mathbf{Z}^d$ such that for all $\varepsilon > 0$ and $x \in \mathbf{R}^d$, $\theta_\varepsilon * f_\varepsilon^\#(x)$ differs from $f_\varepsilon^\#(x)$ by at most the sum of terms $|f_\varepsilon^\#(\varepsilon m + x) - f_\varepsilon^\#(x)|, m \in F$. As $\|T_{\varepsilon e_k} f_\varepsilon^\# - f_\varepsilon^\#\| = \|T_{\varepsilon e_k} f_\varepsilon - f_\varepsilon\| = O(\varepsilon)$ for each k, it follows that $\|\theta_\varepsilon * f_\varepsilon^\# - f_\varepsilon^\#\| = O(\varepsilon)$. $\qquad\square$

We shall now consider the resolvents $(I + H(\varepsilon_n)^{(*)})^{-1}, (\varepsilon_n)$ a sequence tending to 0. Let $f_{\varepsilon_n}, g_{\varepsilon_n} \in L^2(G(\varepsilon_n)^{(*)})$ with $\|g_{\varepsilon_n}\| \leq 1, (I + H(\varepsilon_n)^{(*)}) f_{\varepsilon_n} = g_{\varepsilon_n}$. Then $\|f_{\varepsilon_n}\| \leq 1$ also.

Lemma 2. *With the notations as above, $(f_{\varepsilon_n}^\#)$ is relatively compact in $L^2(\mathbf{R}^d)$.*

Proof. The equation $((I + H(\varepsilon_n)^{(*)}) f_{\varepsilon_n}, f_{\varepsilon_n}) = (g_{\varepsilon_n}, f_{\varepsilon_n})$ leads to the estimates

$$\|f_{\varepsilon_n}\| \leq 1, \quad \|V_{\varepsilon_n}^{1/2} f_{\varepsilon_n}\| \leq 1, \quad \|D_k(\varepsilon_n)^+ f_{\varepsilon_n}\|_b \leq 1 \quad (1 \leq k \leq d)$$

for all $b > 0$. By Lemma 1, $(f_{\varepsilon_n}^\#)$ is relatively compact in $L^2(C(b))$ for any $b > 0$. Since

$$\int_{\mathbf{R}^d \setminus C(b)} |f_{\varepsilon_n}^\#|^2 dx \leq \left(\inf_{x \notin C(b)} V(x) \right)^{-1} \longrightarrow 0$$

as $b \longrightarrow \infty$ uniformly in n we see that $(f_{\varepsilon_n}^\#)$ is relatively compact in $L^2(\mathbf{R}^d)$. This proves the Lemma. $\qquad\square$

A formal functional analytic argument is all that is now needed to prove the approximation of H by $H(\varepsilon_n)^{(*)}$. Let $\mathcal{H}$ be a Hilbert space; $K, K_n (n = 1, 2, \ldots)$ self-adjoint operators on $\mathcal{H}$ with $0 \leq K, K_n \leq I$, K compact; we suppose that $K_n \longrightarrow K$ strongly as $n \longrightarrow \infty$. Further there are closed subspaces $L_n \subset \mathcal{H}$ stable under K_n such that (i) K_n is compact on L_n and $K_n = I$ on $L_n^\perp$. (ii) $P_{L_n} \longrightarrow I$ strongly where P_{L_n} is the orthogonal projection on L_n. (iii) If $K_n f_n = g_n$ where $f_n \in L_n$ and $\|f_n\| \leq 1$, the sequence (g_n) is relatively compact. We then have the following.

Lemma 3. *Let $1 \geq k_1 > k_2 > \cdots > 0$ be the positive eigenvalues of K and S_j the eigensubspace corresponding to k_j. Then (i) if J is a compact subset of $(0, 1]$ not* *containing any eigenvalues of K, then no eigenvalue of $K_n\big|_{L_n}$ belongs to J if n is large enough* (ii) *if J is a compact neighborhood of k_j not containing any $k_i, i \neq j$, all the eigenvalues of $K_n\big|_{L_n}$ that belong to J converge to k_j; if S_{nj} is the span of the corresponding eigenspaces,* $\dim(S_{nj}) = \dim(S_j)$ *for n large enough, and there is an orthonormal basis of S_{nj} that converges to an orthonormal basis of S_j.*

Proof. If $f_r (r = 1, 2, \ldots)$ are unit vectors in L_{n_r} such that $K_{n_r} f_r = k'_r f_r$ where $k'_r \longrightarrow k' > 0$ as $r \longrightarrow \infty$, (f_r) is relatively compact by our assumptions, and for any limit point f^* of (f_r) we have $K f^* = k' f^*$. So k' must be one of the k_j. this proves the first statement and the first part of the second; and further shows that $\dim(S_{nj}) \leq \dim(S_j)$ if n is large enough. The remaining assertions are proved by induction on j. Assume that the result is true for indices $< j$. It is enough to prove that if f is a unit vector in S_j and f_{nj} is its projection on S_{nj}, $f_{nj} \longrightarrow f$. Fix $k, k_j > k > k_{j+1}$ and let f_n be the projection of f on L_n. Then $f_n = f_{n1} + \cdots + f_{nj} + g_n$ where g_n is the projection of f_n on the span of the eigenspaces of $K_n\big|_{L_n}$ of eigenvalues $\leq k$. Now $f_n \longrightarrow f$ while the induction hypothesis shows that $f_{nr} \longrightarrow 0, r < j$. Passing to a subsequence we may assume that $f_{nj} \longrightarrow f^*$ so that $g_n \longrightarrow g^*$ and $f = f^* + g^*$. But $g^* = f - f^* \in S_j$ and $(Kg^*, g^*) \leq k(g^*, g^*)$, showing that $g^* = 0$. Hence $f^* = f$. This proves the Lemma. $\qquad\qquad\square$

It is now enough to take $K_n = (I + H(\varepsilon_n)^{(*)})^{-1}, K = (I + H)^{-1}$ and L_n as the image of $L^2(G(\varepsilon_n)^{(*)})$ to obtain the main theorem of this section.

Theorem 4. *Let (ε_n) be a sequence tending to 0 and $* = p, s, 0$. Let $0 < h_1 < h_2 < \ldots$ be the eigenvalues of H and T_j the eigensubspace corresponding to h_j. Then* (i) *if J is a compact subset of $[0, \infty)$ not containing any eigenvalues of H, then no eigenvalue of $H(\varepsilon_n)^{(*)}$ belongs to J if n is large enough;* (ii) *if J is a compact neighborhood of h_j not containing any $h_i, i \neq j$, all the eigenvalues of $H(\varepsilon_n)^{(*)}$ that belong to J converge to h_j; if T_{nj} is the span of the corresponding eigenspaces,* $\dim(T_{nj}) = \dim(T_j)$ *for n large enough, and there is an orthonormal basis of T_{nj} that converges to an orthonormal basis of T_j.*

3. Convergence of Random Walks on Lattices to Brownian Motion

3.1. In this section we shall write $\varepsilon = c^{-1/2}$ where c is a real parameter ≥ 1 which will go to ∞. We write

$$\mathbf{Z}_c = \mathbf{Z}c^{-1/2}, \quad \mathbf{Z}_c^d = (\mathbf{Z}_c)^d, \quad j_c = jc^{-1/2} \quad (j \in \mathbf{Z}^d)$$

and consider the Markov process $\{X_c(t)\}$ with state space $\mathbf{Z}_c^d$ and infinitesimal generator $Q_c = \frac{1}{2}\Delta(\varepsilon)$. Thus, for $j = (j_1, \ldots, j_d), k = (k_1, \ldots, k_d) \in \mathbf{Z}^d$,

$$Q_c(j_c, k_c) = \begin{cases} c/2, & \text{if for some } s, k_r = j_r \pm \delta_{rs} \\ -cd, & \text{if } k = j \\ 0, & \text{in all other cases .} \end{cases}$$

The Q_c are elements of the Banach algebra **M** of matrices $M = (M(j_c, k_c))_{j,k \in \mathbf{Z}}$ with

$$\|M\| = \sup_j \sum_k |M(j_c, k_c)| < \infty \ .$$

It therefore makes sense to consider the one parameter group in **M** of matrices

$$p_c(t) = \exp tQ_c \ .$$

It is a standard result in the theory of Markov chains that $(p_c(t))_{t \geq 0}$ is a stochastic semigroup and therefore defines a Markov chain [7], and that the associated probability measures are defined on the space of right continuous step functions on $[0, \infty)$ with values in $\mathbf{Z}_c^d$. Since we intend to vary c it is necessary to view these measures as defined on the single larger space of all right continuous maps of $[0, \infty)$ into $\mathbf{R}^d$ whose discontinuities are only of the first kind, namely the *Skorohod space* $D[0, \infty)^d$ [8, 9], and, when we want to consider a fixed time t, in the Skorohod space $D[0, t]^d$. We wish to prove the following basic limit theorem.

Theorem 1. *Let* $x, y \in \mathbf{R}^d$ *and* $t > 0$ *be fixed. Let* $a_c, b_c \in \mathbf{Z}_c^d$ *vary in such a way that* $a_c \longrightarrow x, b_c \longrightarrow y$ *as* $c \longrightarrow \infty$. *Let* $\mathbf{P}_{c,a_c,b_c}$ *be the probability measure on* $D[0, t]^d$ *corresponding to the process* X_c *that starts from* a_c *at time* 0 *and is conditioned to reach* b_c *at time* t. *Then, in the sense of convergence of probability measures in* $D[0, t]^d$,

$$\mathbf{P}_{c,a_c,b_c} \Longrightarrow W_{x,y}^d$$

where $W_{x,y}^d$ *is the probability measure defined by the Brownian motion process in* $D[0, t]^d$ *that starts from* x *at time* 0 *and is conditioned to reach* y *at time* t.

Here convergence of measures is the usual weak convergence (cf. [8, 9]). The conditional Brownian measures (*Brownian bridges*) are defined in the usual way to depend continuously on x and y (cf. [10]). When $d = 1$ we omit the superscript d. For arbitrary d we have

$$W_{x,y}^d = W_{x_1,y_1} \times \cdots \times W_{x_d,y_d} \qquad (x = (x_1, \ldots, x_d), y = (y_1, \ldots, y_d) \in \mathbf{R}^d) \ .$$

They are of course defined on the space $C[0, t]^d$ of continuous paths but we shall view them as defined on the larger space $D[0, t]^d$.

Theorem 1 already contains in itself a *uniformity* of the convergence. Let us suppose that x and y vary in compact sets of $\mathbf{R}^d$ and that $a_c = a_c(x)$ and $b_c = b_c(y)$ depend on x and y. The following corollary is immediate.

Corollary 2. *Suppose that the convergences* $a_c(x) \longrightarrow x, b_c(y) \longrightarrow y$ *are uniform as* $c \longrightarrow \infty$ *when* x, y *vary over compact sets. Then the convergence*

$$\mathbf{P}_{c,a_c(x),b_c(y)} \Longrightarrow W_{x,y}^d$$

is uniform in the following sense. If g is any bounded continuous functional on $D[0,t]^d$, the convergence

$$\int g\,d\mathbf{P}_{c,a_c(x),b_c(y)} \longrightarrow \int g\,dW^d_{x,y}$$

is uniform when x,y vary over compact sets of $\mathbf{R}^d$.

3.2. The first step is to calculate the transition probabilities for the X_c-process. Under the Fourier transform that takes $f \in L^2(\mathbf{Z}^d_c)$ to the function $\hat{f}$ in $L^2((S^1)^d)$ given by

$$\hat{f} = \sum_j f(j_c)e^{ij.\theta} \qquad (j.\theta = j_1\theta_1 + \cdots + j_d\theta_d)$$

Q_c goes over to the operator of multiplication by $-c\sum_{1\le k\le d}(1-\cos\theta_k)$, so that $\exp tQ_c$ goes over to the operator of multiplication by $\exp\{-ct\sum_{1\le k\le d}(1-\cos\theta_k)\}$. This implies that the transition probability $p_c(t:j_c,k_c)$ of going from state j_c to state k_c in time t is given by

$$p_c(t:j_c,k_c) = p_c(t:k-j), \quad p_c(t:r) = \prod_{1\le k\le d} p_c(t:r_k) \qquad (r=(r_1,\ldots,r_d)\in \mathbf{Z}^d)$$

where

$$p_c(t:r) = p_c(t:-r) = \frac{1}{2\pi}\int_{-\pi}^{\pi} e^{-ct(1-\cos\theta)}e^{ir\theta}d\theta \quad (r\in\mathbf{Z}) .$$

It follows from this that X_c is a process of independent increments. It is also clear from the product structure of the transition probabilities that the components of X_c are mutually independent. The basic result on p_c is the following.

Lemma 1. *We have the following.*

(i) $p_c(s:r) > 0$ for all $r \in \mathbf{Z}^d$ and $s > 0$.

(ii) *For any fixed $s > 0$ there is a constant $B > 0$ such that*

$$p_c(s:r) \le Bc^{-d/2} \text{ for all } r \in \mathbf{Z}^d, c \ge 1 .$$

(iii) *For fixed $s > 0$,*

$$p_c(s:r) = \frac{1}{(2\pi cs)^{d/2}}\left(e^{-|r_c|^2/2s} + \eta(r:c)\right)$$

where $|\cdot|$ is the usual Euclidean norm in $\mathbf{R}^d$, and $\eta(r:c) \longrightarrow 0$ as $c \longrightarrow \infty$, uniformly for $r \in \mathbf{Z}^d$.

Proof. It is clearly enough to prove these when $d = 1$.

(i) We have $p'_c(0:\pm 1) = c/2$ while $p_c(0:\pm 1) = 0$. So $p_c(s/n:\pm 1) > 0$ if $n \gg 1$, s being fixed. Also $p_c(s:0) > 0$ for all $s \ge 0$. Hence, if $r > 0$, choosing $n > r$,

$$p_c(s:r) > p_c(s/n:1)^r p_c(s/n:0)^{n-r} > 0 .$$

(ii) For a suitably small $a > 0$ we can find $b > 0$ such that for all $r \in \mathbf{Z}, c \geq 1$,

$$p_c(s : r) \leq \frac{1}{2\pi} \int_{|\theta| \leq a} e^{-cs\theta^2/4} d\theta + \frac{1}{2\pi} \int_{a \leq |\theta| \leq \pi} e^{-bcs} d\theta .$$

This gives (ii).

(iii) We refine the argument in (ii) but work with the new variable $z = \theta\sqrt{cs}$. Then

$$\sqrt{2\pi cs} p_c(s : r) = \frac{1}{\sqrt{2\pi}} \int_{-\pi\sqrt{cs}}^{\pi\sqrt{cs}} e^{-cs(1-\cos(z/\sqrt{cs}))} e^{irz/\sqrt{cs}} dz$$

$$= \frac{1}{\sqrt{2\pi}} \int_{-\infty}^{\infty} e^{-z^2/2 + irz/\sqrt{cs}} dz + \eta$$

$$= e^{\frac{-r_c^2}{2s}} + \eta .$$

To prove that $\eta \to 0$ as $c \to \infty$ uniformly in $r \in \mathbf{Z}$ we split the integrals involved in the definition of η into the regions where $|z| < a\sqrt{cs}$ and $a\sqrt{cs} \leq |z| \leq \pi\sqrt{cs}$. We omit the details. This proves (iii).

For later use we collect the following information on the moments of the X_c-process. Let E_0 denote the expectation value of the 1-dimensional X_c-process starting from 0 at time 0. It is obvious from the space symmetry that $E_0(X_c(t)^k) = 0$ for odd k. Since

$$E_0(X_c(t)^k) = i^k c^{-k/2} f^{(k)}(0), \qquad f(\theta) = e^{-ct(1-\cos\theta)}$$

we find, remembering that the process has independent increments,

$$E_0\left((X_c(s_1) - X_c(s_2))^2\right) = s_1 - s_2, \quad (0 < s_2 < s_1)$$
$$E_0\left((X_c(s_1) - X_c(s_2))^2(X_c(s_2) - X_c(s_3))^2\right) = (s_1 - s_2)(s_2 - s_3)$$
$$(0 < s_3 < s_2 < s_1)$$

By expanding the exponential and noting that $(1 - \cos\theta)^r$ vanishes to the order $2r$ at $\theta = 0$ we find that for each $t > 0$ and $k \geq 1$,

$$\sup_{c \geq 1} E_0\left(X_c(t)^{2k}\right) < \infty .$$

3.3. Convergence of the unconditioned process. In view of the independence of the components of the processes in question it is clearly enough to prove Theorem 3.1.1 when $d = 1$. So we shall assume that $d = 1$ from now on till the end of this section. Let $\mathbf{P}_{c,a_c}$ be the probability measure on $D[0,t]$ defined by the X_c-process starting from $a_c \in \mathbf{Z}_c$ at time 0.

Proposition 1. Let W_x be the probability measure of the Wiener process starting from x at time 0. If $a_c \longrightarrow x$ as $c \longrightarrow \infty$, then

$$\mathbf{P}_{c,a_c} \Longrightarrow W_x \quad (c \longrightarrow \infty) .$$

Proof. We first verify the convergence of the finite dimensional distributions at time points $t_i, 0 < t_1 < t_2 < \cdots < t_N < t$:

$$\mathbf{P}_{c,a_c}^{t_1,t_2,\ldots,t_N} \Longrightarrow W_x^{t_1,t_2,\ldots,t_N} .$$

If $J_i, 1 \leq i \leq N$ are compact intervals of $\mathbf{R}, \mathbf{P}_{c,a_c}(X_c(t_i) \in J_i, 1 \leq i \leq N)$ is equal to

$$\sum_{b_{ic} \in J_i, 1 \leq i \leq N} p_c(t_1 : b_1 - a)p_c(t_2 - t_1 : b_2 - b_1)\ldots p_c(t_N - t_{N-1} : b_N - b_{N-1}) .$$

The uniform estimate for $p_c(t : r)$ given in Lemma 3.1.1 easily implies that this sum converges to

$$\int_{z_i \in J_i, 1 \leq i \leq N} p(t_1 : z_1 - x)p(t_2 - t_1 : z_2 - z_1)\ldots p(t_N - t_{N-1} : z_N - z_{N-1})dz_1 dz_2 \ldots dz_N$$

where

$$p(s : u) = (2\pi s)^{-1/2} \exp\{-u^2/2s\} .$$

The limit is clearly

$$W_x(X_c(t_i) \in J_i, 1 \leq i \leq N) .$$

To complete the proof we must show that the measures $\mathbf{P}_{c,a_c}$ are relatively compact. This is immediate from the criterion of Chentsov [11] because, for $0 < s_3 < s_2 < s_1 \leq t$, we have the uniform bound

$$E_{\mathbf{P}_{c,a_c}}\{(X_c(s_1) - X_c(s_2))^2(X_c(s_2) - X_c(s_3))^2\} \leq (s_1 - s_3)^2$$

for all $c \geq 1$. This proves the proposition. $\square$

3.4. *Convergence of the conditioned processes.* To get the convergence of the conditioned processes from the convergence of the unconditioned processes established in the preceding paragraph we should divide by the probability of the conditioning event, namely, Prob $\{X_c(t) = b_c\}$, and we can take care of this using the precise asymptotics of this probability proved in (iii) of Lemma 3.2.1.

We shall fix $t > 0$ and prove that if $a_c \longrightarrow x, b_c \longrightarrow y$, then $\mathbf{P}_{c,a_c,b_c} \Longrightarrow W_{x,y}$. The method of proof is the same as before; first we establish as in the previous case that the finite dimensional distributions converge, and then verify the relative compactness. The latter is somewhat harder to do in this case. The idea for doing it is as follows. It is a question of showing that if A is the set where the modulus of continuity of the path is large, then $\mathbf{P}_{c,a_c,b_c}(A)$ is small uniformly in c. If the modulus of continuity is computed over the interval $[0, 3t/4]$, this is easy to deduce from the convergence of the unconditioned process as we have to only divide by the probability of the conditioning event, and this is easy because the conditioning takes place at t, which is "far away" from $3t/4$. We also have to show the uniform smallness when the modulus of continuity is calculated on $[t/4, t]$; but this can be deduced from the previous case by using the time reversal $s \longrightarrow t - s$ that takes $\mathbf{P}_{c,a_c,b_c}$ to $\mathbf{P}_{c,b_c,a_c}$. We now proceed to give a few more details of this argument.

Let us write, for any $X \in D[0,t]$ and any $\delta > 0$,

$$\omega(X : \delta) = \sup_{0 \leq s' < s < s'' \leq t, |s' - s''| \leq \delta} \min\left(|X(s) - X(s')|, |X(s) - X(s'')|\right)$$

and for any $\eta > 0$ let

$$A(\delta, \eta) = \{X : \omega(X : \delta) > \eta\} .$$

The compactness criterion [8, 9] comes down to showing that for any $\eta > 0$,

$$\lim_{\delta \to 0} \mathbf{P}_{c,a_c,b_c}(A(\delta, \eta)) = 0 \quad \text{uniformly in } c$$

as $\delta \longrightarrow 0$. Let ω_1 and ω_2 be the analogues of ω for the intervals $[0, 3t/4]$ and $[t/4, t]$ respectively. If $\delta < t/4$ and s', s'' are two points in $[0, t]$ with $|s' - s''| \leq \delta$, then they are both in one of the two intervals. Hence

$$\omega(X : \delta) = \max\{\omega_1(X : \delta), \omega_2(X : \delta)\}$$

so that

$$A(\delta, \eta) = A_1(\delta, \eta) \cup A_2(\delta, \eta), \quad A_j(\delta, \eta) = \{\omega_j(X : \delta) > \eta\}$$

Hence it is enough to prove that

$$\lim_{\delta \to 0} \mathbf{P}_{c,a_c,b_c}(A_j(\delta, \eta)) = 0 \quad \text{uniformly in } c \quad (j = 1, 2)$$

Case of $A_1(\delta, \eta)$. Write A_1 for $A_1(\delta, \eta)$. Then

$$\mathbf{P}_{c,a_c,b_c}(A_1) = \frac{\mathbf{P}_{c,a_c}(A_1 \cap (X(t) = b_c))}{\mathbf{P}_{c,a_c}(X(t) = b_c)} .$$

Considering the values of X at $\frac{3t}{4}$ and we see that the numerator is

$$\sum_j \mathbf{P}_{c,a_c}(A_1 \cap (X(3t/4) = j_c)) \cdot \mathbf{P}_{c,0}(X(t/4) = b_c - j_c)$$

which, in view of the Markov property, is majorized by

$$\mathbf{P}_{c,a_c}(A_1) \cdot \sup_k p_c\left(\frac{t}{4} : k\right) .$$

Hence

$$\mathbf{P}_{c,a_c,b_c}(A_1) \leq \mathbf{P}_{c,a_c}(A_1) \cdot \frac{\sup_k p_c\left(\frac{t}{4} : k\right)}{p_c(t : b - a)} .$$

By Lemma 3.2.1, for fixed $s > 0$, if $k_c \longrightarrow u$, then $\sqrt{c}p_c(s : k) \geq \gamma > 0$ for some constant $\gamma > 0$, while, $\sup_m \sqrt{c}p_c(s : m) \leq \gamma' < \infty$ for some other constant γ'. Hence, for some constant $B > 0$ independent of c, δ, η,

$$\mathbf{P}_{c,a_c,b_c}(A_1) \leq B\mathbf{P}_{c,a_c}(A_1) .$$

But the right side goes to 0 as $\delta \longrightarrow 0$ uniformly in c because of the convergence $\mathbf{P}_{c,a_c} \Longrightarrow W_x$ that we have already established. Hence the left side goes to 0 as $\delta \longrightarrow 0$ uniformly in c.

Case of $A_2(\delta, \eta)$. As we remarked earlier we use time reversal. Since the elements of the Skorohod space are in general only right continuous we define the time reversal map as follows:

$$x^*(s) = x(t - s - 0), \qquad 0 \le s \le t .$$

It is easy to see that the map $x \longrightarrow x^*$ is Borel and involutive. For any probability measure P on $D[0, t]$ let P^* be the probability measure defined by $P(E) = P^*(E^*)$. It is then easily seen from the stochastic continuity of the processes and the symmetry $p_c(t : r) = p_c(t : -r)$ that

$$\mathbf{P}_{c,a_c,b_c}{}^* = \mathbf{P}_{c,b_c,a_c} .$$

If we now define $\omega_j{}^*$ as the analogue of ω_j where $X(s')$ etc are replaced by $X(s' - 0)$ etc, it is then obvious that

$$A_2(\delta, \eta)^* = \{\omega_1{}^*(X : \delta) > \eta\}$$

and further that

$$\{\omega_1{}^*(X : \delta) > \eta\} \subset \{\omega_1(X : 2\delta) > \eta\} .$$

From these remarks we find that

$$\mathbf{P}_{c,a_c,b_c}(A_2(\delta, \eta)) = \mathbf{P}_{c,b_c,a_c}(A_2(\delta, \eta)^*) \le \mathbf{P}_{c,b_c,a_c}(A_1(2\delta, \eta)) .$$

and the proof for the case of A_2 is now reduced to the case of A_1. This completes the proof of the theorem.

4. Convergence of Traces: Continuum Limit of Infinite Lattices

4.1. We shall now prove that in the infinite lattice approximation the traces of the dynamical operators converge. Let $\mathcal{V}$ be the class of potentials on $\mathbf{R}^d$ with the following properties.

  (i) V is continuous and ≥ 0;
  (ii) $V(x)/\log|x| \longrightarrow \infty$ as $|x| \longrightarrow \infty$.
       It is not difficult to show (see also the remarks at the end of the section) that in this case we have
  (iii) $e^{t(\frac{1}{2}\Delta - V)}$ is of trace class for every $t > 0$.

When $d = 1$, the harmonic oscillator potential $V(x) = x^2$ is in $\mathcal{V}$ and so is any potential V on $\mathbf{R}^d$ for which $V(x) \ge |x|^2$ for $|x|$ sufficiently large. We shall prove

Proposition 1. *Let $V \in \mathcal{V}$. Then*

  (i) *$e^{t(Q_c - V)}$ is of trace class on $L^2(\mathbf{Z}_c^d)$ for each $t > 0$.*

(ii) *For each $t > 0$, as $c \longrightarrow \infty$, we have*

$$\text{Tr}\left(e^{t(Q_c - V)}\right) \longrightarrow \text{Tr}\left(e^{t(\frac{1}{2}\Delta - V)}\right) .$$

4.2. Recall that under the Fourier transform $L^2(\mathbf{Z}_c^d) \longrightarrow L^2((S^1)^d)$ the operator Q_c goes over to the operator of multiplication by $-c\sum_{1 \leq k \leq d}(1 - \cos\theta_k)$, thus $Q_c \leq 0$. Hence, $Q_c - V$, self adjoint on the domain of multiplication by V, is ≤ 0. As $e^{t(Q_c - V)} \geq 0$ for each $t > 0$ it will be of trace class if and only if

$$\sum_{j \in \mathbf{Z}^d} e^{t(Q_c - V)}(j_c, j_c) < \infty .$$

This fact will emerge during the course of the proof itself.

By the Feynman–Kac formula for the X_c-process

$$e^{t(Q_c - V)}(j_c, k_c) = \int e^{-\int_0^t V(X(s))ds} d\mathbf{P}_{c,j_c,k_c} \cdot p_c(t : k - j) \geq 0 .$$

Lemma 1. *There is a constant $B_t > 0$ such that for all $c \geq 1$,*

$$\sup_j e^{t(Q_c - V)}(j_c, j_c) \leq B_t c^{-d/2} .$$

Furthermore, let

$$g(x) = \frac{1}{(2\pi t)^{d/2}} \int e^{-\int_0^t V(X(s))ds} dW_{x,x}^d$$

Then

$$\text{Tr}\left(e^{t(\frac{1}{2}\Delta - V)}\right) = \int_{\mathbf{R}^d} g(x)dx .$$

Proof. The Feynman–Kac formula shows that

$$\sup_j e^{t(Q_c - V)}(j_c, j_c) \leq p_c(t : 0)$$

and the first assertion follows from Lemma 3.2.1. We observe next by the Feynman–Kac formula again that $e^{t(\frac{1}{2}\Delta - V)}$ is an integral operator on $L^2(\mathbf{R}^d)$ with kernel

$$K_V(x, y) = K_{V,t}(x, y) = \int e^{-\int_0^t V(X(s))ds} dW_{x,y} \cdot p(t : x - y) .$$

Since the operator is ≥ 0, the trace evaluation follows from Mercer's theorem [12].

Lemma 2. *Suppose $c \longrightarrow \infty$. If $j_c \longrightarrow x, k_c \longrightarrow y$, then,*

$$e^{t(Q_c - V)}(j_c, k_c)c^{d/2} \longrightarrow K_V(x, y) .$$

If the convergence of j_c to x and k_c to y are uniform when x and y vary over compact sets of $\mathbf{R}^d$, the above convergence is also uniform over compact sets for x, y.

Proof. Since, by Lemma 3.2.1,

$$c^{d/2} p_c(t : k - j) = \frac{1}{(2\pi t)^{d/2}} \left(e^{-|k_c - j_c|^2/2t} + o(1) \right)$$

with uniform $o(1)$, it is a question of showing that

$$\int e^{-\int_0^t V(X(s))ds} \, d\mathbf{P}_{c, j_c, k_c} \longrightarrow \int e^{-\int_0^t V(X(s))ds} \, dW_{x,y}^d .$$

This follows from Theorem 3.1.1 since the integrand is a continuous functional on $D[0, t]^d$.

To prove Proposition 1 we must prove that as $c \longrightarrow \infty$,

$$\sum_{j \in \mathbf{Z}} e^{t(Q_c - V)}(j_c, j_c) \longrightarrow \int_{\mathbf{R}^d} g(x) dx$$

where the statement includes the assertion that the left side is finite for all $c > 0$.

Lemma 3. *For any $m > 0$,*

$$\sum_{|j_c| \leq m} e^{t(Q_c - V)}(j_c, j_c) \longrightarrow \int_{|x| \leq m} g(x) dx .$$

Proof. Define the function g_c on $\mathbf{R}^d$ by

$$g_c(u) = e^{t(Q_c - V)}(j_c, j_c) c^{d/2} \qquad \text{if } j_r c^{-1/2} \leq u_r < (j_r + 1) c^{-1/2} \text{ for } 1 \leq r \leq d .$$

It then follows from Lemma 2 that g_c converges uniformly to g so that

$$\int_{|x| \leq m} g_c dx \longrightarrow \int_{|x| \leq m} g dx$$

for any $m > 0$. Since

$$\int_{|x| \leq m} g_c dx = \sum_{|j_c| \leq m} e^{t(Q_c - V)}(j_c, j_c) + O(c^{-1/2})$$

the lemma is proved.

To finish the proof of Proposition 1 it is now sufficient to prove that

$$\lim_{m \to \infty} \sum_{|j_c| \geq m} e^{t(Q_c - V)}(j_c, j_c) = 0 \quad \text{uniformly in } c \qquad (*) .$$

Lemma 4. *There exists a constant $B > 0$ such that for all integers $p \geq 1$, and $|j_c| \geq p$ and all $c \geq 1$, we have the estimate*

$$e^{t(Q_c - V)}(j_c, j_c) \leq Bc^{-d/2} \int_{p-1}^{p} \left(e^{-\frac{t}{2}V^*\left(\frac{x}{2}\right)} + \frac{1}{x^4} \right) dx$$

where $V^*(x) = \inf_{|y| \geq x} V(y)(x \geq 0)$.

Proof. Before starting the proof we recall the classical fact ([13], pp. 21–24) which itself is a consequence of P. Levy's inequality for sums of independent random variables: if $X(t)$ is a separable stochastic process with right continuous paths and independent increments, then

$$P\left(|X(s) - X(0)| > 2a \text{ for some } s \in [0, u]\right) \leq \frac{2}{a^{2k}} \sup_{s,t \in [0,u]} E\left(X(s) - X(t)\right)^{2k}.$$

Applying this to each component of X_c it follows that for each $u > 0$ and integer $k \geq 1$ there is a constant $B_k = B_k(u) > 0$ such that for all $a > 0, c \geq 1, j_c \in \mathbf{Z}_c^d$,

$$\mathbf{P}_{c,j_c}\left(|X(s) - X(0)| > 2a \text{ for some } s \in [0, u]\right) \leq B_k a^{-2k}.$$

With this said, we shall prove the lemma. We write

$$e^{t(Q_c - V)}(j_c, j_c) = I_1 + I_2$$

where

$$I_i = \int_{R_i} e^{-\int_0^t V(X(s))ds} \chi_{j_c}(X(t)) d\mathbf{P}_{c,j_c} \quad (i = 1, 2) .$$

Here χ_{j_c} is the characteristic function of the singleton $\{j_c\}$ and $R_i (i = 1, 2)$ are the regions defined by

$$R_1 = \{X | X(0) = j_c, |X(s)| \geq p/2 \text{ for all } s \in [0, t/2]\}$$
$$R_2 = \{X | X(0) = j_c, |X(s)| < p/2 \text{ for some } s \in [0, t/2]\} .$$

To estimate I_1 we note that for $X \in R_1$,

$$\int_0^t V(X(s))ds \geq \int_0^{\frac{t}{2}} V(X(s))ds \geq tV^*(p/2)/2 .$$

Hence

$$I_1 \leq e^{-tV^*(p/2)/2} \mathbf{P}_{c,j_c}(R_1 \cap (X(t) = j_c))$$
$$\leq e^{-tV^*(p/2)/2} \mathbf{P}_{c,j_c}(R_1) \sup_k p_c(t/2 : k)$$
$$\leq Bc^{-d/2} e^{-tV^*(p/2)/2}$$

where $B > 0$ is a constant > 0 independent of c, j_c, p. For I_2 we replace the integrand by 1 and observe that for $X \in R_2$, we have $|X(s) - X(0)| > p/2$ for some $s \in [0, t/2]$. Hence we get

$$I_2 \leq \mathbf{P}_{c,j_c}(R_2) \cdot \sup_k p_c(t/2 : k)$$

$$\leq Bc^{-d/2} \mathbf{P}_{c,j_c}(|X(s) - X(0)| > p/2 \text{ for some } s \in [0, t/2])$$

$$\leq Bc^{-d/2} \frac{1}{p^4}.$$

This proves the lemma.

To get $(*)$ we sum over j_c in the range $p \leq |j_c| \leq p + 1$, the number of terms in the sum being $O(c^{d/2})$, and then over $p = m, m+1, m+2, \ldots$. Thus

$$\sum_{|j_c| \geq m+1} e^{t(Q_c - V)}(j_c, j_c) \leq B \int_m^\infty \left(e^{-\frac{1}{2}V^*(\frac{x}{2})} + \frac{1}{x^4} \right) dx .$$

Since $V^*(x)/\log x \longrightarrow \infty$ the integrand is $O(x^{-4})$ and so the right side goes to zero as $m \longrightarrow \infty$. As it is independent of c we are finished with the proof of the Proposition.

Remark. It is a consequence of the above proof that the integral of $K_{V,t}(x, x)$ over $\mathbf{R}^d$ is finite. This integral is the Hilbert–Schmidt norm of $K_{V,t/2}$ since the kernels are continuous, symmetric and we have a semi-group. So $K_{V,t}$ is of trace class for every $t > 0$. Also the finiteness of the integral on the diagonal can be established by splitting the integration as in the proof of Lemma 4 and estimating the parts directly.

5. Convergence of Traces: Continuum Limit of Finite Lattices

5.1. We shall now truncate our infinite lattice $\mathbf{Z}_c^d$ to finite grids $\mathbf{Z}_c^{*d} = (\mathbf{Z}_c^*)^d$ where

$$\mathbf{Z}_c^* = \{k_c \| k| \leq N(c)\}$$

and $N(c)$ are integers that tend to ∞ as $c \longrightarrow \infty$. We assume that for some constant $a > 0$,

$$N(c) \geq a \cdot c .$$

We write $\|x\| = \max_{1 \leq r \leq d} |x_r|, x \in \mathbf{R}^d$ and suppose that for suitable integers $N^*(c)$,

$$N^*(c) < N(c), \quad N^*(c)/N(c) \longrightarrow 1 \quad (c \longrightarrow \infty)$$

we have

$$Q_c^*(j_c, k_c) = Q_c(j_c, k_c), \qquad \|j\| \leq N^*(c), \|k\| \leq N(c)$$

and further that the Q_c^* generate Markov processes. We may suppose

$$N^*(c) \geq (3/4)N(c) .$$

We intend to prove

Proposition 1. *Suppose $V \in \mathcal{V}$. Then, for each $t > 0$, as $c \longrightarrow \infty$,*

$$\mathrm{Tr}\left(e^{t(Q_c^* - V)}\right) = \sum_{\|j\| \leq N(c)} e^{t(Q_c^* - V)}(j_c, j_c) \longrightarrow \mathrm{Tr}\left(e^{t(\frac{1}{2}\Delta - V)}\right) .$$

5.2. We write $\mathbf{P}^*_{c, j_c}$ for the probability measure of the Markov process corresponding to the generator Q_c^* starting from j_c at time 0. We have the Feynman–Kac formula

$$e^{t(Q_c^* - V)}(j_c, k_c) = \int e^{-\int_0^t V(X(s))ds} \chi_{k_c}(X(t)) d\mathbf{P}^*_{c, j_c} .$$

To prove the proposition it is enough, in view of Proposition 4.1.1, to show that

$$\sum_{\|j\| \leq N(c)} e^{t(Q_c^* - v)}(j_c, j_c) - \sum_{j \in \mathbf{Z}^d} e^{t(Q_c - V)}(j_c, j_c) \longrightarrow 0 \qquad (*)$$

as $c \longrightarrow \infty$. Since $N(c)c^{-1/2} \longrightarrow \infty$ and

$$\lim_{m \to \infty} \sum_{\|j_c\| > m} e^{t(Q_c - V)}(j_c, j_c) = 0 \quad \text{uniformly in } c$$

we have

$$\lim_{c \to \infty} \sum_{\|j\| > N(c)/2} e^{t(Q_c - V)}(j_c, j_c) = 0 .$$

So, to prove $(*)$, it is enough to prove the following two limit relations:

$$\lim_{c \to \infty} \sum_{\|j\| \leq N(c)/2} \left| e^{t(Q_c^* - V)}(j_c, j_c) - e^{t(Q_c - V)}(j_c, j_c) \right| = 0 \quad \text{(I)}$$

$$\lim_{c \to \infty} \sum_{\|j\| > N(c)/2} e^{t(Q_c^* - V)}(j_c, j_c) = 0 . \quad \text{(II)}$$

We have already noted that $e^{t(Q_c^* - V)}(j_c, k_c)$ is always ≥ 0 by the Feynman–Kac formula. Since each sum has $O(N(c)^d)$ terms, it is enough to verify the following replacements for (I) and (II):

$$\sup_{\|j\| \leq N(c)/2} \left| e^{t(Q_c^* - V)}(j_c, j_c) - e^{t(Q_c - V)}(j_c, j_c) \right| = o(N(c)^{-d}) \quad \text{(I')}$$

$$\sup_{\|j\| \geq N(c)/2} e^{t(Q_c^* - V)}(j_c, j_c) = o(N(c)^{-d}) . \quad \text{(II')}$$

5.3. *Proof of* (I'). The key to proving both (I') and (II') is the following lemma. Let

$$W_{c, j_c} = \left\{ X \in D[0, t]^d \mid X(0) = j_c, \|X(u)\| \leq N^*(c)c^{-1/2} \text{ for all } u \in [0, t] \right\} .$$

Lemma 1. *For* $\|j\| \leq N^*(c)$ *the measures* $\mathbf{P}_{c,j_c}$ *and* $\mathbf{P}^*_{c,j_c}$ *coincide on the σ-algebra of subsets of* W_{c,j_c}. *If* W'_{c,j_c} *denotes the complement of* W_{c,j_c}, *then there are constants* $B > 0, c_1 \geq 1$ *such that for all* $c \geq c_1$ *and* $\|j\| \leq N(c)/2$,

$$\mathbf{P}_{c,j_c}(W'_{c,j_c}) = \mathbf{P}^*_{c,j_c}(W'_{c,j_c}) \leq BN(c)^{-(d+1)}.$$

Proof. The first assertion is more or less obvious if we note that the Markov measures are determined by the waiting times and the jump probabilities. The relation between Q_c and Q^*_c implies that the jump probabilities from j_c to k_c are the same as long as $\|j_c\| \leq N^*(c)$. This shows that the measures coincide on W_{c,j_c} (see [7]).

For the second assertion note that both measures take the same value for W_{c,j_c} and so for its complement also; this common value is the probability under $\mathbf{P}_{c,j_c}$ that X, starting from j_c at time 0, visits some $k_c, \|k\| > N^*(c)$, during $[0, t]$. So, if $\|j\| \leq N(c)/2$, then,

$$\mathbf{P}_{c,j_c}(W'_{c,j_c}) \leq \mathbf{P}_{c,j_c}\left\{\|X(s) - X(0)\| > N(c)c^{-1/2}/4 \text{ for some } s \in [0, t]\right\}$$
$$\leq BN(c)^{-(d+1)}$$

where B is a constant independent of c, j_c. This proves the lemma.

Lemma 2. *We have, uniformly for* $\|j\| \leq N(c)/2$,

$$\sum_{\|k\| \leq N(c)} \left|\left(e^{t(Q^*_c - V)} - e^{t(Q_c - V)}\right)(j_c, k_c)\right| \leq BN(c)^{-(d+1)}$$

where $B > 0$ *is a constant independent of* c, j_c.

Proof. It follows from Lemma 1 that

$$\left|\left(e^{t(Q^*_c - V)} - e^{t(Q_c - V)}\right)(j_c, k_c)\right|$$

is majorized by

$$\left|\int_{W'_{c,j_c} \cap \{X(t) = k_c\}} e^{-\int_0^t V(X(s))ds} \chi_{j_c}(X(t))d(\mathbf{P}_{c,j_c} - \mathbf{P}^*_{c,j_c})\right|$$

which is

$$\leq (\mathbf{P}_{c,j_c} + \mathbf{P}^*_{c,j_c})(W'_{c,j_c} \cap \{X(t) = k_c\}).$$

Hence

$$\sum_{\|k\| \leq N(c)} \left|\left(e^{t(Q^*_c - V)} - e^{t(Q_c - V)}\right)(j_c, k_c)\right| \leq (\mathbf{P}_{c,j_c} + \mathbf{P}^*_{c,j_c})(W'_{c,j_c})$$

$$\leq BN(c)^{-(d+1)}.$$

The proof of (I') is now quite simple. We have, for $\|j\| \le N(c)/2$,

$$\left| \left(e^{t(Q_c^*-V)} - e^{t(Q_c-V)} \right)(j_c, j_c) \right| \le 2BN(c)^{-(d+1)}$$

where the O is uniform in $\|j\| \le N(c)/2$. This proves (I').

5.4. *Proof of* (II'). We consider $j, \|j\| \ge N(c)/2$. We split the integral expressing $e^{t(Q_c^*-V)}(j_c, j_c)$ into two parts, over the region R_1 and over its complement R_1', where

$$R_1 = \left\{ X \mid X(0) = j_c, |X(u)| \ge N(c)c^{-1/2}/4 \text{ for all } u \in [0, t/2] \right\} .$$

Lemma 1. *We have, as* $c \longrightarrow \infty$, *uniformly for* $N(c)/2 \le \|j\| \le N(c)$,

$$\int_{R_1} e^{-\int_0^t V(X(s))ds} \chi_{j_c}(X(t)) dP_{c,j_c}^* \le o(N(c)^{-d}) .$$

Proof. For $X \in R_1$,

$$\int_0^t V(X(s))ds \ge \int_0^{t/2} V(X(s))ds \ge t(V^*(N(c)c^{-1/2}/4)/2$$

so that

$$\int_{R_1} e^{-\int_0^t V(X(s))ds} \chi_{j_c}(X(t)) dP_{c,j_c}^* \le e^{-t(V^*(N(c)c^{-1/2}/4)/2}.$$

For any $b > 0$, $V^*(x) \ge b \log x$ for all sufficiently large $x > 0$, and so the right side of the above estimate is $O(N(c)^{-tb/4})$ which is $o(N(c)^{-d})$ if $tb > 4d$.

Lemma 2. *We have, as* $c \longrightarrow \infty$, *uniformly for* $N(c)/2 \le \|j\| \le N(c)$,

$$\int_{R_1'} e^{-\int_0^t V(X(s))ds} \chi_{j_c}(X(t)) dP_{c,j_c}^* \le o(N(c)^{-d}) .$$

Proof. The integral is obviously majorized by $\mathbf{P}_{c,j_c}^*(R_1')$ and so we must estimate this probability. We have

$$R_1' = \left\{ X | X(0) = j_c, \|X(u)\| < N(c)c^{-1/2}/4 \text{ for some } u \in [0, t/2] \right\} .$$

Let

$$M_c = \left\{ k_c \mid \|k_c\| = [N(c)/2]c^{-1/2} \right\}$$

and let us introduce the stopping time τ of the first visit of the X_c^*-process to the set M_c. Then, writing A for the event

$$A = \left\{ X \mid \tau < t/2, |X(\tau + u)| < N(c)c^{-1/2}/4 \text{ for some } u \in [0, t/2] \right\}$$

we have

$$\sup_{j_c} \mathbf{P}^*_{c,j_c}(R'_1) \leq \sup_{j_c} \mathbf{P}^*_{c,j_c}(A) = \sup_{j_c} \sum_{k_c} \mathbf{P}^*_{c,j_c}(A \cap (X(\tau) = k_c)) \ .$$

In view of the strong Markov property ([7], p. 232), this is majorized by

$$\sup_{j_c} \sum_{k_c} \mathbf{P}^*_{c,j_c}(\tau < t/2, X(\tau) = k_c) \mathbf{P}^*_{c,k_c}(\|X(u)\| < N(c)c^{-1/2}/4$$

$$\text{for some } u \in [0, t/2]) \ .$$

Hence

$$\sup_{j_c} \mathbf{P}^*_{c,j_c}(R'_1) \leq \sup_{k_c} \mathbf{P}^*_{c,k_c}(\|X(u)\| < N(c)c^{-1/2}/4 \text{ for some } u \in [0, t/2]) \ .$$

Let

$$C = \left\{ \|X(u)\| < N(c)c^{-1/2}/4 \text{ for some } u \in [0, t/2] \right\}$$

$$A = \left\{ \|X(u)\| \leq N^*(c)c^{-1/2} \text{ for all } u \in [0, t/2] \right\}$$

$$B = \left\{ \|X(u)\| > N^*(c)c^{-1/2} \text{ for some } u \in [0, t/2] \right\}$$

then

$$\mathbf{P}^*_{c,k_c}(C) = \mathbf{P}^*_{c,k_c}(C \cap A) + \mathbf{P}^*_{c,k_c}(C \cap B) \ .$$

By Lemma 3.1

$$\mathbf{P}^*_{c,k_c}(C \cap A) = \mathbf{P}_{c,k_c}(C \cap A) \leq \mathbf{P}_{c,k_c}(C), \quad \mathbf{P}^*_{c,k_c}(C \cap B) \leq \mathbf{P}^*_{c,k_c}(B) = \mathbf{P}_{c,k_c}(B)$$

so that

$$\mathbf{P}^*_{c,k_c}(C) \leq \mathbf{P}_{c,k_c}(C) + \mathbf{P}_{c,k_c}(B) \ .$$

Furthermore,

$$(X(0) = k_c) \cap (B \cup C) \subset \left\{ X \mid \|X(u) - X(0)\| > [N(c)/4]c^{-1/2} \right.$$

$$\left. \text{for some } u \in [0, t/2] \right\} \ .$$

Hence

$$\mathbf{P}^*_{c,k_c}(C) \leq 2\mathbf{P}_{c,k_c}(\|X(u) - X(0)\| > [N(c)/4]c^{-1/2} \text{ for some } u \in [0, t/2])$$

$$\leq BN(c)^{-(d+1)}$$

for a constant $B > 0$ independent of c, k_c, M_c. Lemmas 1 and 2 imply (II') and thus the proof of Proposition 1 is finished.

6. Convergence of Dynamical Groups in Trace Norm

6.1. Let us denote by $\mathcal{H}_2$ the Hilbert space of operators of Hilbert-Schmidt class on $L^2(\mathbf{R}^d)$ with scalar product and norm given by

$$(S, T)_2 = \text{Tr}(ST^\dagger), \quad \|\|T\|\|_2^2 = \text{Tr}\,(TT^\dagger) \ .$$

We also consider the trace norm $||| \cdot |||_1$; we have

$$||| AB |||_1 \leq ||| A |||_2 ||| B |||_2 .$$

Recall the class $\mathcal{V}$ of potentials defined in Sec. 4. We consider the Hamiltonians $H(\varepsilon)^{(\bullet)} = -\Delta(\varepsilon)^{(\bullet)} + V_\varepsilon$ where the notation is as before except that for the truncated Laplacian we use $2Q_c^*$ as in Sec. 5, with Q_c^* symmetric so that the Hamiltonians are self-adjoint.

Theorem 1. *Let $V \in \mathcal{V}$. Then for any $t > 0$,*

$$||| e^{-tH(\varepsilon)^{(\bullet)}} - e^{-tH} |||_1 \longrightarrow 0 \quad (\varepsilon \longrightarrow 0) .$$

Proof. We do this first with respect to $||| \cdot |||_2$. We note that for this it is sufficient to prove that

$$\mathrm{Tr}\left(e^{-tH(\varepsilon)^{(\bullet)}} \right) \longrightarrow \mathrm{Tr}\left(e^{-tH} \right) \quad (*)$$

for each $t > 0$. For this means that

$$||| e^{-tH(\varepsilon)^{(\bullet)}} |||_2^2 = \mathrm{Tr}(e^{-2tH(\varepsilon)^{(\bullet)}}) \longrightarrow \mathrm{Tr}(e^{-2tH}) = ||| e^{-tH} |||_2^2 .$$

By Corollary to Proposition 2.2 and the density of the operators of finite rank in $\mathcal{H}_2$ we have

$$e^{-tH(\varepsilon)^{(\bullet)}} \longrightarrow e^{-tH}$$

weakly in $\mathcal{H}_2$ while the norms also converge. Hence we have convergence in $\mathcal{H}_2$. The convergence $(*)$ is now immediate from the results of Secs. 4 and 5 except for the Schwinger Hamiltonians. However for $H(\varepsilon)^{(s)}$ we use as before the comparison with the periodic Hamiltonians. Indeed it is clear from the inequality

$$\mathrm{Tr}(e^{-tH(\varepsilon)^{(s)}}) \leq \mathrm{Tr}(e^{-tH(\varepsilon)^{(p)}})$$

that

$$\mathrm{Tr}\left(e^{-tH} \right) = \lim_\varepsilon \mathrm{Tr}\left(e^{-tH(\varepsilon)^{(p)}} \right) \geq \limsup_\varepsilon \mathrm{Tr}\left(e^{-tH(\varepsilon)^{(s)}} \right) \geq \liminf_\varepsilon \mathrm{Tr}\left(e^{-tH(\varepsilon)^{(s)}} \right) .$$

On the other hand, $e^{-tH(\varepsilon)^{(s)}} \longrightarrow e^{-tH}$ strongly. So if $f_1, f_2, \ldots,$ is an orthonormal basis for $L^2(\mathbf{R}^d)$,

$$\mathrm{Tr}\left(e^{-tH(\varepsilon)^{(s)}} \right) \geq \sum_{1 \leq j \leq k} \left(e^{-tH(\varepsilon)^{(s)}} f_j, f_j \right) \longrightarrow \sum_{1 \leq j \leq k} \left(e^{-tH} f_j, f_j \right)$$

so that, letting $k \longrightarrow \infty$,

$$\liminf_\varepsilon \mathrm{Tr}\left(e^{-tH_N^{(s)}} \right) \geq \mathrm{Tr}\left(e^{-tH} \right)$$

which proves ($*$). For furnishing the proof with $|||\cdot|||_1$ we compare trace norms at t with Hilbert–Schmidt norms at $t/2$ and use the relation

$$|||A^2 - B^2|||_1 \leq |||A + B|||_2 \cdot |||A - B|||_2 + 2|||B|||_2 \cdot |||A - B|||_2$$

which follows from the identity

$$A^2 - B^2 = (A + B)(A - B) + (A - B)B - B(A - B)$$

6.2. The trace convergence implies of course convergence in the uniform operator topology and hence the convergence of eigenvalues and eigenfunctions. The results in Secs. 4 and 5 allow us, as we shall show now, to refine the L^2-convergence of the eigenfunctions to uniform convergence on compacta at least *for the stochastic Hamiltonians* (this means that the Laplacian is approximated by $\Delta(\varepsilon)$ for the infinite lattice and by the stochastic Q_c^* in the finite grid). We treat first the case of the infinite lattice approximation $H(\varepsilon)$. The proof for it is based on the following two simple lemmas.

Lemma 1. *Fix $t > 0$. Then there is a constant $C = C(t) > 0$ such that for all $f \in L^2(\mathbf{R}^d)$ and all ε,*

$$\|e^{-tH(\varepsilon)}f\|_\infty \leq C\|f\|_2, \qquad \|e^{-tH}f\|_\infty \leq C\|f\|_2 .$$

Proof. The kernel K_V that defines e^{-tH} as an integral operator satisfies

$$0 \leq K_V(x, y) \leq (2\pi t)^{-d/2}e^{-|x-y|^2/2t}$$

so that

$$|(e^{-tH}f)(x)| \leq (4\pi t)^{-d/4}\|f\|_2 .$$

Further

$$\|e^{-tH(\varepsilon)}f\|_\infty^2 \leq \sup_j \sum_k |e^{-tH(\varepsilon)}(j_c, k_c)|^2 \cdot \|f\|_2^2 \cdot \varepsilon^{-d} .$$

But, as $e^{-tH(\varepsilon)}$ is self-adjoint, Lemma 4.2.1 gives

$$\sum_k |e^{-tH(\varepsilon)}(j_c, k_c)|^2 = e^{-2tH(\varepsilon)}(j_c, j_c) \leq B_{2t}\varepsilon^d .$$

Lemma 2. *Fix $t > 0$. Then, for any $f \in L^2(\mathbf{R}^d)$,*

$$e^{-tH(\varepsilon)}f \longrightarrow e^{-tH}f$$

uniformly on compacta.

Proof. In view of Lemma 1 it is enough to do this for a dense set of f, say $f \in C_c(\mathbf{R}^d)$. Now it is obvious that $e^{-tH(\varepsilon)}$ is the integral operator whose defining kernel is $K(\varepsilon)$ where

$$K(\varepsilon)(x,y) = e^{-tH(\varepsilon)}(j_c, k_c)\varepsilon^{-d} \quad (x \in R(j_c), y \in R(k_c))$$

where for $r \in \mathbf{Z}^d, R(r_c)$ is the cube centered at r_c of sides ε. Lemma 4.2.2 implies that $K(\varepsilon) \longrightarrow K_V$ uniformly on compact sets of $\mathbf{R}^d \times \mathbf{R}^d$. The present lemma is now immediate. $\qquad\qquad\square$

Let f_j be an orthonormal sequence of eigenfunctions of H and $f(\varepsilon)_j$ be a corresponding family for $H(\varepsilon)$ converging in $L^2(\mathbf{R}^d)$.

Proposition 3. *We have*

$$f(\varepsilon)_j \longrightarrow f_j$$

uniformly on compact sets of $\mathbf{R}^d$.

Proof. It is enough to show that

$$e^{-tH(\varepsilon)}f(\varepsilon)_j \longrightarrow e^{-tH}f_j$$

uniformly on compacta. But

$$e^{-tH(\varepsilon)}f(\varepsilon)_j - e^{-tH}f_j = e^{-tH(\varepsilon)}(f(\varepsilon)_j - f_j) + (e^{-tH(\varepsilon)}f_j - e^{-tH}f_j)$$

and the result follows from Lemmas 1 and 2.

We shall now treat the case of the finite grid stochastic approximation. It is clearly enough to establish the analogues of the two lemmas above for $H(\varepsilon)^{(*)}$ in place of $H(\varepsilon)$.

Lemma 1. *Fix $t > 0$. Then for any compact set $E \subset \mathbf{R}^d$ there is a constant $C = C(E,t) > 0$ such that for all $f \in L^2(\mathbf{R}^d)$,*

$$\|e^{-tH(\varepsilon)^{(*)}}f\|_{E,\infty} \leq C\|f\|_2$$

where $\|\cdot\|_{E,\infty}$ is the L^∞-norm on E.

Proof. As before it is enough to show that for any $a > 0, u > 0$ there is a constant $B_{a,u} > 0$ such that for all ε

$$\sup_{|j_c| \leq a} e^{-uH(\varepsilon)^{(*)}}(j_c, j_c) \leq B_{a,u}\varepsilon^d .$$

This is immediate from Lemma 5.3.2 and Lemma 4.2.1.

Lemma 2. *We have, for all* $f \in L^2(\mathbf{R}^d)$ *and stochastic* $H(\varepsilon)^{(*)}$,

$$e^{-tH(\varepsilon)^{(*)}} f \longrightarrow e^{-tH} f$$

uniformly on compacta.

Proof. It is a question of showing that

$$e^{-tH(\varepsilon)^{(*)}} f - e^{-tH(\varepsilon)} f \longrightarrow 0$$

uniformly on compact sets for any $f \in C_c(\mathbf{R}^d)$. This comes to showing that for each $a > 0$

$$\sup_{\substack{|j_c| \le a \\ |k_c| \le a}} \sum \left| e^{-tH(\varepsilon)^{(*)}}(j_c, k_c) - e^{-tH(\varepsilon)}(j_c, k_c) \right| = o(1)$$

This is immediate from Lemma 5.3.2.

The following is now clear.

Proposition 3. *Denoting by* $f(\varepsilon)^*_j$ *an orthonormal sequence of eigenfunctions for stochastic* $H(\varepsilon)^{(*)}$ *converging in* L^2-*norm to* f_j, *we have*

$$f(\varepsilon)^*_j \longrightarrow f_j$$

uniformly on compacta.

References

[1] J. Schwinger, *Proc. Nat. Acad. Sci. USA* **46** (1960) 570–579, 1401–1405.
[2] H. Weyl, *Group Theory and Quantum Mechanics*, Dover, 1931.
[3] H. Kreiss, "Difference approximations for boundary and eigenvalue problems for ordinary differential equations", *Math. Comp.* **26** (1972) 605–624.
[4] M. Reed, and B. Simon, *Methods of Modern Mathematical Physics*, II, Academic Press, 1975.
[5] R. Narasimhan, *Analysis on Real and Complex Manifolds*, North Holland, 1968.
[6] M. Reed, and B. Simon, *Methods of Modern Mathematical Physics*, I, Academic Press, 1980.
[7] D. Freedman, *Markov Chains*, Springer Verlag, 1971.
[8] P. Billingsley, *Weak Convergence of Probability Measures*, Wiley, 1968.
[9] K. R. Parthasarathy, *Probability Measures on Metric Spaces*, Academic Press, 1967.
[10] B. Simon, *Functional Integration and Quantum Physics*, Academic Press, 1979.
[11] N. N. Chentsov, *Theory of Probab. App.* 1 (1956) 140–144.
[12] F. Riesz and B. Sz. Nagy, *Lecons d'Analyse Fonctionnelle*, Budapest, 1952.
[13] S. R. S. Varadhan, *Diffusion Problems and Partial Differential Equations*, Tata Institute Lectures, Bombay, 1980.

Ann. Inst. Henri Poincaré, (In Memoriam C. KIPNIS)

Vol. 31, n° 1, 1995, p. 273-285.   *Probabilités et Statistiques*

Self Diffusion of a tagged particle in equilibrium for asymmetric mean zero random walk with simple exclusion

by

S. R. S. VARADHAN (*)

Courant Institute of Mathematical Sciences,
New York University, New York, U.S.A.

Dedicated to the memory of Claude Kipnis.

ABSTRACT. – We consider a tagged particle in a simple exclusion model where the probability distribution of the jump sizes has zero mean, but is not necessarily symmetric. We establish for the tagged particle, a central limit theorem under the usual scaling.

RÉSUMÉ. – Nous considérons une particule marquée dans un modèle d'exclusion simple où la loi de probabilité des sauts est de moyenne nulle mais pas nécessairement symétrique. Nous démontrons pour le déplacement de la particule marquée un théorème central limite avec le changement d'échelle usuel.

1. INTRODUCTION

We consider a random walk with simple exclusion. This means that we have a collection of particles in $Z^d, d \geq 1$ with at most one particle per site. Each particle waits for an exponential time with mean 1 and at the end of this time picks a random site to jump to. The probability that a

(*) Supported by NSF grant DMS-9201222 and ARO grant DAAL03-92-G-0317.
Classification A.M.S.: 60 K 35, 60 F 17.

Annales de l'Institut Henri Poincaré - Probabilités et Statistiques - 0246-0203
Vol. 31/95/01/$ 4.00/© Gauthier-Villars

particle, located at x, picks the site y to jump to is given by $p(y - x)$. However the jump can be executed only if the site y is free. Otherwise the particle remains in the original site and waits for a new exponential time. All the particles are doing this simultaneously and independently of each other. Since we are dealing with processes in continuous time ties will never occur and this describes the evolution completely.

For a more mathematical definition of the model we start with the state space Ω consisting of functions η on Z^d taking values either 0 or 1. If $\eta(x) = 1$, then there is a particle at x and if $\eta(x) = 0$, the site x is free. We next define certain transformations on the space Ω.

$$(\sigma_{x,y}\eta)(a) = \eta(a) \quad if \quad a \neq x \ or \ y$$
$$(\sigma_{x,y}\eta)(x) = \eta(y)$$
$$(\sigma_{x,y}\eta)(y) = \eta(x)$$

The transformation $\sigma_{x,y}$ for two sites x, y in Z^d with $x \neq y$ could be either a particle jumping from x to y or one jumping from y to x. Of course if $\eta(x) = \eta(y)$ then $\sigma_{x,y}$ does nothing and is identity. Given $p(x) \geq 0$ with $p(0) = 0$ and $\Sigma p(x) = 1$, i.e. the distribution of jumps, we can define a formal infinitesimal generator acting on suitable functions F on Ω

$$(LF)(\eta) = \Sigma_{x,y}\eta(x)(1 - \eta(y))p(y - x)[F(\sigma_{x,y}\eta) - F(\eta)]$$

It is known [3] that this generator defines a good Markov Process on Ω. For each ρ in $0 \leq \rho \leq 1$, we have on Ω the Bernoulli product measure P_ρ with $P_\rho[\eta(x) = 1] = \rho$ for all x in Z^d. Each P_ρ is an invariant measure for our evolution. Corresponding to each ρ we have a system of interacting particles with initial distribution P_ρ which is in equilibrium. Of $p(x)$ we assume that it is zero outside a finite set. The case when $p(x) = p(-x)$ is the symmetric case and then the process is reversible with respect to any one of the invariant measures. But in this article we will assume only that

$$\Sigma < x, \theta > p(x) = 0$$

or that the jumps have mean zero. Now the process is not necessarily reversible. We will also assume an irreducibility condition namely that $\{x : p(x) > 0\}$ generates the whole group Z^d.

If we start our evolution under the assumption that there is a particle at 0 and that at other sites we have a random Bernoulli configuration with density ρ, we can tag the particle at 0 as it moves around and denote its location at time t by z_t. We are interested in proving a Brownian scaling limit for z_t, i.e. to establish that $\lambda^{-\frac{1}{2}}z_{\lambda t}$ converges as $\lambda \to \infty$ to a

Brownian motion in distribution. In [2], with Claude Kipnis we proved that this was indeed so in the symmetric case. The case $p(1) = p(-1) = \dfrac{1}{2}$ in one dimension is special and leads to Brownian motion with zero variance. In all other cases it is nondegenerate. In this article we will extend these results to the asymmetric mean zero case.

2. EVOLUTION OF THE TAGGED PARTICLE

In order to follow the motion of a tagged particle the state space has to be changed. The state space Ω cannot distinguish between particles. We shall find it convenient to describe the current state of the system by giving the location of the tagged particle and the environment around the tagged particle. In other words the state space is $\hat{\Omega} = Z^d \times \Omega_0$. Here Ω_0 is the space of configurations on $Z^d - \{0\}$. In order to describe the evolution in the space $\hat{\Omega}$ we need to describe a transformation τ_x that acts on Ω_0 for each $x \neq 0$.

$$(\tau_x \eta)(a) = \eta(x + a) \quad if \quad a \neq 0 \quad or \quad -x$$
$$(\tau_x \eta)(-x) = 0$$

$\tau_x \eta$ is meaningful only if $\eta(x) = 0$ and describes the effect of the tagged particle, which by definition is always at zero, jumping to x. We relocate the origin at x the new location of the tagged particle. The transformations $\sigma_{x,y}$ for $x, y \neq 0$ are well defined on Ω_0. The evolution of the tagged particle on $\hat{\Omega}$ is governed by the following generator

$$(\hat{L}F)(z, \eta) = \Sigma_{x,y\neq0}\eta(x)(1 - \eta(y))p(y - x)[F(z, \sigma_{x,y}\eta) - F(z, \eta)]$$
$$+ \Sigma_x(1 - \eta(x))p(x)[F(z + x, \tau_x\eta) - F(z, \eta)].$$

If F were a function of η only, then

$$(L_0 F)(\eta) = \Sigma_{x,y\neq0}\eta(x)(1 - \eta(y))p(y - x)[F(\sigma_{x,y}\eta) - F(\eta)]$$
$$+ \Sigma_x(1 - \eta(x))p(x)[F(\tau_x\eta) - F(\eta)].$$

In other words if (z_t, η_t) is the tagged system , η_t by itself is a Markov Process with generator L_0 . The Bernoulli measures P_ρ^0 on Ω_0 defined by

$$P_\rho^0[\eta(x) = 1] = \rho \quad for \ all \ x \ \in Z^d - \{0\}$$

are invariant measures for L_0. $\hat{L}$ on the other hand does not have an invariant measure because the z_t part will wander away in Z^d.

Our initial distribution on $\hat{\Omega}$ is $\delta_0 \times P_\rho^0$, *i.e.* we start from $z = 0$ and η is in equilibrium . Then under the $\hat{L}$ evolution η is always in equilibrium. z_t is the location of the tagged particle at time t and the scaling limit is to be established for it.

3. OUTLINE OF PROOF

A direct calculation yields

$$\hat{L}z = \Sigma_x(1 - \eta(x))xp(x)$$

or equivalently for each θ in R^d

$$\hat{L} < z, \theta > = \Sigma_x(1 - \eta(x)) < x, \theta > p(x)$$
$$= < \psi(\eta), \theta > .$$

Therefore,

$$< z_t, \theta > = \int_0^t < \psi(\eta_s), \theta > ds + < M(t), \theta > .$$

Here $M(t)$ is a Martingale with stationary increments. The basic idea in [2] was to replace the first term on the right in the above equation by a Martingale term and an error term.

$$\int_0^t < \psi(\eta_s), \theta > ds = < N(t), \theta > + < E(t), \theta > .$$

Our ability to do so depended on the following procedure.

Step 1.

Solve the equation

$$\lambda u_\lambda - L_0 u_\lambda = \psi$$

For simplicity let us take just one component of ψ . Based on an estimate

$$|E^{P_\rho^0}[\psi(\eta)G(\eta)]| \le C_\rho[D_\rho(G)]^{\frac{1}{2}}$$

where $D_\rho(u) = - < L_0 u, u >_\rho$ is the Dirichlet form of u, we get for the solution u_λ

$$E[\lambda u_\lambda^2 + D_\rho(u_\lambda)] = < \psi, u_\lambda >_\rho$$
$$\leq C_\rho [D_\rho(u_\lambda)]^{\frac{1}{2}}.$$

This gives us bounds

$$\lambda E u_\lambda^2 \leq C_\rho^2$$
$$D_\rho(u_\lambda) \leq C_\rho^2$$

holding uniformly as $\lambda \to 0$.

Step 2.

Using either spectral theory or a more direct argument one establishes in fact that

$$\lambda E[u_\lambda^2] \to 0 \quad as \quad \lambda \to 0$$

and

$$D_\rho(u_{\lambda_1} - u_{\lambda_2}) \to 0 \quad as \quad \lambda_1, \lambda_2 \to 0.$$

This was enough to reduce the Brownian scaling property for z_t to the same thing for a Martingale with stationary increments which is elementary and standard.

Step 3.

We prove compactness by establishing the tightness of the distributions of

$$\int_0^t \psi(\eta_s) ds$$

under Brownian scaling.

Step 4.

Establish the nondegeneracy of the Brownian motion obtained as the scaling limit.

4. DETAILS OF PROOF

Due to the asymmetry one has to make changes in the proof along the way.

Step 1.

$$\psi(\eta) = \Sigma(1 - \eta(x)) < x, \theta > p(x)$$
$$= \Sigma_{x \neq 0} \eta(x) \omega_\theta(x)$$

where $\omega_\theta(x)$ are weights with $\Sigma \omega_\theta(x) = 0$. We can therefore rewrite

$$\psi(\eta) = \Sigma_{x,y \neq 0} [\eta(x) - \eta(y)] \tilde{\omega}(x, y)$$

for some weights $\tilde{\omega}(x, y)$. Because of irreducibility we can assume that $\tilde{\omega}(x, y) \neq 0$ only if $p(y - x) > 0$. Since $\eta(x) - \eta(y) = -((\sigma_{x,y}\eta)(y) - (\sigma_{x,y}\eta)(x))$ we can write,

$$E[G(\eta)\psi(\eta)] = \frac{1}{2} E[\Sigma \tilde{\omega}(x, y)[\eta(x) - \eta(y)][G(\sigma_{x,y}\eta) - G(\eta)]$$
$$\leq C[D_1(G)]^{1/2}.$$

where

$$D_1(G) = E[\Sigma_{x,y \neq 0} p(y - x)[G(\sigma_{x,y}\eta) - G(\eta)]^2.$$

Here and in what follows E is expectation relative to P_ρ^0 for some fixed $0 < \rho < 1$. If we denote by

$$D_2(G) = E[\Sigma_x [G(\tau_x\eta) - G(\eta)]^2]$$

then the full Dirichlet form for L_0 is

$$D_0(G) = D_1(G) + D_2(G)$$

Step 2.

If now one solves the equation

$$\lambda u_\lambda - L_0 u_\lambda = \psi$$

one can get the estimates

$$\lambda E u_\lambda^2 + D_0(u_\lambda) = E\psi u_\lambda \leq C[D_1(u_\lambda)]^{\frac{1}{2}},$$
$$\lambda E u_\lambda^2 \leq C, \tag{4.1}$$
$$D_0(u_\lambda) \leq C.$$

We can as before take a subsequence of u_λ converging to a limit u in H_1. Here H_1 is the completion of nice functions with respect to the

Dirichlet norm $D_0(u)$. It contains a lot of generalized functions and consists technically of limits of equivalence classes of functions modulo constants. Let us suppose that the following estimate holds. For all functions F and G,

$$E[FL_0G] \le C[D_0(F)]^{\frac{1}{2}}[D_0(G)]^{\frac{1}{2}} \qquad (4.2)$$

then L_0 will be a bounded operator from H_1 into H_{-1} and $\lambda u_\lambda \to 0$ weakly in H_{-1}. Therefore

$$L_0 u = -\psi \quad in \quad H_{-1}$$

From this we obtain,

$$D_0(u) = <u, \psi>$$

where $<,>$ is the pairing between H_1 and H_{-1}, the latter being the formal dual of the former. Taking limits as $\lambda \to 0$ in (4.1) if $d_1 = lim\ \lambda E u_\lambda^2$ and $d_2 = lim\ D_0(u_\lambda)$ along suitable subsequences then

$$d_1 + d_2 = <u, \psi> = D_0(u)$$

By lower semicontinuity $D_0(u) \le d_2$. Therefore $d_1 = 0$ and $d_2 = D_0(u)$. This proves the strong convergence of u_λ in H_1 and also establishes

$$lim\ \lambda E u_\lambda^2 = 0.$$

It is now routine to prove the uniqueness of the limit point. So the inequality (4.2) which is obvious with $C = 1$ in the symmetric case is the crucial step in the nonsymmetric case. We shall prove it in the last section. The rest of the details are identical to the symmetric case.

Step 3.

Whereas in [2], for the symmetric case, compactness was a consequence of the estimate in step 1, in our more general context we need to use some special properties of ψ. We start with the estimate

$$\frac{1}{t}logE\left\{exp\left[\lambda \int_0^t \psi(\eta_s)ds\right]\right\} \le \sup_{||G||_2=1} [\lambda E[\psi G^2] - D_0(G)]$$

As in step 1

$$E[G^2\psi] = \frac{1}{2}E\Sigma[\eta(x) - \eta(y)]\tilde{\omega}(x,y)[G^2(\sigma_{x,y}\eta) - G^2(\eta)]$$

and

$$E[(\eta(x) - \eta(y))(G^2(\sigma_{x,y}\eta) - G^2(\eta))]$$
$$= E[[\eta(x) - \eta(y)][G(\sigma_{x,y}\eta) - G(\eta)][G(\sigma_{x,y}(\eta) + G(\eta)]]$$
$$\leq E[|G(\sigma_{x,y}\eta) - G(\eta)||G(\sigma_{x,y}\eta) + G(\eta)|]$$
$$\leq [E[G(\sigma_{x,y}\eta) - G(\eta)]^2]^{\frac{1}{2}}[E[G(\sigma_{x,y}\eta) + G(\eta)]^2]^{\frac{1}{2}}$$
$$\leq C[E[G(\sigma_{x,y}\eta) - G(\eta)]^2]^{\frac{1}{2}}.$$

We have used the facts that $|\eta(x) - \eta(y)| \leq 1$ and $EG^2(\sigma_{x,y}\eta) = EG^2(\eta) = 1$. Therefore

$$\sup_{||G||_2=1}[\lambda E\psi G^2 \quad - D_0(G)] \leq \sup_{||G||_2=1}[C\lambda(D_0(G))^{\frac{1}{2}} - D_0(G)]$$
$$\leq \frac{A\lambda^2}{2}$$

In other words

$$E\left[exp\left[\lambda\int_s^t \psi(\eta_s)ds\right]\right] \leq exp\left[A(t-s)\frac{\lambda^2}{2}\right].$$

Compactness is now a consequence of Garsia-Rodemick-Rumsey estimate which can be found in [4].

Step 4

We now establish the nondegeneracy of the limiting Brownian motion. From the outline of the proof it is clear that what we need to establish is that the martingale produced by $\int_0^t < \psi(\eta_s), \theta > ds$ namely $< N(t), \theta >$ cannot cancel the martingale $< M(t), \theta >$. There are two classes of martingales generated by the underlying Poisson events: jumps that involve the tagged particle, identified through τ_x and those involving other particles identified through $\sigma_{x,y}$ with $x, y \neq 0$. These are orthogonal martingales because they relate to non simultaneous Poisson jumps. If $N(t)$ were to totally cancel $M(t)$, which involves only the first type of jumps, its orthogonal projection on to the second type of martingales should go to zero. This can happen only if for the corresponding u_λ, $D_1(u_\lambda) \to 0$ as $\lambda \to 0$. But if $D_1(u_\lambda) \to 0$ then it follows from our estimates that $D_0(u_\lambda) \to 0$ as well and this forces $N(t)$ to be identically zero. With a small bit of extra calculation one can turn this into a lower bound for the variance of the limiting Brownian motion.

With only the basic estimate (4.2) remaining to be proved in the next section we have now established our main result.

THEOREM 4.1. – *If z_t is the location of the tagged particle in Z^d then as $k \to \infty$ the distribution of the stochastic process $\dfrac{z_{kt}}{\sqrt{k}}$ converges in Skorohod space to a Brownian motion on R^d with a nondegenerate covariance matrix.*

5. THE BASIC ESTIMATE

For the operator L_0 acting on functions defined on Ω_0

$$(L_0 F)(\eta) = \Sigma_x (1 - \eta(x))[F(\tau_x \eta) - F(\eta)]p(x)$$
$$+ \Sigma_{x,y \neq 0} \eta(x)(1 - \eta(y))p(y - x)[F(\sigma_{x,y}\eta) - F(\eta)]$$

we wish to establish the following estimate.

THEOREM 5.1. - *For functions $F(\eta)$ and $G(\eta)$*

$$| < F, L_0 G >_\rho | \leq C_\rho [D_0(F)]^{\frac{1}{2}} [D_0(G)]^{\frac{1}{2}}$$

for some constant C_ρ depending only on ρ.

Here $<, >_\rho$ represents the inner product in $L_2(P_\rho^0)$ and $D_0(F)$ is the Dirichlet form

$$D_0(F) = \frac{1}{2} E^{P_\rho^0} \{\Sigma_x (1 - \eta(x))p(x)[F(\tau_x \eta) - F(\eta)]^2$$
$$+ \Sigma_{x,y \neq 0} \eta(x)(1 - \eta(y))p(y - x)[F(\sigma_{x,y}\eta) - F(\eta)]^2\}$$

Before we prove Theorem 5.1 let us make some observations.

1. If L_0 were symmetric in $L_2(P_\rho^0)$ then the inequality is valid with $C_\rho = 1$.

2. Suppose A is a Markov generator on a finite state space with an invariant measure μ and A is not symmetric. Let $\bar{A}$ be the symmetrization $\dfrac{A + A^*}{2}$ in $L_2(\mu)$. Since A and $\bar{A}$ have the same range, if we have $Ag = h$ we also have for some $\bar{g}$, $\bar{A}\bar{g} = h$ and the map $g \to \bar{g}$ is well defined and the two Dirichlet forms are related by

$$D(\bar{g}) \leq CD(g)$$

Let us first prove a similar estimate for the untagged system L on Ω.

$$(LF)(\eta) = \Sigma_{x,y} \eta(x)(1 - \eta(y))p(y - x)[F(\sigma_{x,y}\eta) - F(\eta)]$$

LEMMA 5.2. – *For any two functions F,G on Ω and for any ρ*

$$| < F, LG >_\rho | \leq C_\rho [D(F)]^{1/2}[D(G)]^{1/2}.$$

Proof. – The proof depends on the following considerations. Suppose $a_1, a_2, \cdots a_k$ are k points in Z^d such that $a_1 + a_2 + \cdots + a_k = 0$, then the probability distribution $\pi(x)$ on Z^d defined by $\pi(a_j) = \frac{1}{k}$ for $j = 1, 2 \cdots k$ and $\pi(x) = 0$ for all other x has clearly mean 0. The sequence $y_0 = 0, y_1 = a_1, y_2 = a_1 + a_2, \cdots, y_k = a_1 + a_2 + \cdots + a_k = 0$ defines a cycle $C = \{y_0, y_1, \cdots, y_k\}$ and a π can be associated to each such cycle by taking $a_j = y_j - y_{j-1}$. Moreover one can assume that the cycle has no double points. Otherwise the cycle decomposes into two or more cycles and π_C is a convex combination of π_{C_i} corresponding to the component cycles. We can therefore limit ourselves to irreducible cycles. If $C_1, C_2, \cdots, C_l$ are l irreducible cycles and $w_1, w_2, \cdots, w_l$ are nonnegative weights adding up to 1, the convex combination $p(x) = \Sigma w_i \pi_{C_i}(x)$ is a mean zero probability distribution on Z^d. We have the converse

LEMMA 5.3. – *Any $p(x)$ with finite support and mean 0 has a representation*

$$p(x) = \Sigma w_i \pi_{C_i}(x)$$

for some weights $w_1, w_2, \cdots, w_l$ and irreducible cycles $C_1, C_2, \cdots, C_l$.

Proof. – A proof of this lemma can be found in [5].

We continue with the proof of lemma 5.2. Each cycle C introduces an operator A_C on Ω by

$$(A_C F)(\eta) = (1/k)\Sigma_{i=0}^{k-1}\eta(y_i)(1 - \eta(y_{i+1}))[F(\sigma_{y_i,y_{i+1}}\eta) - F(\eta)]$$

One can verify that A_C has P_ρ for an invariant measure and is in general nonreversible unless $k = 2$. For any cycle C we can consider the cycle $C + x$ described by $x, x + y_1, \cdots, x + y_k = x$ starting and ending at x rather than at 0 and define

$$(A_{C+x}F)(\eta) = (1/k)\Sigma_{i=0}^{k-1}\eta(x + y_i)(1 - \eta(x + y_{i+1}))$$
$$[F(\sigma_{x+y_i,x+y_{i+1}}\eta) - F(\eta)]$$

We can make a translation invariant generator on Ω by defining

$$L_C = \Sigma_{x \in Z^d} A_{C+x}$$

L_C is a typical generator for asymmetric mean zero random walk with simple exclusion. A consequence of lemma 5.3. is the representation

LEMMA 5.4. – *Our generator L for asymmetric mean zero simple exclusion has a representation*

$$L = \Sigma_{i=0}^l w_i L_{C_i}$$

for some $C_1, C_2, \cdots, C_l$ *which are irreducible cycles.*

Now we can complete the proof of lemma 5.2. The Dirichlet form for L is easily seen to be weighted sum of the forms for each C_i. Therefore there is no loss of generality in assuming $l = 1$ or $L = A_C$ for some irreducible C.

$$L = \Sigma_x A_{C+x}$$

The Dirichlet form for L_C is the sum of Dirichlet forms for each A_{C+x}. By Schwartz's inequality it is enough to prove the estimate for each A_{C+x} and this is essentially observation 2. The idea is that instead of detailed balance that produces reversibility we now have local balance that yields bounds.

Now we have to extend these considerations to the tagged system. We need another basic idea. Suppose $(\Omega, \mathcal{F}, P)$ is a probability space and $T_1, T_2, \cdots, T_k$ are k measure preserving transformations. Let us assume $T_k \cdots T_1 = I$, and define

$$(LF)(\omega) = \Sigma_{i=0}^k [F(T_i \omega) - F(\omega)].$$

Our claim is that our basic estimate is valid in this context.

LEMMA 5.5. – *We have for all F and G,*

$$|E[FLG]| \le C[D(F)]^{1/2}[D(F)]^{1/2}.$$

Proof.

$$\int FLG dP = \Sigma \int F(\omega)[G(T_i \omega) - G(\omega)] dP$$

$$= \Sigma \int F(T_{i-1} \cdots T_1 \omega)[G(T_i \cdots T_1 \omega) - G(T_{i-1} \cdots T_1 \omega)] dP$$

$$= \Sigma \int [F(T_{i-1} \cdots T_1 \omega) - F(\omega)]$$

$$[G(T_i \cdots T_1 \omega) - G(T_{i-1} \cdots T_1 \omega)] dP$$

because

$$\Sigma_i F(\omega)[G(T_i \cdots T_1 \omega) - G(T_{i-1} \cdots T_1 \omega)]$$
$$= F(\omega)[G(T_k \cdots T_1 \omega) - G(\omega)] = 0.$$

Rest is Schwartz.

Finally we return to the proof of Theorem 5.1. The operator

$$(L_0 F)(\eta) = \Sigma p(x)(1 - \eta(x))[F(\tau_x \eta) - F(\eta)]$$
$$+ \Sigma p(y - x)\eta(x)(1 - \eta(y))[F(\sigma_{x,y}\eta) - F(\eta)]$$

decomposes into cycles and we can assume without loss of generality that $p(x) = \pi_C(x)$ for some irreducible C. The problem is that 0 plays a special role and translates $C + x$ of the cycle C that touch 0 create trouble. We can write

$$L_0 F = L_1 F + L_2 F$$

where $L_2 F$ involves only jumps of the untagged particles and in addition to full cycles that are estimated without any trouble there are incomplete cycles because they go through the origin. L_1 on the other hand involves only jumps of the tagged particle.

Let us remark that

$$\eta(x)(1 - \eta(y))[F(\sigma_{x,y}\eta) - F(\eta)] = (1 - \eta(y))[F(\sigma_{x,y}\eta) - F(\eta)]$$

so that the terms of L_1 and L_2 look similar. Let our cycle be $0 = y_0, y_1, \cdots, y_k = 0$ with $a_i = y_i - y_{i-1}$ for $i = 1, \cdots, k$. Then ignoring constants 1/k and disregarding the full cycles,

$$(L_1 F)(\eta) = \Sigma_i (1 - \eta(a_i))[F(\tau_{a_i}\eta) - F(\eta)]$$
$$(L_2 F)(\eta) = \Sigma_{(x,y)\in\Delta}(1 - \eta(y))[F(\sigma_{x,y}\eta) - F(\eta)]$$

The rest of the proof can be best described in words. Suppose there is an empty site in the loop in front of the tagged particle, *i.e.* $\eta(a_1) = 0$. Then the tagged particle can move, that is we can apply τ_{a_1}. Now the empty site created at the origin is really at $-a_1 = (a_2 + \cdots + a_k)$ because the origin has shifted with the tagged particle. $-a_1$ is the last site of the loop that starts from the new origin *i.e.* the old a_1. One can effect $\sigma_{a_2 + \cdots + a_{k-1}, a_2 + \cdots + a_k}$ and get a free site at $a_2 + \cdots + a_{k-1}$. We can proceed in this fashion till we get a free site at a_2. Now the tagged particle can jump to a_2 and the whole process starts again. After several steps the tagged particle will return to its original starting point with an empty site in front and all other particles in exactly the same position that we started from. This takes exactly $n = k(k - 1)$ steps and we use up every term of L_1 and L_2 exactly once. In other words

$$(L_1 + L_2)(F)(\eta) = \Sigma \chi_{E_j}[F(S_j \eta) - F(\eta)]$$

Annales de l'Institut Henri Poincaré - Probabilités et Statistiques

where $E_{j+1} = S_j S_{j-1} \cdots S_1 E_1$ and $S_n S_{n-1} \cdots S_1 = I$ on E_1. We can use lemma 5.5 at this point and we are done.

6. REMARKS

The basic estimate (4.2) for the untagged system was derived and used by Lin Xu in his New York University PhD dissertation to establish a hydrodynamic scaling limit for mean zero random walks with simple exclusion. A central limit theorem for the position of a tagged particle in the case of asymmetric one dimensional simple exclusion (non zero mean) was considered in [1] by C. Kipnis.

ACKNOWLEDGEMENTS

The author wishes to thank F. Rezakhanlou for valuable discussions.

REFERENCES

[1] C. KIPNIS, Central limit theorems for infinite series of queues and applications to simple exclusion, *Ann. Prob.* Vol. **14**, 1986, pp. 397-408.
[2] C. KIPNIS and S. R. S. VARADHAN, Central limit theorem for additive functionals of reversible Markov processes and application to simple exclusions, *Comm. Math. Phys.* Vol. **104**, 1986, pp. 1-19.
[3] T. LIGGETT, *Interacting Particle Systems*, Springer-Verlag, 1985.
[4] D. W. STROOCK and S. R. S. VARADHAN, *Multidimensional Diffusion Processes*, Springer-Verlag, Berlin-Heidelberg-New York, 1979.
[5] L. XU, *Ph. D Dissertation,* New York University, 1993.

(Manuscript received April 9, 1994;
revised August 14, 1994.)

Communications on Pure and Applied Mathematics, Vol. LIII, 0972–1006 (2000)

Diffusive Limit of a Tagged Particle in Asymmetric Simple Exclusion Processes

SUNDER SETHURAMAN
Iowa State University

S. R. S. VARADHAN
Courant Institute

AND

HORNG-TZER YAU
Courant Institute

Dedicated to Dr C. R. Rao on his eightieth birthday.

Abstract

Invariance principles are proved under diffusive scaling for the centered position of a tagged particle in the simple exclusion process with asymmetric nonzero drift jump probabilities in dimensions $d \geq 3$. The method of proof is by martingale techniques which rely on the fact that symmetric random walks are transient in high dimensions. © 2000 John Wiley & Sons, Inc.

1 Introduction

In the study of interacting particle systems, an important component is the motion of a tagged particle as it moves through other particles that are viewed as indistinguishable among themselves. An initial step is to understand the behavior of a tagged particle in equilibrium, in particular, the fluctuation behavior. The simple exclusion processes are a natural class of models where one can study this problem. In the case of simple exclusion processes on $\mathbb{Z}^d$ when the underlying random walk has a symmetric distribution, this question is answered in [1] for $d = 1$ nearest-neighbor walks, and in [6] for all other distributions and dimensions. It was generalized to the case of random walks with mean 0 in [11]. When the mean is nonzero, the only case studied previously is the asymmetric nearest-neighbor $d = 1$ case in [5]. Here we consider the general case where the mean is not zero. We need to assume, however, that *the dimension d of the lattice $\mathbb{Z}^d$ is at least* 3.

If X is a countable set and $p(x, y)$ is the transition probability matrix of a Markov chain, then one can define the simple exclusion process on X with transition probabilities $p(\cdot, \cdot)$ on X. This is actually a Markov process whose state space is the set $\Omega(X)$ of all possible subsets of X. The set $A \subset X$ signifies that there are particles present at every point of A and that the sites in $X \setminus A$ are empty. One can also view $\Omega(X)$ as the set of maps $\eta : X \rightarrow \{0,1\}$ where $\eta(x) = 1$ means that $x \in A$ or that there is a particle at the site x. $\eta(x) = 0$ signifies that the site is empty. Each particle waits for an exponential time and then jumps to a new site or at least tries to. The new site y is selected with probability $p(x,y)$ where the current site is x. If the site y is empty, then the jump is completed and things start afresh. If the site is occupied with a particle already there, the jump is forbidden and the original particle remains at x. Again things start afresh. All the particles are doing this simultaneously, and because we are dealing with continuous time, there will be no ties to resolve. However, it requires some work to make sure that the process is well-defined, especially if there are an infinite number of particles to begin with.

Suppose $u : \Omega \rightarrow \mathbb{R}$ is a function that depends only on a finite number of coordinates, i.e., on $\{\eta(x) : x \in F\}$ for some finite set F; then the infinitesimal generator of the process is defined by

$$(\mathcal{L}u)(\eta) = \sum_{x,y} \eta(x)(1 - \eta(y))p(x,y)[u(\eta^{x,y}) - u(\eta)]$$

where

(1.1)
$$\eta^{x,y}(z) = \begin{cases} \eta(z) & \text{if } z \neq x \text{ or } y \\ \eta(y) & \text{if } z = x \\ \eta(x) & \text{if } z = y. \end{cases}$$

Under some mild conditions on $p(\cdot,\cdot)$ (for instance, it is sufficient to assume that $\sup_y \sum_x p(x,y) < \infty$), there is a well-defined stochastic process starting from any initial configuration in Ω. The details can be found in [8]. This is referred to as the simple exclusion process on X with transition probability $p(\cdot,\cdot)$. A special case of particular interest is when $p(\cdot,\cdot)$ is doubly stochastic, i.e.,

$$\sum_x p(x,y) = 1 \quad \text{for all } y \quad \text{in addition to} \quad \sum_y p(x,y) = 1 \quad \text{for all } x.$$

In this case the process is always well-defined. A more special case is when $p(\cdot,\cdot)$ is symmetric. This case is referred to as symmetric simple exclusion.

In the doubly stochastic case, the uniform measure on X is a σ-finite invariant measure for the Markov chain. One can verify that, for any $0 \leq \rho \leq 1$, the Bernoulli product measure P_ρ on Ω defined by $P_\rho[\eta(x) = 1] = \rho$ for every x, with $\{\eta(x)\}$ being mutually independent for different x, is an ergodic invariant measure for the evolution on Ω with generator L. If $p(\cdot,\cdot)$ is symmetric, the evolution is reversible with respect to each P_ρ and has for its Dirichlet form the quantity

$$D^\rho(u) = \frac{1}{2} \int_\Omega \sum_{x,y} p(x,y)[u(\eta^{x,y}) - u(\eta)]^2 dP_\rho.$$

Our interest is mainly in the case where $X = \mathbb{Z}^d$ for some $d \geq 3$ and $p(x,y) = p(y-x)$ for some probability distribution $p(\cdot)$ on $\mathbb{Z}^d$. We can assume without loss of generality that $p(0) = 0$. We are interested in the case when

$$(1.2) \qquad\qquad m = \sum_x x p(x) \neq 0.$$

We shall assume for simplicity that $p(x) = 0$ outside a finite set F, although it will not matter that much. We shall start the process in equilibrium, that is to say, with the initial distribution being some P_ρ but conditioned to have a particle at the origin 0, which will be tagged. As the system evolves, we wish to follow the trajectory of the tagged particle. It is convenient to change our coordinates in $\mathbb{Z}^d$ so that the tagged particle is always seen at the origin. In other words, our description of the current state consists of the position of the tagged particle, which we denote by z, and the environment seen from the tagged particle, which can be viewed as a point in $\Omega_0 = \{\eta : \mathbb{Z}^d \setminus \{0\} \to \{0, 1\}\}$.

There are two types of motion. When an untagged particle jumps, it is from some x to a y, neither of which can be 0. When the tagged particle jumps from 0 to an empty site x, the origin shifts with it, so what we see is a shift of the environment by $-x$. The tagged particle at 0 is not part of the shift, and so we always end up with $-x$ being empty. In other words, if we define on the set $\eta(x) = 0$, the map τ_x,

$$(\tau_x \eta)(y) = \begin{cases} \eta(x+y) & \text{for } y \neq 0,\ y \neq -x \\ 0 & \text{for } y = -x, \end{cases}$$

then the generator of our process is given by $\mathcal{L} = \mathcal{L}^{\mathrm{sh}} + \mathcal{L}^{\mathrm{ex}}$ where

$$(\mathcal{L}^{\mathrm{sh}} u)(z, \eta) = \sum_x (1 - \eta(x)) p(x) [u(z + x, \tau_x \eta) - u(z, \eta)]$$

and

$$(\mathcal{L}^{\mathrm{ex}} u)(z, \eta) = \sum_{x,y \neq 0} \eta(x)(1 - \eta(y)) p(y - x) [u(z, \eta^{x,y}) - u(z, \eta)].$$

The environment by itself is a Markov process, and the generator is given by $L = L^{\mathrm{sh}} + L^{\mathrm{ex}}$ where

$$(1.3) \qquad (L^{\mathrm{sh}} u)(\eta) = \sum_x (1 - \eta(x)) p(x) [u(\tau_x \eta) - u(\eta)]$$

$$(1.4) \qquad (L^{\mathrm{ex}} u)(\eta) = \sum_{x,y \neq 0} \eta(x)(1 - \eta(y)) p(y - x) [u(\eta^{x,y}) - u(\eta)].$$

We have adopted the convention that the generators associated with the original process are denoted by script $\mathcal{L}$, while those associated with the environment process are denoted by L. The Bernoulli product measure P_ρ restricted to points in $\mathbb{Z}^d \setminus \{0\}$ is an ergodic invariant measure for the environment process (see proposition 3 of [9]). An elementary computation shows that

$$(1.5) \qquad\qquad \mathcal{L}z = g(\eta) = \sum_x x p(x)(1 - \eta(x))$$

so that

$$z(t) - z(0) - \int_0^t g(\eta(s))ds = M(t)$$

is a martingale with stationary increments. One can see almost surely [9],

$$\lim_{t\to\infty} \frac{z(t)}{t} = \lim_{t\to\infty} \frac{1}{t} \int_0^t g(\eta(s))ds = \int_{\Omega_0} g(\eta)dP_\rho = m(1-\rho).$$

Let

(1.6) $$\xi(t) = \frac{1}{\sqrt{t}}(z(t) - m(1-\rho)t);$$

in this article we prove a functional central limit theorem for ξ. As remarked earlier, the symmetric case for nearest-neighbor $d = 1$ walks was considered in [1], and all the other symmetric cases in [6]. The more general case $m = 0$ was covered in [11]. Also, the $d = 1$ asymmetric nearest-neighbor $m \neq 0$ case was done in [5]. We are now interested in the case when $m \neq 0$ in high dimensions, which requires different methods. Our main result is the following theorem:

THEOREM 1.1 *As $\alpha \to \infty$, the distribution of $(1/\sqrt{\alpha})(z(\alpha t) - m(1 - \rho)\alpha t)$ converges weakly in Skorohod space to a nondegenerate Brownian motion with covariance $\mathbf{C}(\rho)$ given in formula (2.23).*

In the next section, we outline the proof of this theorem. Our proof follows the approach in [6] of considering the environment process as seen by the tagged particle and to explore the associated martingales. The key step in this approach is an estimate of the resolvent equation. The methods used here for solving the resolvent equation are similar to the methods used for solving the fluctuation-dissipation equation of the hydrodynamical limit of the simple exclusion processes in [7] and of lattice gases in [2]. Notice that both the fluctuation-dissipation equation and the resolvent equation require estimates on the Green's function of the generator, though in somewhat different contexts. More technical comments can be found at the end of the next section.

2 Outline of the Proof of Theorem 1.1

The tagged particle process has a lot of martingales associated with it. They are of the form

(2.1) $$M_x(t) = N_x(t) - \int_0^t p(x)(1 - \eta(s,x))ds$$

and

(2.2) $$M_{x,y}(t) = N_{x,y}(t) - \int_0^t p(y-x)\eta(s,x)(1 - \eta(s,y))ds,$$

corresponding to the number of jumps $N_x(t)$ of size x for the tagged particle or the number of jumps $N_{x,y}(t)$ of untagged particles from x to y. The compensators of their squares are given by

$$d\langle [M_{x,y}(t)]^2 \rangle = p(y-x)\eta(t,x)(1-\eta(t,y))dt,$$
$$d\langle [M_x(t)]^2 \rangle = p(x)(1-\eta(t,x))dt.$$

These are the basic martingales, and every other martingale is a combination of these. For example, the position $z(t)$ of the tagged particle satisfies

$$z(t) = \sum_x xN_x(t) = \sum_x xM_x(t) + \int_0^t \sum_x xp(x)(1-\eta(s,x))ds.$$

Subtracting the term $m(1-\rho)t$, we have

$$(2.3) \qquad \xi(t) = z(t) - m(1-\rho)t = \xi^{(1)}(t) + \int_0^t g(\eta(s))ds$$

where

$$(2.4) \qquad \xi^{(1)}(t) = \sum_x xM_x(t)$$

and g is the vector-valued function given by (1.5), i.e.,

$$g(\eta) = \sum_x xp(x)(\rho - \eta(x)).$$

The problem now reduces to proving the central limit theorem for the centered additive functional

$$A(t) = \int_0^t g(\eta(s))ds.$$

We would like to do it by the martingale method, which means finding a square integrable martingale $\xi^{(2)}(t)$ such that

$$A(t) = \xi^{(2)}(t) + B(t)$$

with $B(\cdot)$ becoming negligible under rescaling. Then the central limit theorem for square integrable martingales with stationary increments applies to the sum $\xi^{(1)}(t) + \xi^{(2)}(t)$, and this will establish the result. We may write the generator of the process that describes the environment of the tagged particle as the sum of its symmetric and antisymmetric pieces,

$$L = L_{\text{sym}} + L_{\text{skew}}.$$

Associated with the symmetric part, we have the Dirichlet form

$$(2.5) \qquad D^\rho(u) = D^{ex,\rho}(u) + D^{sh,\rho}(u)$$

where

$$(2.6) \qquad D^{ex,\rho}(u) = \frac{1}{2} \int_{\Omega_0} \sum_{x,y \neq 0} a(y-x)[u(\eta^{x,y}) - u(\eta)]^2 dP_\rho$$

and

$$(2.7) \qquad D^{sh,\rho}(u) = \frac{1}{2} \int_{\Omega_0} \sum_x a(x)(1-\eta(x))[u(\tau_x\eta) - u(\eta)]^2 dP_\rho$$

with $a(x) = \frac{1}{2}[p(x)+p(-x)]$. The associated Dirichlet norm is $\|\cdot\|_1 = \sqrt{D_\rho(u)}$. The dual norm $\|\cdot\|_{-1}$ is defined by

$$\|g\|_{-1} = \inf\left[C : \left|\int gu\, dP_\rho\right| \le C\|u\|_1 \text{ for all local functions } u\right].$$

We define $\mathbb{H}_1$ and $\mathbb{H}_{-1}$ as the Hilbert spaces generated by local functions with respect to the norms $\|\cdot\|_1$ and $\|\cdot\|_{-1}$. We shall drop the index ρ, which is fixed throughout the whole paper.

2.1 Step 1: $\mathbb{H}_{-1}$ Estimate and Tightness

Our first step is to show that for the function g defined in (1.1), $\|g\|_{-1} < \infty$. This follows from the next lemma.

LEMMA 2.1 *Let f be a mean zero local function, i.e., a function that depends only on a finite number of coordinates. Then for $d \ge 3$, there is a bound*

$$|\langle f, v\rangle| \le C_f \|v\|_1.$$

Lemma 2.1 will be proved in Section 3. Recall that $\sqrt{t}\,\xi$ is a sum of a martingale $\xi^{(1)}$ and $A(t)$. The martingale $\xi^{(1)}$ satisfies Doob's inequality. Hence the tightness of ξ in Skorohod space is deduced from the following general theorem:

THEOREM 2.2 *Let P be a stationary and ergodic Markov process with marginal μ on some state space. Let A and A^* be the infinitesimal generators in $L_2(\mu)$ of the original process and its time reversal, respectively. Let $S = \frac{1}{2}[A+A^*]$ be the symmetrized operator defined on $\mathrm{dom}(A) \cap \mathrm{dom}(A^*)$ with the associated Dirichlet form*

$$(2.8) \qquad D(u) = \langle -Su, u\rangle.$$

Let f be a function on the state space belonging to $L_2(\mu)$ such that for some sequence $u_n \in \mathrm{dom}(A) \cap \mathrm{dom}(A^*)$ we have

$$\|Su_n - f\|_{L_2(\mu)} \to 0$$

while

$$(2.9) \qquad \sup_n D(u_n) \le C.$$

Then for all $T \ge 0$

$$E^P\left\{\sup_{0\le t\le T}\left|\int_0^t f(x(s))ds\right|^2\right\} \le 8CT$$

with the same constant C as in equation (2.9).

PROOF: We deal with both the forward and backward filtrations.

$$u_n(x(t)) - u_n(x(s)) - \int_s^t (Au_n)(x(\sigma))d\sigma = M_n^+(t) - M_n^+(s)$$

are martingales adapted to the forward filtration for $t \geq s$.

$$u_n(x(s)) - u_n(x(t)) + \int_s^t (A^*u_n)(x(\sigma))d\sigma = M_n^-(s) - M_n^-(t)$$

are martingales adapted to the backward filtration for $s \leq t$. In any case

$$\int_s^t (Su_n)(x(\sigma))d\sigma = \frac{1}{2}[M_n^+(t) - M_n^+(s) + M_n^-(s) - M_n^-(t)].$$

Since

$$E^P\{[M_n^+(t) - M_n^+(s)]^2\} = E^P\{[M_n^-(t) - M_n^-(s)]^2\} \leq 2C|t-s|,$$

Doob's inequality applied separately to the two martingales provides an estimate that is uniform in n. Since $Su_n \to f$ in $L_2(\mu)$, we are done.   □

Remark. If $\text{dom}(A) \cap \text{dom}(A^*)$ is large enough to be a core for the Dirichlet space, from an estimate of the form

$$|\langle f, u \rangle| \leq \sqrt{C}\sqrt{D(u)}$$

valid for all $u \in \text{dom}(A) \cap \text{dom}(A^*)$ we can construct by standard resolvent techniques a sequence u_n with the desired properties, with the same exact constant C appearing in equation (2.9). In our case of tagged versions of simple exclusion processes, the space of local functions, included in all domains, is always large enough to form a core.

2.2 Step 2: Resolvent Estimate

Consider the resolvent equation

(2.10) $$\lambda u_\lambda - Lu_\lambda = f.$$

The key input for the proof of Theorem 1.1 is the following estimate for the solution of the resolvent equation (2.10) to be proved in Section 6.

THEOREM 2.3 *For any local f with mean 0 and $d \geq 3$,*

(2.11) $$\sup_\lambda \|Lu_\lambda\|_{-1} < \infty, \qquad \sup_\lambda \|\lambda u_\lambda\|_{-1} < \infty.$$

2.3 Step 3: Approximation via the Resolvent Equation

THEOREM 2.4 *If f is a local function with mean 0, then for the solution u_λ of the resolvent equation we have*

(2.12) $$\lim_{\lambda \to 0} \|u_\lambda - w\|_1 = 0$$

for some $w \in \mathbb{H}_1$ and

(2.13) $$\lim_{\lambda \to 0} \lambda \|u_\lambda\|_0^2 = 0.$$

PROOF: Multiplying the resolvent equation by u_λ and integrating, we get

$$(2.14) \qquad \lambda \|u_\lambda\|_0^2 + \|u_\lambda\|_1^2 = \langle u_\lambda, f \rangle.$$

From the estimate of Lemma 2.1, we see that $\sup_{0 \leq \lambda \leq 1} \|u_\lambda\|_1 < \infty$. Hence we can choose a subsequence $\lambda_n \to 0$ such that along this subsequence $u_{\lambda_n} = u_n$ has a weak limit w in the Dirichlet space $\mathbb{H}_1$. Because $\lambda \|u_\lambda\|_0^2$ is bounded, $\lambda u_\lambda \to 0$ as $\lambda \to 0$ in $L_2 = \mathbb{H}$ and consequently, from equation (2.11), converges to 0 in $\mathbb{H}_{-1}$. Together with the resolvent equation, this implies that Lu_n converges weakly to $-f$ in $\mathbb{H}_{-1}$. By standard functional analysis there are convex combinations v_n of $u_1, \ldots, u_n$ such that v_n and Lv_n converge strongly to w and $-f$ in $\mathbb{H}_1$ and $\mathbb{H}_{-1}$, respectively. It is easy to see, from equation (2.11), that $\|v_n\|_1 \to \|w\|_1$ and $\langle v_n, Lv_n \rangle$ converges to $-\langle w, f \rangle$, thus proving

$$\langle w, f \rangle = \|w\|_1^2.$$

From equation (2.14), we have, in particular,

$$(2.15) \qquad \limsup_{n \to \infty} \|u_{\lambda_n}\|_1^2 \leq \limsup_{n \to \infty} [\lambda_n \|u_{\lambda_n}\|_0^2 + \|u_{\lambda_n}\|_1^2] = \langle w, f \rangle = \|w\|_1^2.$$

Since $\| \cdot \|_1$ is lower semicontinuous and $u_{\lambda_n} \to w$ weakly,

$$\|w\|_1^2 \leq \liminf_{n \to \infty} \|u_{\lambda_n}\|_1^2.$$

We have thus proved that the equality holds in equation (2.15). This in turns implies that u_{λ_n} converges to w strongly. The proof of Theorem 2.4 is completed by showing that the limits w_1 and w_2 along any two subsequences λ_n and λ_n' are necessarily equal. To see this, we observe that $L(u_{\lambda_n} - u_{\lambda_n'}) \to 0$ weakly in $\mathbb{H}_{-1}$ while $u_{\lambda_n} - u_{\lambda_n'} \to w_1 - w_2$ strongly in $\mathbb{H}_1$, proving that

$$\|u_{\lambda_n} - u_{\lambda_n'}\|_1^2 = -\langle L(u_{\lambda_n} - u_{\lambda_n'}), (u_{\lambda_n} - u_{\lambda_n'}) \rangle \to 0.$$

$\square$

We now have the following martingale decomposition theorem:

THEOREM 2.5 *There is a square integrable martingale $M(t)$ with stationary increments and an additive functional $\Omega(t)$ such that*

$$\int_0^t f(\eta(s))ds = M(t) + \Omega(t)$$

with

$$(2.16) \qquad T^{-1}EM(T)^2 \leq \|f\|_{-1}^2$$

and

$$(2.17) \qquad \lim_{T \to \infty} \frac{1}{T} E[[\Omega(T)]^2] = 0.$$

PROOF: The proof is based on Theorem 2.4. From Ito's formula

$$u_\lambda(\eta(t)) - u_\lambda(\eta(0)) = \int_0^t Lu_\lambda(\eta(s))ds + M_\lambda(t),$$

where M_λ denotes the martingale part with quadratic variation given by $EM_\lambda(t)^2 = t\|u_\lambda\|_1^2$. From the resolvent equation (2.10),

$$\int_0^t f(\eta(s))ds = M_\lambda(t) + \Omega_\lambda(t)$$

where

$$\Omega_\lambda(t) = u_\lambda(\eta(t)) - u_\lambda(\eta(0)) - \int_0^t \lambda u_\lambda(\eta(s))ds.$$

By Theorem 2.4, the martingale part converges as $\lambda \to 0$, to, say, $M(t)$. Clearly, $M(t)$ satisfies the estimate (2.16). Hence we obtain

$$\int_0^t f(\eta(s))ds = M(t) + \Omega(t)$$

where

$$\Omega(t) = \Omega_\lambda(t) + [M_\lambda(t) - M(t)].$$

If we pick $\lambda = 1/T$, then together with (2.13)

$$\frac{1}{T}E^P[(\Omega(T))^2] \leq \frac{12}{T}\|u_{1/T}\|_0^2 + \frac{4}{T}E^P\left[[M_{1/T}(T) - M(T)]^2\right]$$

$$= \frac{12}{T}\|u_{1/T}\|_0^2 + 4E^P\left[[M_{1/T}(1) - M(1)]^2\right] \to 0$$

as $T \to \infty$. This concludes the proof. $\qquad\square$

We can now apply Theorem 2.5 to our setting with f taken to be the function g defined in equation (1.5). Hence

(2.18) $$\xi(t) = \xi^{(1)}(t) + \xi^{(2)}(t) + \Omega(t)$$

where $\xi^{(1)}$ is defined in equation (2.4), $\xi^{(2)}(t)$ is the martingale obtained from applying Theorem 2.5, and $\Omega(t)$ is the error term satisfying equation (2.17). Hence, up to a negligible error, we have two square integrable martingales $\xi_1(t)$ and $\xi_2(t)$ with stationary increments adapted to the environment process. From the martingale convergence theorem (see theorem 3.2 of [3] where condition (b) can be checked in this situation), $\xi(\alpha t)/\sqrt{\alpha}$, as $\alpha \uparrow \infty$, converges to a Brownian motion with the covariance matrix $\mathbf{C}(\rho)$ characterized by

$$\langle \mathbf{C}(\rho)a, a\rangle = E^{P_\rho}\{\langle \xi_1(1) + \xi_2(1), a\rangle^2\}.$$

Hence our final task is to estimate the variance.

2.4 Step 4: Bounds on the Variance

For upper bounds on the variance, we can estimate each term separately, where, as for lower bounds, we have to worry about possible cancellations. We start with the upper bound. From the definition of $\xi^{(1)}$ in (2.4), we have immediately

$$(2.19) \qquad E^{P_\rho}\{\langle \xi_1(1), a\rangle^2\} = (1-\rho)\sum_x p(x)\langle x, a\rangle^2.$$

An upper bound of $\xi^{(2)}$ can also be easily obtained from equation (2.16)

$$(2.20) \qquad E\{\langle \xi_2(1), a\rangle^2\} \leq C\rho(1-\rho)\sum_x p(x)\langle x, a\rangle^2.$$

Therefore, we have the upper bound that follows:

THEOREM 2.6 $\langle \mathbf{C}(\rho)a, a\rangle \leq C(1-\rho)\sum_x p(x)\langle x, a\rangle^2.$

The lower bound can be reduced to the symmetric case. The limiting variance of the tagged particle in the asymmetric case is bounded below by the limiting variance of the tagged particle in the corresponding symmetric case. A lower bound for the symmetric case was derived in [10].

THEOREM 2.7 $\langle \mathbf{C}a, a\rangle \geq C_3(1-\rho)\langle a, a\rangle.$

PROOF: The martingale $\xi^{(2)}(t)$ can be expressed in terms of the solution w obtained in Therorem 2.4

$$(2.21) \qquad \xi^{(2)}(t) = \sum_x w_x M_x(t) + \sum_{x,y} w_{x,y} M_{x,y}(t)$$

where $\{M_x(\cdot)\}$ and $\{M_{x,y}(\cdot)\}$ are the basic martingales introduced in equations (2.1) and (2.2). The functions $\{w_x(\cdot)\}$ and $\{w_{x,y}(\cdot)\}$ are the limits of $u_{\lambda,x} = (u_\lambda(\tau_x\eta) - u_\lambda(\eta))$ and $u_{\lambda,x,y} = (u_\lambda(\eta^{x,y}) - u_\lambda(\eta))$, respectively, in $L_2(P_\rho)$. The convergence of u_λ to w in $\mathbb{H}_1$ guarantees the convergence to 0 of

$$E^{P_\rho}\left[\sum_x p(x)(1-\eta(x))(u_{\lambda,x} - w_x)^2 + \sum_{x,y\neq 0} p(y-x)(u_{\lambda,x,y} - w_{x,y})^2\right]$$

as $\lambda \to 0$. From (2.4), $\xi^{(1)}$ has the representation

$$(2.22) \qquad \xi^{(1)}(t) = \sum_x x M_x(t).$$

Combining equations (2.21) and (2.22), we get the representation

$$\xi^{(1)}(t) + \xi^{(2)}(t) = \sum_x (w_x + x) M_x(t) + \sum_x w_{x,y} M_{x,y}(t),$$

from which we can easily deduce that

$$(2.23) \qquad \langle \mathbf{C}(\rho)a, a\rangle = E^{P_\rho}\left[\sum_x p(x)(1-\eta(x))(w_x^a + \langle x, a\rangle)^2 + \sum_{x,y} p(y-x)w_{x,y}^2\right].$$

Here we have denoted by w^a the solution that corresponds to $f = \langle a, g \rangle$. The right-hand side of formula (2.23), being essentially a Dirichlet form, is clearly insensitive to replacing $p(\cdot)$ by its symmetrized version. The corresponding formula for the symmetric case, proved in [6], involves solving a different equation and therefore a different w, but for the symmetric case the correct w is given by a variational formula that minimizes the expression on the right-hand side of formula (2.23). Therefore the quantity $\langle C(\rho)a, a \rangle$ is bounded below by the corresponding value for the symmetric case for which a lower bound of the form claimed in Theorem 2.7 was proved in [10].   □

We have based our proof on estimating the resolvent equation (2.10) with the key estimate equation (2.11). In fact, we can also base it on the following estimate: *For any $\varepsilon > 0$ there is a local function u_ε such that*

$$\|Lu_\varepsilon - f\|_{-1} \le \varepsilon.$$

From this estimate we can obtain a result analogous to Theorem 2.4 ([7] or section 6 of [4]) and thus the martingale decomposition Theorem 2.5. This was the estimate established in [2, 7] for the fluctuation-dissipation equation. In a sense these two estimates on the Green's functions can be substituted for each other in many contexts; see section 6 of the lecture [4].

3 Estimates Related to Simple Exclusion Processes

Simple exclusion processes describe dynamics of infinitely many simple random walks on $\mathbb{Z}^d$ with the exclusion that no two particles are allowed to occupy the same site. The simple random walk on $\mathbb{Z}^d$ has the generator given by

$$(Af)(x) = \sum_y p(y-x)[f(y) - f(x)].$$

The corresponding symmetric generator is

(3.1) $$(Sf)(x) = \sum_y a(y-x)[f(y) - f(x)]$$

where

$$p(x) = a(x) + b(x),$$

$a(\cdot)$ and $b(\cdot)$ being the symmetric and asymmetric components of $p(\cdot)$, respectively. Note that

$$|b(x)| = \frac{1}{2}|p(x) - p(-x)| \le \frac{1}{2}[p(x) + p(-x)] = a(x).$$

The Dirichlet form is

(3.2) $$\mathcal{D}(f) = \frac{1}{2}\sum_{x,y} a(y-x)[f(y) - f(x)]^2.$$

Since we are mainly interested in the tagged particle process, the state space is $\mathbb{Z}^d \setminus \{0\}$ rather than $\mathbb{Z}^d$. The random walk on $\mathbb{Z}^d \setminus \{0\}$ can be viewed as the random walk on $\mathbb{Z}^d$ with jumps into the origin disallowed. The main result in this section was obtained in collaboration with C. Landim and was reported in [7] by a purely analytic proof. Here we give a more probabilistic proof and for more general processes.

We can consider the general setting of an irreducible transient Markov process on a countable state space X that is symmetric with respect to the counting or uniform measure on X. The generator and the Dirichlet form are given by (3.1) and (3.2) with obvious interpretation. There is a subset $E \subset X$ that is excluded, and transitions into E are disallowed. On the state space $Y = X \setminus E$ we have the induced generator

$$(\tilde{S}u)(x) = \sum_{y \in Y} a(x,y)[u(y) - u(x)]$$

for $x \in Y$. The corresponding Dirichlet form is

$$\tilde{D}(u) = \frac{1}{2} \sum_{x,y \in Y} a(x,y)[u(y) - u(x)]^2.$$

For the special case of a random walk, E consists of just the origin. We shall keep the set E general because we want to apply it to other settings as well. We shall assume that the basic process on X is transient, which means that the Green's function exists, i.e.,

$$g(x,y) = \int_0^\infty p(t,x,y)dt < \infty \quad \text{for all } x,y \in X$$

where $p(t,x,y)$ is the transition probability function. The Green's function $g(x,y)$ satisfies

$$g(x,y) = g(y,x) \le g(x,x) \quad \text{for all } x,y \in X$$

and

$$(Sg(x,\cdot))(y) = \sum_z [g(x,z) - g(x,y)]a(y,z) = -\delta_x(y) \quad \text{for all } x,y \in X$$

where $\delta_x(y) = 1$ if $x = y$ and 0 otherwise.

We want to show that the induced process on Y with generator $\tilde{S}$ is again transient and compare its Green's function $\tilde{g}(x,y)$ to the original Green's function $g(x,y)$. We will assume that the probability

$$\theta(x) = P_x\{x(t) \text{ visits } E \text{ for some } t \ge 0\} < 1$$

for each $x \in Y$. This is clearly satisfied for any random walk in $d \ge 3$ with $E = \{0\}$. It is well-known that in the transient case there is an estimate of the form

$$|u(x)| \le C(x)\sqrt{D(u)}$$

valid uniformly for all functions u that vanish outside a finite set and by completion for all functions that belong to the Dirichlet space $\mathbb{H}_1$, which we recall was defined as the completion of the space of finitely supported functions under the norm $\|u\|_1 = \sqrt{\mathcal{D}(u)}$. A precise estimate on the constant $C(x)$ is given in the following lemma:

LEMMA 3.1 *Suppose $V(x)$ is a nonnegative, compactly supported function. Then*

$$\sum_x u^2(x)V(x) \leq \sup_x \left[\sum_y g(x,y)V(y)\right] \mathcal{D}(u)$$

for all u. In particular, taking V to be 1 at x and 0 elsewhere,

$$|u(x)| \leq \sqrt{g(x,x)}\sqrt{\mathcal{D}(u)}.$$

PROOF: Let us define $W(x) = \sum_y g(x,y)V(y)$ and $C = \sup_x W(x)$. Since $0 \leq W \leq C$ and $V \geq 0$,

$$\sum_x u^2(x)V(x) \leq C\sum_x u^2(x)\frac{V(x)}{W(x)}$$

$$= -C\sum_x \frac{u^2(x)}{W(x)}(SW)(x)$$

$$= \frac{C}{2}\sum_{x,y}\left[\frac{u^2(y)}{W(y)} - \frac{u^2(x)}{W(x)}\right][W(y) - W(x)]a(x,y)$$

$$= \frac{C}{2}\sum_{x,y}\left[u^2(y) + u^2(x) - u^2(x)\frac{W(y)}{W(x)} - u^2(y)\frac{W(x)}{W(y)}\right]a(x,y)$$

$$\leq \frac{C}{2}\sum_{x,y}[u^2(y) + u^2(x) - 2u(x)u(y)]a(x,y) = C\mathcal{D}(u).$$

$\square$

Notice that $\theta(x)$ solves the equation

$$\begin{cases} (S\theta)(x) = \sum_{y\in X} a(x,y)[\theta(y) - \theta(x)] = 0 & \text{for } x \in Y \\ \theta(x) = 1 & \text{for } x \in E. \end{cases}$$

Therefore

$$(\tilde{S}\theta)(x) = -\sum_{y\in E} a(x,y)[\theta(y) - \theta(x)] = -\sum_{y\in E} a(x,y)[1 - \theta(x)] = -A(x)(1 - \theta(x))$$

where

(3.3) $$A(x) = \sum_{y\in E} a(x,y).$$

Because of irreducibility, $A(x)$ cannot vanish identically on Y. Thus $\theta(x)$ is a bounded, nonconstant superharmonic function, and so the process is transient. We will actually assume that

$$(3.4) \qquad \sup_{x \in Y} \theta(x) = \beta < 1.$$

The following lemma is a quantified version of the transience:

LEMMA 3.2 *For the Green's function $\tilde{g}(x,y)$, we have*

$$\sum_{y \in Y} \tilde{g}(x,y)A(y) \leq \frac{\theta(x)}{(1-\beta)}.$$

PROOF: Since

$$-(\tilde{S}\theta)(x) \geq (1-\beta)A(x),$$

the lemma follows from the maximum principle. $\qquad\square$

LEMMA 3.3 *Let $U \geq 0$ be supported on Y, that is, it vanishes on E. Then*

$$(3.5) \qquad \begin{aligned} \sup_{x \in Y} \sum_{y \in Y} \tilde{g}(x,y)U(y) &\leq \frac{1}{(1-\beta)} \sup_{x \in X} \sum_{y \in X} g(x,y)U(y) \\ &= \frac{1}{(1-\beta)} \sup_{x \in Y} \sum_{y \in Y} g(x,y)U(y) \end{aligned}$$

Furthermore, for all $x,y \in Y$,

$$(3.6) \qquad \tilde{g}(x,y) \leq g(x,y) + \frac{1}{(1-\beta)} g(y,y)\theta(x)$$

and

$$(3.7) \qquad \tilde{g}(x,x) \leq \frac{1}{(1-\beta)} g(x,x).$$

PROOF: Let

$$W(x) = \sum_{y \in X} g(x,y)U(y).$$

The function W is nonnegative and solves $SW = -U$. A computation shows that for $x \in Y$,

$$\begin{aligned} (\tilde{S}W)(x) &= (SW)(x) - \sum_{y \in E} a(x,y)[W(y) - W(x)] \\ &\leq -U(x) + \sum_{y \in E} a(x,y)W(x) = -U(x) + A(x)W(x) \leq -U(x) + CA(x) \end{aligned}$$

where $C = \sup_{x \in Y} W(x)$. Then

$$(3.8) \qquad \sum_{y \in Y} \tilde{g}(x,y)U(y) \leq W(x) + C\sum_{y \in Y} \tilde{g}(x,y)A(y) \leq W(x) + \frac{C\theta(x)}{(1-\beta)}.$$

Taking the supremum over x, we get

$$\sup_{x \in Y} \sum_{y \in Y} \tilde{g}(x,y)U(y) \leq C + \frac{C\beta}{(1-\beta)} = \frac{C}{(1-\beta)},$$

which proves equation (3.5). Taking $U(x)$ to be $\delta_y(x)$, we have $W(x) = g(x,y)$ and $\sup_x W(x) = g(y,y)$. Hence equation (3.6) follows from equation (3.8). Taking $x = y$, we get equation (3.7). □

Remark. If $a(x,y)$ is local, then we can do better if we define $\bar{E} = \bigcup_{x \in E} \{y : a(x,y) > 0\}$ and take $C = \sup_{x \in \bar{E}} W(x)$.

We can combine Lemmas 3.1 and 3.2 to obtain the next lemma.

LEMMA 3.4 *For any function u*

$$\sum_{x \in Y} A(x)u^2(x) \leq \frac{\beta}{(1-\beta)} \tilde{D}(u).$$

PROOF: According to Lemma 3.1

$$\sum_{x \in Y} A(x)u^2(x) \leq C\tilde{D}(u)$$

where the constant can be taken to be

$$C = \sup_{x \in Y} \sum_{y \in Y} \tilde{g}(x,y)A(y).$$

In Lemma 3.2 we established that for $x \in Y$

$$\sum_{y \in Y} \tilde{g}(x,y)A(y) \leq \frac{\theta(x)}{(1-\beta)}.$$

Taking the supremum over x, clearly we can take C to be $\beta/(1-\beta)$. □

The following results concerning two random walks will be useful:

LEMMA 3.5 *For the symmetric random walk on $\{\mathbb{Z}^d \setminus \{0\}\} \times \{\mathbb{Z}^d \setminus \{0\}\}$ with generator*

$$\sum_z a(z)[f(x_1 + z, x_2) - f(x_1, x_2)] + \sum_z a(z)[f(x_1, x_2 + z) - f(x_1, x_2)],$$

the probability $\theta(x_1, x_2)$ of hitting the excluded set $E_2 = \{x_1 = 0\} \cup \{x_2 = 0\} \cup \{x_1 = x_2\}$ has the property

$$\sup_{(x_1, x_2) \in (\mathbb{Z}^d \times \mathbb{Z}^d) \setminus E_2} \theta(x_1, x_2) := \alpha < 1.$$

PROOF: If we denote by $\delta(x)$ the probability that the continuous time random walk in $\mathbb{Z}^d$ hits 0 for some $t \geq 0$, then $\theta(x_1, x_2)$ can be bounded by

$$\theta(x_1, x_2) \leq \delta(x_1) + \delta(x_2) + \delta(x_1 - x_2).$$

We know that $\sup_{x \neq 0} \delta(x) < 1$ and $\delta(x) \to 0$ as $x \to \infty$. This is enough to conclude that $\theta(x_1, x_2)$ stays away from 1 near ∞, and since it is strictly less than 1 for each (x_1, x_2), we are done. $\qquad\square$

The number α is a constant of the underlying random walk and $0 < \alpha < 1$. The same α is also a bound for the hitting probability of 0 for a single random walk on $\mathbb{Z}^d$. From Lemma 3.3 we get for the random walk on $\mathbb{Z}^d$ that excludes the origin the following bound for the Green's function:

$$g_0(x,x) \le \frac{1}{(1-\alpha)} g(x,x) = \frac{g(0,0)}{(1-\alpha)}.$$

We now consider the symmetric simple exclusion processes on $\mathbb{Z}^d \setminus \{0\}$. The state space Ω consists of maps η from the countable set $\mathbb{Z}^d \setminus \{0\}$ into $\{0,1\}$. Fix $0 < \rho < 1$ and denote the Bernoulli product measure with density ρ by P_ρ. For $x \in \mathbb{Z}^d$, we define

(3.9) $$\xi_x(\eta) = \frac{\eta(x) - \rho}{\sqrt{\rho(1-\rho)}}.$$

For $A \subset \mathbb{Z}^d \setminus \{0\}$ we define

(3.10) $$\begin{cases} \xi_A(\eta) = \prod_{x \in A} \xi_x(\eta) & \text{if } A \text{ is nonempty} \\ \xi_A(\eta) = 1 & \text{if } A \text{ is empty.} \end{cases}$$

Then $\{\xi_A(\cdot)\}$ is an orthonormal basis for $\mathbf{H} = L_2(\Omega)$. It comes naturally graded as $\mathbf{H} = \bigoplus_{n \ge 0} H_n$ where H_n is the span of ξ_A over sets of cardinality n.

If we write

$$u = \sum_A \tilde{u}(A) \xi_A,$$

then the Dirichlet form can be calculated explicitly, and we obtain

$$D^{\mathrm{ex}}(u) = \frac{1}{2} \sum_{n \ge 1} \sum_{|A|=n} \sum_{x,y} a(x,y) [\tilde{u}(A^{x,y}) - \tilde{u}(A)]^2 := D^{\mathrm{ex}}(\tilde{u})$$

where

(3.11) $$A^{x,y} = \begin{cases} A & \text{if either } x,y \in A \text{ or } x,y \notin A \\ (A \setminus x) \cup y & \text{if } x \in A \text{ and } y \notin A \\ (A \setminus y) \cup x & \text{if } x \notin A \text{ and } y \in A. \end{cases}$$

The computation depends on the simple observation that, for all x,y in X, $\xi_A(\eta^{x,y}) = \xi_{A^{x,y}}(\eta)$.

Let us denote by $\mathcal{X}$ the space of all finite subsets of $\mathbb{Z}^d \setminus \{0\}$, and we write $\mathcal{X}$ as the natural union $\bigcup \mathcal{X}_n$ of spaces of subsets of cardinality n. We will consider functions on $\mathcal{X}$ or $\mathcal{X}_n$, and we assume initially that they are zero outside a finite set of points in $\mathcal{X}$ or $\mathcal{X}_n$. Then $\tilde{u}$ is a function on $\mathcal{X}$, and we can write it as $\tilde{u} = \sum_n \tilde{u}_n$ with $\tilde{u}_n$ being the restriction of $\tilde{u}$ to $\mathcal{X}_n$. Clearly,

$$D^{\mathrm{ex}}(\tilde{u}) = \sum_{n \geq 1} D_n^{\mathrm{ex}}(\tilde{u}_n),$$

where for $f : \mathcal{X}_n \to R$

$$D_n^{\mathrm{ex}}(f) = \frac{1}{2} \sum_{A \in \mathcal{X}_n} \sum_{x,y} a(x,y)[f(A^{x,y}) - f(A)]^2$$

$$= \sum_{\substack{A \in \mathcal{X}_n \\ }} \sum_{\substack{x \in A \\ y \notin A}} a(x,y)[f((A \setminus x) \cup y) - f(A)]^2.$$

The Dirichlet form D^{sh} associated with the shift operator L^{sh} is not graded naturally according to our decomposition. A related Dirichlet form allowing the tagged particle to jump to an occupied site,

(3.12)
$$\hat{D}^{\mathrm{sh}}(u) = \frac{1}{2} \int \sum_x a(x)[u(\tau_x \eta) - u(\eta)]^2 \, dP_\rho,$$

is naturally graded with

$$\hat{D}_n^{\mathrm{sh}}(u) = \hat{D}_n^{\mathrm{sh}}(\tilde{u}_n) = \frac{1}{2} \sum_{x, |A|=n} a(x)[\tilde{u}_n(\tau_x A) - \tilde{u}_n(A)]^2.$$

We shall prove later on an estimate of $\hat{D}^{\mathrm{sh}}$ in terms of D^{sh} and D^{ex}.

We need the following concept later on: Suppose $n > m$ and we are given a function g on $\mathcal{X}_n$. The function f on $\mathcal{X}_m$ is defined by $f \geq 0$ and

$$f^2(B) = \sum_{\substack{A : A \supset B \\ |A|=n}} g^2(A).$$

Notice that if we consider g as a wave function, then f^2 is simply the m point function associated with the wave function. A very important relation between f and g is the connection between their kinetic energies, especially the special case $m = 1$, which leads to the semiclassical limit for the kinetic energy of Bose gases. Notice that in our setting there is an interaction between particles, namely, the exclusion rule. Nevertheless, it is still valid.

LEMMA 3.6 *For any $n > m$*

$$D_m^{\mathrm{ex}}(f) \leq \binom{n-1}{m-1} D_n^{\mathrm{ex}}(g).$$

In particular, for $m = 1$ we have for the function

$$f^2(x) = \sum_{A : A \ni x} g^2(A),$$

that

$$D_1^{ex}(f) \le D_n^{ex}(g).$$

PROOF: If B_1 and B_2 are two subsets of size m that differ by one point, namely, $B_2 = B_1 \setminus x \cup y$, then one has the following obvious estimate

$$|f(B_1) - f(B_2)|^2 \le \sum_{\substack{A:|A|=n \\ A \supset B_1 \\ x \in A, y \notin A}} |g(A \setminus x \cup y) - g(A)|^2.$$

One multiplies this inequality by $a(x,y)$ and sums over everything in sight. One has to be careful and count the number of times a term like $|g(A \setminus x \cup y) - g(A)|^2$ occurs in the summation on the right-hand side. That number is clearly the number of different subsets of A of size m that include a given $x \in A$, and this yields the combinatorial prefactor $\binom{n-1}{m-1}$. □

The following lemma is the main result of this section:

LEMMA 3.7 *Let $n \ge 2$. For any set A in $\mathcal{X}_n$, i.e., a subset of $\mathbb{Z}^d \setminus \{0\}$ of cardinality n, that consists of n distinct points $x_1, x_2, \ldots, x_n$, we define*

$$W_1(A) = \sum_i a(x_i) \quad and \quad W_2(A) = \sum_{i \ne j} a(x_i - x_j).$$

Then for any u defined on $\mathcal{X}_n$ that is in the Dirichlet space,

$$\sum_A W_1(A)u^2(A) \le \frac{\alpha}{(1-\alpha)} D_n^{ex}(u)$$

and

$$\sum_A [W_1(A) + W_2(A)]u^2(A) \le \frac{\alpha}{(1-\alpha)} n D_n^{ex}(u).$$

PROOF: Define v on $\mathcal{X}_1 = \mathbb{Z}^d \setminus 0$ by

$$v^2(x) = \sum_{A:A \ni x} u^2(A).$$

By Lemma 3.6

$$D_1^{ex}(v) \le D_n^{ex}(u).$$

By definition of v,

$$\sum_{|A|=n} W_1(A)u^2(A) = \sum_{x \ne 0} a(x)v^2(x).$$

By Lemma 3.4 and the remark at the end of Lemma 3.6,

$$\sum_{x \ne 0} a(x)v^2(x) \le \frac{\alpha}{(1-\alpha)} D_n^{ex}(v) = \frac{\alpha}{(1-\alpha)} D_n^{ex}(u).$$

This proves the estimate involving W_1.

To prove the second inequality, define the function w on $\mathcal{X}_2$ by

$$w^2(B) = \sum_{A:A \supset B} u^2(A).$$

We can view w as a symmetric function on $(\mathbb{Z}^d \times \mathbb{Z}^d) \setminus E_2$ and compute its Dirichlet form,

$$D_2^{ex}(w) = \frac{1}{2} \sum_{\substack{x_1,x_2,y_1 \in \mathbb{Z}^d \setminus \{0\} \\ x_1 \neq y_1}} a(y_1 - x_1)[w(y_1,x_2) - w(x_1,x_2)]^2$$

$$+ \frac{1}{2} \sum_{\substack{x_1,x_2,y_2 \in \mathbb{Z}^d \setminus \{0\} \\ x_2 \neq y_2}} a(y_2 - x_2)[w(x_1,y_2) - w(x_1,x_2)]^2;$$

here the factor $\frac{1}{2}$ appears because each set B is counted twice. Again by Lemma 3.6

(3.13) $$D_2^{ex}(w) \leq (n-1)D_n^{ex}(u).$$

Clearly,

$$\sum_A [W_1(A) + W_2(A)]u^2(A) = \frac{1}{2} \sum_{\substack{x_1,x_2 \in \mathbb{Z}^d \setminus \{0\} \\ x_1 \neq x_2}} H(x_1,x_2)w^2(x_1,x_2)$$

where

$$H(x_1,x_2) = a(x_1) + a(x_2) + 2a(x_1 - x_2)$$

is the sum of all the transition rates into the excluded set for the two random walks. An easy calculation using Lemma 3.5 establishes

$$\sum_B H(B)w^2(B) = \frac{1}{2} \sum_{x_1 \neq x_2} H(x_1,x_2)w^2(x_1,x_2) \leq \frac{1}{2} \frac{\alpha}{(1-\alpha)} 2D_2^{ex}(w).$$

This fact together with equation (3.13) proves the lemma. $\qquad\square$

Another important Markovian evolution, referred to as Glauber dynamics, will be useful later on. We now collect some of its properties here. The Glauber dynamics has each site flipping back and forth between 0 and 1 as a two-state Markov process. The flip rates from $1 \to 0$ and $0 \to 1$ are taken to be $1/\rho$ and $1/(1-\rho)$, respectively, so that the generator for a single site is

$$(Gv)(0) = \frac{1}{(1-\rho)}(v(1) - v(0)), \quad (Gv)(1) = \frac{1}{\rho}(v(0) - v(1)),$$

with the corresponding Dirichlet form, $[v(0) - v(1)]^2$. For the full process on Ω, we need to define the flip operator σ_x at site x,

$$\begin{cases} (\sigma_x \eta)(y) = \eta(y) & \text{if } x \neq y \\ (\sigma_x \eta)(x) = 1 - \eta(x) & \text{for } x \in X. \end{cases}$$

The Glauber generator then has the Dirichlet form

$$D_g(u) = \sum_x D_{g,x}(u)$$

where

(3.14)
$$D_{g,x}(u) = \int_\Omega [u(\sigma_x \eta) - u(\eta)]^2 dP_\rho .$$

Our next lemma is an estimate of $D_{g,x}(u)$ in terms of $D^{ex}(u)$ in the transient case. The transience is important because in the Glauber dynamics "particles" are not conserved, whereas in exclusion processes they are conserved. The only way to kill a particle in the exclusion model is to send it to infinity, and transience plays an important role in that.

LEMMA 3.8 *Assume transience. Let* $u : \Omega \to \mathbb{R}$ *depend on a finite set of coordinates. Then for any* $x \in X$,

$$\int_{\eta(x)=1} [u(\sigma_x \eta) - u(\eta)]^2 dP_\rho \le \frac{1}{(1-\rho)} g(x,x) D^{ex}(u)$$

and

$$\int_{\eta(x)=0} [u(\sigma_x \eta) - u(\eta)]^2 dP_\rho \le \frac{1}{\rho} g(x,x) D^{ex}(u) .$$

By adding the two inequalities,

$$D_{g,x}(u) \le \frac{1}{\rho(1-\rho)} g(x,x) D^{ex}(u) .$$

PROOF: An elementary calculation shows that for a function

$$u(\eta) = \sum_A \tilde{u}(A) \xi_A(\eta)$$

we have

$$\int_{\eta(x)=1} [u(\sigma_x \eta) - u(\eta)]^2 dP_\rho = \frac{1}{(1-\rho)} \sum_{A:A \ni x} \tilde{u}^2(A),$$

$$\int_{\eta(x)=0} [u(\sigma_x \eta) - u(\eta)]^2 dP_\rho = \frac{1}{\rho} \sum_{A:A \ni x} \tilde{u}^2(A) .$$

Adding the two equalities, we get

$$D_{g,x}(u) = \frac{1}{\rho(1-\rho)} \sum_{A:A \ni x} \tilde{u}^2(A) .$$

424

If we define $\tilde{u}_n$ to be the restriction of $\tilde{u}$ to $\mathcal{X}_n$,

$$\sum_{A:A \ni x} \tilde{u}^2(A) = \sum_n v_n^2(x) \quad \text{where} \quad v_n^2(x) = \sum_{\substack{A:|A|=n \\ A \ni x}} \tilde{u}_n^2(A).$$

From Lemma 3.1 $v_n^2(x) \leq g(x,x)D_1(v_n)$, and by Lemma 3.6 $D_1^{\text{ex}}(v_n) \leq D_n^{\text{ex}}(\tilde{u}_n)$. The lemma follows. $\square$

As corollaries of previous lemmas, we prove Lemma 2.1.

PROOF OF LEMMA 2.1: Clearly, to get a bound of the form

$$|\langle f, v \rangle| \leq C_f \sqrt{\sum_{x \in F} D_{g,x}(v)},$$

where f depends only on the coordinates x from F, is just finite-dimensional matrix algebra. We can now use Lemma 3.8 to complete the proof. $\square$

Next we obtain an estimate of $\hat{D}^{\text{sh}}$. We already know that $D^{\text{sh}}(u) \leq \hat{D}^{\text{sh}}(u)$, since the $\hat{D}^{\text{sh}}$ form is stronger than the D^{sh} form because it corresponds to the tagged particle exchanging places with an untagged particle, which is not allowed. However, we could make the particle at x disappear, let the tagged particle jump to x, which is now empty, and have our old disappeared particle reappear at $-x$ and accomplish our goal. This means that the $\hat{D}^{\text{sh}}$ form can be estimated in terms of the D^{sh} form and the Glauber forms at x and $-x$. More precisely,

$$\int_{\eta(x)=1} [u(\tau_x \eta) - u(\eta)]^2 dP_\rho$$

$$\leq 3 \int_{\eta(x)=1} \left[[u(\sigma_{-x}\tau_x\sigma_x\eta) - u(\tau_x\sigma_x\eta)]^2 + [u(\tau_x\sigma_x\eta) - u(\sigma_x\eta)]^2 \right.$$

$$\left. + [u(\sigma_x\eta) - u(\eta)]^2 \right] dP_\rho$$

$$= 3\frac{\rho}{(1-\rho)} \int_{\eta(-x)=0} [u(\sigma_{-x}\eta) - u(\eta)]^2 + 3\frac{\rho}{(1-\rho)} \int_{\eta(x)=0} [u(\tau_x\eta) - u(\eta)]^2 dP_\rho$$

$$+ 3 \int_{\eta(x)=1} [u(\sigma_x\eta) - u(\eta)]^2 dP_\rho .$$

Note that the maps σ_x are not measure preserving when $\rho \neq \frac{1}{2}$, and we therefore pick up the factor $\rho/(1-\rho)$. We can now use Lemma 3.8 to conclude that for some constant C not depending on ρ

(3.15) $$\hat{D}^{\text{sh}}(u) \leq \frac{C}{1-\rho}[D^{\text{ex}}(u) + D^{\text{sh}}(u)].$$

4 Some Estimates on the Generator

Recall the generator $L = L^{\text{ex}} + L^{\text{sh}}$ of the environment process defined in equations (1.3) and (1.4). Note also that $L_2 = \mathbf{H} = \bigoplus_{n \geq 0} H_n$ where H_n is spanned by the orthonormal basis $\{\xi_A : |A| = n\}$ with ξ_A defined in equation (3.10). Notice that L is a bounded operator on each H_n and maps it into $H_{n-1} \oplus H_n \oplus H_{n+1}$. There are thus bounded operators $B_{n,n-1} : H_n \to H_{n-1}$, $B_{n,n} : H_n \to H_n$, and $B_{n,n+1} : H_n \to H_{n+1}$ such that for $u \in H_n$

$$Lu = B_{n,n-1}u + B_{n,n}u + B_{n,n+1}u.$$

To compute these operators explicitly, we start with the action of L^{ex} on ξ_A,

$$L^{\text{ex}}\xi_A = \sum_{x,y \neq 0} p(y-x)\eta(x)(1-\eta(y))[\xi_{A^{x,y}} - \xi_A].$$

Recall $p(x) = a(x) + b(x)$, the sum of its symmetric and asymmetric parts. Hence we obtain

$$L^{\text{ex}}\xi_A = \frac{1}{2}\sum_{x,y \neq 0} a(y-x)[\xi_{A^{x,y}} - \xi_A] + \frac{1}{2}\sum_{x,y \neq 0} b(y-x)(\eta(x) - \eta(y))[\xi_{A^{x,y}} - \xi_A].$$

To proceed further, we write

$$\eta(x) - \eta(y) = \sqrt{\rho(1-\rho)}[\xi_x - \xi_y]$$

and use the rule

$$\xi_x^2 = 1 + \frac{1-2\rho}{\sqrt{\rho(1-\rho)}}\xi_x$$

to obtain

$$[\eta(x) - \eta(y)][\xi_{A^{x,y}} - \xi_A] = \begin{cases} 2\sqrt{\rho(1-\rho)}[\xi_{A\cup y} - \xi_{A\setminus x}] + \dfrac{2\rho-1}{\sqrt{\rho(1-\rho)}}[\xi_A + \xi_{A\setminus x\cup y}] & \text{if } x \in A, y \notin A, \\[2ex] 2\sqrt{\rho(1-\rho)}[\xi_{A\setminus y} - \xi_{A\cup x}] - \dfrac{2\rho-1}{\sqrt{\rho(1-\rho)}}[\xi_A + \xi_{A\setminus y\cup x}] & \text{if } x \notin A, y \in A, \\[2ex] 0 & \text{in all other cases.} \end{cases}$$

This yields

(4.1)

$$
L^{ex}\xi_A = \frac{1}{2} \sum_{\substack{x,y\neq 0 \\ x\in A, y\notin A}} a(y-x)[\xi_{A\setminus x\cup y} - \xi_A] + \frac{1}{2} \sum_{\substack{x,y\neq 0 \\ x\notin A, y\in A}} a(y-x)[\xi_{A\setminus y\cup x} - \xi_A]
$$

$$
+ \frac{2\rho-1}{2} \sum_{\substack{x,y\neq 0 \\ x\in A, y\notin A}} b(y-x)[\xi_{A\setminus x\cup y} + \xi_A]
$$

$$
- \frac{2\rho-1}{2} \sum_{\substack{x,y\neq 0 \\ x\notin A, y\in A}} b(y-x)[\xi_{A\setminus y\cup x} + \xi_A]
$$

$$
+ \sqrt{\rho(1-\rho)} \sum_{\substack{x,y\neq 0 \\ x\in A, y\notin A}} b(y-x)\xi_{A\cup y} - \sqrt{\rho(1-\rho)} \sum_{\substack{x,y\neq 0 \\ x\notin A, y\in A}} b(y-x)\xi_{A\cup x}
$$

$$
- \sqrt{\rho(1-\rho)} \sum_{\substack{x,y\neq 0 \\ x\in A, y\notin A}} b(y-x)\xi_{A\setminus x} + \sqrt{\rho(1-\rho)} \sum_{\substack{x,y\neq 0 \\ x\notin A, y\in A}} b(y-x)\xi_{A\setminus y}.
$$

We have arranged the terms in equation (4.1) so that the first line is a symmetric piece of $B_{n,n}$, the second two are an asymmetric piece again of $B_{n,n}$, and the last two lines are parts of $B_{n,n+1}$ and $B_{n,n-1}$. In fact, there is a symmetry relative to the interchange of x and y and the first and second terms on each line are equal. We use this to rewrite

$$
L^{ex}\xi_A =
$$

(4.2)
$$
\sum_{\substack{x,y\neq 0 \\ x\in A, y\notin A}} a(y-x)[\xi_{A\setminus x\cup y} - \xi_A] + (2\rho-1) \sum_{\substack{x,y\neq 0 \\ x\in A, y\notin A}} b(y-x)[\xi_{A\setminus x\cup y} + \xi_A]
$$

$$
+ 2\sqrt{\rho(1-\rho)} \sum_{\substack{x,y\neq 0 \\ x\in A, y\notin A}} b(y-x)\xi_{A\cup y} - 2\sqrt{\rho(1-\rho)} \sum_{\substack{x,y\neq 0 \\ x\in A, y\notin A}} b(y-x)\xi_{A\setminus x}.
$$

Now we compute in a similar fashion the term $L^{sh}\xi_A$. Although the tagged particle cannot jump to a site where there is already a particle present, it is convenient to extend the definition of τ_x to the set $\{x : \eta(x) = 1\}$ by allowing the tagged particle to jump to the site $-x$, which now becomes the origin, and move the particle that was originally at $-x$ back to where the tagged particle was, which is now x. This defines a version of τ_x with $\tau_x\xi_A = \xi_{\tau_x A}$ where

$$
\tau_x A = \begin{cases} A+x & \text{if } -x \notin A \\ (A+x)\setminus 0\cup x & \text{if } -x \in A. \end{cases}
$$

Note that $x \in \tau_x A$ if and only if $-x \in A$. After careful calculation we obtain

(4.3)
$$
\begin{aligned}
L^{\mathrm{sh}}\xi_A &= (1-\rho) \sum_{-x \notin A} a(x)[\xi_{\tau_x A} - \xi_A] + \rho \sum_{-x \in A} a(x)[\xi_{\tau_x A} - \xi_A] \\
&\quad + (1-\rho) \sum_{-x \notin A} b(x)[\xi_{\tau_x A} + \xi_A] + \rho \sum_{-x \in A} b(x)[\xi_{\tau_x A} + \xi_A] \\
&\quad + \sqrt{\rho(1-\rho)} \sum_{x \notin A} p(x)\xi_{A \cup x} - \sqrt{\rho(1-\rho)} \sum_{-x \notin A} p(x)\xi_{(A+x) \cup x} \\
&\quad + \sqrt{\rho(1-\rho)} \sum_{x \in A} p(x)\xi_{A \setminus x} - \sqrt{\rho(1-\rho)} \sum_{-x \in A} p(x)\xi_{(A+x) \setminus 0}.
\end{aligned}
$$

Again the first line is the symmetric part of $B_{n,n}$, followed by its asymmetric part and then the pieces of $B_{n,n+1}$ and $B_{n,n-1}$, respectively.

Putting together L^{ex} and L^{sh}, we have

(4.4)
$$
\begin{aligned}
B_{n,n}\xi_A &= \sum_{\substack{x,y \neq 0 \\ x \in A, y \notin A}} a(y-x)[\xi_{A \setminus x \cup y} - \xi_A] + (1-\rho) \sum_{-x \notin A} a(x)[\xi_{\tau_x A} - \xi_A] \\
&\quad + \rho \sum_{-x \in A} a(x)[\xi_{\tau_x A} - \xi_A] + (2\rho - 1) \sum_{\substack{x,y \neq 0 \\ x \in A, y \notin A}} b(y-x)[\xi_{A \setminus x \cup y} + \xi_A] \\
&\quad + (1-\rho) \sum_{-x \notin A} b(x)[\xi_{\tau_x A} + \xi_A] + \rho \sum_{-x \in A} b(x)[\xi_{\tau_x A} + \xi_A]
\end{aligned}
$$

(4.5)
$$
B_{n,n+1}\xi_A = \sqrt{\rho(1-\rho)} \left[2 \sum_{\substack{x,y \neq 0 \\ x \in A, y \notin A}} b(y-x)\xi_{A \cup y} + \sum_{x \notin A} p(x)\xi_{A \cup x} - \sum_{-x \notin A} p(x)\xi_{(A+x) \cup x} \right],
$$

(4.6)
$$
B_{n+1,n}\xi_A = \sqrt{\rho(1-\rho)} \left[-2 \sum_{\substack{x,y \neq 0 \\ x \in A, y \notin A}} b(y-x)\xi_{A \setminus x} + \sum_{x \in A} p(x)\xi_{A \setminus x} - \sum_{-x \in A} p(x)\xi_{(A+x) \setminus 0} \right].
$$

First, we will provide some estimates on $B_{n,n+1}$ and $B_{n+1,n}$. If we separate out the odd and even parts,

$$
B_{n,n+1} = \sqrt{\rho(1-\rho)}[B^{\mathrm{odd}}_{n,n+1} + B^{\mathrm{even}}_{n,n+1}]
$$

where the odd and even parts are given by

$$B^{\text{odd}}_{n,n+1}(\xi_A) = \left[2 \sum_{\substack{x,y\neq 0 \\ x\in A, y\notin A}} b(y-x)\xi_{A\cup y} + \sum_{x\notin A} b(x)\xi_{A\cup x} - \sum_{-x\notin A} b(x)\xi_{(A+x)\cup x}\right],$$

(4.7)

$$B^{\text{even}}_{n,n+1}(\xi_A) = \left[\sum_{x\notin A} a(x)\xi_{A\cup x} - \sum_{-x\notin A} a(x)\xi_{(A+x)\cup x}\right].$$

The even part of $B_{n+1,n}$ is the adjoint of the even part of $B_{n,n+1}$, and for the odd part an extra change of sign is involved. So if we bound the odd and even pieces of $B_{n,n+1}$ separately, the dual bounds are valid for $B_{n+1,n}$. Our basic estimate is the following:

LEMMA 4.1 *For all $n \geq 1$ and $u \in H_n$ and $v \in H_{n+1}$,*

$$|\langle B_{n,n+1}u, v\rangle| \leq C\sqrt{\rho(1-\rho)}\sqrt{D^{\text{ex}}_{n+1}(v)}\left[\sqrt{n}\sqrt{D^{\text{ex}}_n(u)} + \sqrt{\hat{D}^{\text{sh}}_n(u)}\right]$$

and

$$|\langle B_{n+1,n}v, u\rangle| \leq C\sqrt{\rho(1-\rho)}\sqrt{D^{\text{ex}}_n(u)}\left[\sqrt{n}\sqrt{D^{\text{ex}}_{n+1}(v)} + \sqrt{\hat{D}^{\text{sh}}_{n+1}(v)}\right]$$

where C is a constant that is independent of n and ρ.

PROOF: We shall only prove the first estimate. The second one follows in a similar way or by invoking duality. Write functions u and v in H_n and H_{n+1}, respectively, with expansions $u = \sum_{|A|=n} \tilde{u}(A)\xi_A$ and $v = \sum_{|D|=n+1} \tilde{v}(D)\xi_D$. From equation (4.7), the estimate for $B^{\text{odd}}_{n,n+1}$ reduces to the estimation of

$$\left|2 \sum_{|A|=n} \sum_{x\in A, y\notin A} b(y-x)\tilde{u}(A)\tilde{v}(A\cup y)\right|$$

$$+ \left|\sum_{|A|=n} \sum_{x\notin A} b(x)\tilde{u}(A)\tilde{v}(A\cup x) - \sum_{|A|=n} l\sum_{-x\notin A} b(x)\tilde{u}(A)\tilde{v}((A+x)\cup x))\right|.$$

The first term is estimated by

$$\left| 2 \sum_{|D|=n+1} \sum_{x,y\in D} b(y-x)\tilde{u}(D\setminus y)\tilde{v}(D) \right|$$

(4.8)

$$= \left| \sum_{|D|=n+1} \sum_{x,y\in D} b(y-x)(\tilde{u}(D\setminus y) - \tilde{u}(D\setminus x))\tilde{v}(D) \right| \quad (b \text{ is odd})$$

$$\leq \sum_{|D|=n+1} \sum_{x,y\in D} a(y-x)|\tilde{u}(D\setminus y) - \tilde{u}(D\setminus x)||\tilde{v}(D)| \quad (\text{because } |b| \leq a)$$

$$\leq \left[\sum_{|D|=n+1} \sum_{x,y\in D} a(y-x)|\tilde{u}(D\setminus y) - \tilde{u}(D\setminus x)|^2 \right]^{\frac{1}{2}}$$

$$\times \left[\sum_{|D|=n+1} \sum_{x,y\in D} a(y-x)|\tilde{v}(D)|^2 \right]^{\frac{1}{2}}.$$

The first expression

$$\left[\sum_{|D|=n+1} \sum_{x,y\in D} a(y-x)|\tilde{u}(D\setminus y) - \tilde{u}(D\setminus x)|^2 \right]$$

is seen to equal $2D_{ex,n}(u)$. In view of Lemma 3.7, the second expression is bounded by $CnD_{ex,n+1}(v)$.

We turn to the second term. This leads to

(4.9)

$$\left| \sum_{\substack{x,D \\ x\in D, |D|=n+1}} b(x)[\tilde{u}(D\setminus x) - \tilde{u}((D\setminus x)-x)]\tilde{v}(D) \right|$$

$$\leq \left[\sum_{\substack{x,D \\ x\in D, |D|=n+1}} a(x)[\tilde{u}(D\setminus x) - \tilde{u}((D\setminus x)-x)]^2 \right]^{\frac{1}{2}} \left[\sum_{\substack{x,D \\ x\in D, |D|=n+1}} a(x)[\tilde{v}(D)]^2 \right]^{\frac{1}{2}}$$

$$= \left[\sum_{\substack{x,A \\ x\notin A, |A|=n}} a(x)[\tilde{u}(A) - \tilde{u}(A-x)]^2 \right]^{\frac{1}{2}} \left[\sum_{\substack{x,D \\ x\in D, |D|=n+1}} a(x)[\tilde{v}(D)]^2 \right]^{\frac{1}{2}}.$$

We have used the inequality $|b(x)| \leq a(x)$ in the last step and used $A = D \setminus x$ as the summation variable. The first expression is estimated by

$$\sum_{\substack{x,A \\ x \notin A, |A|=n}} a(x)[\tilde{u}(A) - \tilde{u}(A - x)]^2 = \sum_{\substack{x,A \\ x \notin A, |A|=n}} a(x)[\tilde{u}(\tau_x A) - \tilde{u}(A)]^2$$

$$\leq \sum_{\substack{x,A \\ x \notin A, |A|=n}} a(x)[\tilde{u}(\tau_x A) - \tilde{u}(A)]^2 = 2\hat{D}_n^{\text{sh}}(\tilde{u})$$

and the second by

$$\sum_{\substack{x,D \\ x \in D, |D|=n+1}} a(x)[\tilde{v}(D)]^2 \leq \frac{\alpha}{(1-\alpha)} D_{n+1}^{\text{ex}}(\tilde{v}).$$

The estimate for $B_{n,n+1}^{\text{even}}$ is similar. So we have proved the first estimate of Lemma 4.1. The second one follows from duality. This concludes the proof of the lemma. □

5 Some Estimates on the Resolvent

The main theorem of this section is the following:

THEOREM 5.1 *Let $0 < \rho < 1$ and a local mean zero function f be given. Then the solution u_λ of the resolvent equation (2.10) satisfies the following estimates: For every $k \geq 0$ there is a constant C_k independent of n and ρ such that*

$$\sup_{\lambda > 0} \sum_n n^{2k} [D_n^{\text{ex}}(u_\lambda) + (1-\rho)\hat{D}_n^{\text{sh}}(u_\lambda)] \leq C_k C_f, \quad \sup_{\lambda > 0} \lambda \sum_n n^{2k} \|u_{\lambda,n}\|_0^2 \leq C_k C_f.$$

PROOF: We first recall standard estimates. Multiplying equation (2.10) by u_λ and integrating,

$$\lambda \|u_\lambda\|_0^2 + D^{\text{ex}}(u_\lambda) + D^{\text{sh}}(u_\lambda) = \langle f, u_\lambda \rangle \leq C_f \sqrt{D^{\text{ex}}(u_\lambda)}.$$

This leads immediately to the estimates

(5.1) $$\sup_{\lambda > 0} \lambda \|u_\lambda\|_0^2 \leq C_f,$$

(5.2) $$\sup_{\lambda > 0} D^{\text{ex}}(u_\lambda) \leq C_f,$$

(5.3) $$\sup_{\lambda > 0} D^{\text{sh}}(u_\lambda) \leq C_f.$$

Because of estimate (3.15), we also have

(5.4) $$\sup_{\lambda > 0} \hat{D}^{\text{sh}}(u_\lambda) \leq \frac{1}{(1-\rho)} C_f.$$

Let us define the operator T as multiplication by a scalar $t(n)$ on each H_n. The sequence $t(n)$ is assumed to be positive, increasing, and eventually constant. Since L is bounded on each H_n and T is a multiple of I except on a finite number of H_n, it is easily verified that T leaves the domain of L invariant and the commutator $[T,L] = TL - LT$ is a bounded operator from $\mathbf{H} \to \mathbf{H}$. For $u = \sum_n u_n$ with $u_n \in H_n$, an explicit calculation yields

$$[TL - LT]u = \sum_n \left[(t(n+1) - t(n))B_{n,n+1}u_n + (t(n-1) - t(n))B_{n,n-1}u_n \right]$$

and

(5.5)
$$\langle [TL - LT]u, Tu \rangle = \sum_n t(n+1)(t(n+1) - t(n))\langle B_{n,n+1}u_n, u_{n+1} \rangle$$
$$+ \sum_n t(n-1)(t(n-1) - t(n))\langle B_{n,n-1}u_n, u_{n-1} \rangle.$$

From Lemma 4.1, we have

$$|\langle [LT - TL]u, Tu \rangle| \leq C \sum_n t(n+1)|t(n+1)$$
$$- t(n)| \sqrt{\rho(1-\rho)} \sqrt{D_{n+1}^{\mathrm{ex}}(u)} \left[\sqrt{n} \sqrt{D_n^{\mathrm{ex}}(u)} + \sqrt{\hat{D}_n^s(u)} \right]$$
$$+ C \sum_n t(n-1)|t(n-1)$$
$$- t(n)| \sqrt{\rho(1-\rho)} \sqrt{D_{n-1}^{\mathrm{ex}}(u)} \left[\sqrt{n} \sqrt{D_n^{\mathrm{ex}}(u)} + \sqrt{\hat{D}_n^{\mathrm{sh}}(u)} \right].$$

Since $D_n^{\mathrm{ex}}(u) = t(n)^{-2} D_n^{\mathrm{ex}}(Tu)$, the last term is equal to

$$C \sum_n \left| \frac{t(n+1)}{t(n)} - 1 \right| \sqrt{\rho(1-\rho)} \sqrt{D_{n+1}^{\mathrm{ex}}(Tu)} \left[\sqrt{n} \sqrt{D_n^{\mathrm{ex}}(Tu)} + \sqrt{\hat{D}_n^{\mathrm{sh}}(Tu)} \right]$$
$$+ \sum_n \left| \frac{t(n-1)}{t(n)} - 1 \right| \sqrt{\rho(1-\rho)} \sqrt{D_{n-1}^{\mathrm{ex}}(Tu)} \left[\sqrt{n} \sqrt{D_n^{\mathrm{ex}}(Tu)} + \sqrt{\hat{D}_n^{\mathrm{sh}}(Tu)} \right].$$

We now make the choice of $t(n)$ such that for every n, $t(n)$ satisfies

(5.6)
$$\frac{C(1+\sqrt{n})}{2} \left\{ \left| \frac{t(n+1)}{t(n)} - 1 \right| + \left| \frac{t(n-1)}{t(n)} - 1 \right| \right\} \leq \delta$$

where δ will be chosen soon. This can be achieved by making $t(n) = n_0^k$ for $1 \leq n \leq n_0$ and $t(n) = n^k$ for $n_0 \leq n \leq n_1$ and $t(n) = n_1^k$ for $n \geq n_1$. Whereas the choice of n_0 will be governed by δ and k, n_1 can be totally arbitrary and the estimates will be uniform in n_1. In the end n_1 can be allowed to go to infinity, providing us with estimates for the choice of $t(n) = (n_0 \vee n)^k$.

432

From the factor $\sqrt{\rho(1-\rho)}$, we keep only $\sqrt{(1-\rho)}$ and use it only with $\hat{D}_n^{sh}(u)$ terms. Then

$$|\langle [LT - TL]u, Tu \rangle|$$

(5.7)
$$\leq \delta \sum_n [D_{n-1}^{ex}(Tu) + D_{ex,n}(Tu) + D_{n+1}^{ex}(Tu) + (1-\rho)\hat{D}_n^{sh}(Tu)]$$

$$= \delta[3D^{ex}(Tu) + (1-\rho)\hat{D}^{sh}(Tu)] \leq \delta CD(Tu)$$

where the constant C comes from equation (3.15). We pick δ so that $C\delta < \frac{1}{4}$. Let us remark that the estimates depend on T only through δ.

From the resolvent equation we have

$$\lambda Tu_\lambda - LTu_\lambda = [T,L]u_\lambda + Tf.$$

Multiply both sides by Tu_λ and integrate. From equation (5.7) and Lemma 5.1,

$$\lambda \|Tu_\lambda\|_0^2 + D^{env}(Tu_\lambda) \leq \frac{1}{4}D^{env}(Tu_\lambda) + C_{Tf}\sqrt{D^{ex}(u_\lambda)}.$$

Clearly this is sufficient to give the estimates

$$\sup_\lambda \lambda \|Tu_\lambda\|_0^2 \leq C_{Tf}, \quad \sup_\lambda \sum_n [t(n)]^2 [D_n^{ex}(u_\lambda) + (1-\rho)\hat{D}_n^{sh}(u)] \leq C_0 C_{Tf}^2.$$

Since f is local, C_{Tf} is easily controlled. $\square$

6 Proof of Theorem 1.1

We first rewrite $B_{n,n}$ as follows:

LEMMA 6.1 *The operator $B_{n,n}$ from $H_n \rightarrow H_n$ can be rewritten as*

$$B_{n,n}\xi_A = \sum_{\substack{x,y\neq 0 \\ x\in A, y\notin A}} q(y-x)[\xi_{A\setminus x\cup y} - \xi_A] + (1-\rho)\sum_{-x\notin A} p(x)[\xi_{\tau_x A} - \xi_A]$$

$$+ \rho \sum_{-x\in A} p(x)[\xi_{\tau_x A} - \xi_A]$$

where $q(x) = q_\rho(x) = a(x) + (2\rho-1)b(x) = \rho p(x) + (1-\rho)p(-x)$.

PROOF: We have to compare with the earlier expression and make sure that the difference is zero. For this we need to show that

$$(2\rho-1)\sum_{\substack{x,y\neq 0 \\ x\in A, y\notin A}} b(y-x) + (1-\rho)\sum_{-x\notin A} b(x) + \rho \sum_{-x\in A} b(x) = 0.$$

If we use the antisymmetry of $b(\cdot)$ as well as its consequence $\sum_x b(x) = 0$, the above relation is easily seen to be true. $\square$

Our next step is to make $B_{n,n}$ look more like a convolution operator. The full translation symmetry is not available because the set $\mathcal{X}_n$ corresponds only to distinct nonzero n-tuples. A subset $A \subset \mathbb{Z}^d \setminus 0$ of cardinality n is really an equivalence class of $n!$ points in $(\mathbb{Z}^d)^n$. All the functions that we consider on $(\mathbb{Z}^d)^n$ will be symmetric under permutation. Let us denote by $\mathcal{G}_n \subset (\mathbb{Z}^d)^n$ the collection of distinct ordered nonzero n-tuples. A function on $\mathcal{X}_n$ can be considered as a symmetric function on $\mathcal{G}_n$ and then extended to all of $(\mathbb{Z}^d)^n$ by defining it to be 0 on the complement $(\mathbb{Z}^d)^n \setminus \mathcal{G}_n$. We will decompose this complement into three parts:

$$\mathcal{B}_n^1 = \{(x_1,\ldots,x_n) : x_i \neq 0 \text{ for } 1 \leq i \leq n \text{ and } x_i = x_j \text{ for just one pair}\},$$
$$\mathcal{B}_n^2 = \{(x_1,\cdots,x_n) : x_i \neq x_j \text{ for } 1 \leq i \neq j \leq n \text{ and for just one } i, \ x_i = 0\},$$

and

$$\mathcal{B}_n^3 = \mathcal{G}_n^c \setminus (\mathcal{B}_n^1 \cup \mathcal{B}_n^2).$$

We want to replace the operator $B_{n,n}$ acting on functions defined on $\mathcal{G}_n$ by the following operator $C_{n,n}$ of convolution type acting on the space of functions defined on all of $(\mathbb{Z}^d)^n$. Note that $\mathcal{B}_n^1$ and $\mathcal{B}_n^2$ are the boundary of $\mathcal{G}_n$, and transitions are possible from $\mathcal{G}_n$ only into $(\mathcal{G}_n \cup \mathcal{B}_n^1 \cup \mathcal{B}_n^2)$.

$$(C_{n,n}v)(x_1,\ldots,x_n)$$
$$= \sum_{x,j} q(x)[v(x_1,\ldots,x_j+x,\ldots,x_n) - v(x_1,\ldots,x_n)]$$
$$+ (1-\rho)\sum_x p(x)[v(x_1-x,\ldots,x_j-x,\ldots,x_n-x) - v(x_1,\ldots,x_n)].$$

A comparison has to be made with $B_{n,n}$. Given a function u on $\mathcal{X}_n$, we can view it as a symmetric function defined on $\mathcal{G}_n$ and extend it to all of $(\mathbb{Z}^d)^n$ as a symmetric function by making it zero outside $\mathcal{G}_n$. We will abuse the notation somewhat and not distinguish between the three versions of the same function on $\mathcal{X}_n$, on $\mathcal{G}_n$, and on $(\mathbb{Z}^d)^n$. Since we will deal only with symmetric functions, it will not matter. We can also extend in the same fashion the function $f = B_{n,n}u$.

$$(B_{n,n}u)(x_1,\ldots,x_n)$$
$$= \sum_j \sum_{(x_1,\ldots,y,\ldots,x_n)\in\mathcal{G}_n} q(y-x_j)[u(x_1,\ldots,y,\ldots,x_n) - u(x_1,\ldots,x_n)]$$
$$+ (1-\rho)\sum_{x\notin\{x_1,\ldots,x_n\}} p(x)[u(x_1-x,\ldots,x_n-x) - u(x_1,\ldots,x_n)]$$
$$+ \rho\sum_j p(x_j)[u(x_1-x_j,\ldots,-x_j,\ldots,x_n-x_j) - u(x_1,\ldots,x_n)].$$

We define h by

$$C_{n,n}u = B_{n,n}u + h$$

and try to estimate h in terms of u. First, let us compute h explicitly. On $\mathcal{B}_n^3$ the function h is identically zero. For a point in $\mathcal{B}_n^2$, consisting of n distinct points $x_1, \ldots, x_n$, exactly one of which is zero,

$$h(x_1, \ldots, x_{n-1}, 0) = \sum_{x_n : (x_1, \ldots, x_n) \in \mathcal{G}_n} q(x_n) u(x_1, \ldots, x_{n-1}, x_n)$$

$$+ (1 - \rho) \sum_{x \neq x_1, \ldots, x_n} p(x) u(x_1 - x, \ldots, x_n - x).$$

On $\mathcal{B}_n^1$ where a typical point is $(x_1, \ldots, x_{n-1}, x_{n-1})$ with distinct nonzero $x_1, \ldots, x_{n-1}$,

$$h(x_1, \ldots, x_{n-1}, x_{n-1}) = 2 \sum_{x_n : (x_1, \ldots, x_n) \in \mathcal{G}_n} q(x_n - x_{n-1}) u(x_1, \ldots, x_{n-1}, x_n) :$$

Finally on $\mathcal{G}_n$,

$$h(x_1, \ldots, x_{n-1}, x_n)$$

$$= - \Bigg[\Big[\sum_{i \neq j} q(x_i - x_j) + \sum_i q(-x_i) \Big] u(x_1, \ldots, x_n) - (1 - \rho) \Big[\sum_j p(x_j) \Big] u(x_1, \ldots, x_n)$$

$$- \rho \sum_j p(x_j) [u(x_1 - x_j, \ldots, -x_j, \ldots, x_n - x_j) - u(x_1, \ldots, x_n)] \Bigg].$$

There are two new Dirichlet forms. The forms $D_n^{\mathrm{ex}}(\cdot)$ that we already saw on functions defined on $\mathcal{X}_n$ as well as the new Dirichlet forms

$$\bar{D}_n(u) = \frac{1}{2} \sum_{\substack{i, x_i' \\ x_1, \ldots, x_n}} a(x_i - x_i') [u(x_1, \ldots, x_i, \ldots x_n) - u(x_1, \ldots, x_i', \ldots x_n)]^2,$$

which corresponds to n free random walks, and

$$\bar{D}_n^{\mathrm{sh}}(u) = \frac{1}{2} \sum_{x, x_1, \ldots, x_n} a(x) [u(x_1 - x, \ldots, x_n - x) - u(x_1, \ldots, x_n)]^2,$$

which corresponds to shifts. We have the following estimates:

LEMMA 6.2 *If u is any symmetric function of $x_1, \ldots, x_n$ on $(\mathbb{Z}^d)^n$*

$$D_n^{\mathrm{ex}}(u) \leq \frac{1}{n!} \bar{D}_n(u).$$

If, in addition, $u \equiv 0$ outside $\mathcal{G}_n$, then for some constant C independent of n and u,

$$\frac{1}{n!} \bar{D}_n(u) \leq C D_{\mathrm{ex}, n}(u) \quad \text{and} \quad \left| \frac{1}{n!} \bar{D}_n^{\mathrm{sh}}(u) - \hat{D}_n^{\mathrm{sh}}(u) \right| \leq C D_n^{\mathrm{ex}}(u).$$

PROOF: The first part is obvious. The factorial is just the number of times each term is counted. Clearly

$$\bar{D}_n(u) \leq n! D_n^{\mathrm{ex}}(u) + 2(n-1)! \sum_{x_1, \ldots, x_n} a(x_n - x_{n-1}) u^2(x_1, \ldots, x_n),$$

and by Lemma 3.7

$$\sum_{x_1,\ldots,x_n} \sum_{i \neq j} a(x_i - x_j) u^2(x_1,\ldots,x_n) \leq C n D_n^{\mathrm{ex}}(u),$$

and now the second part follows. The third part also follows from Lemma 3.7 and is nearly identical:

$$\left| \frac{1}{n!} \bar{D}_n^{\mathrm{sh}}(u) - \hat{D}_n^{\mathrm{sh}}(u) \right| \leq C \sum_{A \in \mathcal{X}_n} \left(\sum_{x \in A} a(x) \right) [u(A)]^2,$$

where we have identified A with $(x_1,\ldots,x_n)$. $\qquad \square$

We now return to the estimation of h. If we define

$$\langle h, w \rangle = \frac{1}{n!} \sum h(x_1,\ldots,x_n) w(x_1,\ldots,x_n),$$

then $|\langle h, w \rangle|$ is less than

(6.1)
$$\begin{aligned}
|\langle h, w \rangle| &\leq C \left[\frac{1}{n!} \sum_{x_1,\ldots,x_n} A(x_1,\ldots,x_n) u^2(x_1,\ldots,x_n) \right]^{\frac{1}{2}} \\
&\quad \times \left[\frac{1}{n!} \sum_{x_1,\ldots,x_n} A(x_1,\ldots,x_n) w^2(x_1,\ldots,x_n) \right]^{\frac{1}{2}} \\
&\leq C n [D_n^{\mathrm{ex}}(u)]^{\frac{1}{2}} \left[\frac{1}{n!} \bar{D}_n(w) \right]^{\frac{1}{2}}.
\end{aligned}$$

We have defined

$$A(x_1,\ldots,x_n) = \sum_{i \neq j} a(x_i - x_j) + \sum_j a(x_j).$$

We have used the following facts: The symmetric part q is also a, and p and q are dominated by $2a$.

The next lemma is a simple consequence of Fourier analysis.

LEMMA 6.3 *Let u be a symmetric function of n variables satisfying*

$$\lambda u - C_{n,n} u = v$$

where v satisfies the bound

$$\left| \sum_{x_1,\ldots,x_n} v(x_1,\ldots,x_n) w(x_1,\ldots,x_n) \right| \leq C [\bar{D}_n(w)]^{\frac{1}{2}}.$$

Then λu satisfies the same bound with the same constant,

$$\left| \sum_{x_1,\ldots,x_n} \lambda u(x_1,\ldots,x_n) w(x_1,\ldots,x_n) \right| \leq C [\bar{D}_n(w)]^{\frac{1}{2}}.$$

PROOF: Denoting by $\mathbb{T}_d^n$ the n-fold product of the d-torus and by $\hat{u}(\theta)$ and $\hat{v}(\theta)$ the Fourier transforms of u and v, respectively, we have

$$[\lambda + \Phi(\theta) + i\Psi(\theta)]\hat{u}(\theta) = \hat{v}(\theta)$$

where

$$\Phi(\theta) + i\Psi(\theta) = \sum_j \sum_x q(x)[1 - \cos(\theta_j x) - i\sin(\theta_j x)]$$

$$+ (1 - \rho) \sum_x p(x)\left[1 - \cos\left(x\sum_j \theta_j\right) + i\sin\left(x\sum_j \theta_j\right)\right].$$

Φ and Ψ are real, and $\Phi(\theta) \geq 0$. Moreover,

$$\|v\|_{-1}^2 = \int_{\mathbb{T}_d^n} \frac{|\hat{v}(\theta)|^2}{H(\theta)} d\theta \quad \text{where } H(\theta) = \sum_j \sum_x a(x)[1 - \cos(\theta_j x)]$$

and

$$\|\lambda u\|_{-1}^2 = \int_{\mathbb{T}_d^n} \frac{|\lambda \hat{u}(\theta)|^2}{H(\theta)} d\theta = \int_{\mathbb{T}_d^n} \frac{|\hat{v}(\theta)|^2}{H(\theta)} \frac{|\lambda|^2}{|\lambda + \Phi(\theta) + i\Psi(\theta)|^2} d\theta$$

$$\leq \int_{\mathbb{T}_d^n} \frac{|\hat{v}(\theta)|^2}{H(\theta)} d\theta,$$

which proves the lemma.   □

PROOF OF THEOREM 1.1: We are now ready to prove our main theorem. We start with a solution u_λ of (2.10) with a local f. If we decompose and write $u = \sum_n u_n$ and $f = \sum_n f_n$ with u_n and f_n from H_n, following the notation of (4.5) and (4.6),

$$\lambda u_{\lambda,n} - B_{n,n} u_{\lambda,n} = f_n + B_{n-1,n} u_{\lambda,n-1} + B_{n+1,n} u_{\lambda,n+1} = g_n.$$

As before, we rewrite

$$\lambda u_{\lambda,n} - C_{n,n} u_{\lambda,n} = g_n + h_n.$$

From (6.1) and Theorem 5.1, we have the estimate

$$|\langle h_n, w\rangle| \leq C_f C_k n^{-k}\left[\frac{1}{n!}\bar{D}(w)\right]^{\frac{1}{2}}.$$

From Lemmas 4.1 and 6.3 and Theorem 5.1, we have

$$|\langle w, B_{n-1,n} u_{\lambda,n-1} + B_{n+1,n} u_{\lambda,n+1}\rangle| \leq C_f C_k n^{-k}\left[\frac{1}{n!}\bar{D}(w)\right]^{\frac{1}{2}}.$$

If f is local, $f_n = 0$ for large enough n. By Lemma 6.4, we now conclude that for all large n,

$$|\langle \lambda u_{\lambda,n}, w\rangle| \leq C_f C_k \, n^{-k}\left[\frac{1}{n!}\bar{D}(w)\right]^{\frac{1}{2}}.$$

For small n where $f_n \neq 0$, we have from the proof of Lemma 2.1 and Lemma 6.3 that

$$|\langle w, f_n \rangle| \leq C_f \sqrt{D_n^e} \leq C_f \left[\frac{1}{n!} \bar{D}(w) \right]^{\frac{1}{2}}$$

and so for small n by Lemma 6.4,

$$|\langle \lambda u_{\lambda,n}, w \rangle| \leq C_f \left[\frac{1}{n!} \bar{D}(w) \right]^{\frac{1}{2}}.$$

Finally, by the use of Lemma 6.3, adding up contributions with a larger constant,

$$\|\lambda u_\lambda\|_{-1} \leq C_f C_k \sum_n n^{-k},$$

and we are done. $\qquad\square$

Acknowledgments. The authors wish to express their thanks to the referee. The original proof of Theorem 2.7 was by direct estimation and the referee's suggestion to bound it below by the corresponding variance for the symmetric situation simplified the proof considerably.

The authors also wish to acknowledge the support of the National Science Foundation through grants DMS-9703811 for Sethuraman, DMS-9503419 for Varadhan, and DMS-9403462 as well as DMS-9703752 for Yau. In addition, Varadhan wants to acknowledge the support of ARO through Grant ARO-DAAH04-95-1-0666, while Yau wishes to acknowledge the support of the David and Lucile Packard Foundation.

Bibliography

[1] Arratia, R. The motion of a tagged particle in the simple symmetric exclusion system on $\mathbb{Z}$. *Ann. Probab.* **11** (1983), no. 2, 362–373.

[2] Esposito, R.; Marra, R.; Yau, H. T. Navier-Stokes equations for stochastic particle systems on the lattice. *Comm. Math. Phys.* **182** (1996), no. 2, 395–456.

[3] Helland, I. S. Central limit theorems for martingales with discrete or continuous time. *Scand. J. Statist.* **9** (1982), no. 2, 79–94.

[4] Jensen, L.; Yau, H.-T. Hydrodynamical scaling limits of simple exclusion models. *Probability theory and applications (Princeton, N.J., 1996)*, 167–225. IAS/Park City Math. Ser., 6. Amer. Math. Soc., Providence, R.I., 1999.

[5] Kipnis, C. Central limit theorems for infinite series of queues and applications to simple exclusion. *Ann. Probab.* **14** (1986), no. 2, 397–408.

[6] Kipnis, C.; Varadhan, S. R. S. Central limit theorem for additive functionals of reversible Markov processes and applications to simple exclusions. *Comm. Math. Phys.* **104** (1986), no. 1, 1–19.

[7] Landim, C.; Yau, H. T. Fluctuation-dissipation equation of asymmetric simple exclusion processes. *Probab. Theory Related Fields* **108** (1997), no. 3, 321–356.

[8] Liggett, T. M. *Interacting particle systems.* Grundlehren der Mathematischen Wissenschaften, 276. Springer, New York-Berlin, 1985.

[9] Saada, E. A limit theorem for the position of a tagged particle in a simple exclusion process. *Ann. Probab.* **15** (1987), no. 1, 375–381.

[10] Varadhan, S. R. S. Regularity of self-diffusion coefficient. *The Dynkin Festschrift*, 387–397. Progr. Probab., 34. Birkhäuser Boston, Boston, 1994.

[11] Varadhan, S. R. S. Self-diffusion of a tagged particle in equilibrium for asymmetric mean zero random walk with simple exclusion. *Ann. Inst. H. Poincaré Probab. Statist.* **31** (1995), no. 1, 273–285.

SUNDER SETHURAMAN   S. R. S. VARADHAN
Iowa State University   Courant Institute
Department of Mathematics   251 Mercer Street
400 Carver Hall   New York, NY 10012
Ames, IA 50011   E-mail: varadhan@cims.nyu.edu
E-mail: sethuram@math.iastate.edu

HORNG-TZER YAU
Courant Institute
251 Mercer Street
New York, NY 10012
E-mail: yau@cims.nyu.edu

Received March 1999.

Vol. 10 (2005), Paper no. 36, pages 1221-1235.

Journal URL
http://www.math.washington.edu/~ejpecp/

A Martingale Proof of Dobrushin's Theorem for Non-Homogeneous Markov Chains [1]

S. Sethuraman and **S.R.S. Varadhan**

Department of Mathematics, Iowa State University
Ames, IA 50011, USA
Email: sethuram@iastate.edu

and

Courant Institute, New York University
New York, NY 10012, USA
Email: varadhan@cims.nyu.edu

Abstract. In 1956, Dobrushin proved an important central limit theorem for non-homogeneous Markov chains. In this note, a shorter and different proof elucidating more the assumptions is given through martingale approximation.

Keywords and phrases: non-homogeneous Markov, contraction coefficient, central limit theorem, martingale approximation

AMS Subject Classification (2000): Primary 60J10; secondary 60F05.

Submitted to EJP on March 21, 2005. Final version accepted on September 6, 2005.

[1]Research supported in part by NSF/DMS-0071504, NSA-H982300510041 and NSF/DMS-0104343.

1 Introduction and Results

Nearly fifty years ago, R. Dobrushin proved in his thesis [2] an important central limit theorem (CLT) for Markov chains in discrete time that are not necessarily homogeneous in time. Previously, Markov, Bernstein, Sapagov, and Linnik, among others, had considered the central limit question under various sufficient conditions. Roughly, the progression of results relaxed the state space structure from 2 states to an arbitrary set of states, and also the level of asymptotic degeneracy allowed for the transition probabilities of the chain.

After Dobrushin's work, some refinements and extensions of his CLT, some of which under more stringent assumptions, were proved by Statuljavicius [16] and Sarymsakov [13]. See also Hanen [6] in this regard. A corresponding invariance principle was also proved by Gudinas [4]. More general references on non-homogeneous Markov processes can be found in Isaacson and Madsen [7], Iosifescu [8], Iosifescu and Theodorescu [9], and Winkler [18].

We now define what is meant by "degeneracy." Although there are many measures of "degeneracy," the measure which turns out to be most useful to work with is that in terms of the contraction coefficient. This coefficient has appeared in early results concerning Markov chains, however, in his thesis, Dobrushin popularized its use, and developed many of its important properties. [See Seneta [14] for some history.]

Let $(\mathbf{X}, \mathcal{B}(\mathbf{X}))$ be a Borel space, and let $\pi = \pi(x, dy)$ be a Markov transition probability on $(\mathbf{X}, \mathcal{B}(\mathbf{X}))$. Define the contraction coefficient $\delta(\pi)$ of π as

$$
\begin{aligned}
\delta(\pi) &= \sup_{x_1, x_2 \in \mathbf{X}} \|\pi(x_1, \cdot) - \pi(x_2, \cdot)\|_{\mathrm{Var}} \\
&= \sup_{\substack{x_1, x_2 \in \mathbf{X} \\ A \in \mathcal{B}(\mathbf{X})}} |\pi(x_1, A) - \pi(x_2, A)| \\
&= \frac{1}{2} \sup_{\substack{x_1, x_2 \in \mathbf{X} \\ \|f\|_{L^\infty} \le 1}} |\int f(y)[\pi(x_1, dy) - \pi(x_2, dy)]|.
\end{aligned}
$$

Also, define the related coefficient $\alpha(\pi) = 1 - \delta(\pi)$.

Clearly, $0 \le \delta(\pi) \le 1$, and $\delta(\pi) = 0$ if and only if $\pi(x, dy)$ does not depend on x. It makes sense to call π "non-degenerate" if $0 \le \delta(\pi) < 1$. We use the standard convention and denote by $\mu\pi$ and πu the transformations induced by π on countably additive measures and bounded measurable functions respectively,

$$
(\mu\pi)(A) = \int \mu(dx)\,\pi(x, A) \qquad \text{and} \qquad (\pi u)(x) = \int \pi(x, dy)\,u(y).
$$

One can see that $\delta(\pi)$ has the following properties.

$$
\delta(\pi) = \sup_{\substack{x_1, x_2 \in \mathbf{X} \\ u \in \mathcal{U}}} |(\pi u)(x_1) - (\pi u)(x_2)| \tag{1.1}
$$

with $\mathcal{U} = \{u : \sup_{y_1,y_2} |u(y_1) - u(y_2)| \leq 1\}$. It is the operator norm of π with respect to the Banach (semi-) norm $\mathrm{Osc}(u) = \sup_{x_1,x_2} |u(x_1) - u(x_2)|$, namely the oscillation of u. In particular, for any transition probabilities π_1, π_2 we have

$$\delta(\pi_1 \, \pi_2) \leq \delta(\pi_1) \, \delta(\pi_2) \tag{1.2}$$

where $\pi_1\pi_2$ is the two-step transition probability $\pi_1\pi_2(x,\cdot) = \int \pi_1(x,dy)\pi_2(y,\cdot)$.

By a non-homogeneous Markov chain of length n on state space $(\mathbf{X}, \mathcal{B}(\mathbf{X}))$ corresponding to transition operators $\{\pi_{i,i+1} = \pi_{i,i+1}(x,dy) : 1 \leq i \leq n - 1\}$ we mean the Markov process P on the product space $(\mathbf{X}^n, \mathcal{B}(\mathbf{X}^n))$,

$$P[X_{i+1} \in A | X_i = x] = \pi_{i,i+1}(x, A),$$

where $\{X_i : 1 \leq i \leq n\}$ are the canonical projections. In particular, under the initial distribution $X_1 \sim \mu$, the distribution at time $k \geq 1$ is $\mu\pi_{1,2}\pi_{2,3}\cdots\pi_{k-1,k}$. For $i < j$ we will define

$$\pi_{i,j} = \pi_{i,i+1}\pi_{i+1,i+2}\cdots\pi_{j-1,j}.$$

We denote by $E[Z]$ and $V(Z)$ the expectation and variance of the random variable Z with respect to P.

Dobrushin's theorem concerns the fluctuations of an array of non-homogeneous Markov chains. For each $n \geq 1$, let $\{X_i^{(n)} : 1 \leq i \leq n\}$ be n observations of a non-homogeneous Markov chain on $\mathbf{X}$ with transition matrices $\{\pi_{i,i+1}^{(n)} = \pi_{i,i+1}^{(n)}(x,dy) : 1 \leq i \leq n - 1\}$ and initial distribution $\mu^{(n)}$. Let also

$$\alpha_n = \min_{1 \leq i \leq n-1} \alpha\big(\pi_{i,i+1}^{(n)}\big).$$

In addition, let $\{f_i^{(n)} : 1 \leq i \leq n\}$ be real valued functions on $\mathbf{X}$. Define, for $n \geq 1$, the sum

$$S_n = \sum_{i=1}^{n} f_i^{(n)}(X_i^{(n)}).$$

Theorem 1.1 *Suppose that for some finite constants C_n,*

$$\sup_{1 \leq i \leq n} \sup_{x \in \mathbf{X}} |f_i^{(n)}(x)| \leq C_n.$$

Then, if

$$\lim_{n \to \infty} C_n^2 \alpha_n^{-3} \left[\sum_{i=1}^{n} V\big(f_i^{(n)}(X_i^{(n)})\big) \right]^{-1} = 0, \tag{1.3}$$

we have the standard Normal convergence

$$\frac{S_n - E[S_n]}{\sqrt{V(S_n)}} \Rightarrow \mathrm{N}(0,1). \tag{1.4}$$

Also, there is an example where the result is not true if condition (1.3) is not met.

In [2], Dobrushin also states the direct corollary which simplifies some of the assumptions.

Corollary 1.1 *When the functions are uniformly bounded, i.e.* $\sup_n C_n = C < \infty$ *and the variances are bounded below, i.e.* $V(f_i^{(n)}(X_i^{(n)})) \geq c > 0$, *for all* $1 \leq i \leq n$ *and* $n \geq 1$, *then we have the convergence (1.4) provided*

$$\lim_{n \to \infty} n^{1/3} \alpha_n = \infty.$$

We remark that in [2] (e.g. Theorems 3, 8) there are also results where the boundedness condition on $f_i^{(n)}$ is replaced by integrability conditions. As these results follow from truncation methods and Theorem 1.1 for bounded variables, we only consider Dobrushin's theorem in the bounded case.

Also, for the ease of the reader, and to be complete, we will discuss in the next section an example, given in [2] and due to Bernstein and Dobrushin, of how the convergence (1.4) may fail when the condition (1.3) is not satisfied.

We now consider Dobrushin's methods. The techniques used in [2] to prove the above results fall under the general heading of the "blocking method." The condition (1.3) ensures that well-separated blocks of observations may be approximated by independent versions with small error. Indeed, in many remarkable steps, Dobrushin exploits the Markov property and several contraction coefficient properties, which he himself derives, to deduce error bounds sufficient to apply CLT's for independent variables. However, in [2], it is difficult to see, even at the technical level, why condition (1.3) is natural.

The aim of this note is to provide a different, shorter proof of Theorem 1.1 which explains more why condition (1.3) appears in the result. The methods are through martingale approximations and martingale CLT's. These methods go back at least to Gordin [3] in the context of homogeneous processes, and have been used by others in mostly "stationary" situations (e.g. Kifer [10], Kipnis and Varadhan [11], Pinsky [12], and Wu and Woodroofe [19]). The approximation with respect to the non-homogeneous setting of Theorem 1.1 makes use of three ingredients: (1) negligibility estimates for individual components, (2) a law of large numbers (LLN) for conditional variances, and (3) lower bounds for the variance $V(S_n)$. Negligibility bounds and a LLN are well known requirements for martingale CLT's (cf. Hall-Heyde [5, ch. 3]), and in fact, as will be seen, the sufficiency of condition (1.3) is transparent in the proofs of these two components (Lemma 3.2, and Lemmas 3.3 and 3.4). The variance lower bounds which we will use (Proposition 3.2) were as well derived by Dobrushin in his proof. However, using some martingale and spectral tools, we give a more direct argument for a better estimate.

We note also, with this martingale approximation, that an invariance principle for the partial sums holds through standard martingale propositions, Hall-Heyde [5], among other results. In fact, from the martingale invariance principle, it : ould be possible to derive Gudynas's theorems [4] although this is not done here.

We now explain the structure of the article. In section 2, we give the Bernstein-Dobrushin example of a Markov chain with anomalous behavior. In section 3, we state a martingale

CLT and prove Theorem 1.1 assuming a lower bound on the variance $V(S_n)$. Last, in section 4, we prove this variance estimate.

2   Bernstein-Dobrushin Example

Here, we summarize the example in Dobrushin's thesis, attributed to Bernstein, which shows that condition (1.3) is sharp.

Example 2.1 Let $\mathbf{X} = \{1, 2\}$, and consider the 2×2 transition matrices on $\mathbf{X}$,

$$Q(p) = \begin{pmatrix} 1 - p & p \\ p & 1 - p \end{pmatrix}$$

for $0 \le p \le 1$. The contraction coefficient $\delta(Q(p))$ of $Q(p)$ is $|1 - 2p|$. Note that $\delta(Q(p)) = \delta(Q(1 - p))$. The invariant measures for all the $Q(p)$ are the same $\mu(1) = \mu(2) = \frac{1}{2}$. We will be looking at $Q(p)$ for p close to 0 or 1 and the special case of $p = \frac{1}{2}$. However, when p is small, the homogeneous chains behave very differently under $Q(p)$ and $Q(1 - p)$. More specifically, when p is small there are very few switches between the two states whereas when $1 - p$ is small it switches most of the time. In fact, this behavior can be made more precise (see Dobrushin [1], or from direct computation). Let $T_n = \sum_{i=1}^{n} \mathbf{1}_{\{1\}}(X_i)$ count the number of visits to state 1 in the first n steps.

Case A. Consider the homogeneous chain under $Q(p)$ with $p = \frac{1}{n}$ and initial distribution $\mu(1) = \mu(2) = \frac{1}{2}$. Then,

$$\frac{T_n}{n} \Rightarrow G \quad \text{and} \quad \lim_{n \to \infty} n^{-2} V(T_n) = V_A \tag{2.1}$$

where $0 < V_A < \infty$ and G is a non-degenerate distribution supported on $[0, 1]$.

Case B. Consider the homogeneous chain run under $Q(p)$ with $p = 1 - \frac{1}{n}$ and initial distribution $\mu(1) = \mu(2) = \frac{1}{2}$. Then,

$$T_n - \frac{n}{2} \Rightarrow F \quad \text{and} \quad \lim_{n \to \infty} V(T_n) = V_B \tag{2.2}$$

where $0 < V_B < \infty$ and F is a non-degenerate distribution.

Let a sequence $\alpha_n \to 0$ with $\alpha_n \ge n^{-\frac{1}{3}}$ be given . To construct the anomalous Markov chain, it will be helpful to split the time horizon $[1, 2, \ldots, n]$ into roughly $n\alpha_n$ blocks of size α_n^{-1}. We interpose a $Q(\frac{1}{2})$ between any two blocks that has the effect of making the blocks independent of each other. More precisely let $k_i^{(n)} = i[\alpha_n^{-1}]$ for $1 \le i \le m_n$ where $m_n = [n/[\alpha_n^{-1}]]$. Also, define $k_0^{(n)} = 0$, and $k_{m_n+1}^{(n)} = n$.

Define now, for $1 \leq i \leq n$,

$$\pi_{i,i+1}^{(n)} = \begin{cases} Q(\alpha_n) & \text{for } i = 1, 2, \ldots, k_1^{(n)} - 1 \\ Q(\frac{1}{2}) & \text{for } i = k_1^{(n)}, k_2^{(n)}, \ldots, k_{m_n}^{(n)} \\ Q(1 - \alpha_n) & \text{for all other } i. \end{cases}$$

Consider the non-homogeneous chain with respect to $\{\pi_{i,i+1}^{(n)} : 1 \leq i \leq n-1\}$ starting from equilibrium $\mu^{(n)}(0) = \mu^{(n)}(1) = \frac{1}{2}$. ¿From the definition of the chain, one observes, as $Q(\frac{1}{2})$ does not distinguish between states, that the process in time horizons $\{(k_i^{(n)} + 1, k_{i+1}^{(n)}) : 0 \leq i \leq m_n\}$ are mutually independent. For the first time segment 1 to $k_1^{(n)}$, the chain is in regime A, while for the other segments, the chain is in case B.

Once again, let us concentrate on the number of visits to state 1. Denote by $T^{(n)} = \sum_{i=1}^n \mathbf{1}_{\{1\}}(X_i^{(n)})$ and $T^{(n)}(k, l) = \sum_{i=k}^l \mathbf{1}_{\{1\}}(X_i^{(n)})$ the counts in the first n steps and in steps k to l respectively. It follows from the discussion of independence above that

$$T^{(n)} = \sum_{i=0}^{m_n} T^{(n)}(k_i^{(n)} + 1, k_{i+1}^{(n)})$$

is the sum of independent sub-counts where, additionally, the sub-counts for $1 \leq i \leq m_n - 1$ are identically distributed, the last sub-count perhaps being shorter. Also, as the initial distribution is invariant, we have $V(\mathbf{1}_{\{1\}}(X_i^{(n)})) = 1/4$ for all i and n. Then, in the notation of Corollary 1.1, $C = 1$ and $c = 1/4$.

¿From (2.1), we have that

$$V(T^{(n)}(1, k_1^{(n)})) \sim \alpha_n^{-2} V_A \quad \text{as } n \uparrow \infty.$$

Also, from (2.2) and independence of m_n sub-counts, we have that

$$V(T^{(n)}(k_1^{(n)} + 1, n)) \sim n\alpha_n V_B \quad \text{as } n \uparrow \infty.$$

¿From these calculations, we see if $n^{1/3}\alpha_n \to \infty$, then $\alpha_n^{-2} << n\alpha_n$, and so the major contribution to $T^{(n)}$ is from $T^{(n)}(k_1^{(n)} + 1, n)$. However, since this last count is (virtually) the sum of m_n i.i.d. sub-counts, we have that $T^{(n)}$, properly normalized, converges to $N(0, 1)$, as predicted by Dobrushin's Theorem 1.1.

On the other hand, if $\alpha_n = n^{-1/3}$, we have $\alpha_n^{-2} = n\alpha_n$, and count $T^{(n)}(1, k_1^{(n)})$, independent of $T^{(n)}(k_1^{(n)}, n)$, also contributes to the sum $T^{(n)}$. After centering and scaling, then, $T^{(n)}$ approaches the convolution of a non-trivial non-normal distribution and a normal distribution, and therefore is not Gaussian.

3 Proof of Theorem 1.1

The CLT for martingale differences is a standard tool. We quote the following form of the result implied by Corollary 3.1 in Hall and Heyde [5].

Proposition 3.1 *For each $n \geq 1$, let $\{(W_i^{(n)}, \mathcal{G}_i^{(n)}) : 0 \leq i \leq n\}$ be a martingale relative to the nested family $\mathcal{G}_i^{(n)} \subset \mathcal{G}_{i+1}^{(n)}$ with $W_0^{(n)} = 0$. Let $\xi_i^{(n)} = W_i^{(n)} - W_{i-1}^{(n)}$ be their differences. Suppose that*

$$\max_{1 \leq i \leq n} \|\xi_i^{(n)}\|_{L^\infty} \to 0 \quad \text{and}$$
$$\sum_{i=1}^n E[(\xi_i^{(n)})^2 | \mathcal{G}_{i-1}^{(n)}] \to 1 \quad \text{in } L^2.$$

Then,

$$W_n^{(n)} \Rightarrow N(0,1).$$

The first and second limit conditions are the so called "negligibility" assumption on the sequence, and LLN for conditional variances mentioned in the introduction.

Consider now the non-homogeneous setting of Theorem 1.1. To simplify notation, we will assume throughout that the functions $\{f_i^{(n)}\}$ are mean-zero, $E[f_i^{(n)}(X_i^{(n)})] = 0$ for $1 \leq i \leq n$ and $n \geq 1$. Define

$$Z_k^{(n)} = \sum_{i=k}^n E[f_i^{(n)}(X_i^{(n)}) | X_k^{(n)}]$$

so that

$$Z_k^{(n)} = \begin{cases} f_k^{(n)}(X_k^{(n)}) + \sum_{i=k+1}^n E[f_i^{(n)}(X_i^{(n)}) | X_k^{(n)}] & \text{for } 1 \leq k \leq n-1 \\ f_n^{(n)}(X_n^{(n)}) & \text{for } k = n. \end{cases} \quad (3.1)$$

Remark 3.1 Before going further, we remark that the sequence $\{Z_k^{(n)}\}$ can be thought of as a type of "Poisson-resolvent" sequence often seen in martingale approximations Namely, when the array $\{X_i^{(n)}\}$ is formed from the sequence $\{X_i\}$, $f_i^{(n)} = f$ for all i and n, and the chain is homogeneous, $P_n = P$ for all n, then indeed $Z_k^{(n)}$ reduces to $Z_k^{(n)} = f(X_k) + \sum_{i=1}^{n-k}(P^i f)(X_k)$ which approximates the Poisson-resolvent solution $\sum_{i=0}^\infty (P^i f)(X_k) = [(I - P)^{-1} f](X_k)$ usually used to prove the CLT in this case (cf. p. 145-6 Varadhan [17]).

Returning to the full non-homogeneous setting of Theorem 1.1, by rearranging terms in (3.1), we obtain for $1 \leq k \leq n-1$ that

$$f_k^{(n)}(X_k^{(n)}) = Z_k^{(n)} - E[Z_{k+1}^{(n)} | X_k^{(n)}] \quad (3.2)$$

which for $2 \leq k \leq n-1$ further equals $[Z_k^{(n)} - E[Z_k^{(n)} | X_{k-1}^{(n)}]] + [E[Z_k^{(n)} | X_{k-1}^{(n)}] - E[Z_{k+1}^{(n)} | X_k^{(n)}]]$. Then, we have the decomposition,

$$\begin{aligned} S_n &= \sum_{k=1}^n f_k^{(n)}(X_k^{(n)}) \\ &= \sum_{k=2}^n [Z_k^{(n)} - E[Z_k^{(n)} | X_{k-1}^{(n)}]] + Z_1^{(n)} \end{aligned} \quad (3.3)$$

and so in particular $V(S_n) = \sum_{k=2}^{n} V(Z_k^{(n)} - E[Z_k^{(n)}|X_{k-1}^{(n)}]) + V(Z_1^{(n)})$. Let us now define the scaled differences

$$\xi_k^{(n)} = \frac{1}{\sqrt{V(S_n)}}[Z_k^{(n)} - E[Z_k^{(n)}|X_{k-1}^{(n)}]] \qquad (3.4)$$

and the martingale $M_k^{(n)} = \sum_{l=2}^{k} \xi_l^{(n)}$ with respect to $\mathcal{F}_k^{(n)} = \sigma\{X_l^{(n)} : 1 \leq l \leq k\}$ for $2 \leq k \leq n$. The plan to obtain Theorem 1.1 will now be to approximate $S_n/\sqrt{V(S_n)}$ by $M_n^{(n)}$ and use Proposition 3.1. Condition (1.3) will be a sufficent condition for "negligibility" (Lemma 3.2) and "LLN" (Lemmas 3.3 and 3.4) with regard to Proposition 3.1.

Lemma 3.1 *We have, for $1 \leq i < j \leq n$,*

$$\|\pi_{i,j} f_j^{(n)}\|_{L^\infty} \leq 2C_n(1-\alpha_n)^{j-i} \quad \text{and} \quad \mathrm{Osc}(\pi_{i,j}(f_j^{(n)})^2) \leq 2C_n^2(1-\alpha_n)^{j-i}$$

and, for $1 \leq l < i < j \leq n$,

$$\mathrm{Osc}(\pi_{l,i}(f_i^{(n)} \pi_{i,j} f_j^{(n)})) \leq 6C_n^2 (1-\alpha_n)^{i-l}(1-\alpha_n)^{j-i}.$$

Proof. As $\|f_j^{(n)}\|_{L^\infty} \leq C_n$ its oscillation $\mathrm{Osc}(f_j^{(n)}) \leq 2C_n$. From definition of $\delta(\cdot)$ (cf. (1.1)) and (1.2),

$$\mathrm{Osc}(\pi_{i,j} f_j^{(n)}) \leq \mathrm{Osc}(f_j^{(n)})\, \delta(\pi_{i,j}) \leq 2C_n(1-\alpha_n)^{j-i}.$$

Because $E[(\pi_{i,j} f_j^{(n)})(X_i^{(n)})] = E[f_j^{(n)}(X_j^{(n)})] = 0$, the first bound follows as

$$\|\pi_{i,j} f_j^{(n)}\|_{L^\infty} \leq \mathrm{Osc}(\pi_{i,j} f_j^{(n)}) \leq 2C_n(1-\alpha_n)^{j-i}.$$

The second bound is analogous. For the third bound, write

$$\begin{aligned}
\mathrm{Osc}(\pi_{l,i}(f_i^{(n)} \pi_{i,j} f_j^{(n)})) &\leq (1-\alpha_n)^{i-l}\, \mathrm{Osc}(f_i^{(n)} \pi_{i,j} f_j^{(n)}) \\
&\leq (1-\alpha_n)^{i-l}\Big[\mathrm{Osc}(f_i^{(n)})\, \|\pi_{i,j} f_j^{(n)}\|_{L^\infty} \\
&\qquad\qquad + \|f_i^{(n)}\|_{L^\infty}\, \mathrm{Osc}(\pi_{i,j} f_j^{(n)})\Big] \\
&\leq 6C_n^2 (1-\alpha_n)^{i-l}(1-\alpha_n)^{j-i}.
\end{aligned}$$

$\square$

We now state a lower bound for the variance proved in the next section. For comparison, we remark that in [2] the bound $V(S_n) \geq (\alpha_n/8) \sum_{i=1}^{n} V(f_i^{(n)}(X_i^{(n)}))$ is given (see also section 1.2.2 [9]).

Proposition 3.2 *For $n \geq 1$,*

$$V(S_n) \geq \frac{\alpha_n}{4} \sum_{i=1}^{n} V\big(f_i^{(n)}(X_i^{(n)})\big). \tag{3.5}$$

The next estimate shows that the asymptotics of $S_n/\sqrt{V(S_n)}$ depend only on the martingale approximant $M_n^{(n)}$, and that the differences $\xi_k^{(n)}$ are negligible.

Lemma 3.2 *Under condition (1.3), we have that*

$$\lim_{n \to \infty} \sup_{1 \leq k \leq n} \frac{\|Z_k^{(n)}\|_{L^\infty}}{\sqrt{V(S_n)}} = 0.$$

Proof. By Lemma 3.1,

$$\|Z_k^{(n)}\|_{L^\infty} \leq \sum_{i=k}^{n} \|E[f_i^{(n)}(X_i^{(n)})|X_k^{(n)}]\|_{L^\infty} \leq 2C_n \sum_{i=k}^{n}(1 - \alpha_n)^{i-k} \leq 2C_n \alpha_n^{-1}.$$

Then, by Proposition 3.2,

$$\sup_{1 \leq k \leq n} \frac{\|Z_k^{(n)}\|_{L^\infty}}{\sqrt{V(S_n)}} \leq 4C_n \bigg(\alpha_n^3 \sum_{i=1}^{n} V(f_i^{(n)}(X_i^{(n)}))\bigg)^{-1/2}$$

which in turn by (1.3) is $o(1)$. $\qquad\square$

The next two lemmas help prove the LLN part of Proposition 3.1 for array $\{M_k^{(n)}\}$. By the oscillation of a random variable η we mean $\mathrm{Osc}(\eta) = \sup_{\omega,\omega'}|\eta(\omega) - \eta(\omega')|$.

Lemma 3.3 *Let $\{Y_l^{(n)} : 1 \leq l \leq n\}$ and $\{\mathcal{G}_l^{(n)} : 1 \leq l \leq n\}$, for $n \geq 1$, be respectively an array of non-negative variables and σ-fields such that $\sigma\{Y_1^{(n)}, \ldots, Y_l^{(n)}\} \subset \mathcal{G}_l^{(n)}$. Suppose that*

$$\lim_{n \to \infty} E\bigg[\sum_{l=1}^{n} Y_l^{(n)}\bigg] = 1 \quad \text{and} \quad \sup_{1 \leq i \leq n}\|Y_i^{(n)}\|_{L^\infty} \leq \epsilon_n$$

where $\lim_{n \to \infty} \epsilon_n = 0$. In addition, assume

$$\lim_{n \to \infty} \sup_{1 \leq l \leq n-1} \mathrm{Osc}\bigg(E\bigg[\sum_{j=l+1}^{n} Y_j^{(n)}|\mathcal{G}_l^{(n)}\bigg]\bigg) = 0.$$

Then,

$$\lim_{n \to \infty} \sum_{l=1}^{n} Y_l^{(n)} = 1 \quad \text{in } L^2.$$

Proof. Write

$$E\big[\big(\sum_{l=1}^{n} Y_l^{(n)}\big)^2\big] = \sum_{l=1}^{n} E\big[(Y_l^{(n)})^2\big] + 2\sum_{l=1}^{n-1} E\big[Y_l^{(n)}\big(\sum_{j=l+1}^{n} Y_j^{(n)}\big)\big].$$

The first sum on the right-hand side is bounded as follows. ¿From non-negativity,

$$\sum_{l=1}^{n} E\big[(Y_l^{(n)})^2\big] \le \epsilon_n \sum_{l=1}^{n} E\big[Y_l^{(n)}\big] = \epsilon_n \cdot (1+o(1)) \to 0 \quad \text{as} \quad n \uparrow \infty.$$

For the second sum, write

$$\sum_{l=1}^{n-1} E\big[Y_l^{(n)}\big(\sum_{j=l+1}^{n} Y_j^{(n)}\big)\big] = \sum_{l=1}^{n-1} E\big[Y_l^{(n)} E\big[\sum_{j=l+1}^{n} Y_j^{(n)}|\mathcal{G}_l^{(n)}\big]\big].$$

¿From the oscillation assumption, we have that

$$\sup_{1\le l\le n-1} \sup_{\omega} \big|E\big[\sum_{j=l+1}^{n} Y_j^{(n)}|\mathcal{G}_l^{(n)}\big](\omega) - E\big[\sum_{j=l+1}^{n} Y_j^{(n)}\big]\big| = o(1).$$

Therefore,

$$2\sum_{l=1}^{n-1} E\big[Y_l^{(n)}\big(\sum_{j=l+1}^{n} Y_j^{(n)}\big)\big] = 2\sum_{l=1}^{n-1} E\big[Y_l^{(n)}\big]E\big[\sum_{j=l+1}^{n} Y_j^{(n)}\big] + o(1)\cdot \sum_{l=1}^{n-1} E\big[Y_l^{(n)}\big]$$

$$= \big(\sum_{l=1}^{n} E\big[Y_l^{(n)}\big]\big)^2 - \sum_{l=1}^{n} E\big[(Y_l^{(n)})^2\big] + o(1)$$

$$= 1 + o(1)$$

finishing the proof. $\qquad\square$

To apply later this result to $v_j^{(n)} = E[(\xi_j^{(n)})^2|\mathcal{F}_{j-1}^{(n)}]$ measureable with respect to $\mathcal{G}_j^{(n)} = \mathcal{F}_{j-1}^{(n)}$ for $2 \le j \le n$ we will need the following oscillation estimate.

Lemma 3.4 *Under condition (1.3), we have*

$$\sup_{2\le l\le n-1} \mathrm{Osc}\Big(E\big[\sum_{j=l+1}^{n} v_j^{(n)}|\mathcal{F}_{l-1}^{(n)}\big](\omega)\Big) = o(1).$$

Proof. From the martingale property, $E[\xi_r^{(n)}\xi_s^{(n)}|\mathcal{F}_u^{(n)}] = 0$ for $r > s > u$, (3.4) and (3.2), we have

$$
\begin{aligned}
E\Big[\sum_{j=l+1}^{n} v_j^{(n)}|\mathcal{F}_{l-1}^{(n)}\Big] &= E\Big[\sum_{j=l+1}^{n} (\xi_j^{(n)})^2 \,|\mathcal{F}_{l-1}^{(n)}\Big] \\
&= E\Big[\Big(\sum_{j=l+1}^{n} \xi_j^{(n)} \Big)^2 |X_{l-1}^{(n)}\Big] \\
&= V(S_n)^{-1} E\Big[\Big(\sum_{j=l+1}^{n} f_j^{(n)}(X_j^{(n)}) - E[Z_{l+1}^{(n)}|X_l^{(n)}]\Big)^2 |X_{l-1}^{(n)}\Big] \\
&= V(S_n)^{-1} E\Big[\Big(\sum_{j=l+1}^{n} f_j^{(n)}(X_j^{(n)})\Big)^2 |X_{l-1}^{(n)}\Big] \\
&\quad - V(S_n)^{-1} E\big[E[Z_{l+1}^{(n)}|X_l^{(n)}]^2|X_{l-1}^{(n)}\big]. \tag{3.6}
\end{aligned}
$$

By Lemma 3.2, the last term in (3.6) is bounded $\sup_{2\le l\le n-1} V(S_n)^{-1}\|Z_{l+1}^{(n)}\|_{L^\infty}^2 = o(1)$, and so its oscillation is also uniformly $o(1)$.

To estimate oscillation of the first term on right-side of (3.6), we write

$$
\mathrm{Osc}\Big(V(S_n)^{-1} E\big[(\sum_{j=l+1}^{n} f_j^{(n)}(X_j^{(n)}))^2|X_{l-1}^{(n)}\big] \Big) \tag{3.7}
$$

$$
\le V(S_n)^{-1} \sum_{l+1\le j,m\le n} \mathrm{Osc}\Big(E\big[f_j^{(n)}(X_j^{(n)})f_m^{(n)}(X_m^{(n)})|X_{l-1}^{(n)}\big] \Big).
$$

But, for $l + 1 \le j \le m \le n$, we have from Lemma 3.1 that

$$
\mathrm{Osc}\Big(E\big[f_j^{(n)}(X_j^{(n)})f_m^{(n)}(X_m^{(n)})|X_{l-1}^{(n)}\big] \Big) \le 6C_n^2 (1 - \alpha_n)^{j-l+1}(1 - \alpha_n)^{m-j}.
$$

Then, (3.7) is bounded, uniformly in l, on order $V(S_n)^{-1}C_n^2\alpha_n^{-2}$ which from Proposition 3.2 and (1.3) is $o(1)$. $\qquad\square$

Proof of Theorem 1.1. From Lemma 3.2, we need only show that $M_n^{(n)}/\sqrt{V(S_n)} \Rightarrow N(0,1)$. This will follow from martingale convergence (Proposition 3.1) as soon as we show (1) $\sup_{2\le k\le n}\|\xi_k^{(n)}\|_{L^\infty} \to 0$ and (2) $\sum_{k=2}^{n} E[(\xi_k^{(n)})^2|\mathcal{F}_{k-1}^{(n)}] \to 1$. However, (1) follows from the negligibility estimate Lemma 3.2, and (2) from LLN Lemmas 3.3 and 3.4 since "negligibility" (1) holds and $\sum_{k=2}^{n} E[(\xi_k^{(n)})^2] = 1+o(1)$ (from variance decomposition after (3.3) and Lemma 3.2). $\qquad\square$

4 Proof of Variance Lower Bound

Let λ be a probability measure on $\mathbf{X} \times \mathbf{X}$ with marginals α and β respectively. Let $\pi(x_1, dx_2)$ and $\widehat{\pi}(x_2, dx_1)$ be the corresponding transition probabilities in the two directions so that $\alpha\pi = \beta$ and $\beta\widehat{\pi} = \alpha$.

Lemma 4.1 *Let $f(x_1)$ and $g(x_2)$ be square integrable with respect to α and β respectively. If*

$$\int f(x_1)\alpha(dx_1) = \int g(x_2)\beta(dx_2) = 0$$

then,

$$\left| \int f(x_1)g(x_2)\lambda(dx_1, dx_2) \right| \leq \sqrt{\delta(\pi)} \, \|f\|_{L_2(\alpha)} \|g\|_{L_2(\beta)}.$$

Proof. Let us construct a measure on $\mathbf{X} \times \mathbf{X} \times \mathbf{X}$ by starting with λ on $\mathbf{X} \times \mathbf{X}$ and using reversed $\widehat{\pi}(x_2, dx_3)$ to go from x_2 to x_3. The transition probability from x_1 to x_3 defined by

$$Q(x_1, A) = \int \pi(x_1, dx_2)\widehat{\pi}(x_2, A)$$

satisfies $\delta(Q) \leq \delta(\pi)\delta(\widehat{\pi}) \leq \delta(\pi)$. Moreover $\alpha Q = \alpha$ and the operator Q is self-adjoint and bounded with norm 1 on $L_2(\alpha)$. Then, if f is a bounded function with $\int f(x)\alpha(dx) = 0$ (and so $E_\alpha[Q^n f] = 0$), we have for $n \geq 1$,

$$\|Q^n f\|_{L_2(\alpha)} \leq \|Q^n f\|_{L_\infty} \leq (\delta(Q))^n \mathrm{Osc}(f). \tag{4.1}$$

Hence, as bounded functions are dense, on the subspace of functions, $M = \{f \in L_2(\alpha) : \int f(x)\alpha(dx) = 0\}$, the top of the spectrum of Q is less than $\delta(Q)$ and so $\|Q\|_{L_2(\alpha,M)} \leq \delta(Q)$. Indeed, suppose the spectral radius of Q on M is larger than $\delta(Q) + \epsilon$ for $\epsilon > 0$, and $f \in M$ is a non-trivial bounded function whose spectral decomposition is with respect to spectral values larger than $\delta(Q) + \epsilon$. Then, $\|Q^n f\|_{L_2(\alpha)} \geq \|f\|_{L_2(\alpha)}(\delta(Q) + \epsilon)^n$ which contradicts the bound (4.1) when $n \uparrow \infty$. [cf. Thm. 2.10 [15] for a proof in discrete space settings.]

Then,

$$\|\widehat{\pi} f\|_{L_2(\beta)}^2 = \langle \pi\widehat{\pi} f, f \rangle_{L_2(\alpha)} = \langle Q f, f \rangle_{L_2(\alpha)} \leq \|Q\|_{L_2(\alpha,M)} \|f\|_{L_2(\alpha)}^2 \leq \delta(Q)\|f\|_{L_2(\alpha)}^2.$$

Finally,

$$\left| \int f(x_1)g(x_2)\lambda(dx_1, dx_2) \right| = |\langle \widehat{\pi} f, g \rangle_{L_2(\beta)}| \leq \sqrt{\delta(\pi)} \, \|f\|_{L_2(\alpha)} \|g\|_{L_2(\beta)}.$$

$\square$

Lemma 4.2 *Let $f(x_1)$ and $g(x_2)$ be square integrable with respect to α and β respectively. Then,*

$$E\big[\big(f(x_1) - g(x_2)\big)^2\big] \geq (1 - \delta(\pi))\, V\big(f(x_1)\big)$$

as well as

$$E\big[\big(f(x_1) - g(x_2)\big)^2\big] \geq (1 - \delta(\pi))\, V\big(g(x_2)\big).$$

Proof. To get lower bounds, we can assume without loss of generality that f and g have mean 0 with respect to α and β respectively. Then by Lemma 4.1

$$
\begin{aligned}
E\big[\big(f(x_1) - g(x_2)\big)^2\big] &= E\big[[f(x_1)]^2\big] + E\big[[g(x_2)]^2\big] - 2E\big[f(x_1)g(x_2)\big] \\
&\geq E\big[[f(x_1)]^2\big] + E\big[[g(x_2)]^2\big] - 2\sqrt{\delta(\pi)}\,\|f\|_{L_2(\alpha)}\|g\|_{L_2(\beta)} \\
&\geq (1 - \delta(\pi))\|f\|_{L_2(\alpha)}^2.
\end{aligned}
$$

The proof of the second half is identical. □

Proof of Proposition 3.2. Applying Lemma 4.2 to the Markov pairs $\{(X_k^{(n)}, X_{k+1}^{(n)}) : 1 \leq k \leq n-1\}$ with $f(X_k^{(n)}) = E[Z_{k+1}^{(n)}|X_k^{(n)}]$ and $g(X_{k+1}^{(n)}) = Z_{k+1}^{(n)}$, we get

$$E\big[\big(Z_{k+1}^{(n)} - E[Z_{k+1}^{(n)}|X_k^{(n)}]\big)^2\big] \geq \alpha_n E\big[\big(Z_{k+1}^{(n)}\big)^2\big].$$

On the other hand from (3.2), for $1 \leq k \leq n-1$, we have

$$
\begin{aligned}
V(f_k^{(n)}(X_k^{(n)})) &\leq E\big[\big(f_k^{(n)}(X_k^{(n)})\big)^2\big] \\
&\leq 2E\big[\big(Z_k^{(n)}\big)^2\big] + 2E\big[\big(E[Z_{k+1}^{(n)}|X_k^{(n)}]\big)^2\big] \\
&\leq 2E\big[\big(Z_k^{(n)}\big)^2\big] + 2E\big[\big(Z_{k+1}^{(n)}\big)^2\big].
\end{aligned}
$$

Summing over k, and noting $f_n^{(n)}(X_n^{(n)}) = Z_n^{(n)}$ and variance decomposition after (3.3),

$$
\begin{aligned}
\sum_{k=1}^n V(f_k^{(n)}(X_k^{(n)})) &\leq 4\sum_{k=1}^n E\big[\big(Z_k^{(n)}\big)^2\big] \\
&\leq \frac{4}{\alpha_n}\left[\sum_{k=1}^{n-1} E\big[\big(Z_{k+1}^{(n)} - E[Z_{k+1}^{(n)}|X_k^{(n)}]\big)^2\big] + E[(Z_1^{(n)})^2]\right] = \frac{4}{\alpha_n}V(S_n).
\end{aligned}
$$

□

Acknowledgement. We would like to thank the referees for their comments.

References

[1] Dobrushin, R. (1953) Limit theorems for Markov chains with two states. (Russian) *Izv. Adad. Nauk SSSR* **17:4** 291-330.

[2] Dobrushin, R. (1956) Central limit theorems for non-stationary Markov chains I,II. *Theory of Probab. and its Appl.* **1** 65-80, 329-383.

[3] Gordin, M.I. (1969) The central limit theorem for stationary processes. *Soviet Math. Dokl.* **10** 1174-1176.

[4] Gudynas, P. (1977) An invariance principle for inhomogeneous Markov chains. *Lithuanian Math. J.* **17** 184-192.

[5] Hall, P. and Heyde, C.C. (1980) *Martingale Limit Theory and Its Application.* Academic Press, New York.

[6] Hanen, A. (1963) Théorèmes limites pour une suite de châines de Markov. *Ann. Inst. H. Poincaré* **18** 197-301.

[7] Isaacson, D.L., and Madsen, R.W. (1976) *Markov Chains. Theory and Applications.* John Wiley and Sons, New York.

[8] Iosifescu, M. (1980) *Finite Markov Processes and Their Applications.* John Wiley and Sons, New York.

[9] Iosifescu, M., and Theodorescu, R. (1969) *Random Processes and Learning. Springer,* Berlin.

[10] Kifer, Y. (1998) Limit theorems for random transformations and processes in random environments. *Trans. Amer. Math. Soc.* **350** 1481-1518.

[11] Kipnis, C., Varadhan, S. R. S. (1986) Central limit theorem for additive functionals of reversible markov processes. *Commun. Math. Phys.* **104** 1-19.

[12] Pinsky, M. (1991) *Lectures on Random Evolution.* World Scientific, Singapore.

[13] Sarymsakov, T.A. (1961) Inhomogeneous Markov chains. *Theor. Probability Appl.* **6** 178-185.

[14] Seneta, E. (1973) On the historical development of the theory of finite inhomogeneous Markov chains. *Proc. Cambridge Philos. Soc.* **74** 507-513.

[15] Seneta, E. (1981) *Non-negative Matrices and Markov Chains.* Second Edition, Springer-Verlag, New York.

[16] Statuljavicius, V. (1969) Limit theorems for sums of random variables connected in Markov chains.(Russian) *Litovsk. Mat. Sb.* **9** 345-362; ibid. **9**, 635-672; ibid. **10** 161–169.

[17] Varadhan, S.R.S. (2001) *Probability Theory.* Courant Lecture Notes **7**, American Mathematical Society, Providence, R.I.

[18] Winkler, G. (1995) *Image Analysis, Random Fields and Dynamic Monte Carlo Methods. A Mathematical Introduction.* Applications of Mathematics **27**, Springer-Verlag, Berlin.

[19] Wu, Wei Biao, Woodroofe, M. (2004) Martingale approximations for sums of stationary processes. *Ann. Probab.* **32** 1674–1690.

Review Articles

Diffusion Processes

S. R. S. Varadhan

1. Brownian motion

Brownian motion is the prototypical example of a diffusion process. The standard Brownian motion on R is a stochastic process $x(t)$ defined for $0 \leq t < \infty$. It has independent increments, i.e. the increments over disjoint intervals are stochastically independent. The distribution of the increment $x(t) - x(s)$, over the interval $[s, t]$ with $s \leq t$, is normally distributed with mean 0 and variance $t - s$. The Brownian motion is usually normalized so that $x(0) = 0$ with probability 1. It can be equivalently defined as a Gaussian Process with mean 0 and covariance $E[x(s)x(t)] = s \wedge t$. The joint distribution of $x(t_1), \ldots, x(t_n)$ at n time points $0 < t_1 < \cdots < t_n < \infty$ is multivariate normal with mean 0 and covariance $C_{i,j} = t_i \wedge t_j$. Its distribution on R^n has the joint density

$$
p_n(t_1, \ldots, t_n; x_1, \ldots, x_n) = \left[\prod_{j=1}^{n} (2\pi(t_j - t_{j-1}))^{-\frac{1}{2}} \right]
$$
$$
\times \exp\left[-\sum_{j=1}^{n} \frac{(x_j - x_{j-1})^2}{2(t_j - t_{j-1})} \right] \tag{1.1}
$$

with respect to the Lebesgue measure. Here we adopt the convention that $t_0 = x_0 = 0$. This defines a consistent family of finite dimensional distributions, and by Kolmogorov's theorem defines a stochastic process. It is a theorem Norbert Wiener that this stochastic process can be supported on the space $\Omega = C[0, \infty)$ of continuous functions on $[0, \infty)$. The space Ω comes with a natural σ-field of Borel sets $\mathscr{F}$ and canonical sub-σ-fields $\mathscr{F}_t$ of events generated by $[x(s) : 0 \leq s \leq t]$. Clearly $\mathscr{F}_s \subset \mathscr{F}_t$ for $s \leq t$ and $\mathscr{F} = \sigma(\cup_t \mathscr{F}_t)$, i.e. the smallest σ-field containing $\mathscr{F}_t$ for all t. The Wiener measure P_0 on Ω is the unique measure such that for every positive integer n, time points $0 < t_1 < \cdots < t_n$

$$
P_0[x(t_1), \ldots, x(t_n) \in A] = \int_A p_n(t_1, \ldots, t_n; x_1, \ldots, x_n) dx_1 \ldots dx_n
$$

for all Borel sets $A \subset R^n$. It has the following additional properties.

1. If we denote by H_{α,C,T_1,T_2} the set of functions $x(\cdot)$ that satisfy a local Hölder condition of the form

$$|x(s) - x(t)| \leq C|t - s|^{\alpha}$$

for $T_1 \leq s \leq t \leq T_2$ then for every $\alpha \geq 1/2$, $T_1 < T_2$ and $C < \infty$

$$P_0[H_{\alpha,C,T_1,T_2}] = 0 \ .$$

2. On the other hand for every $\alpha < 1/2$, and $T_1 < T_2 < \infty$,

$$\lim_{C \to \infty} P_0[H_{\alpha,C,T_1,T_2}] = 1 \ .$$

3. If D_t is the set of paths that are differentiable at the point t, then

$$P_0[\cup_t D_t] = 0$$

4. For any positive integer k, if $\{j/2^k; j \geq 0\}$ is the discrete set of points with spacing $1/2^k$, then

$$P_0\left[\lim_{k \to \infty} \sum_{1 \leq j \leq [2^k t]} \left(x\left(\frac{j}{2^k}\right) - x\left(\frac{j-1}{2^k}\right)\right)^2 = t \quad \text{for all } t \geq 0\right] = 1 \quad (1.2)$$

It is clear from these properties that a typical Brownian path, being nowhere differentiable, is not of bounded variation in any finite interval. However it has a definite 'quadratic variation' in any interval that is almost surely equal to the length of the interval.

One of the ways of understanding the behavior of a Brownian path is to start with a random walk $\{S_n : n \geq 1\}$ which, for each n, is the sum

$$S_n = X_1 + X_2 + \cdots + X_n$$

of the first n terms of a sequence $\{X_i\}$ of independent identically distributed random variables that take the value ± 1 with probability $1/2$ each. While the central limit theorem asserts that for large n, the distribution $S_n/\sqrt{n}$ is asymptotically the standard normal, Brownian motion is to be thought of as the limit in distribution of $S_{[nt]}/\sqrt{n}$ as a function of t.

The Wiener measure P_0 or the Brownian motion $x(\cdot)$ is not only the prime example of a diffusion process but it is also the building block out of which other diffusion processes are constructed. A discussion of Brownian motion and its properties can be found, among other places, in the following books (Stroock, 1993; Revuz and Yor, 1999). Usually one refers to the measure P_0 as the Wiener measure and a random path $x(\cdot)$ that is 'distributed' according to the Wiener measure as Brownian motion.

Brownian motion derives its name from the botanist Robert Brown who observed in 1828, that, in water, pollen from plants dispersed in an irregular swarming motion. From that time many scientists have examined the Brownian phenomenon including Einstein. Norbert Wiener was the first one who essentially constructed the measure corresponding to Brownian motion, and proved the

Hölder continuity of it paths in Wiener (1923) and Wiener (1924). It was the first rigorous example of integration in an infinite dimensional function space.

Just as any multivariate normal random vector can be generated by a linear transformation $y = Tx$ from one with covariance equal to the Identity matrix, one can start with any orthonormal basis $\{f_j(\cdot)\}$ in $L_2[0, \infty]$ and represent the Brownian motion as

$$x(t) = \sum_j Z_j F_j(t)$$

where $\{Z_j\}$ are independent standard normal random variables and

$$F_j(t) = \int_0^t f_j(s)\mathrm{d}s \ .$$

While formally the expansion is always valid its convergence is far more delicate. Such expansions have been considered for Brownian motion and more generally for Gaussian processes in Itô and Nisio (1968).

2. Brownian motion as a Markov process

The probability density p_n in (1.1) can be rewritten as

$$p_n(t_1, \ldots, t_n; x_1, \ldots, x_n) = \prod_{j=1}^n p(t_j - t_{j-1}, x_j - x_{j-1}) \tag{2.1}$$

where

$$p(t, x) = \frac{1}{\sqrt{2\pi t}} \exp\left[-\frac{x^2}{2t}\right] \tag{2.2}$$

In other words the process P_0 is a Markov process with transition probability

$$P_0[x(t) \in A | x(\sigma) : 0 \le \sigma \le s] = \int_A p(t - s, y - x(s))\mathrm{d}y$$

$$= \int_A q(s, x(s), t, y)\mathrm{d}y$$

where $q(s, x; t, y) = p(t - s, y - x)$.

We can in fact initialize the Brownian motion to start from any point x_0 at time 0 to get P_{x_0} which is just the distribution of $x_0 + x(\cdot)$ under P_0. The transition probability density as we saw depends only on the differences $y - x$ and $t - s$. It is because Brownian motion as a Markov process is homogeneous in time and has independent increments, i.e. is it is invariant under translations in space as well as time.

In the theory of Markov processes the concept of stopping times is extremely important. These are random variables $\tau(\omega) \ge 0$ that are $\mathscr{F}$ measurable that have the additional property that for every $t \ge 0$, the set $[\omega : \tau(\omega) \le t]$ is $\mathscr{F}_t$ measurable. Examples of stopping times are the first time some thing happens, for instance

$$\tau(\omega) = \inf[t : x(t) \geq a]$$

In order to determine if $\tau(\omega) \leq t$, we only need to observe the path up to time t. On the other hand, the last time some thing happens like

$$\sup\left[t : \int_0^t V(x(s))\mathrm{d}s \leq 0\right]$$

are not stopping times. An important fact is that for Brownian motion as well as for most other diffusion processes, the Markov property extends to stopping times and this is called the strong Markov property and was first considered by Hunt (1956).

Associated with any stopping time τ there is a natural σ-field $\mathscr{F}_\tau$ defined by

$$\mathscr{F}_\tau = \{A : A \cap [\tau \leq t] \in \mathscr{F}_t \quad \text{for all } t \geq 0\}$$

Roughly speaking if we observe the process only up to time τ and then stop, then $\mathscr{F}_\tau$ represents the information that is available. The strong Markov property for Brownian motion, first established by G. A. Hunt [H], states that with respect to any P_{x_0}, the future increments $y_\tau(t) = x(\tau + t) - x(\tau)$ is again a Brownian motion, i.e. distributed according to P_0 and more over is stochastically independent of the events in $\mathscr{F}_\tau$. This can be restated in the form

$$P_{x_0}[x(\tau + t) \in A | \mathscr{F}_\tau] = P_{x(\tau)}[x(t) \in A]$$

An easy consequence of this is the reflection principle of Bachelier that computes for $\ell \geq 0$,

$$P_0\left[\sup_{0 \leq s \leq t} x(s) \geq \ell\right] = 2P[x(t) \geq \ell] = \sqrt{\frac{2}{\pi t}} \int_\ell^\infty \exp\left[-\frac{x^2}{2t}\right]\mathrm{d}x$$

3. Semigroups and generators

An important functional analytical tool in the study of Markov processes is the notion of semigroups associated with them. For any bounded measurable function f on R, we define $T_t f$ by

$$(T_t f)(x) = E[f(x(t)|x(0) = x] = \int_R f(y)p(t, y - x)\mathrm{d}y$$

The Chapman–Kolmogorov equations which in our case reduce to

$$\int_R p(t, y - x)p(s, z - y)\mathrm{d}y = p(t + s, z - x)$$

yield

$$T_t T_s f = T_s T_t f = T_{t+s} f$$

for all f and $s, t \geq 0$. Such semigroups are characterized by their infinitesimal generators

$$A = \frac{\mathrm{d} T_t}{\mathrm{d} t}\bigg|_{t=0}$$

and can be recovered in some sense as

$$T_t = \exp[tA]$$

For our Brownian motion semigroup it is an easy calculation to derive that for any smooth function f

$$(Af)(x) = \lim_{t \to 0} \frac{1}{t} \int [f(x+y) - f(x)]p(t,y)\mathrm{d}y = \frac{1}{2}\frac{\mathrm{d}^2 f(x)}{\mathrm{d}x^2}$$

It is therefore natural to expect connections between the operator $1/2D_x^2$ and Brownian motion. The first such relation is that the transition probability $p(t,x)$ satisfies the heat equation

$$\frac{\partial p}{\partial t} = \frac{1}{2}\frac{\partial^2 p}{\partial x^2}$$

and consequently

$$(T_t f)(x) = u(t,x) = \int f(y)p(t, y-x)\mathrm{d}y$$

solves the Cauchy problem

$$\frac{\partial u}{\partial t} = \frac{1}{2}\frac{\partial^2 u}{\partial x^2}$$

with the initial condition $\lim_{t \to 0} u(t,x) = f(x)$.

Taking Laplace transforms in t, for $\lambda > 0$, the resolvent equations

$$\lambda u - \tfrac{1}{2}u_{xx} = f$$

are solved by

$$u(x) = \int_0^\infty \mathrm{e}^{-\lambda t}(T_t f)(x)\mathrm{d}t = E^{P_x}\left[\int_0^\infty \mathrm{e}^{-\lambda t}f(x(t))\mathrm{d}t\right]$$

The Feynman–Kac formula provides a representation for solutions of

$$\frac{\partial u}{\partial t} = \frac{1}{2}\frac{\partial^2 u}{\partial x^2} + V(x)u$$

with the initial condition $\lim_{t\to 0} u(t,x) = f(x)$ in the form

$$u(t,x) = E^{P_x}\left[f(x(t))\exp\left[\int_0^t V(x(s))\mathrm{d}s\right]\right]$$

4. Stochastic integrals

Since $x(t)$ is almost surely a continuous function of t we can define $Y = \int_0^\infty f(t)x(t)\mathrm{d}t$ for smooth functions f with compact support and will give us a Gaussian random variable with mean 0 and variance $\iint f(s)f(t)s \wedge t\, \mathrm{d}s\, \mathrm{d}t$. In fact, if f is a smooth function we can even define $\int f(t)\mathrm{d}x(t)$ as $-\int f'(t)x(t)\mathrm{d}t$ after an integration by part and calculate

$$E\left[\left|\int f(t)\mathrm{d}x(t)\right|^2\right] = \iint f'(s)f'(t)s \wedge t\, \mathrm{d}s\, \mathrm{d}t = \int [f(t)]^2\, \mathrm{d}t$$

By completion in $L_2[P]$, we can now define $\int f(t)\mathrm{d}x(t)$ for functions f in $L_2[0,\infty]$ and will give us a mean zero normal random variable with variance $\|f\|_2^2$. This was carried out already by Wiener and yields the following result known as the Cameron–Martin formula [see Cameron and Martin (1953)].

Let $F(t)$ be of the form $F(t) = \int_0^t f(s)\mathrm{d}s$ for some f in $L_2[0,\infty]$. Let $x(\cdot)$ be Brownian Motion distributed according to the Wiener measure P_0. Then the distribution P_F of $y(\cdot) = x(\cdot) + F(\cdot)$ is absolutely continuous with respect to P_0 and has the Radon–Nikodym derivative

$$\frac{\mathrm{d}P_F}{\mathrm{d}P_0}(\omega) = \exp\left[\int_0^\infty f(t)\mathrm{d}x(t) - \frac{1}{2}\int_0^\infty [f(t)]^2\, \mathrm{d}t\right] .$$

Formally,the relative density of a Gaussian vector ξ with mean μ and covariance C with respect the Gaussian with mean 0 and the same covariance is given by

$$\exp\left[\langle \xi, C^{-1}\mu\rangle - \tfrac{1}{2}\langle \mu, C^{-1}\mu\rangle\right]$$

and our formula is just a special case.

In the definition of Wiener's stochastic integral the integrand f is nonrandom, i.e. does depend on the path ω. Itô extended Wiener's definition in a far reaching way. Since the calculation is essentially one involving two moments, in an approximation of the form

$$S = \sum_j f(t_j)(x(t_{j+1}) - x(t_j))$$

if we let $f(t) = f(t,\omega)$ depend on ω, but insist that for every t, $f(t,\omega)$ be measurable with respect to $\mathscr{F}_t$ then the increment $x(t_{j+1}) - x(t_j)$ is independent of $f(t_j,\omega)$. One still gets

$$E[S] = 0, \quad E[S^2] = E\left[\sum_j [f(t_j, \omega)]^2\right]$$

leading to Itô's definition of the stochastic integral for the class of 'progressively measurable' functions $f(t, \omega)$ satisfying

$$E\left[\int_0^\infty [f(t, \omega)]^2 dt\right] < \infty$$

Since we do not necessarily have to integrate all the way to ∞, it suffices that f is progressively measurable and satisfies

$$E\left[\int_0^t [f(s, \omega)]^2 ds\right] < \infty \quad \text{for all } t < \infty$$

Itô's stochastic integral

$$y(t) = \int_0^t f(s, \omega) dx(s)$$

is well defined under these conditions and has the following properties.

(1) The process $y(t)$ is almost surely continuous in t and is a progressively measurable martingale relative to $\mathscr{F}_t$.

(2) If τ is any stopping time, $y(\tau \wedge t)$ is also a martingale and

$$y(\tau \wedge t) = \int_0^t [\mathbf{1}_{\tau > s}(\omega) f(s, \omega)] dx(s) = \int_0^t f_\tau(s, \omega) dx(s) \ .$$

(3) The process $z(t) = y^2(t) - \int_0^t [f(s, \omega)]^2 ds$ is a progressively measurable martingale.

(4) If f is in addition uniformly bounded, then

$$\exp\left[y(t) - \frac{1}{2}\int_0^t [f(s, \omega)]^2 ds\right]$$

is again a martingale.

5. Itô calculus

The stochastic integral $y(t) = \int_0^t f(s, \omega) dx(s)$ can be thought of as $dy = f dx$ and the first step in developing any calculus is the chain rule. What is $d\phi(y(t))$? The problem exists already if $f = 1$ i.e. what is $d\phi(x(t))$? If we take $\phi(x) = x^2$ one might think that $dx^2(t) = 2x(t) dx(t)$ or

$$x^2(t) - x^2(0) = x^2(t) = \int_0^t 2x(t) dx(t)$$

This is clearly wrong because the left hand side has expectation t while the right hand side has expectation 0. A better guess is

$$x^2(t) - x^2(0) = x^2(t) = \int_0^t 2x(t)dx(t) + t$$

This can indeed be seen to be correct by using (1.2). More generally Itô's formula asserts that

$$f(x(t)) - f(x(0)) = \int_0^t f'(x(s))dx(s) + \frac{1}{2}\int_0^t f''(x(s))ds$$

Since $x(\cdot)$ is not of bounded variation but has finite quadratic variation, formally $(dx)^2 = dt$ and $(dx)^i(dt)^j = 0$ if $i + 2j \geq 3$. We have to take an additional term in the Taylor expansion. We can now compute

$$d\phi(t,y(t)) = \phi_t dt + \phi_y dy + \frac{1}{2}\phi_{yy}(dy)^2 = [\phi_t + \frac{1}{2}\phi_{yy}f^2]dt + \phi_y f \, dx$$

or

$$\phi(t,y(t)) - \phi(0,y(0))$$
$$= \int_0^t \left[\phi_s(s,y(s)) + \frac{1}{2}\phi_{yy}(s,y(s))[f(s,\omega)]^2\right]ds$$
$$+ \int_0^t \phi_y(s,y(s))f(s,\omega)dx(s)$$

In particular if $f = 1$

$$\phi(t,x(t)) - \phi(0,x(0)) = \int_0^t \left[\phi_s(s,x(s)) + \frac{1}{2}\phi_{xx}(s,x(s))\right]ds$$
$$+ \int_0^t \phi_x(s,x(s))dx(s)$$

One can use Itô's formula to establish a connection with the infinitesimal generator $1/2D_x^2$. For example if $u(t,x)$ solves

$$u_t = \frac{1}{2}D_x^2 u$$

with $u(0,x) = f(x)$, Itô's formula for $v(t,x) = u(T-t,x)$ yields

$$dv(t,x(t)) = \int_0^t v_x(s,x(s))dx(s)$$

with no dt term and therefore $v(t,x(t))$ is a martingale. Equating expectations at times 0 and T we get

$$v(0,x) = u(T,x) = E^{P_x}[v(T,x(T))] = E^{P_x}[f(x(T))] = \int f(y)p(T,y-x)dy$$

Itô's formula in particular provides a proof of the uniqueness of the Cauchy problem for the heat equation!

A more interesting example is to consider a smooth function $u = u(t, x)$ that solves

$$u_t + \tfrac{1}{2}D_x^2 u + V(x, t)u = 0 \quad \text{in } [0, T] \times [a, b]$$

with $u(T, x) = f(x)$. Then by a similar calculation

$$u(t, x(t)) \exp\left[\int_0^t V(s, x(s)) \, ds\right]$$

is a martingale until the exit time τ from the interval $[a, b]$, and we get

$$u(0, x) = E^{P_x}[u(\tau \wedge T, x(\tau \wedge T))] \exp\left[\int_0^{\tau \wedge T} V(s, x(s)) \, ds\right]$$

expressing $u(0, x)$ in terms of the boundary values of u along $t = T$ and $0 < t < T, x = a$ or b. Itô's theory of stochastic integrals and stochastic differential equations appeared in [Itô (1942)]. We can now find a treatment in various texts in probability, engineering and in recent times finance.

6. Brownian motion with a drift

Brownian motion is characterized as a process with independent increments with the distribution of the increment $x(t) - x(s)$ being normal with mean 0 and variance $t - s$. We can build in a drift by requiring the increment to have a normal distribution with mean $\mu(t - s)$ and variance $t - s$. Such a process is very easily defined as $y(t) = x(t) + \mu t$. It is again a Markov process with a transition probability density

$$q_\mu(s, x, t, y) = \frac{1}{\sqrt{2\pi(t - s)}} \exp\left[-\frac{(y - x - \mu(t - s))^2}{2(t - s)}\right]$$

Symbolically $dy = dx + \mu dt$ and there are analogous results. But the basic difference is that the infinitesimal generator $1/2D_x^2$ has now to be replaced by $1/2D_x^2 + \mu D_x$.

Often, in considering stochastic models, we come across situations where the drift is a restoring force that is proportional to the displacement. Formally

$$dy = -\mu y \, dt + dx \ . \tag{6.1}$$

Such an example will not be a process with independent increments, but will still be Markov. We can solve (6.1) explicitly with initial condition $y(0) = y$, to obtain

$$y(t) = e^{-\mu t}y(0) + \int_0^t e^{-\mu(t-s)} \, dx(s) \tag{6.2}$$

From the representation (6.2), we see that given $y(0) = y$, $y(t)$ is again normally distributed, but now with mean $e^{-\mu t}y$ and variance $1/2\mu(1 - e^{-2\mu t})$. We therefore have a time homogeneous Markov process with transition probability density

$$q(s, x, t, y) = \frac{1}{\sqrt{\frac{\pi}{\mu}(1 - e^{-2\mu(t-s)})}} \exp\left[-\frac{\mu(y - e^{-\mu(t-s)}x)^2}{(1 - e^{-2\mu(t-s)})}\right]$$

As $t \to \infty$ this has a limit which is a stationary density for the process and is in fact the normal distribution with mean 0 and variance $1/2\mu$. For this Markov process, the generator is $1/2D_y^2 - \mu y D_y$. The way to compute the generator quickly is by applying Itô's formula

$$du(y(t)) = u'dy(t) + \frac{1}{2}u''(dy(t))^2 = u'[-\mu y dt] + u'dx + \frac{1}{2}u'' dt$$

$$= \left[\frac{1}{2}D_y^2 u - \mu y D_y u\right] dt + D_y u\, dx$$

The generator is given by the coefficient of the dt term. More generally we can have models of the form $dy = b(y)dt + dx$, written as

$$y(t) = y(0) + x(t) + \int_0^t b(y(s))ds$$

If $b(y)$ satisfies a Lipschitz condition in y, i.e., a condition of the form $|b(y) - b(z)| \leq C|y - z|$ which is valid for some constant C for all y and z, the equation can be solved uniquely by Picard iteration. The solution $y(t)$ will be a function $y(t) = \phi(t, y(0), \omega)$ that will be a measurable function of ω relative to the σ-field $\mathscr{F}_t$. The autonomous nature of the equation means that

$$\phi(t, y(0), \omega) = \phi(t - s, y(s), \theta_s\omega)$$

where $\theta_s\omega$ is the new Brownian path $x(\sigma) - x(s)$ for $\sigma \geq s$. This guarantees that $y(t)$ is still a Markov process and Itô's formula computes the generator as $1/2D_y^2 + b(y)D_y$. The transition probability is no longer explicit. It is of course given by

$$q(s, x, t, A) = P_0[\phi(t - s, x, \omega) \in A]$$

Hopefully this has a density. The conditional distribution of the increment $z = y(t + h) - y(t)$ of a Brownian motion with drift $b(y)$ in an interval $[t, t + h]$ is roughly normal with mean $hb(y(t))$ and variance h. The relative density with respect to the distribution of the increment of Brownian motion over the same interval is given by

$$\exp\left[b(y(t))z - \frac{h}{2}[b(y(t))]^2\right]$$

This suggests the formula

$$\exp\left[\int_0^t b(x(s))\mathrm{d}x(s) - \frac{1}{2}\int_0^t [b(x(s))]^2\,\mathrm{d}s\right]$$

for the Radon–Nikodym derivative $(\mathrm{d}P_{b(\cdot),x})/\mathrm{d}P_x$ of the Brownian motion with drift $b(\cdot)$ relative to Brownian motion starting at the same point x, but without any drift. This formula, known some times as Girsanov's formula, can be proved again using Itô's formula, at least when b is bounded. It can become quite technical if b is unbounded for reasons that have to do with the possible explosion of the process, i.e. becoming infinite at a finite time. The formula in a sense is still true, but needs to be carefully interpreted. See Stroock and Varadhan (1997) for a general discussion on explosion.

7. One-dimensional diffusions

The idea of a one dimensional diffusion process is to have a Markov process $x(t)$ with continuous paths such that given the past history $[x(s) : 0 \le s \le t]$, the future increment $x(t + h) - x(t)$ has a conditional distribution that for small h is approximately normal with mean hb and variance ha. The mean b and the variance a can be functions $b(t, x(t))$ and $a(t, x(t))$ respectively. The case $b \equiv 0$, $a \equiv 1$ is clearly Brownian motion. If $b(t, x) \equiv \mu$ and $a \equiv 1$ we have the Brownian motion with a constant drift. If a and b are purely functions of t we have a Gaussian process with independent increments, with the increment over $[s, t]$ being normally distributed with mean $\int_s^t b(\sigma)\mathrm{d}\sigma$ and variance $\int_s^t a(\sigma)\mathrm{d}\sigma$. The case when a and b are functions of x will give us the time homogeneous case. Another interesting class of examples are given by $a = a(t)$ and $b = c(t)x + d(t)$. These are the Gauss Markov processes.

The aim is to explore the relationship between the processes and the coefficients that define them. There are several possible avenues to make the connection.

a. Stochastic differential equations

If we want to generate a Gaussian random variable Y with mean b and variance a we can do that by means of a linear transformation $Y = \sigma X + b$ from a standard normal X, with the choice of $\sigma = \sqrt{a}$. With this analogy in mind, in Itô (1942), Itô's treatment of the problem was to construct the increment of a 'Diffusion Process' $y(t)$ by

$$y(t + h) - y(t) \simeq b(t, y(t))h + \sigma(t, y(t))(x(t + h) - x(t))$$

where $x(t)$ is a standard Brownian motion. Here $\sigma(t, y) = \sqrt{a(t, y)}$. This is formally written as

$$\mathrm{d}y = \sigma\,\mathrm{d}x + b\,\mathrm{d}t$$

Mathematically one tries to solve the above in integrated form

$$y(t) = y(0) + \int_0^t b(s, y(s)) ds + \int_0^t \sigma(s, y(s)) dx(s) \tag{7.1}$$

Let us, following Itô, make the following assumptions.

(A) The functions σ and b satisfy a Lipschitz condition.

$$|\sigma(t, x) - \sigma(t, y)| + |b(t, x) - b(t, y)| \leq C|x - y|$$

for some constant C.

(B) They satisfy linear growth conditions

$$|\sigma(t, x)| + |b(t, x)| \leq C(1 + |x|)$$

for some constant C.

Then within the class of almost surely continuous (in s) progressively measurable functions $\xi(s, \omega)$ on the Wiener space $(\Omega, \mathscr{F}_s, P_0)$ that satisfy

$$\sup_{0 \leq s \leq t} E^{P_0}[(\xi(s, \omega))^2] \leq C(t) < \infty$$

for every t, there exists a unique solution $y(\cdot, \cdot)$ to (7.1). Moreover the solution is a strong Markov process with transition probability function $p(s, y, t, A)$ described below. The equation can be solved for $t \geq s$ in the form

$$y(t) = x + \int_s^t b(v, y(v)) dv + \int_s^t \sigma(v, y(v)) dx(v)$$

and if we denote the solution by $y(s, x, t, \omega)$, then

$$p(s, y, t, A) = P_0[y(s, x, t, \omega) \in A]$$

This gives us a family $Q_{s,x}$ of probability measures on the space of paths $\Omega_{s,x}$ that start from x at time s.

It can be shown [see McKean (1969)] that under additional conditions of smoothness on the coefficients σ and b the solutions $y(s, x, t, \omega)$ depend smoothly on x and in fact for any smooth function $f(x)$

$$u(s, x) = \int f(y) p(s, x, t, dy) = E^{Q_{s,x}}[f(x(t))]$$

is a smooth function of s and x and satisfies the equation

$$\frac{\partial u}{\partial s} + \frac{1}{2} a(s, x) \frac{\partial^2 u}{\partial x^2} + b(s, x) \frac{\partial u}{\partial x} = 0 \tag{7.2}$$

with the boundary condition $u(t, x) = f(x)$. The relation between a, b and $p(\cdot, \cdot, \cdot, \cdot)$ can also expressed by

$$\lim_{t \to s} \frac{1}{t-s} \int (y-x) p(s,x,t,dy) = b(s,x)$$

$$\lim_{t \to s} \frac{1}{t-s} \int (y-x)^2 p(s,x,t,dy) = a(s,x)$$

$$\lim_{t \to s} \frac{1}{t-s} \int (y-x)^4 p(s,x,t,dy) = 0$$

In fact this was the way Kolmogorov [see Kolmogorov (1931)], formulated the problem of constructing the transition probabilities from given coefficients.

b. Partial differential equations

One can try to construct $p(s,x,t,y)$ as a probability density directly and then construct a Markov process with these transition probability densities. To do this we need to find a nonnegative p that solves the equation

$$\frac{\partial p}{\partial s} + \frac{1}{2} a(s,x) \frac{\partial^2 p}{\partial x^2} + b(s,x) \frac{\partial p}{\partial x} = 0$$

with the boundary condition

$$p(s,x,t,\cdot) \to \delta_x(\cdot)$$

as $s \uparrow t$. Under the assumptions

(C) For some constant C and $0 < \alpha \le 1$

$$|a(s,x) - a(t,y)| + |b(s,x) - b(t,y)| \le C[|s-t|^\alpha + |x-y|^\alpha]$$

(D) For some constants $C < \infty$ and $c > 0$

$$c \le a(s,x) \le C, \quad \text{and} \quad |b(s,x)| \le C$$

the existence uniqueness of p can be established. See for instance Friedman (1964) for the details. From p one constructs the Markov process in a canonical manner via finite dimensional distributions. The almost sure continuity of the paths follows from estimates on p that are obtained along the way.

If we drop the lower bound $a \ge c > 0$ but strengthen the smoothness assumptions by demanding additional regularity on a and b as functions its variables of s and x, one can prove the existence of smooth solutions u for the Equation (7.2), provided the boundary data f is smooth. The value of the solution $u(s,x)$ can be shown for each fixed t to be a nonnegative bounded linear functional of f and a suitable version of Riesz representation theorem will give

$$u(s,x) = \int f(y) p(s,x,t,dy)$$

and we get p in this manner. No matter what assumptions we use the final goal is always to get the family of measures $Q_{s,x}$ on the space of paths. One can show, as one should, that if more than one method works in a given situation they lead to the same measures $Q_{s,x}$.

c. Martingale formulation

No matter how the measures are constructed it is always true that for any smooth f

$$f(x(t)) - f(x(s)) - \int_s^t (L_{v,x}f)(x(v))dv \qquad (7.3)$$

is a martingale for times $t \geq s$ relative to $(\Omega_s, \mathcal{F}_t^s, Q_{s,x})$ and

$$Q_{s,x}[\omega : x(s,\omega) = x] = 1$$

Here Ω_s is the space of continuous maps $\omega = x(\cdot)$ from $[s, \infty]$ into R. $\mathcal{F}_t^s$ is the σ-field generated by $\{x(v) : s \leq v \leq t\}$. And

$$(L_{v,x}f)(x) = \tfrac{1}{2}a(v,x)f''(x) + b(v,x)f'(x)$$

In the martingale formulation, that was introduced in Stroock and Varadhan (1969), for given a and b and starting point (s,x) one asks if $Q_{s,x}$ exists such that (7.3) is a martingale for every smooth f and if such a $Q_{s,x}$ is unique. If we assume that the coefficients are bounded and continuous, existence is easy. Uniqueness is more difficult and under assumptions (A), (B) or (C), (D) uniqueness can be shown. Uniqueness implies that the processes $Q_{s,x}$ are strong Markov processes. In the martingale formulation, by establishing the corresponding Girsanov formula the consideration for any bounded measurable b can be reduced to the case of $b \equiv 0$ provided a has a uniform positive lower bound.

Using the martingale formulation one can establish the following random time change relation in the time homogeneous case. If $V(x)$ is positive function with $0 < c \leq V(x) \leq C < \infty$ we can define stopping times τ_t by the relation

$$\int_0^{\tau_t} \frac{1}{V(x(s))} ds = t$$

and use the stopping times to change the time scale of the process through the following map $\Phi : \Omega \to \Omega$

$$\Phi(\omega)(t) = x(\tau_t(\omega), \omega)$$

If Q is a solution to the martingale problem for $[a(x), b(x)]$ then $Q\Phi_V^{-1}$ solves the martingale problem for $[V(x)a(x), V(x)b(x)]$. One can therefore transfer existence and uniqueness results from one to the other. For the matingale formulation Stroock and Varadhan (1997) is a good source.

In the one dimensional case there are some special results. For instance in the Stochastic Differential Equations formulation the Lipschitz condition can be weakened to a Hölder condition in x of exponent $1/2$. Using a combination of random time change and Girsanov's formula one can reduce the case of any bounded [a, b] with a positive lower bound for a to the case of $a \equiv 1, b \equiv 0$, which is the case of Brownian motion and enjoys both existence and uniqueness. This gives a way of directly constructing such processes from Brownian motion. These results can be found in Dynkin (1959) or Stroock and Varadhan (1997) in full detail.

d. *Limit theorems*

Suppose that we have a Markov chain with transition probability $\pi_h(x, dy)$ and this represents a transition in a time step of duration h. We can construct a piecewise constant stochastic process $X_h(t)$ by defining first, a Markov chain ξ_n^h with transition probability π_h and then making a stochastic process out of it by defining,

$$X_h(t) = \xi_n^h \quad \text{for } nh \leq t < (n+1)h$$

We can, for any initial condition $X_h(0) = x$ define the distribution of this process as a measure $Q_{x,h}$ on the Skorohod space of paths with discontinuities only of the first kind. If one assumes that the limits

$$\lim_{h \to 0} \frac{1}{h} \int (y - x)\pi_h(x, dy) = b(x)$$

$$\lim_{h \to 0} \frac{1}{h} \int (y - x)^2 \pi_h(x, dy) = a(x)$$

and

$$\lim_{h \to 0} \frac{1}{h} \int |y - x|^{2+\delta} \pi_h(x, dy) = 0$$

hold locally uniformly in x for some bounded continuous functions $[a(x), b(x)]$ and that there is unique family Q_x that is a solution to the martingale problem for these coefficients that start at time 0 from the point x, then it follows that

$$\lim_{h \to 0} Q_{h,x} = Q_x$$

in the sense of weak convergence. The basic idea in this approach is that with $\pi_h f$ defined as $\int f(y)\pi_h(x, dy)$,

$$f(X_h(nh)) - f(X_h(0)) - \sum_{j=0}^{n-1} (\pi_h f - f)(X_h((n-1)h))$$

is a martingale with respect to $(\Omega, \mathscr{F}_{nh}, Q_{h,x})$ and under our assumptions

$$\lim_{h \to 0} \frac{1}{h} (\pi_h f - f) = Lf$$

exists locally uniformly, with

$$(Lf)(x) = \tfrac{1}{2}a(x)f''(x) + b(x)f'(x)$$

This link is enough to establish that any limit point of $Q_{h,x}$ as $h \to 0$ is a solution of the martingale problem for $[a(x), b(x)]$. If we have uniqueness the limit is identified. See for instance Stroock and Varadhan (1997) and Ethier and Kurtz (1986) for results of this type.

8. The multidimensional case

To start with, we note that one can define Brownian motion in d dimensions by just taking d independent Brownian motions and making them the d components of a vector valued process $\{x_j(\cdot)\}$. We get a Markov process (actually independent increments) with transition probability density given by

$$p(s,x,t,y) = \frac{1}{[2\pi(t-s)]^{d/2}} \exp\left[-\frac{\|y-x\|^2}{2(t-s)}\right]$$

This diffusion process has the generator

$$L = \frac{1}{2}\sum_{j=1}^{d} \frac{\partial^2}{\partial x_j^2}$$

and the transition probability density itself is the fundamental solution of the heat equation

$$\frac{\partial p}{\partial t} = \frac{1}{2}\sum_{j=1}^{d} \frac{\partial^2 p}{\partial x_j^2}$$

For our Brownian motion we have taken the identity $\mathcal{I}$ as the covariance matrix. We can instead take any positive definite matrix A for the covariance and if $A = \sigma\sigma^*$ with $*$ representing the adjoint operation, we can represent the new process as the linear transform $y(t) = \sigma x(t)$. The new transition probabilities will be

$$p(s,x,t,y) = \frac{1}{[2\pi(t-s)]^{d/2}} \frac{1}{[\det A]^{1/2}} \exp\left[-\frac{\langle(y-x), A^{-1}(y-x)\rangle}{2(t-s)}\right]$$

and the new generator

$$L = \frac{1}{2}\sum_{i,j=1}^{d} a_{i,j}\frac{\partial^2}{\partial x_i \partial x_j}$$

We can add a constant drift vector vector b and then the corresponding transition probability density and generator will be given respectively by

472

$$p(s,x,t,y) = \frac{1}{[2\pi(t-s)]^{d/2}} \frac{1}{[\det A]^{1/2}}$$

$$\times \exp\left[-\frac{\langle(y-x-b(t-s)), A^{-1}(y-x-b(t-s))\rangle}{2(t-s)}\right]$$

$$L = \frac{1}{2}\sum_{i,j=1}^{d} a_{i,j} \frac{\partial^2}{\partial x_i \partial x_j} + \sum_j b_j \frac{\partial}{\partial x_j}$$

The general time dependent diffusion process in d dimensions corresponds to the generator

$$L = \frac{1}{2}\sum_{i,j=1}^{d} a_{i,j}(t,x) \frac{\partial^2}{\partial x_i \partial x_j} + \sum_j b_j(t,x) \frac{\partial}{\partial x_j}$$

and can be constructed by any of the methods outlined in the one dimensional case. The essential intuition is that we have a vector valued process $x(t)$ such that the conditional distribution of $x(t+h) - x(t)$, given the past history $\{x(s) : 0 \le s \le t\}$, is approximately normally distributed in R^d with mean $hb(t,x(t))$ and covariance $ha(t,x(t))$. Here $b(t,x) = \{b_j(t,x)\}$ is an R^d valued function on $[0,\infty] \times R^d$ while $a(t,x) = a_{i,j}(t,x)$ is a symmetric positive semidefinite $d \times d$ matrix for each $(t,x) \in [0,\infty) \times R^d$.

Just as in the one dimensional case if we assume that $a(t,x)$ can be written as $\sigma(t,x)\sigma(t,x)^*$ and that both $\sigma(t,x)$ and $b(t,x)$ satisfy a Lipschitz condition of the form

$$\sum_j |b_j(t,x) - b_j(t,y)| + \sum_{i,j} |\sigma_{i,j}(t,x) - \sigma_{i,j}(t,y)| \le C|x-y|$$

as well as a bound

$$\sum_j |b_j(t,x)| + \sum_{i,j} |\sigma_{i,j}(t,x)| \le C(1+|x|)$$

with a constant C that does not depend on t, then we can solve the stochastic differential equation

$$dy(t) = \sigma(t,y(t))dx(t) + b(t,y(t))dt$$

and construct a map from the d dimensional Brownian motion to the d dimensional diffusion $y(\cdot)$ corresponding to $[a(\cdot,\cdot), b(\cdot,\cdot)]$.

If we assume that $a(t,x)$ is uniformly elliptic, i.e. for some constants $0 < c < C < \infty$ we have

$$c\sum_{j=1}^{d} \xi_j^2 \le \sum_{i,j=1}^{d} a_{i,j}(t,x)\xi_i\xi_j \le C\sum_{j+1}^{d} \xi_j^2 \,,$$

$b_j(\cdot, \cdot)$ are uniformly bounded by some constant C, and that $a(\cdot, \cdot)$ and $b(\cdot, \cdot)$ satisfy a Hölder condition

$$\sum_j |b_j(s,x) - b_j(t,y)| + \sum_{i,j} |a_{i,j}(s,x) - a_{i,j}(t,y)| \leq C[|t - s|^\alpha + |x - y|^\alpha]$$

for some exponent α and constant C, then just as in the one dimensional case we can get a fundamental solution $p(s, x, t, y)$ that can serve as the transition probability density for the diffusion. The heat operator is of course replaced by $\partial/\partial s + L_s$ where

$$L_s = \frac{1}{2} \sum_{i,j=1}^{d} a_{i,j}(s,x) \frac{\partial^2}{\partial x_i \partial x_j} + \sum_{j+1}^{d} b_j(s,x) \frac{\partial}{\partial x_j}$$

The corresponding diffusion processes $Q_{s,x}$ can again be characterized through the martingale formulation as one for which $Q_{s,x}[x(s) = x] = 1$, and

$$f(x(t)) - f(x(s)) - \int_s^t (L_v f)(x(v)) dv$$

is a martingale for every smooth function f.

An important observation is the connection with the Dirichlet problem. We will state it for the Brownian motion and of course there are analogs for the general time homogeneous case.

If G is a connected open set in R^d with a smooth boundary, for example the open ball $B = \{x : |x| < 1\}$ and if P_x is the d dimensional Brownian motion starting from the point $x \in B$ we can define the stopping time

$$\tau(\omega) = \inf[t : x(t) \notin B]$$

By the continuity of paths one can see that $|x(\tau)| = 1$ and it is not difficult to see that $P_x[\omega : \tau(\omega) < \infty] = 1$ for $x \in B$. Then for any continuous data f on $|x| = 1$

$$u(x) = E^{P_x}[f(x(\tau))]$$

solves the Dirichlet problem

$$\sum_j \frac{\partial^2 u}{\partial x_j^2} = 0$$

in B and for $b \in \partial B$,

$$\lim_{\substack{x \to b \\ x \in B}} u(x) = f(b)$$

Although, Kolmogorov [see Kolmogorov (1931)], introduced the connection between Markov processes and parabolic partial differential equations it was not till much later that advances in PDE Friedman (1964) made it possible to use the results from PDE for constructing diffusions whereas Itô had made considerable progress using stochastic differential equations. For additional results on one dimensional diffusions Dynkin (1959) is an excellent source.

9. Diffusions with reflection

If we try to construct the one dimensional Brownian motion on the half line rather than the full line we run into trouble. The Brownian path eventually reaches 0 and wants to get out to the negative side. We must do some thing to keep it nonnegative. We may decide to give it enough of a kick to keep it non-negative. This is easier to see in the discrete setting of a symmetric random walk or a gambler's ruin situation. Every time the gambler loses his entire fortune (i.e. reaches 0) we provide him with enough to make one more bet (i.e. move him from 0 to 1). He may lose it again in which case we make another contribution. Every time the random walk comes to zero, at the next step it is moved to 1.

The continuous analog is the set of relations $y(t) = x(t) + F(t)$, $F(t)$ is non-decreasing and continuous in t, $y(t) \geq 0$ and $x(t)$ is a given Brownian path. We are interested in the minimal F that achieves this. Alternately F is allowed to increase only when $y(t) = 0$. It turns out that such a pair of function $y(t), F(t)$ exist and is unique for every continuous path $x(t)$ and in fact are given explicitly by

$$F(t) = -\left[\inf\left[0, \inf_{0 \leq s \leq t} x(s)\right]\right]$$

and

$$y(t) = x(t) + F(t)$$

Moreover $y(t)$ is a Markov process whose distribution as a process is the same as that of $|x(t)|$. The corresponding expectations

$$u(t, y) = E[f(y(t))|y(0) = y]$$

now solve the equation

$$\frac{\partial u}{\partial t} = \frac{1}{2} \frac{\partial^2 u}{\partial y^2} \quad \text{on } [0, \infty) \times [0, \infty)$$

with the Neumann boundary condition $u_y(t, 0) = 0$ for all $t > 0$.

There are analogs for general diffusions in R^d. It becomes more complicated because the push into the region from the boundary which can be administered in only one direction in the one dimensional case can now be administered in any nontangential direction pointing towards the interior. Wentzell in (1959) wrote down the most general boundary condition that can arise. The one dimensional case for general diffusions was considered by Feller in (1957). For a martingale formulation of the problem of constructing diffusions with boundary conditions see Stroock and Varadhan (1971).

References

Cameron, R. H. and W. T. Martin (1953). The transformation of Wiener integrals by non linear transformations. *Trans. Amer. Math. Soc.* **75**, 552–575.

Dynkin, E. B. (1959). One dimensional continuous strong Markov processes. *Theor. Prob. Appl.* **4**, 3–54.

Ethier, S. N. and T. G. Kurtz (1986). *Markov processes. Characterization and convergence.* Wiley Series in Probability and Mathematical Statistics: John Wiley & Sons, Inc.

Feller, W. (1957). Generalized second order differential operators and their lateral conditions. *Illinois J. Math.* **1**, 495–504.

Friedman, A. (1964). *A Partial Differential Equations of Parabolic Type.* Prentice Hall, New Jersey.

Hunt, G. A. (1956). Some theorems concerning Brownian motion. *Trans. Amer. Math. Soc.* **81**, 294–319.

Itô, K. (1942). Differential equations determining a Markoff process (original Japanese title: Zenkoku Sizyo Sugaku Danwakai-si). *J. Pan-Japan Math. Coll.* 1077.

Itô, K. and M. Nisio (1968). On the Convergence of Sums of Independent Banach Space Valued Random Variables, Osaka. *J. Math.* **5**, 35–48.

Kolmogorov, A. N. (1931). Uber die Analytischen Methoden in der Wahrschienlichtkeitsrechnung. *Math. Ann.* **104**, 415–458.

McKean, H. P. (1969). *Stochastic Integrals.* Academic Press.

Revuz, D. and M. Yor (1999). *Continuous Martingale and Brownian Motion.* 3rd edn., Springer-Verlag.

Stroock, D. W. (1993). Probability Theory. An Analytic View. Cambridge University Press.

Stroock, D. W. and S. R. S. Varadhan (1997). *Multidimensional Diffusion Processes.* 2nd edn., Springer-Verlag.

Stroock, D. W. and S. R. S. Varadhan (1969). Diffusion Processes with continuous coefficients, I, II. *Comm. Pure Appl. Math.* **XXII**, 345–400, 479–530.

Stroock, D. W. and S. R. S. Varadhan (1971). Diffusion Processes with boundary conditions. *Comm. Pure Appl. Math.* **XXIV**, 147–225.

Wentzel, A. D. (1959). On the boundary conditions for multidimensional diffusion processes. *Theor. Prob. Appl.* **4**, 164–177.

Wiener, N. (1923). N. Differential-Space. *J. Math. Phys.* **2**, 131–174.

Wiener, N. (1924). The average value of a functional. *Proc. Lon. Math. Soc.* **22**, 454–467.

Bull. Amer. Math. Soc. (N.S.) 40 (2003) 89-97

STOCHASTIC ANALYSIS AND APPLICATIONS

S. R. S. VARADHAN

1. Introduction

The world we live in has never been very predictable, and randomness has always been part of our lives. There is ample evidence that our ancestors did enjoy playing games of chance, and the early development of probability theory had to do with these games. The study of probability has always been motivated by potential applications. Besides gambling, the stimulus has come from life insurance, pricing of annuities, statistical modeling of errors in observations, genetics, etc.

By 1900, there had developed a fairly well understood body of work, although it was not within an axiomatic framework. This prompted Hilbert to pose the axiomatization of probability theory as one of his famous problems. Although it was not until 1933 in [13] that Kolmogorov axiomatized probability theory by making it part of measure theory, the subject continued to blossom in the early part of the twentieth century at the hands of people like Lévy, Khintchin, and Wiener.

2. Infinite dimensional analysis

Wiener, in [27], was the first person to construct a measure, corresponding to what we now call Brownian motion, on the space of R^d valued continuous functions on $[0, 1]$, thereby making it possible to integrate legitimately in function spaces.

The basic ingredient in the construction of the Wiener measure is the kernel

$$p(t, x, y) = \frac{1}{(2\pi t)^{\frac{d}{2}}} \exp\left[-\frac{\|y - x\|^2}{2t}\right]$$

and its relationship to the Wiener measure P through the formula

$$P[x(\cdot) : x(t_1) \in A_1, \ldots, x(t_n) \in A_n]$$
$$= \int_{A_1} \cdots \int_{A_n} p(t_1, 0, x_1)p(t_2 - t_1, x_1, x_2) \cdots p(t_n - t_{n-1}, x_{n-1}, x_n)dx_1 \cdots dx_n.$$

The function p is of course the fundamental solution of the heat equation

$$\frac{\partial p}{\partial t} = \frac{1}{2}\Delta p,$$

and this creates a basic connection between the Wiener measure and the Laplace operator. Kolmogorov observed, in [12], that this relationship can be extended to a large class of probability measures (Markov or diffusion processes) that generalize

Received by the editors September 7, 2001, and, in revised form, February 21, 2002.
2000 *Mathematics Subject Classification*. Primary 60-02.

the Wiener measure on the one hand, and a large class of differential operators (second order parabolic) that generalize the heat equation on the other:

$$\frac{\partial p}{\partial t} = \frac{1}{2} \sum_{i,j} a_{i,j}(x) \frac{\partial^2 p}{\partial x_i \partial x_j} + \sum_j b_j(x) \frac{\partial p}{\partial x_j}$$

or

$$\frac{\partial p}{\partial t} = \frac{1}{2} \sum_{i,j} \frac{\partial^2}{\partial y_i \partial y_j} [a_{i,j}(y)p] - \sum_j \frac{\partial}{\partial y_j} [b_j(y)p].$$

These are referred to respectively as Kolmogorov's backward and forward equations. Itô, in 1944, described a mapping of the space of trajectories that will directly transform the Wiener measure into the more general measure that corresponds to a given operator

$$\mathcal{L} = \frac{1}{2} \sum_{i,j} a_{i,j}(x) \frac{\partial^2}{\partial x_i \partial x_j} + \sum_j b_j(x) \frac{\partial}{\partial x_j}.$$

This transformation is defined in [9] by means of a system of stochastic differential equations

$$dy_i(t) = \sum_j \sigma_{i,j}(y(t))dx_j(t) + b_i(y(t))dt$$

with $y(0) = z$, where

$$\sigma(x)\sigma^*(x) \equiv a(x).$$

One big advantage of Itô's construction is that it bypasses the PDE, thereby opening up the possibility of saying something about the solution of the PDE by studying the properties of the SDE.

It is not as simple as it sounds. The functions $x_j(t)$ are not of bounded variation (almost surely with respect to the Wiener measure), and therefore even in integrated form it is a task to make sense of the dx integrals

$$y_i(t) = z_i + \int_0^t \sum_j \sigma_{i,j}(y(s))dx_j(s) + \int_0^t b_i(y(s))ds.$$

Itô developed his theory of stochastic integration to take care of this sticky point, including a stochastic calculus with special rules

$$dx_i(t)dx_k(t) = \delta_{i,k}dt,$$
$$dx_i(t)dt = 0,$$

and

$$dy_i(t)dy_j(t) = a_{i,j}(y(t))dt.$$

In integrated form the identity

$$f(y(t)) - f(y(0)) = \int_0^t \langle \sigma^*(y(s))\nabla f(y(s)), dx(s) \rangle + \int_0^t (\mathcal{L}f)(y(s))ds$$

is known as Itô's formula. An advantage of Itô's approach is that the matrix of coefficients $\{a_{i,j}(x)\}$ can be semi-definite. Often in PDE, nondegeneracy is needed.

Another big advantage of Itô's approach is the possibility of dealing with infinite systems just as easily as finite ones. With suitable regularity and proper definition of norms, this can be realized. The PDE, on the other hand, is hard to study directly in infinite dimensions.

By a combination of methods from analysis (PDE) and probability (SDE), diffusion processes in finite dimensional spaces were thoroughly studied during 1950–1980.

The Wiener measure provides us with a reasonable integration theory in the infinite dimensional space $\Omega = C_0[[0,1]; R^d]$. However it is not invariant with respect to translations on Ω, and in fact is not even quasi-invariant with respect to all translations. It is however quasi-invariant with respect to translations from the following dense subspace $\mathcal{H} \subset \Omega$:

$$\mathcal{H} = \left\{ h : h(t) = \int_0^t f(s)ds, \int_0^1 \|f(s)\|^2 ds < \infty \right\}.$$

One can therefore talk about the smoothness of a function, defined only almost everywhere, in the direction $h \in \mathcal{H}$. The gradient operator ∇_h exists, and it makes sense to talk of $\|\nabla_h u\|_p$ existing in the L_p sense. Since $\mathcal{H}$ has a natural Hilbert structure, it also makes sense to talk about the squared gradient

$$\sum_i \|\nabla_{e_i} u\|_2^2,$$

where $\{e_i\}$ is an orthonormal basis of $\mathcal{H}$. There are natural Sobolev spaces and corresponding spaces of distributions. Since we are in infinite dimensions, no amount of L_p regularity will yield continuity, but a very nice theory of differential analysis can still be built up. This was initiated by Malliavin in [16] and developed further by others, including Kusuoka and Stroock in [14], [15], and [20].

A particularly interesting byproduct is Hörmander's theory of hypoellipticity, which can be explained in terms of differential calculus on Wiener space and an integration by parts formula. Using Itô's theory, one can write the fundamental solution $p(t, z, y)$ of

$$\frac{\partial}{\partial t} - \mathcal{L}$$

as

$$E^P[\delta(y_z(t) - y)].$$

The idea is to differentiate as many times as we need to with respect to z and y, integrate by parts to get rid of unwanted derivatives, and express the answers as Wiener integrals and estimate them.

The general method bears the name of Malliavin calculus and is an important tool in the study of regularity of maps defined as stochastic integrals from one infinite dimensional space to another. Clearly the study of infinite dimensional spaces locally modeled after Wiener spaces will rely heavily on such a calculus.

One area from where problems are likely to come is nonlinear analysis in an infinite dimensional setting.

Probability theory has always generated its problems by its contact with other areas. There are very few problems that are generated by its own internal structure. This is partly because, once stripped of everything else, a probability space is essentially the unit interval with Lebesgue measure. If you have seen one you have seen them all.

3. LARGE DISCRETE RANDOM STRUCTURES

Perhaps the most studied example of a discrete random structure is the Ising model [8]. In its simplest version we have a finite lattice $\Lambda_N \subset Z^d$, a state space

$\Omega_N = \{-1, 1\}^{\Lambda_N}$ of maps $\xi(\cdot) : \Lambda_N \to \{-1, 1\}$, and a probability measure depending on a small number of parameters. Given the two parameters J and H, the Ising measure on Ω_N is the probability measure defined by

$$p^g_{N,J,H}(\xi) = \frac{1}{Z^g_N(J,H)} \exp[-JE^g_N(H, \xi)],$$

where

$$E^g_N(H, \xi) = \sum_{\substack{x,y \in \Lambda_N \\ |x-y|=1}} |\xi(x) - \xi(y)|^2 + 2 \sum_{\substack{x \in \Lambda_N \\ y \notin \Lambda_N \\ |x-y|=1}} |\xi(x) - g(y)|^2 + H \sum_{x \in Z_N} \xi(x).$$

The interesting questions concern the behavior of the measures p_N on Ω_N as $N \to \infty$ and the absence or persistence of the influence of the boundary conditions g, after the infinite volume limit has been taken.

This is an area that has been very well studied as a statistical model for phase transitions. While a lot is known, there are still some basic issues that are unresolved, even in the simplest model. There are issues of universality, etc., that are still open. The infinite volume limits of these types of models provide natural examples of random fields in Z^d that are correlated. The recent results concerning surface tension and the Wulff construction are excellent examples of interesting detailed analysis of such systems in the strongly correlated region.

The subject of random graphs had its originals in a paper of Erdös and Rényi [3]. The basic model is of a graph with n vertices in which some randomly chosen edges are turned on while the others are shut down. The subtle interplay between the probability with which the individual edges are turned on and the qualitative behavior of the resulting graph for large values of n has been a fascinating subject for study ever since the paper of Erdös and Rényi.

Percolation. If we take the standard square lattice in Z^d with $d \geq 2$ and select the nearest neighbor edges independently with probability p, there is a qualitative change in the nature of the resultant infinite graph as p changes. For small p the graph is just a collection of finite clusters, whereas for p large enough, with probability 1, there is an infinite cluster. There are detailed studies regarding the nature of this transition at a critical value $0 < p_c < 1$ of p. See [5].

The random cluster models of Fortuin and Kasteleyn [4] provide a unified point of view, connecting percolation and Ising type models.

Recently there have been exciting developments [24], [1], and [11] in situations that appear at first glance to be different, but lead to the same phenomenon. These have to do with (non-Gaussian) asymmetric fluctuations of certain extremal problems.

1. Random matrix theory. We look at the largest eigenvalue λ_N of a random symmetric matrix with an orthogonally invariant Gaussian distribution.

2. We look at the length l_N of the longest increasing subsequence of a random permutation of $1, 2, \ldots, N$.

3. Random growth models. We have a collection of independent random variables $X_{i,j}$ at each site in Z^2 and a path π connecting $(0,0)$ and (M, N) that moves only either up or to the right, and we define

$$G(M, N) = \max_\pi \sum_{(i,j) \in \pi} X_{i,j}.$$

All these random variables exhibit non-Gaussian asymmetric fluctuations and after rescaling have the same limiting distribution, known as the Tracy-Widom distribution. Its distribution function is given by

$$F(s) = \exp\left[-\int_s^\infty (x-s)[u(x)]\right]^2 dx,$$

where u is the solution of the Painlevé II equation

$$u'' = 2u^3 + xu$$

with $u \simeq \mathrm{Ai}(x)$ as $x \to \infty$.

The theory, as developed by P. Deift and others, depends on exact formulas and a precise analysis of the asymptotic behavior of solutions of associated Riemann-Hilbert problems.

To a probabilist there is a mystery here, and it is very likely that these results will one day be part of a larger class of limit theorems.

Disordered systems. Random environments, etc. One of the earliest examples of a problem involving disorder is a random walk in a random environment. Let us take a family of independent random variables π_j, $-\infty < j < \infty$, with $0 < \pi_j < 1$, and consider a random walk on Z with transition probabilities $p_{x,x+1} = \pi_x$ and $p_{x,x-1} = 1 - \pi_x$. The random choices are made once and fixed for all time. The asymptotic behavior of such a walk can be studied in detail for the one dimensional nearest neighbor case. Other problems of a similar nature range from the difficult to the impossibly difficult.

Roughly speaking, these problems deal with probability measures $\mu_{\mathbf{a}}$, where $\mathbf{a}$ is a large or an infinite set of parameters $\{a_x\}$ indexed by $x \in \Lambda$. If $a_x \equiv a$ (a constant), the situation is well understood. In problems of disorder, $\{a_x\}$ are assumed to be random, chosen independently for each x and with a common distribution. One is interested in properties that are shared by $\mu_{\mathbf{a}}$ for almost all $\mathbf{a}$. A random walk in a random environment is one example. See [21].

The Sherrington-Kirkpatrick or S-K model consists of a family of probability measures $\mu_N = \mu_N^{\beta,h,\{g_{i,j}\}}$ on $\{0,1\}^N$ depending on parameters $\beta, h, \{g_{i,j}\}$:

$$\mu_N\{\epsilon\} = \frac{1}{Z_N} \exp[-\beta H_N(\epsilon)],$$

where

$$H_N(\epsilon) = -\frac{1}{\sqrt{N}} \sum_{1 \le i < j \le N} g_{i,j}\epsilon_i\epsilon_j + h \sum_{1 \le i \le N} \epsilon_i;$$

$\{g_{i,j}\}$ is a realization of i.i.d. standard Gaussian random variables.

If $h = 0$ and $\beta < 1$, the direction of the vector ϵ is rather random, so that

$$E^{\mu_N \times \mu_N}\left[\frac{(\epsilon.\epsilon')^2}{N^2}\right]$$

is small for most $\{g_{i,j}\}$. This is not expected to be true for $\beta > 1$, although what is known is a slightly weaker form of the result for $\beta \gg 1$.

This is just one of several similar problems. See the recent work of Talagrand [22] and [23] for detailed comments.

The similar problem in the case of short range interactions, where

$$H_N(\epsilon) = -\sum_{(x,y):|x-y|=1} g_{x,y}\epsilon_x\epsilon_y + h\sum_x \epsilon_x$$

and the summation extends over nearest neighbors in Z^d, is even less understood. See for instance the article [17] by Newman and Stein.

4. SCALING LIMITS

Another area that has been active is the dynamical behavior of large or infinite interacting systems. The evolution may have conserved quantities, in which case the evolution of slow modes can be studied under hydrodynamical scaling, averaging out the fast modes with respect to appropriate invariant distributions.

Fairly general methods involving entropy, rate of entropy production and relative entropy have been developed.

In the case of systems without conserved quantities, one of the issues is whether the rate of approach to equilibrium slows down as the system size increases. This can be measured in terms of either the spectral gap or the log-Sobolev constant. Conditions for these estimates to hold uniformly are known. They can be used to estimate the accuracy of simulations. See Stroock and Zegarliński [19].

In the presence of conserved quantities, the spectral gaps and the log-Sobolev constants become degenerate with increasing volume, and the rate can be guessed by looking at the slow modes. That this is in fact the case is a useful tool in the analysis of these systems and is known for several models. See Yau [29], or Cancrini and Martinelli [2].

A typical class of models that illustrates the scaling behavior is known as simple exclusion processes. See [25], [26] and [28].

One has, as physical space, either Z^d or Z_N^d, the periodic lattice in Z^d of N^d elements. Each site can have either one particle or no particle. The state is described by $\{\eta(x) : x \in Z_N^d\}$.

The dynamics is specified by the generator

$$(\mathcal{A}f)(\eta) = \sum_{x,y} \eta(x)(1 - \eta(y))p(y - x)[f(\eta^{x,y}) - f(\eta)].$$

The scaling $x \to \frac{x}{N}$ of space imbeds Z_N^d into the d-torus T^d. The micro-state η can be replaced by a macro-state

$$r_N(t,\cdot) = \frac{1}{N^d}\sum_x \delta_{\frac{x}{N}},$$

which as $n \to \infty$ turns into the particle density $\rho(\theta)$.

Assuming that the initial configuration leads to the macro-state $\rho_0(\theta)$, the dynamical behavior of the large system should be describable in terms of $\rho_0(\cdot)$.

There are two cases. If $m = \sum zp(z)$, either $m = 0$ or $m \neq 0$. Let us first look at $m \neq 0$. The special case $d = 1$, $p(1) = 1$ is typical. In this case one needs to speed up the time scale by a factor of N as well, and the micro-state at time t is, with probability nearly 1, close to the macro-state $\rho(t,\theta)$, which is the unique entropic solution of the Burgers equation

$$\rho_t + [\rho(1 - \rho)]_\theta = 0.$$

The secondary issues of probabilities of large deviations can also be studied for this model. See Jensen [10]. Given $\rho(\cdot,\cdot)$, the question of estimating the probability

$$P[r_N(\cdot,\cdot) \simeq \rho(\cdot,\cdot)] = \exp[-N\, I(\rho(\cdot,\cdot)) + o(N)]$$

with an explicit I has a simple answer: $I(\rho(\cdot,\cdot)) = \infty$ unless $\rho(\cdot,\cdot)$ is also a weak solution, and then $I(\rho(\cdot,\cdot))$ is the precise amount by which the entropy condition is violated for the convex functional

$$h(\rho) = \rho \log \rho + (1 - \rho) \log(1 - \rho);$$

that is,

$$I(\rho(\cdot,\cdot)) = \int_0^T \int [[h(\rho)]_t + [g(\rho)]_x]^+ dx dt.$$

In particular, one can show that

$$\lim_{N\to\infty} \frac{1}{N} \log E[\exp\{hN(t)\}] = \begin{cases} \frac{h}{4} & \text{if } h \geq 0, \\ \tanh \frac{h}{4} & \text{if } h \leq 0, \end{cases}$$

and one can compare the fluctuation theory with the large deviation estimate.

The case $m = 0$ exhibits more complex behavior. There exists $A(\rho)$ such that the equality

$$\rho_t = \frac{1}{2} \nabla A(\rho) \nabla \rho$$

describes the evolution of the macroscopic state in the new time scale, which now needs to be speeded up by a factor of N^2. The calculation of $A(\rho)$ is in general rather complicated. However, if p is symmetric, $A(\rho)$ simplifies to $A = \sum z \otimes zp(z)$.

Again the problems of large deviation as well as the motion of a tracer particle have satisfactory solutions. $S(\rho)$ is the self-diffusion coefficient in equilibrium, i.e., $\frac{x(t)}{\sqrt{t}} \simeq \beta$, and β has dispersion $S(\rho)$. Instead of just looking at the evolution of the macro-state, we may wish to consider the random process

$$R_N = \frac{1}{N^d} \sum_j \delta_{x_j}(\cdot),$$

which describes the collective history of all the particles. The limit

$$\lim_{N\to\infty} R_N = Q$$

exists. Both Q and the rate function $I(R)$ in the large deviation estimate

$$P[R_N \simeq R] = \exp[-N^d I(R) + o(N)]$$

have explicit exact descriptions. See Quastel, Rezakhanlou, and Varadhan [18].

The models that we have described have a fairly regular structure. The physical space is the lattice Z^d with connections between nearby sites. The rates are translation invariant.

One can imagine disordered versions of these problems. There have been some studies where the disorder is only in the rates. See [6] and [7]. But a more challenging problem is when the underlying structure itself is some sort of a random graph.

Whereas problems coming from physics present a more regular structure, problems from the social sciences present far more disorder.

Today more sophisticated models are being used in finance, insurance and other areas involving an analysis of uncertainty and risk. If we wish to fully understand how large systems behave and why, one possible approach is to specify the model at the microscopic level and develop methods that allow us to make predictions of macroscopic behavior.

REFERENCES

1. Jinho Baik, Percy Deift, and Kurt Johansson, *On the distribution of the length of the longest increasing subsequence of random permutations*, J. Amer. Math. Soc. **12** (1999), 1119–1178. MR **2000e:**05006

2. N. Cancrini and F. Martinelli, *On the spectral gap of Kawasaki dynamics under a mixing condition revisited. Probabilistic techniques in equilibrium and nonequilibrium statistical physics*, J. Math. Phys. **41** (2000), 1391–1423.

3. P. Erdös and A. Renyi, *On the evolution of random graphs*, Magyar Tud. Akad. Mat. Kutató Int. Közl. **5** (1960), 17–61. MR **23:**A2338

4. C. M. Fortuin and P. W. Kasteleyn, *On the random cluster model. I: Introduction and relation to other models*, Physica **59** (1972), 535–564. MR **50:**12107

5. Geoffrey Grimmett, *Percolation*, 2nd ed., Springer-Verlag, Berlin, 1999. MR **2001a:**60114

6. Alice Guionnet and Boguslaw Zegarliński, *Decay to equilibrium in random spin systems on a lattice*, Comm. Math. Phys. **181** (1996), 703–732. MR **97j:**82105

7. _____, *Decay to equilibrium in random spin systems on a lattice. II*, J. Statist. Phys. **86** (1997), 899–904. MR **98h:**82046

8. E. Ising, *Beitrag zur Theorie des Ferromagnitismus*, Z. Phys. **31** (1926), 253–258.

9. Kiyosi Itô, *Stochastic integral*, Proc. Imp. Acad. Tokyo **20** (1944), 519–524. MR **7:**313c

10. Leif Jensen, *Large deviations of the asymmetric simple exclusion process*, Ph.D. thesis, Courant Inst., New York Univ., New York, 2000.

11. Kurt Johansson, *Shape fluctuations and random matrices*, Comm. Math. Phys. **209** (2000), 437–476. MR **2001h:**60177

12. A. N. Kolmogorov, *Über die analyttischen Methoden in der Wahrscheinlichkeitsrechnung*, Math. Ann. **104** (1931), 415–458.

13. _____, *Grundbegriffe der Wahrscheinlichkeitsrechnung*, Springer-Verlag, Berlin, 1933.

14. S. Kusuoka and D. Stroock, *Applications of the Malliavin calculus. II*, J. Fac. Sci. Univ. Tokyo Sect. IA Math. **32** (1985), 1–76. MR **86k:**60100b

15. _____, *Applications of the Malliavin calculus. III*, J. Fac. Sci. Univ. Tokyo Sect. IA Math. **34** (1987), 391–442. MR **89c:**60093

16. Paul Malliavin, *Stochastic calculus of variation and hypoelliptic operators*, Proc. Internat. Sympos. Stochastic Differential Equations (Res. Inst. Math. Sci., Kyoto Univ., Kyoto, 1976), Wiley, New York, 1978, pp. 195–263. MR **81f:**60083

17. C. M. Newman and D. L. Stein, *Equilibrium pure states and nonequilibrium chaos*, J. Statist. Phys. **94** (1999), 709–722. MR **2000d:**82040

18. Jeremy Quastel, Fraydoun Rezakhnalou, and S. R. S. Varadhan, *Large deviations for the symmetric simple exclusion process in dimensions $d \geq 3$*, Probab. Theory Related Fields **113** (1999), 1–84. MR **2000c:**60162

19. D. Stroock and B. Zegarliński, *On the ergodic properties of Glauber dynamics*, J. Statist. Phys. **81** (1995), 1007–1019. MR **96i:**60109

20. Daniel Stroock, *Applications of the Malliavin calculus. I*, Stochastic Analysis (Katata/Kyoto, 1982), North-Holland Math. Library, vol. 32, North-Holland, Amsterdam, 1984, pp. 271–306. MR **86k:**60100a

21. Alain-Sol Sznitman, *Brownian motion, obstacles and random media*, Springer-Verlag, Berlin, 1998. MR **2001h:**60147

22. Michel Talagrand, *Huge random structures and mean field models for spin glasses*, Proc. Internat. Congr. Math. (Berlin, 1998), Vol. I, Doc. Math. Extra Vol. I (1998), 507–536 (electronic). MR **2000c:**60164

23. _____, *The Sherrington-Kirkpatrick model: a challenge for mathematicians*, Probab. Theory Related Fields **110** (1998), 109–176. MR **99a:**60117

24. Craig A. Tracy and Harold Widom, *Level spacing distributions and the Bessel kernel*, Comm. Math. Phys. **161** (1994), 289–309. MR **95e:**82004

25. S. R. S. Varadhan, *The complex story of simple exclusion*, Itô's Stochastic Calculus and Probability Theory, Springer, Tokyo, 1996, pp. 385–400. MR **98g:**60058

26. _____ , *Large deviations for interacting particle systems*, Perplexing Problems in Probability, 373–383, Progr. Probab., vol. 44, Birkhauser Boston, Boston, MA, 1999, pp. 373–383. MR **2001h:**60044

27. Norbert Wiener, *Differential space*, J. Math. and Phys. **2** (1923), 131–174.

28. Horng-Tzer Yau, *Scaling limit of particle systems, incompressible Navier-Stokes equation and Boltzmann equation*, Proc. Internat. Congr. Math. (Berlin, 1998), Vol. III Doc. Math. Extra Vol. III, 193–202 (electronic). MR **2000h:**82061

29. _____ , *Logarithmic Sobolev inequality for lattice gases with mixing conditions*, Comm. Math. Phys. **181** (1996), 367–408. MR **98e:**82021

COURANT INSTITUTE, NEW YORK UNIVERSITY, NEW YORK, NEW YORK 10012-1110
E-mail address: Varadhan@cims.nyu.edu

Entropy, Princeton Univ. Press, 2003, 199-214

Large Deviations and Entropy

S. R. S. Varadhan

Courant Institute of Mathematical Sciences

We describe how entropy plays a central role in the calculation of probabilities of large deviations. This is illustrated by several examples.

9.1 Where Does Entropy Come From?

The entropy functional $h(p) = h(p_1, \ldots, p_m) = -\sum_j p_j \log p_j$, defined on the simplex of probability distributions on m points, i.e. on the simplex

$$\left\{ p = (p_1, \ldots, p_m) : p_i \geqslant 0, \sum_i p_i = 1 \right\},$$

is usually characterized abstractly as the only function that satisfies certain natural properties (Khinchin 1957). However, for a probabilist the natural way to meet entropy is in a combinatorial computation, while approximating factorials with the use of Stirling's formula (Feller 1957).

Let us a toss the mythical fair coin n times and observe the number of heads. The number of ways of getting k heads is $\binom{n}{k}$. If we use Stirling's approximation for factorials, then we obtain

$$\binom{n}{k} \simeq \frac{\sqrt{2\pi}\,e^{-n}n^{n+1/2}}{\sqrt{2\pi}\,e^{-k}k^{k+1/2}\sqrt{2\pi}\,e^{-(n-k)}(n-k)^{n-k+1/2}}$$

$$= \frac{\sqrt{n}}{\sqrt{2\pi k(n-k)}} \exp\left[n\left[-\frac{k}{n}\log\frac{k}{n} - \frac{n-k}{n}\log\frac{n-k}{n} \right] \right].$$

If we fix the proportion of heads and tails at approximately p and $q = 1 - p$, then

$$\binom{n}{k} \simeq \frac{1}{\sqrt{2\pi npq}} e^{nh(p,q)}.$$

The entropy is the factor in the exponential growth rate of the 'volume,' or in this case the number of distinct outcomes in the space of coin tosses that corresponds to a given number k of heads. The actual probability in the context of tossing a fair coin is

$$P_n(k) \simeq \frac{1}{\sqrt{2\pi npq}} e^{n[h(p,q)-\log 2]}.$$

Let us look at the situation when an 'unfair' coin with probabilities α for heads and $1 - \alpha$ for tails is tossed. Then

$$P_n(\alpha, k) \simeq \frac{1}{\sqrt{2\pi npq}} e^{n[h(p,q)+p\log\alpha+(1-p)\log(1-\alpha)]}.$$

The exponential constant $h(p, q) + p\log\alpha + (1 - p)\log(1 - \alpha)$ is a sum of two terms, both depending on $p = 1 - q = k/n$. The first term $h(p, q)$ is the 'volume' term that we have already seen and does not depend on α. The second term is the 'energy' term and is a function of p and α. The most likely state for k corresponds to the value of p that maximizes the combination $h(p, q) - E(p, q)$, which is $p = \alpha$. Deviations of p away from α are called 'large deviations.' Their probabilities decay exponentially as

$$\exp[-nI(\alpha; p)],$$

where

$$I(\alpha: p) = p\log\frac{p}{\alpha} + (1 - p)\log\frac{1 - p}{1 - \alpha} = E(p, q) - h(p, q). \qquad (9.1)$$

The large-deviation rate function is the rate of decay given by (9.1). $I(\alpha; p) \geqslant 0$ and in fact $I(\alpha; p) > 0$ unless $p = \alpha$.

While it is difficult to extend the definition of the entropy functional $h(p)$ to arbitrary probability distributions, the 'relative entropy' $I_\alpha(p)$ has a definition that makes sense for any pair of probability distributions α and β on an arbitrary measurable space. If $\beta \ll \alpha$, with a Radon–Nikodym derivative

$$f(x) = \frac{d\beta}{d\alpha} \quad \text{and} \quad \int f(x)\log f(x)\,d\alpha(x) < \infty,$$

then

$$I(\alpha; \beta) = \int f(x)\log f(x)\,d\alpha(x) = \int \log f(x)\,d\beta(x). \qquad (9.2)$$

Otherwise, $I_\alpha(\beta) = +\infty$.

There are large-deviation rate functions that at first glance do not seem to come directly from entropy. Cramér's Theorem (Cramér 1938) is an example. Consider a sequence $X_1, X_2, \ldots, X_n, \ldots$ of i.i.d. random variables. Assume that the moment-generating function

$$M(\theta) = E[\exp[\theta X]]$$

is finite for all $\theta \in \mathbb{R}$. Let $\phi(a)$ be the Legendre transform of $\psi(\theta) = \log M(\theta)$:

$$\phi(a) = \sup_\theta[a\theta - \psi(\theta)] = \sup_\theta[a\theta - \log M(\theta)]. \qquad (9.3)$$

Then $\phi(a) \geqslant 0$ and $\phi(a) = 0$ if and only if $a = E[X]$. The 'law of large numbers' tells us that the probabilities

$$P\left[\frac{X_1 + \cdots + X_n}{n} \sim a\right]$$

of large-deviations decay, away from $a = E[X]$. In fact, they do so exponentially, with an explicit rate constant

$$P\left[\frac{X_1 + \cdots + X_n}{n} \sim a\right] \sim \exp[-\phi(a)]$$

with $\phi(a)$ given by (9.3).

The precise formulation of the above relation is as follows. A sequence of probability distributions $\{P_n\}$ on a Polish space (complete separable metric space) $\mathcal{X}$ is said to satisfy a 'large-deviation principle' with rate n and rate function $I(x)$ if the following are valid.

1. The rate function $I(x)$ is lower semicontinuous, nonnegative and has level sets $K_\ell = \{x : I(x) \leqslant \ell\}$ that are compact in $\mathcal{X}$.

2. Moreover,

$$\limsup_{n \to \infty} \frac{1}{n} \log P_n(C) \leqslant -\inf_{x \in C} I(x) \quad \text{for closed } C \subset \mathcal{X}, \qquad (9.4)$$

$$\liminf_{n \to \infty} \frac{1}{n} \log P_n(G) \geqslant -\inf_{x \in G} I(x) \quad \text{for open } G \subset \mathcal{X}. \qquad (9.5)$$

It looks like that, in the context of Cramér's Theorem, Legendre transforms control large-deviation rate functions rather than entropy. But, as we shall see in the next section, this is deceptive.

One of the consequences of the validity of a large-deviation principle is a Laplace asymptotic formula for certain types of integrals. If P_n satisfies a large-deviation principle with a rate function $I(x)$ on a Polish space $\mathcal{X}$ and if $F : \mathcal{X} \to \mathbb{R}$ is a bounded continuous function, then

$$\lim_{n \to \infty} \frac{1}{n} \log E^{P_n}[\exp[nF(x)]] = \sup_{x \in \mathcal{X}}[F(x) - I(x)]. \qquad (9.6)$$

9.2 Sanov's Theorem

In the multinomial case, $X_1, X_2, \ldots, X_n$ are n independent random variables that take values $1, 2, \ldots, m$ with probabilities $\pi_1, \pi_2, \ldots, \pi_m$ and the frequencies $f_1, f_2, \ldots, f_m$ are observed. The probabilities of large deviations for the empirical probabilities $\{f_j/n\}$ are

$$P_n\left[\left\{\frac{f_j}{n}\right\} \sim \{p_j\}\right] = \exp\left[-n\sum_j p_j \log \frac{p_j}{\pi_j} + o(n)\right]$$

$$= \exp[-nI(\{\pi_j\}; \{p_j\}) + o(n)].$$

This extends to the empirical distribution

$$\nu_n = \frac{1}{n}\sum_{r=1}^{n} \delta_{X_r} \qquad (9.7)$$

of n identically distributed random variables on $\mathbb{R}$ with a common distribution μ viewed as a random measure on $\mathbb{R}$. The large-deviation rate function takes the form

$$P[\nu_n \simeq \nu] = \exp[-n\, I(\mu; \nu) + o(n)],$$

where $I(\mu; \nu)$ was defined in (9.2). The connection to Cramér's Theorem is provided by the contraction principle

$$\inf_{\nu:\int x\, d\nu=a} I(\mu; \nu) = \phi(a)$$

and the infimum is attained when

$$d\nu = \frac{e^{\theta x}}{M(\theta)}\, d\mu$$

and θ is chosen so that

$$\psi'(\theta) = \frac{M'(\theta)}{M(\theta)} = a.$$

9.3   What about Markov Chains?

Let $\pi_{i,j}$ be the transition probabilities of a Markov chain on a finite state space with k points. We want to compute the probabilities of large deviations for the empirical distribution ν_n of the proportion of times the chain spends at different sites. The rate function is easily computed as

$$\mathcal{I}_\pi(\nu) = \sup_V \left[\sum_j V(j)\nu(j) - \lambda(V) \right],$$

where the supremum is taken over all vectors $V = \{V(1), V(2), \ldots, V(k)\}$. Here

$$\lambda(V) = \log \rho(V)$$

and $\rho(V)$ is the eigenvalue with the largest modulus of the matrix

$$\pi_{i,j}^V = \pi_{i,j} e^{V(j)}$$

(which is positive because the matrix has nonnegative entries). It is not clear what the spectral radius of nonnegative matrices has to do with entropy. But again there is a connection. Instead of considering just the one-dimensional marginals $\nu_n(j)$, we could consider the frequencies of sites visited at consecutive times,

$$\nu_n^{(2)}(i, j) = \frac{1}{n} \sum_{r=0}^{n-1} \delta_{X_r}(i)\delta_{X_{r+1}}(j), \tag{9.8}$$

where we have adopted the convention that $X_n = X_0$. If F is the state space, then $v_n^{(2)}$ is a probability measure on $F \times F$ such that both marginals are equal to the empirical distribution v_n defined in (9.7). If $\mu(i)$ is the invariant measure for the chain, then $v_n^{(2)} \to \mu^{(2)}$ given by $\mu^{(2)}(i, j) = \mu(i)\pi_{i,j}$. We want to compute the rate function $\mathcal{I}_2(q)$ defined for measures q on $F \times F$ with equal marginals that governs the probabilities of large deviations of the empirical frequencies $v_n^{(2)}(\cdot, \cdot)$ away from $\mu^{(2)}(\cdot, \cdot)$:

$$P[v_n^{(2)} \sim q] = \exp[-n\mathcal{I}_2(q) + o(n)].$$

For each q define q_1 to be its marginal and $\hat{q}$ on $F \times F$ by $\hat{q}(i, j) = q_1(i)\pi_{i,j}$. Then

$$\mathcal{I}_2(q) = I(\hat{q}; q) = \sum_{i,j} q(i, j) \log \frac{q(i, j)}{\hat{q}(i, j)} \tag{9.9}$$

and we are back with our relative entropy. In particular,

$$\inf_{q:q_1=v} \mathcal{I}_2(q) = \mathcal{I}_\pi(v).$$

9.4 Gibbs Measures and Large Deviations

All of this can be understood in terms of large deviations of Gibbs measures in one dimension. Let F be a finite set and $\Omega = F^{\mathbb{Z}}$ be the set of maps $\omega : \mathbb{Z} \to F$ of the set of integers $\mathbb{Z}$ into F. $T : \Omega \to \Omega$ is the shift operator defined by $(T\omega)(n) = \omega(n+1)$. If $\varphi(\omega)$ is a local function of ω, i.e. a function depending only on a finite number of coordinates $\{\omega(j) : |j| \leqslant k\}$, the Gibbs measure P_φ is the (unique) limit as $N \to \infty$ of normalized probability measures $P_{N,\varphi}$ on $F^{[-N, N]}$, defined by

$$P_{N,\varphi}[\omega(-N) \ldots, \omega(N)] \simeq \exp\left[-\sum_{j=-N}^{N} \varphi(T^j \omega)\right].$$

In principle, for finite N, these measures depend on some boundary effects, arising from the edges in the summation $\sum_{j=-N}^{N} \varphi(T^j \omega)$. But, in one dimension as $N \to \infty$, these effects disappear and we get a unique, well-defined, stationary probability measure P_φ on $\Omega = F^{\mathbb{Z}}$ corresponding to any local function φ. A special P, which we denote by P_0, corresponds to $\varphi \equiv 0$ and is the product measure on Ω of the uniform distribution on the finite set F. If Q is any stationary measure on Ω, we have its marginal distribution Q_N on $\Omega_N = F^{[-N, N]}$ with entropy $h_N(Q)$ given by

$$-\sum_{\omega_N \in \Omega_N} Q_N(\omega_N) \log Q_N(\omega_N)$$

and the Kolmogorov–Sinai entropy $H(Q)$ of Q, defined by

$$H(Q) = \lim_{N \to \infty} \frac{1}{2N+1} h_N(Q),$$

can be easily shown to exist (by considerations of subadditivity). The largest possible value for $H(Q)$ is $\log k$ and is achieved only when $Q = P_0$, the product of uniform distributions.

The 'free energy' $\psi_0(g)$ for a local function g (relative to P_0) is defined by

$$\psi_0(g) = \lim_{N \to \infty} \frac{1}{2N+1} \log E^{P_0} \left[\exp \left[\sum_{j=-N}^{N} g(T^j \omega) \right] \right]. \tag{9.10}$$

The empirical process $R_n(\omega)$ is defined by

$$R_n(\omega) = \frac{1}{n} [\delta_\omega + \delta_{T\omega} + \cdots + \delta_{T^{n-1}\omega}]$$

and is nearly stationary for large n. Moreover, the marginals of R_n are precisely the empirical distributions $v_n^{(1)}, v_n^{(2)}, \ldots$ with a negligible correction coming from the edges for large n. By the law of large numbers $R_n(\omega) \to P_0$ and the probabilities of large deviations from the limit P_0 are governed by the rate function

$$I_0(Q) = [\log k - H(Q)].$$

There is also a variational formula for the free energy, arising from the Laplace asymptotic formula (9.6)

$$\psi_0(\varphi) = \sup_{Q \in \mathcal{S}} [E^Q[\varphi(\omega)] - I_0(Q)],$$

where the supremum is taken over all translation-invariant probability measures Q on Ω. It is attained uniquely at some $Q = P_\varphi$ which is called the Gibbs measure with energy φ. The free energy relative to P_φ is defined as

$$\psi_\varphi(g) = \lim_{N \to \infty} \frac{1}{2N+1} \log E^{P_\varphi} \left[\exp \left[\sum_{j=-N}^{N} g(T^j \omega) \right] \right]$$

and is seen to equal

$$\psi_0(\varphi + g) - \psi_0(\varphi) = \sup_{Q \in \mathcal{S}} [E^{P_\varphi}[\varphi(\omega)] - I_\varphi(Q)]$$

with

$$I_\varphi(Q) = \psi_0(\varphi) - [E^Q[\varphi(\omega)] - I_0(Q)].$$

It is now clear that the large deviations of $R_n(\omega)$ under P_φ will be controlled by the rate function $I_\varphi(Q)$. If we are only interested in the marginals $v_n^{(1)}$ or $v_n^{(2)}$, the rate function will have to be contracted to yield

$$I_\varphi^{(1)}(\{p_j\}) = \inf_{Q \in \mathcal{S}_1(\{p_j\})} I_\varphi(Q), \tag{9.11}$$

where the infimum is taken over stationary (i.e. translation-invariant) probability measures Q on Ω with one-dimensional marginal $\{p_j\}$ or

$$I_\varphi^{(2)}(\{q_{i,j}\}) = \inf_{Q \in \mathscr{S}_2(\{q_{ij}\})} I_\varphi(Q), \tag{9.12}$$

where the infimum is taken over stationary processes Q with two-dimensional marginal $\{q_{i,j}\}$.

If $\phi(\omega) = V(\omega(0))$, then P_φ is a product measure with $\pi_j \simeq e^{V(j)}$. The infimum in attained at the product measure $Q = \otimes^{\mathbb{Z}} \{p_j\}$ and $I_\varphi^{(1)}(\{p_j\})$ becomes the relative entropy $I(\{\pi_j\}; \{p_j\})$. We are back at Sanov's Theorem.

If $\phi(\omega) = V(\omega(0), \omega(1))$, then P_φ is Markov with transition a transition probability π that can be determined. Moreover, $\{q_{i,j}\}$ determines a unique Markov process Q_0 and $I_\varphi^{(2)}(\{q_{i,j}\})$, with the infimum attained at Q_0, is seen to equal $I_2(q)$, which was defined in (9.9). See Lanford (1973) for more details.

9.5 Ventcel–Freidlin Theory

There are other large-deviation investigations, e.g. the Ventcel–Freidlin theory (Ventcel and Freidlin 1970). Here the context is very different. Consider a Markov process with generator

$$L_\varepsilon = \tfrac{1}{2}\varepsilon\Delta + b(x) \cdot \nabla$$

acting on smooth functions on $\mathbb{R}^d$. If the paths start from $x \in \mathbb{R}^d$, then they are given by the solution to the SDE

$$x_\varepsilon(t) = x + \int_0^t b(x_\varepsilon(s))\, ds + \sqrt{\varepsilon}\,\beta(t),$$

where $\beta(t)$ is the d-dimensional Brownian motion. As $\varepsilon \to 0$, $x_\varepsilon(t)$ converges to the solution of the ODE

$$x(t) = x + \int_0^t b(x(s))\, ds.$$

We are interested in the probabilities of large deviations, i.e.

$$P[x_\varepsilon(\cdot) \sim f(\cdot)],$$

where f is an arbitrary function $f(\cdot) : [0, T] \to \mathbb{R}^d$ with $f(0) = x$. These probabilities decay like $\exp[-(1/\varepsilon)I(f)]$ and $I(f)$ is calculated as

$$I(f) = \frac{1}{2}\int_0^T \|f'(t) - b(f(t))\|^2\, dt.$$

This is also essentially an entropy calculation. One way to make sure that the solution converges to $f(t)$ is to replace the original generator by the new time-dependent generator

$$L_{\varepsilon, f} = \tfrac{1}{2}\varepsilon\Delta + f'(t) \cdot \nabla$$

corresponding to the SDE

$$y_\varepsilon(t) = x + \int_0^t f'(s)\, ds + \sqrt{\varepsilon}\, \beta(t).$$

The solution $y_\varepsilon(t)$ converges to $f(t)$. A formula, known to probabilists as Girsanov's formula, gives us the Radon–Nikodym derivative (on the σ-field up to time T) of the distribution μ_ε (on path space) of the solution of

$$dz(t) = b(t, z(t))\, dt + \sqrt{\varepsilon}\, d\beta(t)$$

relative to the distribution of λ_ε of $z(\cdot)$ given by $z(t) = \sqrt{\varepsilon}\beta(t)$, as

$$\frac{d\mu_\varepsilon}{d\lambda_\varepsilon}(z(\cdot)) = \exp\left[\frac{1}{\varepsilon}\int_0^T \langle b(t, z(t)), dz(t)\rangle - \frac{1}{2\varepsilon}\int_0^T \|b(t, z(t))\|^2\, dt\right],$$

where $\int_0^T \langle b(t, z(t)), dz(t)\rangle$ is the Itô stochastic integral. The relative entropy of the two measures P_ε of $x_\varepsilon(\cdot)$ and Q_ε of $y_\varepsilon(\cdot)$, which correspond to the choice of $b(t, x) = b(x)$ and $b(t, x) = f'(t)$ respectively, can now be easily computed as

$$I(P_\varepsilon; Q_\varepsilon) = E^{Q_\varepsilon}\left[\frac{1}{\varepsilon}\int_0^T \langle b(x(s)), dx(s)\rangle - \frac{1}{2\varepsilon}\int_0^T \|b(x(s))\|^2\, ds\right.$$
$$\left. - \frac{1}{\varepsilon}\int_0^T \langle f'(s), dx(s)\rangle + \frac{1}{2\varepsilon}\int_0^T \|f'(s)\|^2\, ds\right]$$
$$\simeq \frac{1}{2\varepsilon}\int_0^T |f'(s) - b(f(s))|^2\, ds.$$

9.6 Entropy and Large Deviations

In fact, it is not hard to see that any 'large deviation' has to be related to 'entropy.' If P is a measure and we want to estimate $P(A)$, then we can do it if we can estimate

$$\inf_{Q:Q(A)=1} I(P; Q) = \ell.$$

We have the inequality (Jensen) that says

$$E^Q[g] \le I(P; Q) + \log E^P[\exp[g]]. \tag{9.13}$$

Taking $g = c\chi_A$ in (9.13), with $c > 0$, we get

$$Q(A) \le \frac{1}{c}[I(P; Q) + \log[e^c P(A) + (1 - P(A))]]$$

or

$$\ell \ge c - \log[1 - P(A) + e^c P(A)].$$

Letting $c \to \infty$ we get

$$\ell \ge -\log P(A)$$

or
$$P(A) \leqslant e^{-\ell}.$$

There is a partial converse. If $Q \ll P$ and $E^Q|\log dQ/dP - \ell|$ is small and $Q(A) = q$, then

$$
\begin{aligned}
P(A) &\geqslant \int_A \frac{dP}{dQ}\, dQ = Q(A) \frac{1}{Q(A)} \int_A \exp\left[-\log\frac{dQ}{dP}\right] dQ \\
&\geqslant q \exp\left[\frac{1}{q} \int_A \left[-\log\frac{dQ}{dP}\right] dQ\right] \\
&\geqslant q \exp\left[-\ell - \frac{1}{q} E^Q\left|\log\frac{dQ}{dP} - \ell\right|\right].
\end{aligned}
$$

Large deviations is then essentially estimating entropies, but the new measure Q has to be guessed. In a dynamical situation the dynamics has to be perturbed in order to get a Q with the right property. There are lots of choices and the entropy can be calculated by a Girsanov formula. Then the optimal choice becomes an optimal control problem with entropy as the cost functional (Dupuis and Ellis 1997; Fleming 1985).

Let us illustrate this by means of the examples in Sections 9.2–9.4.

If $X_1, \ldots, X_n, \ldots$ is a sequence of i.i.d. random variables with a common distribution α and we want the empirical distribution $(1/n)\sum_{r=1}^n \delta_{X_r}$ to be close to β, we can achieve this by changing the joint distribution of the $\{X_r\}$ from the product measure $P = \otimes^{\mathbb{Z}}\alpha$ to any stationary process Q with one-dimensional marginals equal to β. The relative entropy $I_n(P; Q)$ of Q with respect to P on the σ-field generated by the first n coordinates grows linearly in n and

$$\lim_{n\to\infty} \frac{1}{n} I_n(P; Q) = I(P; Q)$$

exists and can be evaluated as

$$I(P; Q) = E^Q[h(q_\omega, \alpha)],$$

where q_ω is the conditional distribution of X_1 given the past history $\{X_j : j \leqslant 0\}$ calculated under Q. The control problem reduces to the variational problem

$$I(\beta) = \inf_{Q \in \mathcal{M}_\beta} [I(P; Q)],$$

where $\mathcal{M}_\beta$ is the set of stationary processes Q, with one-dimensional marginals β. Self-consistency requires that for $Q \in \mathcal{M}_\beta$,

$$\int q_\omega\, dQ = \beta.$$

It is not hard to see at this point that, from the convexity of $I(\cdot, \alpha)$, the infimum is attained when $Q = \otimes^{\mathbb{Z}}\beta$, and we arrive at Sanov's Theorem.

The advantage of this somewhat complex view point is that the principle is universal. If we replace the product measure P by a nicely mixing Markov chain with transition probabilities $\pi(x, dy)$, nothing changes except the formula for $\mathcal{I}(P; Q)$, which now reads

$$\mathcal{I}(P; Q) = E^Q[h(q_\omega, p_\omega)],$$

where, by our Markovian assumptions, $p_\omega = \pi(X_{-1}, dy)$. Let us carry out the optimization

$$\inf_{Q \in \mathcal{M}_\beta} \mathcal{I}(P; Q)$$

in two steps. We let Γ_β be the set of bivariate distributions γ with common marginals β and denote by $\mathcal{M}_\gamma^{(2)}$ the set of stationary processes Q such that its bivariate marginal at two consecutive times is γ. Then the infimum above can be carried out as

$$\inf_{\gamma \in \Gamma_\beta} \inf_{Q \in \mathcal{M}_\gamma} \mathcal{I}(P; Q).$$

Any $\gamma \in \Gamma_\beta$ can be disintegrated as $\beta(dx) \otimes \hat{\pi}(x, dy)$ through its marginal and conditional distribution. $\hat{\pi}(\cdot, \cdot)$ is a transition probability with β as an invariant measure, and therefore defines a canonical stationary Markov process Q_γ with (consecutive) bivariate marginals γ and one-dimensional marginals β. One can show that the inner infimum is attained at $Q = Q_\gamma$,

$$\mathcal{I}(Q_\gamma; P) = \mathcal{I}^{(2)}(\gamma) = \int I(\pi(x, \cdot), \hat{\pi}(x, \cdot)) \, \beta(dx),$$

and the rate function becomes

$$\mathcal{I}(\beta) = \inf_{\gamma \in \Gamma_\beta} \mathcal{I}^{(2)}(\gamma).$$

The duality relation

$$\sup_\beta \left[\int V(x) \, \beta(dx) - \mathcal{I}(\beta) \right] = \log \lambda(V),$$

where $\lambda(V)$ is the spectral radius of $\pi(x, dy)e^{V(x)}$, is just a piece of convex analysis and establishes the connection with Section 9.3. It goes more or less like this. We denote by

$$\Gamma = \bigcup_\beta \Gamma_\beta$$

the set of all bivariate distributions with equal marginals:

$$
\sup_V \left[\int V(x)\, d\beta - \mathcal{I}(\beta) \right]
$$

$$
= \sup_\beta \sup_{\gamma \in \Gamma_\beta} \left[\int V(x)\, \gamma(dx, dy) - \mathcal{I}^{(2)}(\gamma) \right]
$$

$$
= \sup_{\gamma \in \Gamma} \left[\int V(x)\, \gamma(dx, dy) - \int \log \frac{\hat{\pi}(x, dy)}{\pi(x, dy)} \gamma(dx, dy) \right]
$$

$$
= \sup_\gamma \inf_U \left[\int [V(x) + U(x) - U(y)]\, \gamma(dx, dy) \right.
$$

$$
\left. - \int \log \frac{\hat{\pi}(x, dy)}{\pi(x, dy)} \gamma(dx, dy) \right]
$$

$$
= \inf_U \sup_\gamma \left[\int [V(x) + U(x) - U(y)]\, \gamma(dx, dy) \right.
$$

$$
\left. - \int \log \frac{\hat{\pi}(x, dy)}{\pi(x, dy)} \gamma(dx, dy) \right]
$$

$$
= \inf_U \left[\sup_x [V(x) + U(x) + \log \int e^{-U(y)} \pi(x, dy)] \right]
$$

$$
= \lambda(V).
$$

Finally, the connection with the example in Section 9.5 is straightforward. We replace $b(x)$ with a new $c(t, x)$ and the SDE is

$$
dx(t) = c(t, x(t))\, dt + \sqrt{\varepsilon}\, \beta(t).
$$

We need c to be such that f is a solution of $\dot{x} = c(t, x(t))$ or $c(t, f(t)) = f'(t)$, which is really what we did.

9.7   Entropy and Analysis

Entropy plays a role in analysis as well. Let $p(t, x, y)$ be the transition probability density (relative to an invariant distribution q) of a Markov process. We have the Markov semigroup

$$
(T_t f)(x) = \int f(y) p(t, x, y) q(dy)
$$

acting on bounded measurable functions and the adjoint semigroup

$$
(T_t^* g)(y) = \int g(x) p(t, x, y) q(dx)
$$

acting on $L_1(q)$. Let L and L^* be the infinitesimal generators of the semigroups T_t and T_t^*, respectively. The function

$$u(t, x) = (T_t f)(x)$$

satisfies the evolution equation

$$\frac{\partial u}{\partial t} = Lu$$

and $g_t = T_t^* g_0$ satisfies the Fokker–Planck (Kolmogorov's forward) equation

$$\frac{\partial g}{\partial t} = L^* g$$

with L^* being the adjoint of L with respect to q, i.e.

$$\int (Lf)g \, dq = \int f(L^* g) \, dq.$$

The H-theorem asserts that the relative entropy $H(t) = \int g_t \log g_t \, dq$ is non-increasing in t. If we know that $H_0 = \int g_0 \log g_0 \, dq < \infty$, then

$$\int_0^\infty \left[-\frac{dH_t}{dt} \right] dt \leqslant H_0,$$

giving us a decay rate on $I(t) = -dH_t/dt$. An elementary calculation shows that

$$-\frac{d}{dt} H_t = -\frac{d}{dt} \int g_t \log g_t \, dq = -\int (L^* g_t) \log g_t \, dq = -\int g_t (L \log g_t) \, dq$$

$$= -2 \int g_t (L \log \sqrt{g_t}) \, dq \geqslant -2 \int g_t \frac{L\sqrt{g_t}}{\sqrt{g_t}} \, dq = 2\mathcal{D}(\sqrt{g_t}).$$

Here $\mathcal{D}$ is the Dirichlet form on $L_2(q)$, defined as

$$\mathcal{D}(g) = \langle g, -Lg \rangle_q = -\int g(x)(Lg)(x)q(dx) = -\int g(x)(L^* g)(x)q(dx),$$

and this provides us with a control

$$\int_0^\infty \mathcal{D}(\sqrt{g_t}) \, dt \leqslant \tfrac{1}{2} H_0.$$

Since usually $D(g) = 0$ only if g is a constant, a small value of the Dirichlet form implies that the state is close to equilibrium. This can be used to show that large systems with multiple equilibria approach local equilibria over large volumes with a length scale that is related to the time scale. If a log-Sobolev estimate is available, then this can be used to establish hydrodynamic scaling results. These typically derive a partial differential equation for the parameters that define the local equilibria as functions of rescaled space and time. We shall illustrate this in the next section. Even if there are no such inequalities available, one can still use the estimate on the Dirichlet form in very many ways (see, for instance, Guo et al. 1988).

9.8 Hydrodynamic Scaling: an Example

Hydrodynamic scaling refers to a wide class of phenomena where a large interacting system with conserved quantities and local interactions evolves over time.
Because of the conserved quantities there will be multiple equilibria, which can
usually be parametrized by the average values of the conserved quantities. The
system will reach local equilibria first before evolving towards the global equilibrium rather slowly. If we speed up time, the local equilibria will be established
over larger domains. At the appropriate rescaling of space and time, the evolution of the complex system can be fully described by describing the parametric
label of the equilibrium as a function of space and time, which is determined
for later times by an evolution equation from its initial values. This evolution
equation, usually a parabolic or hyperbolic transport equation, is called the
hydrodynamic limit of the complex interacting system.

Let us illustrate this by a familiar class of examples called simple exclusion
processes. We will limit ourselves to the one-dimensional case. We start with
the lattice $\mathbb{Z}$. At any given time we have particles at some of the sites and the
associated variables $\eta(x)$, which are 1 if there is a particle at x and 0 otherwise.
The term 'exclusion' signifies that we are not allowed to have more than one
particle at the same time at any site. The particles wait for a random exponential
waiting time (with mean 1) and then pick a random site to jump to. If the particle
is at site x, then the probability of picking site y to jump to is $p(y-x)$. Of course,
$\sum_{z \neq 0} p(z) = 1$. If the site y is selected, the particle jumps to the new site if it
is empty. If the chosen site has a particle and is therefore unavailable, the jump
is disallowed and the particle waits at the original site for a new exponential
time. All of this can be specified by writing down the generator L acting on
functions $f(\eta)$ that depend on the configuration η:

$$(Lf)(\eta) = \sum_{x,y} p(y-x)\eta(x)(1-\eta(y))[f(\eta^{x,y}) - f(\eta)].$$

Here $\eta^{x,y}$ is the new configuration obtained by exchanging the situations at x
and y, i.e.

$$\eta^{x,y}(z) = \begin{cases} \eta(x) & \text{if } z = y, \\ \eta(y) & \text{if } z = x, \\ \eta(z) & \text{otherwise.} \end{cases} \tag{9.14}$$

For simplicity, let us look at the periodic case where the lattice $\mathbb{Z}$ is replaced by
$\mathbb{Z}_N$ of N sites with periodic boundary conditions. Let us also suppose that $p(z)$
is symmetric, i.e. $p(z) = p(-z)$. Let the initial configuration, which depends
on N and could be random, be such that

$$\lim_{N \to \infty} \frac{1}{N} \sum_x J\left(\frac{x}{N}\right) \eta_0(x) = \int_T J(z)\rho_0(z)\,dz \tag{9.15}$$

for some deterministic $0 \leqslant \rho_0(z) \leqslant 1$. In the random case the above limit is interpreted as limit in probability. Here the continuum limit of the rescaled lattice is the circle T, which is viewed as the unit interval with end points identified. Then the evolution has the property that, at time $N^2 t$,

$$\lim_{N \to \infty} \frac{1}{N} \sum_x J\left(\frac{x}{N}\right) \eta_{N^2 t}(x) = \int_T J(z) \rho(t, z) \, dz$$

in probability, where $\rho(\cdot, \cdot)$ is the solution of the heat equation

$$\rho_t = \tfrac{1}{2} \sigma^2 \rho_{xx} \qquad \qquad (9.16)$$

with initial condition $\rho(0, z) = \rho_0(z)$. Here $\sigma^2 = \sum_z z^2 p(z)$. In this context the heat equation (9.16) is called the hydrodynamic limit of the symmetric simple exclusion process.

It is fairly easy to establish this fact by a direct computation of two moments remembering that the speeded-up process has generator $N^2 L$. One expects that in the speeded-up time scale, for any local function $f(\eta)$, denoting by $f_x(\eta)$ its spatial translate by the lattice variable x,

$$\lim_{N \to \infty} \frac{1}{N} \int_0^T \left[\sum_x f_x(\eta_{N^2 s}) \right] ds = \int_0^T \int_T \hat{f}(\rho(t, z)) \, dz \, dt \qquad (9.17)$$

with $\hat{f}(\rho) = E^\rho[f(\eta)]$. Here E^ρ refers to the expectation with respect to the Bernoulli product measure on $\{0, 1\}^{\mathbb{Z}}$ with $E^\rho[\eta(x)] = \rho$. This result will attest to the fact that in the time scale N^2 the system is locally in a Bernoulli state over spatial scales that involve N lattice sites. This can be deduced from the estimate on the Dirichlet form in the following way: if we use the uniform distribution on 2^N sites as our q, then the relative entropy of any initial state with respect to q is at most $N \log 2$. Therefore, in the speeded-up time scale, since the Dirichlet form inherits a factor of N^2,

$$\int_0^\infty \mathcal{D}(\sqrt{g_t}) \, dt \leqslant \frac{\log 2}{2N}.$$

By convexity, if

$$\bar{g} = \frac{1}{T} \int_0^T g_t \, dt,$$

then

$$\mathcal{D}(\sqrt{\bar{g}}) \leqslant \frac{\log 2}{2NT}.$$

We want to use this estimate to show that on a large block of εN sites the conditional distribution under $\bar{q}$ of the configurations given the mean number m of particles in that block is close to the Bernoulli measure μ_m on that block. This can be controlled if we can control the relative entropy of the restriction of $\bar{q}$ to a block of size εN to a mixture of μ_m on that block. We have to use a mixture, because the average density m can be random under $\bar{q}$. Since the Dirichlet form is additive in volume, the restriction of $\bar{q}$ to a typical block of length εN will have a Dirichlet form of about $(\varepsilon \log 2)/N$. The log-Sobolev inequality is known to hold with a constant proportional to the square of the block size. Therefore, the relative entropy of the restriction of $\bar{q}$ to a typical block of size εN will have a relative entropy of order $\varepsilon^3 N$ relative to a suitable mixture of Bernoulli distribution in that block. The specific entropy is roughly ε^3 per site. This can be controlled uniformly in N provided ε is small. This is enough to prove (9.17).

We now turn to the totally asymmetric case where $p(1) = 1$ and $p(z) = 0$ for $z \neq 1$. In this case time is rescaled by a factor of N and we assume that the initial condition (9.15) holds as before. Then

$$\lim_{N \to \infty} \frac{1}{N} \sum_x J\left(\frac{x}{N}\right) \eta_{Nt}(x) = \int_T J(z) \rho(t, z) \, dz, \qquad (9.18)$$

where $\rho(t, z)$ is now a weak solution of Burgers equation

$$\frac{\partial \rho}{\partial t} + \frac{\partial [\rho(1 - \rho)]}{\partial z} = 0 \qquad (9.19)$$

on $[0, T] \times T$ with initial condition

$$\rho(0, z) = \rho_0(z).$$

There is no uniqueness for solutions of Burgers equation (9.19) in the class of weak solutions. See Chapter 6 by Dafermos (2003) in this book. While there is uniqueness within the class of regular solutions, in general, regular solutions do not exist or exist only up to a certain time when shocks develop. The hydrodynamic limit (9.18) is singled out as the unique weak solution satisfying the entropy condition (Rezakhanlou 1991), i.e. for every convex 'entropy' h and the corresponding 'flux' g, defined by

$$h'(\rho)(1 - 2\rho) = g'(\rho),$$

one has

$$\frac{\partial h(\rho)}{\partial t} + \frac{\partial g(\rho)}{\partial z} \leqslant 0 \qquad (9.20)$$

as a distribution. A special case is the entropy

$$h(\rho) = \rho \log \rho + (1 - \rho) \log(1 - \rho)$$

with corresponding flux

$$g(\rho) = \rho - \rho(1 - \rho) \log \frac{\rho}{1 - \rho}.$$

It turns out (Jensen 2000) that this particular entropy controls the probabilities of large deviations. If we want the empirical density to be close to an arbitrary

$\rho(\cdot, \cdot)$ in $[0, T]$, which may not be the entropy solution, this will be a large deviation. The rate function $I(\rho)$ for such a large deviation is finite only if $\rho = \rho(t, z)$ is a weak solution of equation (9.19), and the distribution

$$\mu = \frac{\partial h(\rho)}{\partial t} + \frac{\partial g(\rho)}{\partial z}$$

is a signed measure of bounded variation. While the condition for the weak solution $\rho(\cdot, \cdot)$ to be an entropy solution is $\mu \leqslant 0$, the large-deviation rate function is given by

$$I(\rho(\cdot, \cdot)) = \mu^+[[0, T] \times T].$$

References

Cramér, H. 1938 Sur un noveau théorème—limites de la théorie des probabilités. *Actualités Scientifiques et Industrielles* **736**, 5–23. *Colloque consacré à la théorie des probabilités*, Vol. 3. Hermann, Paris

Dafermos, C. 2003 Entropy for hyperbolic conservation laws. In *Entropy* (ed. A. Greven, G. Keller and G. Warnecke), pp. 107–120. Princeton University Press.

Dupuis, P. and Ellis, R. S. 1997 *A Weak Convergence Approach to the Theory of Large Deviations*. Wiley Series in Probability and Statistics (Probability and Statistics Section). Wiley-Interscience.

Feller, W. 1957 *An Introduction to Probability Theory and Its Applications*, Vol. I, 2nd edn. Wiley, Chapman and Hall.

Fleming, W. H. 1985 A stochastic control approach to some large deviations problems. *Recent Mathematical Methods in Dynamic Programming* (Rome, 1984), pp. 52–66. Lecture Notes in Mathematics, vol. 1119. Springer.

Guo, M. Z., Papanicolaou, G. C. and Varadhan, S. R. S. 1988 Nonlinear diffusion limit for a system with nearest neighbor interactions. *Commun. Math. Phys.* **118**(1), 31–59.

Jensen, L. 2000 Large deviations of the asymmetric simple exclusion process. PhD thesis, New York University.

Khinchin, A. I. 1957 *Mathematical Foundations of Information Theory* (translated by R. A. Silverman and M. D. Friedman). Dover, New York.

Lanford III, O. E. 1973 Entropy and equilibrium states in classical statistical mechanics. In *Statistical Mechanics and Mathematical Problems* (ed. A. Lenard), pp. 1–113. Lecture Notes in Physics, 20. Springer.

Rezakhanlou, F. 1991 Hydrodynamic limit for attractive particle systems on Z^d. *Commun. Math. Phys.* **140**, 417–448.

Ventcel, A. D. and Freidlin, M. I. 1970 Small random perturbations of dynamical systems. *Russ. Math. Surv.* (translation of *Usp. Mat. Nauk*) **25**, 3–55.

The Annals of Probability
2008, Vol. 36, No. 2, 397–419
DOI: 10.1214/07-AOP348
© Institute of Mathematical Statistics, 2008

SPECIAL INVITED PAPER

LARGE DEVIATIONS[1]

BY S. R. S. VARADHAN

New York University

This paper is based on Wald Lectures given at the annual meeting of the
IMS in Minneapolis during August 2005. It is a survey of the theory of large
deviations.

1. Large deviations for sums. The role of "large deviations" is best understood through an example. Suppose that $X_1, X_2, \ldots, X_n, \ldots$ is a sequence of i.i.d.
random variables, for instance, normally distributed with mean zero and variance 1. Then,

$$E[e^{\theta(X_1+\cdots+X_n)}] = E[e^{\theta X_1}]^n = e^{n(\theta^2/2)}.$$

On the other hand,

$$E[e^{\theta(X_1+\cdots+X_n)}] = E[e^{n\theta(S_n/n)}].$$

Since, by the law of large numbers, $\frac{S_n}{n}$ is nearly zero, we have

$$E[e^{\theta(X_1+\cdots+X_n)}] = E[e^{o(n)}] \neq e^{o(n)}.$$

There is, of course, a very simple explanation for this. In computing expectations
of random variables that can assume large values with small probabilities, contributions from such values cannot be ignored. After all, a product of something big
and something small can still be big! In our case, assuming $\theta > 0$, for any $a > 0$,

$$E[e^{\theta S_n}] \geq e^{n\theta a} P\left[\frac{S_n}{n} \geq a\right] = e^{n\theta a} e^{-(na^2/2)+o(n)} = e^{n(a\theta - a^2/2)+o(n)}.$$

Since $a > 0$ is arbitrary,

$$E[e^{\theta S_n}] \geq e^{n \sup_{a>0}(\theta a - a^2/2)+o(n)} = e^{n\theta^2/2+o(n)},$$

which is the correct answer.

The simplest example for which one can calculate probabilities of large devia-

Received April 2007; revised April 2007.
[1] Supported in part by the NSF Grants DMS-01-04343 and DMS-06-04380.
AMS 2000 subject classifications. 60-02, 60F10.
Key words and phrases. Large deviations, rare events.

tions is coin tossing. The probability of k heads in n tosses of a fair coin is

$$P(n, k) = \binom{n}{k} 2^{-n} = \frac{n! 2^{-n}}{k!(n-k)!},$$

which using Stirling's approximation, is

$$\simeq \frac{\sqrt{2\pi} e^{-n} n^{n+1/2} 2^{-n}}{\sqrt{2\pi} e^{-(n-k)} (n-k)^{n-k+1/2} \sqrt{2\pi} e^{-k} k^{k+1/2}},$$

$$\log P(n, k) \simeq -\frac{1}{2} \log(2\pi) \left(n + \frac{1}{2}\right) \log n - \left(n - k + \frac{1}{2}\right) \log(n-k)$$

$$- \left(k + \frac{1}{2}\right) \log k - n \log 2$$

$$= -\frac{1}{2} \log(2\pi) - \frac{1}{2} \log n - \left(n - k + \frac{1}{2}\right) \log\left(1 - \frac{k}{n}\right)$$

$$- \left(k + \frac{1}{2}\right) \log \frac{k}{n} - n \log 2.$$

If $k \simeq nx$, then

$$\log P(n, k) \simeq -n[\log 2 + x \log x + (1 - x) \log(1 - x)] + o(n)$$
$$= -n H(x) + o(n),$$

where $H(x)$ is the Kullback–Leibler information or relative entropy of Binomial$(x, 1 - x)$ with respect to Binomial$(\frac{1}{2}, \frac{1}{2})$.

This is not a coincidence. In fact, if f_i are the observed frequencies in n trials of a multinomial with probabilities $\{p_i\}$ for the individual cells, then

$$P(n, p_1, \ldots, p_k; f_1, \ldots, f_k) = \frac{n!}{f_1! \cdots f_k!} p_1^{f_1} \cdots p_k^{f_k}.$$

A similar calculation using Stirling's approximation yields, assuming $f_i \simeq n x_i$,

$$\log P(n, p_1, \ldots, p_k; f_1, \ldots, f_k) = -n H(x_1, \ldots, x_k; p_1, \ldots, p_k) + o(n),$$

where $H(x, p)$ is again the Kullback–Leibler information number

$$H(x) = \sum_{i=1}^{k} x_i \log \frac{x_i}{p_i}.$$

Any probability distribution can be approximated by one that is concentrated on a finite set and the empirical distribution from a sample of size n will then have a multinomial distribution. One therefore expects that the probability $P(n, \alpha, \beta)$ that the empirical distribution

$$\frac{1}{n} \sum_{i=1}^{n} \delta_{X_i}$$

of n independent observations from a distribution α is close to β should satisfy

$$\log P(n, \alpha, \beta) = -nH(\beta, \alpha) + o(n),$$

where $H(\beta, \alpha)$ is again the Kullback–Leibler information number

$$\int \log \frac{d\beta}{d\alpha} \, d\beta = \int \frac{d\beta}{d\alpha} \log \frac{d\beta}{d\alpha} \, d\alpha.$$

This theorem, proven by Sanov, must be made precise. This requires a formal definition of what is meant by large deviations. We have a family $\{P_n\}$ of probability distributions on some space X which we assume to be a complete separable metric space. There is a sequence of numbers $a_n \to \infty$ which we might as well assume to be n. Typically, P_n concentrates around a point $x_0 \in X$ and, for sets A away from x_0, $P_n(A)$ tends to zero exponentially rapidly in n, that is,

$$\log P_n(A) \simeq -nc(A),$$

where $c(A) > 0$ if $x_0 \notin A$.

We say that a *large deviation principle* holds for a sequence of probability measures P_n defined on the Borel subsets of a Polish (complete separable metric) space X, with a *rate function* $H(x)$, if:

(1) $H(x) \geq 0$ is a lower semicontinuous function on X with the property that $K_\ell = \{x : H(x) \leq \ell\}$ is a compact set for every $\ell < \infty$;

(2) for any closed set $C \subset X$,

$$\limsup_{n \to \infty} \frac{1}{n} \log P_n(C) \leq - \inf_{x \in C} H(x);$$

(3) for any open set $U \subset X$,

$$\liminf_{n \to \infty} \frac{1}{n} \log P_n(U) \geq - \inf_{x \in U} H(x).$$

While condition (1) is not really necessary for the validity of the large deviation principle, it is a useful condition on the rate function that will allow us to reduce the analysis to what happens on compact sets. Rate functions with this property are referred to as "good" rate functions.

For Sanov's theorem, the i.i.d. random variables will be taking values in a complete separable metric space X. Their common distribution will be a probability measure $\alpha \in \mathcal{M} = \mathcal{M}(X)$, the set of probability measures on X. The space $\mathcal{M}$ under weak convergence is a complete separable metric space. The empirical distributions will be random variables with values in $\mathcal{M}$, and P_n will be their distribution, which will therefore be a probability measure on $\mathcal{M}$. The rate function

$$H(\beta, \alpha) = \int \frac{d\beta}{d\alpha} \log \frac{d\beta}{d\alpha} \, d\alpha$$

will be defined to be $+\infty$ unless $\beta \ll \alpha$ and the Radon–Nikodym derivative $f = \frac{d\beta}{d\alpha}$ is such that $f \log f$ is integrable with respect to α.

2. Rate functions, duality and generating functions. We start with i.i.d. random variables and look at the sample mean

$$Y_n = \frac{S_n}{n} = \frac{X_1 + \cdots + X_n}{n}.$$

According to a theorem of Cramér, its distribution P_n satisfies a large deviation principle with a rate function $h(a)$ given by

$$h(a) = \sup_{\theta}[\theta a - \log E[e^{\theta X}]].$$

In general, if a large deviation principle is valid for probability measures P_n on a space X with some rate function $H(x)$, it is not hard to see that, under suitable conditions on the function F (boundedness and continuity will suffice),

$$\frac{1}{n} \log \int e^{nF(x)} dP_n \to \sup_{x}[F(x) - H(x)].$$

The basic idea is just

$$\frac{1}{n} \log \sum e^{na_i} \to \sup_{i} a_i.$$

In other words, the logarithms of generating functions are dual to, or Legendre transforms of, large deviation rate functions. For instance,

$$\log\left[\int e^{V(x)} d\alpha\right] = \sup_{\beta}\left[\int V(x) d\beta - H(\beta, \alpha)\right].$$

If the rate function is convex, as it is in the case of sums of independent identically distributed random variables, the duality relationship is invertible and

$$H(\beta, \alpha) = \sup_{V(\cdot)}\left[\int V(x) d\beta - \log E[e^{V(x)} d\alpha]\right],$$

where the supremum on V in either case is taken over all bounded measurable functions or bounded continuous functions.

3. Markov processes. This relationship for i.i.d. sequences can be extended to the Markovian context. For simplicity, let us assume that we have a finite state space X and transition probabilities $\pi(x, y)$ of a Markov chain on X. Let us suppose that $\pi(x, y) > 0$ for all x, y. If $V(\cdot): X \to R$ is a function on X, then

$$E_x[\exp[V(X_1) + V(X_2) + \cdots + V(X_n)]]$$

can be explicitly evaluated as

$$\sum_{y} \pi_V^n(x, y),$$

where $\pi_V(x, y) = \pi(x, y)e^{V(y)}$ and π_V^n is the nth power of π_V. Since π_V is a matrix with positive entries,

$$\frac{1}{n} \log \sum_y \pi_V^n(x, y) \to \log \lambda_\pi(V),$$

where $\lambda_\pi(V)$ is the principal eigenvalue of π_V. The analog of Sanov's theorem in this context establishes a large deviation result for the empirical distributions of a long sequence $\{x_1, x_2, \ldots, x_n\}$. They belong to the space $\mathcal{M}$ of probability measures on X. Let us denote by P the Markov process with transition probability $\pi(x, y)$ and by Q_n the measure on $\mathcal{M}$ which is the distribution of the empirical distribution $\{x_1, x_2, \ldots, x_n\}$. The large deviation upper bound for Q_n, with a rate function given by

$$H_\pi(q) = \sup_V \left[\sum_x V(x)q(x) - \log \lambda_\pi(V) \right],$$

is an easy consequence of estimates on the generating function.

There is a more direct way of approaching H_π through lower bounds. Let us pretend that our Markov chain exhibits "atypical" behavior and behaves like a different Markov chain, one with transition probability $\tilde{\pi}(x, y)$. In other words, the empirical distribution of visits to different sites of X is close to q, which turns out to be the invariant distribution for a different chain, one with transition probabilities $\tilde{\pi}(x, y)$. We can estimate the probability of this event. Let U be an open set around q in the space $\mathcal{M}$ of probability measures on X. Let A_n be the set of all realizations of $\{x_1, \ldots, x_n\}$ with empirical distributions belonging to U. We can estimate $P(A_n)$, the probability of A_n under the original $\pi(x, y)$ chain, by

$$P(A_n) = \sum_{x_1, x_2, \ldots, x_n \in A_n} \pi(x, x_1) \cdots \pi(x_{n-1}, x_n)$$

$$= \sum_{x_1, x_2, \ldots, x_n \in A_n} \tilde{\pi}(x, x_1) \cdots \tilde{\pi}(x_{n-1}, x_n) \exp\left[-\sum_{i=1}^n \log \frac{\tilde{\pi}(x_{i-1}, x_i)}{\pi(x_{i-1}, x_i)} \right]$$

$$= \int_{A_n} \exp\left[-\sum_{i=1}^n \log \frac{\tilde{\pi}(x_{i-1}, x_i)}{\pi(x_{i-1}, x_i)} \right] d\tilde{P},$$

where $\tilde{P}$ is the Markov chain with transition probability $\tilde{\pi}(x, y)$. By the ergodic theorem,

$$\lim_{n \to \infty} \tilde{P}(A_n) = 1$$

as $n \to \infty$. An application of Jensen's inequality yields

$$\liminf_{n \to \infty} \frac{1}{n} \log Q_n(A) \geq -\sum_{x, y} \tilde{\pi}(x, y) \log \frac{\tilde{\pi}(x, y)}{\pi(x, y)} q(x).$$

We can pick any $\tilde{\pi}$, provided q is invariant for $\tilde{\pi}$, that is, $q\tilde{\pi} = q$ and we will get an upper bound for $H_\pi(q)$. Therefore,

$$H_\pi(q) \leq \inf_{\tilde{\pi}:\tilde{q}=q} \sum_{x,y} \tilde{\pi}(x,y) \log \frac{\tilde{\pi}(x,y)}{\pi(x,y)} q(x).$$

In fact, there is equality here and $H_\pi(q)$ is dual to $\log \lambda_\pi(V)$.

$$H_\pi(q) = \sup_{V(\cdot)} \left[\sum_x V(x)q(x) - \log \lambda_\pi(V) \right],$$

$$\log \lambda_\pi(V) = \sup_{q(\cdot)} \left[\sum_x V(x)q(x) - H_\pi(q) \right].$$

If we are dealing with a process in continuous time, we will have a matrix A of transition rates $\{a(x,y)\}$ with $a(x,y) \geq 0$ for $x \neq y$ and $\sum_y a(x,y) = 0$. The transition probabilities $\{\pi(t,x,y)\}$ will be given by $\pi(t,x,y) = (\exp tA)(x,y)$. If a is symmetric, that is, $a(x,y) = a(y,x)$, then so is $\pi(t,\cdot,\cdot)$. The uniform distribution on X will, in this case, be the invariant measure. The role of $\log \lambda(V)$ in the previous discrete situation will now be played by the principal eigenvalue $\lambda_a(V)$ of $A + V$, where $(A + V)(x,y) = a(x,y) + \delta(x,y)V(y)$. Here, $I = \{\delta(x,y)\}$ is the identity matrix. This is the conjugate of the rate function $H(q)$.

$$H_a(q) = \sup_{V(\cdot)} \left[\sum_x V(x)q(x) - \lambda_a(V) \right],$$

$$\lambda_a(V) = \sup_{q(\cdot)} \left[\sum_x V(x)q(x) - H_a(q) \right].$$

Note that, in the symmetric case, we have the usual variational formula

$$\lambda_a(V) = \sup_{\substack{u(\cdot) \\ \sum_x [u(x)]^2=1}} \left[\sum_x V(x)u^2(x) + \sum_{x,y} a(x,y)u(x)u(y) \right]$$

$$= \sup_{\substack{u(\cdot) \\ \sum_x [u(x)]^2=1}} \left[\sum_x V(x)u^2(x) - \tfrac{1}{2}\sum_{x,y} a(x,y)\left(u(x)-u(y)\right)^2 \right]$$

$$= \sup_{\substack{q(\cdot)\geq 0 \\ \sum_x q(x)=1}} \left[\sum_x V(x)u^2(x) - \tfrac{1}{2}\sum_{x,y} a(x,y)\left(\sqrt{q(x)}-\sqrt{q(y)}\right)^2 \right].$$

It is therefore not very surprising that

$$H_a(q) = \tfrac{1}{2}\sum_{x,y} a(x,y)\left(\sqrt{q(x)}-\sqrt{q(y)}\right)^2.$$

4. Small random perturbations. The exit problem. The context for large deviations is a probability distribution that is nearly degenerate. So far, this has come from the law of large numbers or the ergodic theorem. But it can also come from small random perturbations of a deterministic system.

Consider, for instance, the Brownian motion or Wiener measure on the space of continuous functions $C_0[[0, T]; R]$ that are zero at 0, that is, $x(0) = 0$. We can make the variance of the Brownian motion at time t equal to ϵt instead of t. We can consider $x_\epsilon(t) = \sqrt{\epsilon} x(t)$. Perhaps consider $\epsilon = \frac{1}{n}$ and $x_n(t) = \frac{1}{n} \sum_i y_i(t)$, the average of n independent Brownian motions. In any case, the measure P_ϵ of x_ϵ is nearly degenerate at the path $f(\cdot) \equiv 0$.

We have a large deviation principle for P_ϵ with a rate function

$$H(f) = \tfrac{1}{2} \int_0^T [f'(t)]^2 \, dt.$$

There are various ways of seeing this. We will exhibit two. As a Gaussian process, the generating function for Brownian motion is

$$\log E\left[\exp\left[\int_0^T x(t)g(t)\,dt\right]\right] = \tfrac{1}{2} \int_0^T \int_0^T \min(s, t)g(s)g(t)\,ds\,dt.$$

Its dual is given by

$$\sup_{g(\cdot)}\left[\int_0^T f(t)g(t)\,dt - \tfrac{1}{2}\int_0^T \int_0^T \min(s, t)g(s)g(t)\,ds\,dt\right]$$

$$= \tfrac{1}{2} \int_0^T [f'(t)]^2 \, dt.$$

Or, we can perturb $\sqrt{\epsilon} x(t)$ by $f(t)$ and change the measure from P_ϵ to Q_ϵ that concentrates near f rather that at 0. Of course, there are many measures that do this. Q_ϵ is just one such. The relative entropy is easily calculated for our choice. It is

$$H(Q_\epsilon, P_\epsilon) = \frac{1}{2\epsilon} \int_0^T [f'(t)]^2 \, dt.$$

This is also a lower bound for the rate function $H(f)$. In general, the rate function has a lower bound which is the "entropy cost" of changing the measure to do what we want it to do, namely, to concentrate on the "wrong spot." The cheapest way of achieving this is invariably the actual rate function. Let us now start with an ODE in R^n,

$$dx(t) = b(x(t))\, dt, \qquad x(0) = x_0,$$

and perturb it with a small noise,

$$dx_\epsilon(t) = b(x_\epsilon(t))\, dt + \sqrt{\epsilon} \beta(t), \qquad x(0) = x_0,$$

where $\beta(t)$ is the Brownian motion in R^n. As $\epsilon \to 0$, the distribution P_ϵ of $x_\epsilon(\cdot)$ will concentrate on the unique solution $x_0(\cdot)$ of the ODE. But there will be a large deviation principle with a rate function

$$H(f) = \tfrac{1}{2} \int_0^T \|f'(t) - b(f(t))\|^2 \, dt$$

which will happen if $\sqrt{\epsilon}\beta(\cdot)$ concentrates around $g(\cdot)$ with $g'(\cdot) = f'(\cdot) - b(f(\cdot))$.

One application of this is to the "exit problem" and the resulting interpretation of "punctuated equilibria." Let us take the ODE to be a gradient flow:

$$dx(t) = -(\nabla V)(x(t)) \, dt.$$

The system will move toward a minimum of V. If V has multiple local minimum or valleys surrounded by mountains, the solutions of the ODE could be trapped near a local minimum, depending on the starting point. They will not move from one local minimum to a deeper minimum. On the other hand, with even a small noise, they develop "wanderlust." While, most of the time, they follow the path dictated by the ODE, they will, from time to time, deviate sufficiently to be able to find and reach a lower minimum. If one were to sit at a deeper minimum and wait for the path to arrive, then trace its history back, one would be apt to find that it did not wander at all, but made the most efficient beeline from the previous minimum to this one, as if it were guided by a higher power. The large deviation interpretation of this curious phenomenon is simply that the system will experiment with every conceivable path, with probabilities that are extremely small. The higher the rate function, that is, the less efficient the path, the smaller will be the probability, hence, less frequent the attempts involving that path. Therefore, the first attempt to take the path to its new location will be the most efficient path, one that climbs the lowest mountain pass to get to the new fertile land!

5. Gibbs measures and statistical mechanics.

Let us look at point processes on R. The simplest is the Poisson point process. One can imagine it as starting from a finite interval $[-\frac{n}{2}, \frac{n}{2}]$ and placing $k_n = \rho n$ particles there randomly, independently and uniformly. Their joint density is

$$\frac{1}{n^{k_n}} dx_1 \cdots dx_{k_n}.$$

We can ignore the labeling and think of it as a point process P_n. The density is then

$$\frac{k_n!}{n^{k_n}} dx_1 \cdots dx_{k_n}.$$

As $n \to \infty$, we obtain a point process P which is the Poisson point process with intensity ρ. We can try to modify P_n into Q_n given by

$$dQ_n = \frac{1}{Z_n} \exp\left[-\sum_{i,j} V(x_i - x_j)\right] dP_n,$$

where $V \geq 0$ has compact support. Here, Z_n is the normalizing constant

$$Z_n = \int \exp\left[-\sum_{i,j} V(x_i - x_j)\right] dP_n.$$

The limit

$$\lim_{n\to\infty} \frac{1}{n} \log Z_n = \Psi(V)$$

is called the *free energy*. If Q is any stationary, that is, translation-invariant point process on R, its restriction to $[a, b]$ is a symmetric (i.e., permutation-invariant) probability distribution $q_{[a,b]}$ on

$$\bigcup_{k\geq 0} [a, b]^k.$$

Similarly, the Poisson point process with intensity ρ generates $p_{[a,b]}$, a convex combination of uniform distributions on $[a, b]^k$ with Poisson weights

$$e^{-\rho(b-a)} \frac{\rho^k (b - a)^k}{k!}.$$

The relative entropy $H(q_{[a,b]}, p_{[a,b]}) = H_\ell(Q, P)$ depends only on the length $\ell = (b - a)$ and is superadditive, that is,

$$H_{\ell_1 + \ell_2}(Q, P) \geq H_{\ell_1}(Q, P) + H_{\ell_2}(Q, P).$$

Therefore, the limit

$$\lim_{\ell\to\infty} \frac{1}{\ell} H_\ell(Q, P) = \sup_\ell H_\ell(Q, P) = \widehat{H}(Q, P)$$

exists and is called the *specific relative entropy*. There is the expectation

$$E^Q[V] = \lim_{n\to\infty} \frac{1}{n} E^Q\left[\sum_{x_\alpha, x_\beta \in [-n/2, n/2]} V(x_\alpha - x_\beta)\right].$$

Then,

$$\Psi(V) = -\inf_Q [E^Q[V] + \widehat{H}(Q, P)]$$

over all stationary point processes and the minimizer (unique) is the limit of Q_n. Such considerations play a crucial role in the theory of equilibrium statistical mechanics and thermodynamics [16].

6. Interacting particle systems. Interacting particle systems offer an interesting area where methods of large deviation can be applied. Let us look at some examples. Consider the lattice $\mathbf{Z}_N$ of integers modulo N. We think of them as points arranged uniformly on the circle of unit circumference or the unit interval with endpoints identified. For each $x \in \mathbf{Z}_N$, there corresponds a point $\xi = \frac{x}{N}$ on the interval (really, the circle). There are a certain number of particles in the system that occupy some of the sites. The particles wait for an exponential time and then decide to jump. They pick the adjacent site either to its left or right with equal probability and then jump. In this model, the particles do not interact and, after a diffusive rescaling of space and time, the particles behave like independent Brownian motions. Since there is a large number of particles, by an application of the law of large numbers, the empirical density will be close to a solution $\rho(t, \xi)$ of the heat equation

$$\rho_t = \tfrac{1}{2}\rho_{\xi\xi},$$

with a suitable initial condition, depending on how the particles were initially placed.

We can introduce an interaction in this model, by imposing a limit of at most one particle per site. When the particle decides to jump to a randomly chosen site after a random exponential waiting time, it can jump only when the site is unoccupied. Otherwise, it must wait for a new exponential time before trying to jump again. All of the particles wait and attempt to jump independently of each other. Since we are dealing with continuous-time and exponential distributions, two particle will not try to jump at the same time and we do not have to resolve ties.

One can express all of this by simply writing down the generator. η is a configuration and $X_N = \{0, 1\}^N$ is the set of all configurations. $\eta(x) = 1$ if the site x is occupied and $\eta(x) = 0$ if the site x is free. A point in X_N is just a map $\eta : \mathbf{Z}_N \to \{0, 1\}$. The generator A_N acting on functions $f : X_N \to R$ is given by

$$(A_N f)(\eta) = \tfrac{1}{2} \sum_x \eta(x)[(1 - \eta(x+1))(f(\eta^{x,x+1}) - f(\eta))$$

$$+ (1 - \eta(x-1))(f(\eta^{x,x-1}) - f(\eta))],$$

where $\eta^{x,y}$ is defined by

$$\eta^{x,y}(z) = \begin{cases} \eta(x), & \text{if } z = y, \\ \eta(y), & \text{if } z = x, \\ \eta(z), & \text{otherwise.} \end{cases}$$

This system has some interesting features. The total number $\sum_x \eta(x)$ of particles does not change over time. Particles just diffuse over time. It will take time of order N^2 for the effect to be felt at a distance of order N. The equilibrium distributions $\mu_{N,k}$ are uniform over all possible $\binom{N}{k}$ configurations, where k is the number of particles. In particular, there are multiple equilibria. Such systems can be locally in equilibrium while approaching the global equilibrium rather slowly. For example, we initially place our particles in such a way that one half of the circle C_1 has particle density $\frac{1}{4}$, while the other half C_2 has density $\frac{3}{4}$. The system will locally stay near the two different equilibria at the two intervals for quite some time, with just some smoothing near the edges. If we wait for times of order N^2, that is, time $N^2 t$ with $t > 0$, the density will be close to $\rho(t, \frac{x}{N})$. A calculation that involves applying the speeded up generator $N^2 \mathcal{A}_N$ to expressions of the form

$$\frac{1}{N} \sum J\left(\frac{x}{N}\right) \eta(x)$$

shows that $\rho(t, \xi)$ will be the solution of the heat equation

$$\rho_t(t, \xi) = \tfrac{1}{2} \rho_{\xi\xi}(t, \xi); \qquad \rho(0, \xi) = \tfrac{1}{4} \mathbf{1}_{C_1}(\xi) + \tfrac{3}{4} \mathbf{1}_{C_2}(\xi).$$

In this case, this behavior is the same as if the particles moved independently without any interaction. While this is true at the level of the law of large numbers, we shall see that for large deviation rates, the interaction does matter.

7. Superexponential estimates. Let us consider the process in equilibrium, with $k = \rho N$ particles. We let N, k go to ∞ while ρ is kept fixed. It is easy to see, for any fixed site x, that

$$\mu_{N,k}[\eta(x) = 1] \to \rho$$

and $\{\eta(x_j)\}$ become independent for any finite set of distinct sites. If $f(\eta)$ is a local function and $f_x(\eta) = f(\tau_x \eta)$ its translation, then

$$\frac{1}{N} \sum_x f_x(\eta) \to E^{P_\rho}[f(\eta)]$$

as $N \to \infty$. The probabilities of deviations decay exponentially fast. The exponential rate is given by

$$H(a) = \inf_{Q \in \mathcal{M}(a,\rho)} H(Q, P_\rho),$$

where $\mathcal{M}(a, \rho)$ consists of stationary measures Q with density ρ and $E^Q[f(\eta)] = a$, and P_ρ is Bernoulli with density ρ. $H(Q, P)$ is the specific entropy, calculated as the limit

$$\lim_{n \to \infty} \frac{1}{n} H(Q_n, P_n),$$

where Q_n and P_n are restrictions of Q and P to a block of n sites. This is, of course, an equilibrium calculation done at a fixed time. What about space–time averages

$$\frac{1}{NT} \int_0^T \sum_x f_x(\eta(t)) \, dt?$$

The rate function is now some $H_T(a)$ that depends on T. While this is hard to calculate, one can show that

$$\lim_{T \to \infty} H_T(a) = \inf_{\rho(\cdot) \in \mathcal{P}(a,\rho)} H(P_{\rho(\xi)}, P_\rho) \, d\xi,$$

where $\mathcal{P}(a, \rho)$ consists of density profiles $\rho(\xi)$ that satisfy

$$\int \rho(\xi) \, d\xi = \rho \quad \text{and} \quad \int E^{P_{\rho(\xi)}}[f(\eta)] \, d\xi = a.$$

Note that the rate increases to a finite limit as $T \to \infty$. The reason is that large-scale fluctuations in density can occur with exponentially small probability and these fluctuations do not necessarily decay when T is large. Particles diffuse slowly and it takes times of order N^2 to diffuse through order N sites.

To resolve this difficulty, we define the approximate large-scale empirical density by

$$r_{N,\epsilon}(\xi, \eta) = \frac{1}{N\epsilon} \sum_{x:|x/N-\xi| \le \epsilon} \eta(x)$$

and compare

$$\frac{1}{N^3} \int_0^{N^2 T} \sum_x f_x(\eta(t)) \, dt = Y_{N,f}(T)$$

with

$$\frac{1}{N^3} \int_0^{N^2 T} \sum_x \widehat{f}\left(r_{N,\epsilon}\left(\frac{x}{N}, \eta(t)\right)\right) dt = \widehat{Y}_{N,f,\epsilon}(T),$$

where

$$\widehat{f}(\rho) = E^{P_\rho}[f(\eta)].$$

One can show that

$$\limsup_{\epsilon \to 0} \limsup_{N \to \infty} \frac{1}{N} \log P[|Y_{N,f}(T) - \widehat{Y}_{N,f,\epsilon}(T)| \ge \delta] = -\infty$$

for any $\delta > 0$. In other words, in time scale N^2, the large scale density fluctuations are responsible for all large deviations with a logarithmic decay rate of N. Such estimates, called *superexponential estimates*, play a crucial role in the study of interacting particle systems where approach to equilibrium is slow in large scales.

8. Hydrodynamical limits. If we consider a large system with multiple equilibria, the system can remain in a local equilibrium for a long time. Let the different equilibria be labeled by parameter ρ. If the system lives on a large spatial domain and if ξ denotes a point in the macroscopic scale of space, a function $\rho(\xi)$ could describe a system that is locally in equilibrium, albeit at different ones at macroscopically different points. If t is time measured in a suitably chosen faster scale, then $\rho(\xi) = \rho(t, \xi)$ may evolve gently in time and converge as $t \to \infty$ to a constant ρ that identifies the global equilibrium.

For instance, in our example, let us initially start with $k = N\rho$ particles and distribute them in such a way that we achieve a density profile $\rho_0(\xi)$ as $N \to \infty$. Technically, this means

$$\lim_{N \to \infty} \frac{1}{N} \sum_x J\left(\frac{x}{N}\right) \eta(x) = \int J(\xi) \rho_0(\xi) \, d\xi$$

for bounded continuous test functions J. What will we see after time tN^2, especially when N is large?

The answer is very easy. Consider the sum

$$f(\eta) = \frac{1}{N} \sum_x J\left(\frac{x}{N}\right) \eta(x)$$

and apply the speeded up operator $N^2 A_N$ to it.

$$f(\eta^{x,y}) - f(\eta) = \frac{1}{N} [\eta(x) - \eta(y)] \left[J\left(\frac{y}{N}\right) - J\left(\frac{x}{N}\right) \right],$$

$$(N^2 A_N f)(\eta) = \frac{N}{2} \sum_x \eta(x)(1 - \eta(x+1)) \left[J\left(\frac{x+1}{N}\right) - J\left(\frac{x}{N}\right) \right]$$

$$+ \eta(x)(1 - \eta(x-1)) \left[J\left(\frac{x-1}{N}\right) - J\left(\frac{x}{N}\right) \right]$$

$$= \frac{N}{2} \sum_x [\eta(x)(1 - \eta(x+1)) - \eta(x+1)(1 - \eta(x))]$$

$$\times \left(J\left(\frac{x+1}{N}\right) - J\left(\frac{x}{N}\right) \right)$$

$$= \frac{N}{2} \sum_x [\eta(x) - \eta(x+1)] \left[J\left(\frac{x+1}{N}\right) - J\left(\frac{x}{N}\right) \right]$$

$$= \frac{N}{2} \sum_x \left[J\left(\frac{x+1}{N}\right) + J\left(\frac{x-1}{N}\right) - 2J\left(\frac{x}{N}\right) \right] \eta(x)$$

$$\simeq \frac{1}{2N} \sum_x J''\left(\frac{x}{N}\right) \eta(x),$$

leading to the limiting heat equation

$$\rho_t = \tfrac{1}{2} \rho_{\xi\xi}, \qquad \rho(0, \xi) = \rho_0(\xi).$$

Let us change the problem slightly. Introduce a slight bias. The probabilities to right and left are $\frac{1}{2} \pm \frac{1}{2N} b$, where $b > 0$ is the bias to the right. This introduces an extra term, so that now

$$N^2 A_N \frac{1}{N} \sum_x J\left(\frac{x}{N}\right) \eta(x) = \frac{1}{2N} \sum_x J''\left(\frac{x}{N}\right) \eta(x) + F_N(\eta),$$

where

$$F_N(\eta) = \frac{b}{2N} \sum J'\left(\frac{x}{N}\right)[\eta(x)(1 - \eta(x+1)) + \eta(x)(1 - \eta(x-1))].$$

This is a problem because F_N is nonlinear in η and cannot be replaced by a simple expression involving ρ. If we were locally in equilibrium, F_N would be replaced by

$$G_{N,\epsilon}(\eta) = \frac{b}{N} \sum J'\left(\frac{x}{N}\right) r_{N,\epsilon}\left(\frac{x}{N}, \eta\right)\left(1 - r_{N,\epsilon}\left(\frac{x}{N}, \eta\right)\right),$$

where $r_{N,\epsilon}(\frac{x}{N}, \eta)$ is the empirical density. If we were in a global equilibrium, in the faster time scale,

$$\left| \int_0^T F_N(\eta(t)) - G_{N,\epsilon}(\eta(t)) \, dt \right| \geq \delta,$$

with superexponentially small probability. The importance of superexponential estimates lies in the following elementary, but universal, inequality:

$$Q(A) \leq \frac{2 + H(Q, P)}{\log(1/P(A))}.$$

If $P(A)$ is superexponentially small and $H(Q, P)$ is linear in N, then $Q(A)$ is small. Perturbation of the process by a slight bias of order $\frac{1}{N}$ produces a relative entropy of order $\frac{1}{N^2}$ per site per unit time. With N sites and time of order N^2, this is still only linear in N. Changes in initial conditions can also only contribute linear relative entropy. So, we can carry out the approximation and end up with

$$\rho_t = \tfrac{1}{2}\rho_{\xi\xi} - [b\rho(1 - \rho)]_\xi.$$

Note that without exclusion, the limit would have been

$$\rho_t = \tfrac{1}{2}\rho_{\xi\xi} - [b\rho]_\xi,$$

which is just the CLT when there is a small mean.

9. Large deviations in hydrodynamical limits. We saw earlier that in our model, if we started with an initial profile with density ρ_0, in a speeded up time scale, the system will evolve with a time-dependent profile which is the solution of the heat equation

$$\rho_t = \tfrac{1}{2}\rho_{\xi\xi}.$$

This is, of course, true with probability nearly 1. We can make the initial condition deterministic in order that no initial deviation is possible. Still, our random evolution could produce, with small probability, strange behavior. If we limit our-selves to deviations with only exponentially small probabilities, what deviations are possible and what are their exponential rates?

We can doctor the system with a bias b which is not constant, but is a function $b(t, \frac{x}{N})$ of x and fast time t with $b(t, \xi)$ being a nice function of t and ξ. This will produce a solution of

$$\rho_t(t, \xi) = \tfrac{1}{2} D_\xi^2 \rho(t, \xi) - D_\xi[b(t, \xi)\rho(t, \xi)(1 - \rho(t, \xi))], \qquad \rho(0, \xi) = \rho_0(\xi).$$

This can be done with an entropy cost that can be calculated as

$$\Psi(b) = \tfrac{1}{2} \int_0^T \int |b(t, \xi)|^2 \rho(t, \xi)(1 - \rho(t, \xi)) \, dt \, d\xi,$$

for the duration $[0, T]$. If $\rho(t, \xi)$ is given, then we can minimize $\Psi(b)$ over all compatible b. This is a lower bound for the rate. One can match this bound in the other direction.

10. Large deviations for random walks in random environments. We will start with a probability space (Ω, Σ, P) on which $\mathbf{Z}^d$ acts ergodically as a family τ_z of measure-preserving transformations. We are given $\pi(\omega, z)$, which is a proba-bility distribution on $\mathbf{Z}^d$ for each ω, and is a measurable functions of ω for each z. One can then generate random transition probabilities $\pi(\omega, z', z)$ by defining

$$\pi(\omega, z', z' + z) = \pi(\tau_{z'}\omega, z).$$

For each ω, $\pi(\omega, z', z)$ can serve as the transition probability of a Markov process on $\mathbf{Z}^d$, and the measure corresponding to this process, starting from 0, is denoted by Q^ω. This is, of course, random (it depends on ω and is called the *random walk in the random environment* ω). One can ask the usual questions about this random walk and, in some form, they may be true for almost all ω with respect to P. The law of large numbers, if valid, will take the form

$$P\left[\omega : \lim_{n \to \infty} \frac{S_n}{n} = m(P) \text{ a.e. } Q^\omega\right] = 1.$$

Such statement concerning the almost sure behavior under Q^ω for almost all ω with respect to P are said to deal with the "quenched" version. Sometimes one wishes to study the behavior of the "averaged" or "annealed" measure,

$$\overline{Q} = \int Q^\omega P(d\omega).$$

The law of large numbers is the same because it is equivalent to

$$\overline{Q}\left[\omega : \lim_{n \to \infty} \frac{S_n}{n} = m(P)\right] = 1.$$

On the other hand, questions on the asymptotic behavior of probabilities, like the central limit theorem or large deviations, could be different for the quenched and the averaged cases.

A special environment, which is called the *product environment*, is one in which $\pi(\omega, z', z' + z)$ are independent for different z' and have a common distribution β which is a probability measure on the space $\mathcal{M}$ of all probability measures on $\mathbf{Z}^d$. In this case, the canonical choice for (Ω, Σ, P) is the countable product of $\mathcal{M}$ and the product measure P with marginals β.

There are large deviation results regarding the limits

$$\lim_{n \to \infty} \frac{1}{n} \log Q^\omega \left[\frac{S_n}{n} \simeq a \right] = I(a)$$

and

$$\lim_{n \to \infty} \frac{1}{n} \log \overline{Q} \left[\frac{S_n}{n} \simeq a \right] = \overline{I}(a).$$

The difference between I and $\overline{I}$ has a natural explanation. In the special case of Z, in terms of the large deviation behavior in the one-dimensional random environment, it is related to the following question. If, at time n, we see a particle reach an improbable value na, what does it say about the environment in $[0, na]$? Did the particle behave strangely in the environment or did it encounter a strange environment? It is probably a combination of both.

Large deviation results exists, however, in much wider generality, both in the quenched and the averaged cases. The large deviation principle is essentially the existence of the limits

$$\lim_{n \to \infty} \frac{1}{n} \log E[\exp[\langle \theta, S_n \rangle]] = \Psi(\theta).$$

The expectation is with respect to Q^ω or $\overline{Q}$, which could produce different limits for Ψ. The law of large numbers and central limit theorem involve the differentiability of Ψ at $\theta = 0$, which is harder. In fact, large deviation results have been proven by general subadditivity arguments for the quenched case. Roughly speaking, fixing ω, we have

$$Q^\omega[S_{k+\ell} \simeq (k + \ell)a] \geq Q^\omega[S_k \simeq ka] \times Q^{\tau_{ka}\omega}[S_\ell \simeq \ell a].$$

It is harder for the averaged case. We want to prove that the limits

$$\lim_{n \to \infty} \frac{1}{n} \log \overline{Q} \left[\frac{S_n}{n} \simeq a \right] = -\overline{I}(a)$$

or, equivalently,

$$\lim_{n \to \infty} \frac{1}{n} \log E^{\overline{Q}}[\exp[\langle \theta, S_n \rangle]] = \overline{\Psi}(\theta),$$

exist. The problem is that the measure $\overline{Q}$ is not very nice. As the random walk explores $\mathbf{Z}^d$, it learns about the environment and in the case of the product environment, when it returns to a site that it has visited before, the experience has not been forgotten and leads to long term correlations. However, if we are interested in the behavior $S_n \simeq na$ with $a \neq 0$, the same site is not visited too often and the correlations should rapidly decay.

One can use Bayes' rule to calculate the conditional distribution

$$\overline{Q}[S_{n+1} = S_n + z | S_1, S_2, \ldots, S_n] = q(z|w),$$

where w is the past history of the walk. Before we do that, it is more convenient to shift the origin as we go along so that the current position of the random walk is always the origin and the current time is always 0. An n step walk then looks like $w = \{S_0 = 0, S_{-1}, \ldots, S_{-n}\}$. We pick a z with probability $q(z|w)$. We obtain a new walk of $n + 1$ steps $w' = \{S'_0 = 0, S'_{-1}, \ldots, S'_{-(n+1)}\}$ given by $S'_{-(k+1)} = S_{-k} - z$ for $k \geq 0$. We can now calculate $q(z|w)$. We need to know all the quantities $\{k(w, x, z)\}$, the numbers of times the walk has visited x in the past and jumped from x to $x + z$. It is not hard to see that the a posteriori probability can be calculated as

$$q(z|w) = \frac{\int \pi(z) \Pi_{z'} \pi(z')^{k(w,0,z')} \beta(d\pi)}{\int \Pi_{z'} \pi(z')^{k(w,0,z')} \beta(d\pi)}.$$

While this initially makes sense only for walks of finite length, it can clearly be extended to all transient paths. Note that although we only use $k(w, 0, z')$, in order to obtain the new $k(w', 0, z')$, we would need to know the collection $\{k(w, z, z')\}$.

Now, suppose that R is a process with stationary increments $\{z_j\}$. Then, we can again make the current position the origin and if the process is transient, as it would be if the increments were ergodic and had a nonzero mean a, the conditional probabilities $q(z|w)$ would exist a.e. R and can be compared to the corresponding conditional probabilities $r(z|w)$ under R. The relative entropy

$$H(R) = E^R \left[\sum_z r(z|w) \log \frac{r(z|w)}{q(z|w)} \right]$$

is then well defined.

The function

$$\overline{I}(a) = \inf_{\substack{R: \int z_1 dR = a \\ R \text{ ergodic}}} H(R),$$

defined for $a \neq 0$, extends as a convex function to all of R^d and, with this $\overline{I}$ as rate function, a large deviation result holds.

We now turn to the quenched case. Although a proof using the subadditive ergodic theorem exists, we will provide an alternate approach that is more appealing.

We will illustrate this in the context of Brownian motion with a random drift. Instead of the action of $\mathbf{Z}^d$, we can have $\mathbf{R}^d$ acting on (Ω, Σ, P) ergodically and consider a diffusion on $\mathbf{R}^d$ with a random infinitesimal generator

$$(\mathcal{L}^\omega u)(x) = \tfrac{1}{2}(\Delta u)(x) + \langle b(\omega, x), (\nabla u)(x) \rangle$$

acting on smooth functions on $\mathbf{R}^d$. Here, $b(\omega, x)$ is generated from a map $b(\omega): \Omega \to \mathbf{R}^d$ by the action of $\{\tau_x : x \in \mathbf{R}^d\}$ on ω:

$$b(\omega, x) = b(\tau_x \omega).$$

Again, there is the quenched measure Q^ω that corresponds to the diffusion with generator $\mathcal{L}^\omega$ that starts from 0 at time 0, and the averaged measure that is given by a similar formula. This model is referred to as *diffusion with a random drift*. Exactly the same questions can be asked in this context. We can define a diffusion on Ω with generator

$$\mathcal{L} = \tfrac{1}{2}\Delta + \langle b(\omega), \nabla \rangle,$$

where $\nabla = \{D_i\}$ are the generators of the translation group $\{\tau_x : x \in \mathbf{R}^d\}$. This is essentially the image of lifting the paths $x(t)$ of the diffusion on $\mathbf{R}^d$ corresponding to $\mathcal{L}^\omega$ to Ω by

$$\omega(t) = \tau_{x(t)}\omega.$$

While there is no possibility of having an invariant probability measure on $\mathbf{R}^d$, on Ω, one can hope to find an invariant probability density $\phi(\omega)$, that is, to find $\phi(\omega) \geq 0$ in $L_1(P)$ with $\int \phi \, dP = 1$ which solves

$$\tfrac{1}{2}\Delta\phi = \nabla \cdot (b\phi).$$

If such a ϕ exists, then we have an ergodic theorem for the diffusion process Q^ω corresponding to $\mathcal{L}$ on Ω,

$$\lim_{t \to \infty} \frac{1}{t} \int_0^t f(\omega(s)) \, ds = \int f(\omega)\phi(\omega) \, dP \qquad \text{a.e. } P.$$

This also translates to an ergodic theorem on R^d. If we define the stationary process g by $g(\omega, x) = f(\tau_x \omega)$, then

$$\lim_{t \to \infty} \frac{1}{t} \int_0^t f(\omega, x(s)) \, ds = \int f(\omega)\phi(\omega) \, dP \qquad \text{a.e. } Q^\omega, \text{ a.e. } P,$$

where Q^ω is now the quenched process in the random environment. Since

$$x(t) = \int_0^t b(\omega, x(s)) \, ds + \beta(t),$$

where $\beta(\cdot)$ is the Brownian motion, it is clear that

$$\lim_{t \to \infty} \frac{x(t)}{t} = \int b(\omega)\phi(\omega) \, dP \qquad \text{a.e. } Q^\omega, \text{ a.e. } P,$$

providing a law of large numbers for $x(t)$. While we cannot be sure of finding ϕ for a given b, it is easy to find a b for a given ϕ. For instance, if $\phi > 0$, we could take $b = \frac{\nabla\phi}{2\phi}$. Or, more generally, $b = \frac{\nabla\phi}{2\phi} + \frac{c}{\phi}$ with $\nabla \cdot c = 0$. If we change b to b' which satisfies $\frac{1}{2}\Delta\phi = \nabla \cdot (b'\phi)$, the new process $Q^{b',\omega}$ with drift b' will, in the time interval $[0, t]$, have relative entropy

$$E^{Q^{b',\omega}}\left[\frac{1}{2}\int_0^t \|b(\omega(s)) - b'(\omega(s))\|^2 \, ds\right]$$

and, by the ergodic theorem, one can see that, a.e. with respect to P,

$$\lim_{t\to\infty} \frac{1}{t} E^{Q^{b',\omega}}\left[\frac{1}{2}\int_0^t \|b(\omega(s)) - b'(\omega(s))\|^2 \, ds\right] = \frac{1}{2}\int \|b(\omega) - b'(\omega)\|^2 \phi(\omega) \, dP.$$

Moreover, for almost all ω with respect to P, almost surely with respect to $Q^{b',\omega}$,

$$\lim_{t\to\infty} \frac{x(t)}{t} = \int b'(\omega)\phi(\omega) \, dP.$$

If we fix $\int b'(\omega)\phi(\omega) \, dP = a$, the bound

$$\liminf_{t\to\infty} \frac{1}{t} \log Q^\omega\left[\frac{x(t)}{t} \simeq a\right] \geq -\frac{1}{2}\int \|b - b'\|^2 \phi \, dP$$

is easily obtained. If we define

$$I(a) = \inf_{\substack{b',\phi \\ (1/2)\Delta\phi=\nabla\cdot(b'\phi) \\ \int b'\phi dP=a}} \frac{1}{2}\int \|b - b'\|^2 \phi \, dP,$$

then

$$\liminf_{t\to\infty} \frac{1}{t} \log Q^\omega\left[\frac{x(t)}{t} \simeq a\right] \geq -I(a).$$

Of course, these statements are valid a.e. P. One can check that I is convex and that the upper bound amounts to proving the dual estimate

$$\lim_{t\to\infty} \frac{1}{t} \log E^{Q^\omega}\left[e^{\langle\theta,x(t)\rangle}\right] \leq \Psi(\theta),$$

where

$$\Psi(\theta) = \sup_a[\langle a, \theta\rangle - I(a)].$$

We need a bound on the solution of

$$u_t = \frac{1}{2}\Delta u + \langle b, \nabla u\rangle$$

with $u(0) = \exp[\langle\theta, x\rangle]$. By the Hopf–Cole transformation $v = \log u$, this reduces to estimating

$$v_t = \frac{1}{2}\Delta v + \frac{1}{2}\|\nabla v\|^2 + \langle b, \nabla v\rangle.$$

with $v(0) = \langle \theta, x \rangle$. This can be done if we can construct a subsolution

$$\tfrac{1}{2} \nabla \cdot w + \tfrac{1}{2} \|\nabla w\|^2 + \langle b, w \rangle \le \psi(\theta)$$

on Ω, where $w : \Omega \to \mathbf{R}^d$ satisfies $\int w \, dP = \theta$ and $\nabla \times w = 0$ in the sense that $D_i w_j = D_j w_i$. The existence of the subsolution comes from convex analysis.

$$\psi(\theta) = \sup_{\substack{b', \phi \\ (1/2)\Delta\phi = \nabla \cdot (b'\phi)}} \left[\int \langle b', \theta \rangle \phi \, dP - \tfrac{1}{2} \int \|b - b'\|^2 \phi \, dP \right]$$

$$= \sup_{\phi} \sup_{b'} \inf_{u} \left[\int \langle b', \theta \rangle \phi \, dP - \tfrac{1}{2} \int \|b - b'\|^2 \phi \, dP \right.$$
$$\left. + \tfrac{1}{2} [\Delta u + \langle b', \nabla u \rangle] \phi \, dP \right]$$

$$= \sup_{\phi} \inf_{u} \sup_{b'} \left[\int \langle b', \theta \rangle \phi \, dP - \tfrac{1}{2} \int \|b - b'\|^2 \phi \, dP \right.$$
$$\left. + \int \tfrac{1}{2} [\Delta u + \langle b', \nabla u \rangle] \phi \, dP \right]$$

$$= \sup_{\phi} \inf_{u} \int \sup_{b'} \left[\langle b', \theta + \nabla \cdot u \rangle \phi \, dP - \tfrac{1}{2} \int \|b - b'\|^2 \phi \, dP + \int \tfrac{1}{2} \Delta u \phi \, dP \right]$$

$$= \sup_{\phi} \inf_{u} \int \left[\tfrac{1}{2} \Delta u + \langle b, \theta + \nabla u \rangle + \tfrac{1}{2} \|\theta + \nabla u\|^2 \right] \phi \, dP$$

$$= \sup_{\phi} \inf_{\substack{\nabla \times w = 0 \\ \int w \, dP = \theta}} \int \left[\tfrac{1}{2} \nabla \cdot w + \langle b, w \rangle + \tfrac{1}{2} \|w\|^2 \right] \phi \, dP$$

$$= \inf_{\substack{\nabla \times w = 0 \\ \int w \, dP = \theta}} \sup_{\phi} \int \left[\tfrac{1}{2} \nabla \cdot w + \langle b, w \rangle + \tfrac{1}{2} \|w\|^2 \right] \phi \, dP$$

$$= \inf_{\substack{\nabla \times w = 0 \\ \int w \, dP = \theta}} \sup_{\omega} \left[\tfrac{1}{2} \nabla \cdot w + \langle b, w \rangle + \tfrac{1}{2} \|w\|^2 \right],$$

which proves the existence of a subsolution. One needs to justify the free interchange of sup and inf. In passing from the first line to the second, the restriction on b' and ϕ is replaced by the Lagrange multiplier u.

This can be viewed as showing the existence of a limit as $\epsilon \to 0$ (homogenization) of the solution of

$$u_t^\epsilon = \frac{\epsilon}{2} \Delta u^\epsilon + \frac{1}{2} \|\nabla u^\epsilon\|^2 + \left\langle b\left(\frac{x}{\epsilon}, \omega\right), \nabla u^\epsilon \right\rangle$$

with $u^\epsilon(0, x) = f(x)$. The limit satisfies

$$u_t = \Psi(\nabla u)$$

with $u(0, x) = f(x)$.

11. Homogenization of Hamilton–Jacobi–Bellman equations.

This can be generalized to equations of the form

$$u_t^\epsilon = \frac{\epsilon}{2}\Delta u^\epsilon + H\left(\frac{x}{\epsilon}, \nabla u^\epsilon, \omega\right), \qquad u(0, x) = f(x),$$

where $H(x, p, \omega) = H(\tau_x\omega, p)$ is a stationary process of convex functions in p. By the changes of variables $x = \epsilon y$, $u = \epsilon v$, $t = \epsilon \tau$, this reduces to the behavior of $\epsilon v^\epsilon(\frac{T}{\epsilon}, \frac{x}{\epsilon})$, where v^ϵ solves

$$v_\tau^\epsilon = \tfrac{1}{2}\Delta v^\epsilon + H(y, \nabla v^\epsilon, \omega), \qquad v^\epsilon(0, y) = \epsilon^{-1}f(\epsilon y).$$

From the principle of dynamic programming, we can represent the solution as the supremum of a family of solutions of linear equations,

$$v^\epsilon(\tau, y) = \sup_{b(\cdot, \cdot) \in \mathcal{B}} w^\epsilon\big(b(\cdot, \cdot), T - \tau, y\big),$$

where w^ϵ solves

$$w_\tau^\epsilon + \tfrac{1}{2}\Delta w^\epsilon + \langle b(\tau, y), \nabla w^\epsilon\rangle - L(\tau_y\omega, b(\tau, y)) = 0,$$

$$w^\epsilon(\epsilon^{-1}T, y) = \epsilon^{-1}f(\epsilon y),$$

L being the convex dual of H with respect to the variable p. If b is chosen as $b(\tau_y\omega)$, for suitable choices of $b(\omega) \in \mathcal{C}$ that admit positive integrable solutions ϕ to

$$\tfrac{1}{2}\Delta\phi(\tau_x\omega) = \nabla \cdot b(\tau_x\omega)\phi(\tau_x, \omega),$$

then it is not hard to see that

$$\epsilon w^\epsilon(0, 0) \to f\left(T\int b(\omega)\phi(\omega)\,dP\right) - T\int L(b(\omega), \omega)\phi(\omega)\,dP.$$

This provides a lower bound

$$\liminf_{\epsilon \to 0} u^\epsilon(T, 0) \geq \sup_{b \in \mathcal{C}}\left[f\left(T\int b(\omega)\phi(\omega)\,dP\right) - T\int L(b(\omega), \omega)\phi(\omega)\,dP\right]$$

which can be shown to be an upper bound as well.

12. History and references.   The origin of large deviation theory goes back to Scandinavian actuaries [10] who were interested in the analysis of risk in the insurance industry. For sums of independent random variables, the general large deviations result was established by Cramér in [1]. The result for empirical distributions of independent identically distributed random variables is due to Sanov [18]. The generalization to Markov chains and processes can be found in several papers of Donsker and Varadhan [3–6] and Gärtner [13]. The results concerning small random perturbations of deterministic systems goes back to the work of Varadhan [20], as well as Vencel and Freidlin [12]. Several monographs have appeared on the subject. Lecture notes by Varadhan [21], texts by Deuschel and Stroock [7], Dembo and Zeitouni [2], Schwartz and Weiss [19], Ellis [9], Dupuis and Ellis [8] and, most recently, by Feng and Kurtz [11]. They cover a wide spectrum of topics in large deviation theory. For large deviations in the context of hydrodynamic scaling, there is the text by Kipnis and Landim [14], as well as an exposition by Varadhan [23]. As for large deviations for random walks in a random environment, see [24], as well as references in Zeitouni's article [25]. For a general survey on large deviations and entropy, see [22]. The results on homogenization of random Hamilton–Jacobi–Bellman equations and its application to large deviations has appeared in [15] and [17]. Undoubtedly, there are many more references. A recent Google search on "Large Deviations" produced 3.4 million hits.

REFERENCES

[1] CRAMÈR, H. (1938). Sur un nouveau théorème-limite de la théorie des probabilités. *Actualités Scientifiques et Industrialles* **736** 5–23. *Colloque Consecré à la Théorie des Probabilités* 3. Hermann, Paris.

[2] DEMBO, A. and ZEITOUNI, O. (1998). *Large Deviations Techniques and Applications*, 2nd ed. Springer, New York. MR1619036

[3] DONSKER, M. D. and VARADHAN, S. R. S. (1975). Asymptotic evaluation of certain Markov process expectations for large time. I. *Comm. Pure Appl. Math.* **28** 1–47. MR0386024

[4] DONSKER, M. D. and VARADHAN, S. R. S. (1975). Asymptotic evaluation of certain Markov process expectations for large time. II. *Comm. Pure Appl. Math.* **28** 279–301. MR0386024

[5] DONSKER, M. D. and VARADHAN, S. R. S. (1976). Asymptotic evaluation of certain Markov process expectations for large time. III. *Comm. Pure Appl. Math.* **29** 389–461. MR0428471

[6] DONSKER, M. D. and VARADHAN, S. R. S. (1983). Asymptotic evaluation of certain Markov process expectations for large time. IV. *Comm. Pure Appl. Math.* **36** 183–212. MR0690656

[7] DEUSCHEL, J. D. and STROOCK, D. W. (1989). *Large Deviations*. Academic Press, Boston. MR0997938

[8] DUPUIS, P. and ELLIS, R. S. (1997). *A Weak Convergence Approach to the Theory of Large Deviations*. Wiley, New York. MR1431744

[9] ELLIS, R. S. (2006). *Entropy, Large Deviations, and Statistical Mechanics*. Springer, Berlin. MR2189669

[10] ESSCHER, F. (1932). On the probability function in the collective theory of risk. *Skandinavisk Aktuarietidskrift* **15** 175–195.

[11] FENG, J. and KURTZ, T. G. (2006). *Large Deviations for Stochastic Processes*. Amer. Math. Soc., Providence, RI. MR2260560

[12] FREIDLIN, M. I. and WENTZELL, A. D. (1998). *Random Perturbations of Dynamical Systems*, 2nd ed. Springer, New York. MR1652127

[13] GÄRTNER, J. (1977). On large deviations from an invariant measure (in Russian). *Teor. Verojatnost. i Primenen.* **22** 27–42. MR0471040

[14] KIPNIS, C. and LANDIM, C. (1999). *Scaling Limits of Interacting Particle Systems*. Springer, Berlin. MR1707314

[15] KOSYGINA, E., REZAKHANLOU, F. and VARADHAN, S. R. S. (2006). Stochastic homogenization of Hamilton–Jacobi–Bellman equations. *Comm. Pure Appl. Math.* **59** 1489–1521. MR2248897

[16] LANFORD, O. E. (1973). Entropy and equilibrium states in classical statistical mechanics. In *Statistical Mechanics and Mathematical Problems* (A. Lenard, ed.) 1–113. *Lecture Notes in Phys.* **20**. Springer, Berlin.

[17] LIONS, P. L. and SOUGANIDIS, P. E. (2005). Homogenization of "viscous" Hamilton–Jacobi equations in stationary ergodic media. *Comm. Partial Differential Equations* **30** 335–375. MR2131058

[18] SANOV, I. N. (1957). On the probability of large deviations of random magnitudes (in Russian). *Mat. Sb. N. S.* **42 (84)** 11–44. MR0088087

[19] SHWARTZ, A. and WEISS, A. (1995). *Large Deviations for Performance Analysis*. Chapman and Hall, London. MR1335456

[20] VARADHAN, S. R. S. (1967). Diffusion processes in a small time interval. *Comm. Pure Appl. Math.* **20** 659–685. MR0217881

[21] VARADHAN, S. R. S. (1984). *Large Deviations and Applications*. SIAM, Philadelphia. MR0758258

[22] VARADHAN, S. R. S. (2003). Large deviations and entropy. In *Entropy. Princeton Ser. Appl. Math.* 199–214. Princeton Univ. Press. MR2035822

[23] VARADHAN, S. R. S. (1996). The complex story of simple exclusion. In *Itô's Stochastic Calculus and Probability Theory* 385–400. Springer, Tokyo. MR1439538

[24] VARADHAN, S. R. S. (2003). Large deviations for random walks in a random environment. Dedicated to the memory of Jürgen K. Moser. *Comm. Pure Appl. Math.* **56** 1222–1245. MR1989232

[25] ZEITOUNI, O. (2002). Random walks in random environments. In *Proceedings of the International Congress of Mathematicians* **III** (*Beijing, 2002*) 117–127. Higher Ed. Press, Beijing. MR1957524

COURANT INSTITUTE OF MATHEMATICAL SCIENCES
NEW YORK UNIVERSITY
NEW YORK, NEW YORK 10012
USA
E-MAIL: varadhan@cims.nyu.edu

Contemporary Mathematics
Volume **490**, 2009, 3-10

The Role of Weak Convergence in Probability Theory

S.R.S.Varadhan

1. Introduction

The concept of limit theorems in probability goes way back. The first limit theorem, the weak law of large numbers was proved by Jacob Bernoulli [1] in the early eighteenth century. This was quickly followed by De Moivre [3] who proved the central limit theorem establishing the approximation of the binomial distribution by a normal distribution. Further work by, Laplace, Gauss, Levy, Khintichine, Kolmogorov, Gnedenko and others put limit theorems at the center of probability theory [11].

The connection between random walks and Brownian motion was understood by Levy and others along with the idea that distributions of quantities like the maximum etc, based on random walks, converges to the corresponding distributions derived from Brownian motion. Doob [6] formulated this more precisely in his paper on 'Heuristic approach to the Kolmogorov-Smirnov Theorems'. Donsker [5], in his thesis, established the first general theorem to the effect that Doob's heuristic proof can in fact be justified. However his approach was too dependent on finite dimensional approximations.

At this point the study of stochastic processes as probability distributions on function spaces began. Contributions were made by LeCam [13] in the United States, Kolmogorov [12], Prohorov [15], Skorohod [17] and others in USSR, as well as Varadarajan [20] in India. Alexandrov in the 1940's had studied set functions on topological spaces and now powerful techniques from functional analysis could be used to study stochastic processes as measures on function spaces. A random walk or any stochastic process induces a probability distribution on the space of paths. By interpolation or some such simple device both the approximating and the limiting distributions can be put on the same space of paths. The question then reduces to the investigation of the convergence of a sequence μ_n of probability measures on a space X of paths to a limit μ. It is clear that the measures μ_n, in the case of random walks, look qualitatively different from Brownian paths and hence $\mu_n \perp \mu$. It is not going to be true that $\mu_n(A) \to \mu(A)$ for all measurable sets $A \in X$.

Functional analysis now provides a useful window. The space X of paths comes with a topology. A probability measure μ defines a normalized non-negative linear functional

$$\Lambda_\mu(f) = \int f(x)\mu(dx)$$

on $\mathbf{B} = C(X)$ and they form a convex subset $\mathcal{M}$ in the dual $\mathbf{B}^*$ of $\mathbf{B}$. The natural weak∗ topolgy on this subset is called 'weak convergence' by probabilists, and has proved to be a very useful tool. The following questions arise naturally and were investigated.

1. Which linear functionals $\Lambda(f)$ have the above representation as Λ_μ for some probability measure μ, i.e. when does $\lambda \in \mathcal{M}$ and when is μ uniquely determined by Λ?

2. What can one say about the space $\mathcal{M}$ as a topological space?

3. Is it metrizable? What are its compact subsets?

Viewed in this manner the classical limit theorems tell us that if we place random walks as well as Brownian Motion on the same space $X = C[0, T]$ of continuous paths, then the only possible limit for the measures μ_n coming from random walks is the Wiener measure μ. Weak convergence as elements in $C(X)^*$ would provide justification for the convergence of distributions of continuous functionals on X under μ_n to the corresponding distribution under μ. The issue then is that of compactness.

Prohorov, Skorohod and others in the Russian school worked mostly under the assumption that X is a complete separable metric space. They obtained characterizations of compact subsets of $\mathcal{M}$, and provided useful general sufficient conditions to verify compactness in several useful function spaces. Varadarajan had independently worked out similar results, in a more general context, in his thesis during 1956-57 at the Indisn Statistical Institute. These tools were used by Parthasarathy, Ranga Rao and Varadhan [14] , [21] to study limit theorems in different contexts.

2. The Martingale Problem.

The work of Stroock and Varadhan [18], [19] on the Martingale approach to the study of Markov process is also motivated by these considerations. In approximating diffusion processes by Markov chains we start with an approximation that is valid at the infinitesimal level. If $h > 0$ is the discretized time unit and $\pi_h(x, dy)$ is the single step transition probability, it is natural to assume that

$$\int [f(y) - f(x)]\pi_h(x, dy) = h(Lf)(x) + o(h)$$

where L is the generator of the semigroup associated with the limiting process. Our aim is to show $v_n(x) = \int f(y)\pi_h^n(x, dy) \simeq (e^{n h L}f)(x)$. The standard analytical method is to to solve the evolution equation

$$u_t = Lu, u(0, x) = f(x)$$

and estimate the difference $\Delta_n(x) = v_n(x) - u(n h, x)$. Let $nh = t$.

$$\int u(k h, y)\pi_h(x, dy) \simeq u(k h, x) + h(Lu(k h, \cdot))(x) + o(h)$$

$$= u(k h, x) + hu_t(k h, x)(x) + o(h)$$

$$= u((k + 1) h, x) + o(h)$$

Hence

$$\left| u(n\,h, x) - \int u(0, y)\pi_h^n(x, dy) \right| \le n\,o(h) = o(1)$$

and the proof depends on the regularity of the solution $u(t, x)$. The actual random paths of the Markov chain did not play any role. Only the iterates of π_h were needed.

The martingale method, on the other hand, starts with the measures $P_{h,x}$ on a function space of paths, that corresponds to the given Markov chain starting from x and shows that sequence has a limit P_x. Then if $nh = t$,

$$v_n(x) = E^{P_{h,x}}[f(x(t))] \to E^{P_x}[f(x(t))] = u(t, x)$$

To achieve this one needs to verify compactness and then characterize the possible limit uniquely. It is done as the unique measure on $C[0, T]$ (or $D[0, T]$ which is a space of paths that admits simple jumps), with certain properties:

$$P[x(0) = x] = 1$$

and

$$f(x(t)) - f(x(0)) - \int_0^t (Lf)(x(s))ds = Z_f(t)$$

is a martingale with respect to $(\Omega, \mathcal{F}_t, P)$ for a wide class of functions f. From the definition of a Markov chain, if P_h is the probability distribution of the chain and

$$P_h[x(0) = x] = 1$$

then

$$f(x(n\,h)) - f(x(0)) - \sum_{j=0}^{n-1} \int [f(y) - f(x(j\,h))]\pi_h(x(j\,h), dy) = Z_f^h$$

is a martingale. One now uses ideas from weak convergence to show that $\{P_h\}$ is compact and that any limit satisfies the properties.

3. Large Deviations

In limit theorems one is primarily interested in the behavior of $P_n(A)$ as $n \to \infty$. We look at the larger problem of weak convergence of P_n and establish such convergence to the limit P, in a some topological space X that contains A as a subset. If A is a continuity set for P, i.e $P[\delta A] = 0$, then $P_n(A) \to P(A)$.

In large deviations, we are dealing with a situation where P_n tends to a distribution that is degenerate at some point, i.e. $P_n \to \delta_{x_0}$ for some $x_0 \in X$. Then if $x \notin \bar{A}$, $P_n(A) \to 0$ and we wish to know how fast. In particular we expect the rate to be exponential and we wish to examine

$$\lim_{n \to \infty} \frac{1}{n} \log P_n(A) = -c(A)$$

It will turn out that

$$c(A) = \inf_{x \in A} I(x)$$

for some non-negative function $I(x)$ with $I(x_0) = 0$.

In the theory of weak convergence it is established that in metric spaces weak convergence is equivalent to

$$\limsup_{n \to \infty} P_n(C) \leq P(C) \qquad \text{for closed sets} \quad C$$

$$\liminf_{n \to \infty} P_n(G) \geq P(G) \qquad \text{for open sets} \quad G$$

and for continuity sets i.e sets with $P(A^\circ) = P(\bar{A})$, one has

$$\lim_{n \to \infty} P_n(A) = P(A)$$

An important criterion for compactness of $\{P_n\}$, is the uniform tightness condition: given $\epsilon > 0$, there is a compact set K_ϵ such that

$$\inf_n P_n(K_\epsilon) \geq 1 - \epsilon$$

In large deviation theory a sequence P_n is said to satisfy the large deviation principle with the rate function $I(x)$ if

(3.1) $$\limsup_{n \to \infty} \frac{1}{n} \log P_n(C) \leq - \inf_{x \in C} I(x) \qquad \text{for closed sets} \quad C$$

(3.2) $$\liminf_{n \to \infty} \frac{1}{n} \log P_n(G) \geq - \sup_{x \in G} I(x) \qquad \text{for open sets} \quad G$$

$I(x) : X \to [0, \infty]$ is a lower semi continuous function. There is a similar tightness condition.

1. $I(x)$ has compact level sets, i.e $C_\ell = \{x : I(x) \leq \ell\}$ is compact for each $\ell < \infty$.

2. For any ℓ there is a compact set K_ℓ such that

$$\sup_n P_n(K_\ell^c) \leq e^{-n\ell}$$

Just as weak convergence implies

$$\lim_{n \to \infty} P_n(A) = P(A)$$

for continuity sets, in large deviation theory it follows easily from (3.1) and (3.2) that for sets A such that

$$\inf_{x \in A^\circ} I(x) = \inf_{x \in A} I(x) = \inf_{x \in \bar{A}} I(x)$$

we will have

$$\lim_{n \to \infty} \frac{1}{n} \log P_n(A) = -c(A) = - \inf_{x \in A} I(x)$$

In weak convergence we will have

$$\lim_{n \to \infty} \int f(x) dP_n = \int f(x) dP$$

for bounded continuous functons f. In large deviation theory for such functions the analogous result is

$$\lim_{n \to \infty} \frac{1}{n} \log \int e^{n\,f(x)} dP_n = \sup_x [f(x) - I(x)]$$

In weak convergence when $f_n \to f$, in order to make sure that $\int f_n(x)dP_n \to \int f(x)dP$ when P_n converges weakly to P, one assumes the uniform convergence of f_k to f on compact sets. This is used together with tightness to control

$$\limsup_{k \to \infty} \limsup_{n \to \infty} P_n[|f_k - f| \geq \epsilon] = 0$$

for every $\epsilon > 0$. In large deviation theory super exponential estimates play a similar role. For every $\epsilon > 0$,

(3.3) $$\limsup_{k \to \infty} \limsup_{n \to \infty} \frac{1}{n} \log P_n[|f_k - f| \geq \epsilon] = -\infty$$

The methods used in large deviation theory are often similar to the ones used in studying weak convergence of probability measures. There are several sources [22], [4], [2] that describe these. Let us look at an example.

Schilder's theorem [16]. P_n is the distribution of $x(t) = \frac{1}{\sqrt{n}}\beta(t)$, where β is the standard Brownian motion. There is a large deviation principle on $C[0,1]$ with rate function

$$I(f) = \frac{1}{2} \int_0^1 [f'(t)]^2 dt$$

on functions f with $f(0) = 0$ that have a square integrable derivative. Otherwise $I(f) = \infty$.

The proof proceeds by approximation. If $x_k(\cdot)$ is the piecewise linear approximation with intervals $\frac{1}{k}$.

$$P_n[\|x_k(\cdot) - x(\cdot)\| \geq \epsilon] \leq C\,n\,exp[-\frac{n\,k\,\epsilon^2}{2}].$$

This provides the estimate (3.3) and enables us to interchange the limits on k and n. In a recent book [7] Ellis and Dupuis emphasize the weak convergence aspect of large deviations. If one wants to prove a large deviation upper bound for P_n with rate function $I(x)$, it suffices to show that when ever $Q_n << P_n$ and $Q_n \to \delta_x$ weakly, then

$$\liminf_{n \to \infty} \frac{1}{n} H(Q_n|P_n) \geq I(x).$$

Here $H(Q|P)$ is the relative entropy $\int \frac{dQ}{dP} \log \frac{dQ}{dP} dP$.

4. Scaling Limits

In studying scaling limits of large systems of interacting processes ideas from weak convergence play a crucial role. For simplicity consider as in [8], a family $\{x_i(t)\}$ of processes indexed by points $i = 1, 2, \ldots, N$ arranged (periodically) on the unit circle that satisfy the stochastic differential equation

$$dx_i(t) = [\phi(x_{i-1}(t)) - 2\phi(x_i(t)) + \phi(x_{i+1}(t))]dt + [d\beta_{i,i+1}(t) - d\beta_{i-1,i}(t)].$$

Assume initially that

$$\frac{1}{N} \sum_i \delta_{\frac{i}{N}} x_i(0) = \mu_N(0) \to u_0(x)dx$$

in probability. Then show that

$$\frac{1}{N} \sum_i \delta_{\frac{i}{N}} x_i(N^2 t) = \mu_N(N^2 t) \to u(t, x)dx$$

where $u(t,x)$ satisfies a certain partial differential equation with $u(0,x) = u_0(x)$. Let P_N be the distribution of the process $\mu_N(N^2\cdot)$ on the space $C[[0,T]; \mathcal{M}(S^1)]$. Then,

1. P_N is tight.

2. If P is any limit then P is supported on paths $\mu(t)$, that satisfy

(i) they have a density for every t, i.e $\mu(t,dx) = u(t,x)dx$

(ii) if $\psi' = \phi$ then with

$$h(x) = \sup[\theta\, x - \log \int e^{\theta y - \psi(y)} dy]$$

$u(t,x)$ is a weak solution of

$$u_t = [h(u(t,x))]_{xx}$$

that satisfies

$$\int_{S^1} \int_0^T [[h'(u(t,x))]_x]^2 dx dt < \infty$$

(iii) $u(0,x) = u_0(x)$

3. The solution satisfying (i), (ii) and (iii) is unique.

The measure P therefore has to be the δ-measure at this unique solution.

5. Large Deviations for the Simple Exclusion Process.

The totally asymmetric simple exclusion model (TASEP) is a Markov process of interacting particles [**10**]. These particles live on Z and are restricted to at most one particle per site. The current state then is a map $\eta : Z \to \{0,1\}$. If there is a particle at site x then $\eta(x) = 1$, otherwise it is $\eta(x) = 0$. The generator of the process is given by

$$(\mathcal{L}f)(\eta) = \sum_z \eta(z)(1 - \eta(z+1))[f(\eta^{z,z+1}) - f(\eta)]$$

which means that particles wait for a random exponential time at the end of which they try to jump to the next site on its right. They jump if the site is free. If not they wait again for the next opportunity. If the density of particles is not too high, then they march to the right, in a haphazard fashion. We rescale space by $\frac{1}{N}$ and speed up time by a factor of N. The density profile $\rho(t,x)$ that provides a macroscopic view, at time Nt is defined by

$$\frac{1}{N} \sum_z J(\frac{z}{N}) \eta(Nt,z) \to \int J(x)\rho(t,x)dx.$$

Given $\rho(0,x) = \rho_0(x)$ at time $t = 0$, for $t > 0$, $\rho(t,x)$ is determined as the weak solution of

$$\rho_t + [\rho(1 - \rho)]_x = 0$$

with initial condition $\rho(0,x) = \rho_0(x)$. However the weak solution is not unique and an 'entropy condition' has to be imposed in order to make the solution unique. If ϕ is a convex function and h is determined by $h'(r) = \phi'(r)(1 - 2r)$, then for smooth solutions we will have

$$[\phi(\rho)]_t + [h(\rho)]_x = 0.$$

The entropy entropy condition for weak solutions is that

$$[\phi(\rho)]_t + [h(\rho)]_x \leq 0$$

At a simple shock discontinuity this requires the density on the left to be lower than the density on the right. Non entropic solutions can occur with small probability. For example one particle may not jump for a long time and will hold up every particle behind it, producing a shock, i.e. discontinuity. We want to see how small this probability is. It turns out that the rate function $I(\rho(\cdot, \cdot))$ is infinite unless $\rho(\cdot, \cdot)$ is a weak solution and in such a case the rate function is the total mass of the positive part (can be infinite) of $[\phi(\rho)]_t + [h(\rho)]_x$ viewed as a distribution on $R \times [0, T]$ for the special convex function

$$\phi(r) = r \log r + (1 - r) \log(1 - r).$$

The proof by Jensen [9], [23] depends on the method of Ellis-Dupuis that uses weak convergence. If we perturb the dynamics from $\mathcal{L}$ to

$$\mathcal{L}_c = \sum_z c_z(t, \eta) \eta(z)(1 - \eta(z + 1))[f(\eta^{z, z+1}) - f(\eta)]$$

so that the macroscopic profile converges to $\rho(t, x)$ that may not be entropic then, denoting by Q_n the perturbed process and by P_n the unperturbed one, it suffices to prove that we must necessarily have

$$\liminf_{N \to \infty} \frac{1}{N} H(Q_N | P_N) \geq I(\rho(\cdot, \cdot).)$$

This can be done.

6. Conclusion

I have tried to demonstrate how ideas from 'weak convergence' of probability measures on abstract spaces, that at the first glance appear to be too abstract, do in fact provide an essential tool to study some hard and concrete problems. Of course my view is biased, in that I have found the tool very useful in my own work. I learnt these tools as a graduate student in Calcutta from Varadarajan through my assocaition with Ranga Rao and Parthasarathy. You can not teach an old dog new tricks!

Acknowledgments. This work was supported by a grant from the National Science Foundation DMS-0604380.

References

[1] Bernoulli, J. Ars Conjectandi, Basel, 1713

[2] Dembo, A.; Zeitouni, O. Large deviations techniques and applications. Jones and Bartlett Publishers, Boston, MA, 1993.

[3] DeMoivre, A. The Doctrine of Chances, London, 1718

[4] Deuschel, J.D.; Stroock, D. W. Large deviations. Pure and Applied Mathematics, 137. Academic Press, Inc., Boston, MA, 1989.

[5] Donsker, M. D. Justification and extension of Doob's heuristic approach to the Komogorov-Smirnov theorems. Ann. Math. Statistics 23, (1952). 277–281.

[6] Doob, J. L. Heuristic approach to the Kolmogorov-Smirnov theorems. Ann. Math. Statistics 20, (1949). 393–403.

[7] Dupuis, P.; Ellis, R. S. A weak convergence approach to the theory of large deviations. Wiley Series in Probability and Statistics: Probability and Statistics. A Wiley-Interscience Publication. John Wiley & Sons, Inc., New York, 1997.

[8] Guo, M. Z.; Papanicolaou, G. C.; Varadhan, S. R. S. Nonlinear diffusion limit for a system
     with nearest neighbor interactions. Comm. Math. Phys. 118 (1988), no. 1, 31–59.

[9] Jensen, L. Large deviations of TASEP. PhD thesis, NYU, 2000.

[10] Kipnis, C.; Landim, C. Scaling limits of interacting particle systems. Grundlehren der Mathe-
     matischen Wissenschaften [Fundamental Principles of Mathematical Sciences], 320. Springer-
     Verlag, Berlin, 1999.

[11] Gnedenko, B.V.; Kolmogorov, A.N. Limit Distributions for Sums of Independent Random
     Variables. Moscow, 1949.

[12] Kolmogoroff, A.N. ; Prochorow, Yu.V. Zufllige Funktionen und Grenzverteilungsstze. (Ger-
     man) Bericht ber die Tagung Wahrscheinlichkeitsrechnung und mathematische Statistik in
     Berlin, Oktober, 1954, pp. 113–126. Deutscher Verlag der Wissenschaften, Berlin, 1956.

[13] LeCam, L. Convergence in distribution of stochastic processes. Univ. Calif. Publ. Statist. 2
     (1957), 207–236.

[14] Parthasarathy, K. R.; Ranga Rao, R.; Varadhan, S. R. S. Probability distributions on locally
     compact abelian groups. Illinois J. Math. 7 1963 337–369.

[15] Prohorov, Yu. V. Convergence of random processes and limit theorems in probability theory.
     (Russian) Teor. Veroyatnost. i Primenen. 1 (1956), 177–238.

[16] Schilder, M. Some asymptotic formulas for Wiener integrals. Trans. Amer. Math. Soc. 125
     1966 63–85.

[17] Skorohod, A. V. Limit theorems for stochastic processes. (Russian) Teor. Veroyatnost. i
     Primenen. 1 (1956), 289–319.

[18] Stroock, D.W.; Varadhan, S. R. S. Diffusion processes with continuous coefficients. I. Comm.
     Pure Appl. Math. 22 1969 345–400.

[19] Stroock, Daniel W.; Varadhan, S. R. S. Diffusion processes with continuous coefficients. II.
     Comm. Pure Appl. Math. 22 1969 479–530.

[20] Varadarajan, V. S. Convergence of stochastic processes. Bull. Amer. Math. Soc. 67 1961
     276–280.

[21] Varadhan, S. R. S. Limit theorems for sums of independent random variables with values in
     a Hilbert space. Sankhyā Ser. A 24 1962 213–238.

[22] Varadhan, S. R. S. Large deviations and applications. CBMS-NSF Regional Conference Se-
     ries in Applied Mathematics, 46. Society for Industrial and Applied Mathematics (SIAM),
     Philadelphia, PA, 1984.

[23] Varadhan, S. R. S. Large deviations for the asymmetric simple exclusion process. Stochastic
     analysis on large scale interacting systems, 1–27, Adv. Stud. Pure Math., 39, Math. Soc.
     Japan, Tokyo, 2004.

COURANT INSTITUTE OF MATHEMATICAL SCIENCES, NEW YORK UNIVERSITY, NEW YORK,
NY, 10012, USA. E-MAIL: varadhan@cims.nyu.edu